CELLULAR LIPID BINDING PROTEINS

Cellular Lipid Binding Proteins

Edited by

JAN F.C. GLATZ

Department of Physiology
Cardiovascular Research Institute Maastricht (CARIM)
Maastricht University
P.O. Box 616
6200 MD Maastricht
The Netherlands

Reprinted from *Molecular and Cellular Biochemistry*, Volume 239 (2002)

Kluwer Academic Publishers
Dordrecht / Boston / London

Distributors for North, Central and South America:
Kluwer Academic Publishers
101 Philip Drive
Assinippi Park
Norwell, Massachusetts 02061 USA
Telephone (781) 871-6600
Fax (781) 681-9045
E-mail <kluwer@wkap.com>

Distributors for all other countries:
Kluwer Academic Publishers Group
Distribution Centre
Post Office Box 322
3300 AH Dordrecht, The Netherlands
Telephone 31 78 6392 392
Fax 31 78 6546 474
E-mail <services@wkap.nl>

Electronic Services <http://www.wkap.nl>

Library of Congress Cataloging-in-Publication Data

A C.I.P. Catalogue record for this book is available from the Library of Congress

ISBN 1-4020-7220-1

Printed on acid-free paper

Printed in The Netherlands

Molecular and Cellular Biochemistry:

An International Journal for Chemical Biology in Health and Disease

CONTENTS VOLUME 239, Nos. 1 & 2, October 2002

CELLULAR LIPID BINDING PROTEINS
Dr. Jan F.C. Glatz

Molecular and Cellular Biochemistry **239:** 1, 2002.

Preface

Cellular lipid binding proteins are well established to serve central roles in cellular lipid uptake and metabolism. Membrane-associated fatty acid-binding proteins putatively function in protein-mediated transmembrane transport of fatty acids, while the intracellular trafficking of fatty acids, bile acids, and other lipid ligands may involve their interaction with specific membrane or protein targets, which are unique properties of some but not of all cytoplasmic lipid binding proteins. Importantly, recent studies indicate that the various proteins interacting with lipids not only facilitate their transport in aqueous media but, by interacting with specific targets, also modulate their subsequent biological action or metabolism. In addition, during the last decade evidence has been presented that various metabolic diseases, such as hyperlipidemia, atherosclerosis, insulin resistance and diabetes, are characterized by malfunctioning or deficiencies in cellular lipid binding proteins. Therefore, the notion arises that for better understanding of the action of lipids as signalling compounds and the role of lipids in intermediairy metabolism, it is essential to have detailed knowledge of the interactions between lipids and their cognate binding proteins.

In view of this growing interest in lipid-protein interaction, the 4th International Conference on Lipid Binding Proteins was held in Maastricht, the Netherlands, in June, 2001. The organization of this meeting was prompted by the success of previous conferences held in Maastricht (1989 and 1992) and Minneapolis (1997). The proceedings of these previous three meetings have been published in *Molecular and Cellular Biochemistry*, volume 98 (1990), volume 123 (1993) and volume 192 (1999), respectively. The present focussed issue of *Molecular and Cellular Biochemistry* is a collection of selected papers based on the lectures and posters presented during the conference, and thus provides insight into our current understanding of the significance of these proteins for the functioning of the cell.

Jan F.C. Glatz PhD
Department of Physiology
Cardiovascular Research Institute Maastricht (CARIM)
Maastricht University
Maastricht, The Netherlands

Molecular and Cellular Biochemistry **239**: 3–7, 2002.

Cellular lipid binding proteins as facilitators and regulators of lipid metabolism

Jan F.C. Glatz, Joost J.F.P. Luiken, Marc van Bilsen and Ger J. van der Vusse

Department of Physiology, Cardiovascular Research Institute Maastricht (CARIM), Maastricht University, Maastricht, The Netherlands

Abstract

Evidence is accumulating that cellular lipid binding proteins are playing central roles in cellular lipid uptake and metabolism. Membrane-associated fatty acid-binding proteins putatively function in protein-mediated transmembrane transport of fatty acids, likely coexisting with passive diffusional uptake. The intracellular trafficking of fatty acids, bile acids, and other lipid ligands, may involve their interaction with specific membrane or protein targets, which are unique properties of some but not of all cytoplasmic lipid binding proteins. Recent studies indicate that these proteins not only facilitate but also regulate cellular lipid utilization. For instance, muscle fatty acid uptake is subject to short-term regulation by translocation of fatty acid translocase (FAT)/CD36 from intracellular storage sites to the plasma membrane, and liver-type cytoplasmic fatty acid-binding protein (L-$FABP_c$) functions in long-term, ligand-induced regulation of gene expression by directly interacting with nuclear receptors. Therefore, the properties of the lipid-protein complex, rather than those of the lipid ligand itself, determine the fate of the ligand in the cell. Finally, there are an increasing number of reports that deficiencies or altered functioning of both membrane-associated and cytoplasmic lipid binding proteins are associated with disease states, such as obesity, diabetes and atherosclerosis. In conclusion, because of their central role in the regulation of lipid metabolism, cellular lipid binding proteins are promising targets for the treatment of diseases resulting from or characterised by disturbances in lipid metabolism, such as atherosclerosis, hyperlipidemia, and insulin resistance. (Mol Cell Biochem **239**: 3–7, 2002)

Key words: fatty acid-binding protein, fatty acid translocase (FAT), CD36, bile acid-binding protein, membrane protein, peroxisome proliferator activated receptor

Abbreviations: FABP – fatty acid-binding protein; $FABP_c$ – cytoplasmic FABP; $FABP_{pm}$ – plasma membrane FABP; FAT – fatty acid translocase (CD36); FATP – fatty acid-transport protein; PPAR – peroxisome proliferator activated receptor; CRBP – cellular retinol binding protein; I-BABP – intestinal bile acid-binding protein

Introduction

By virtue of their chemical nature, lipid molecules generally show limited solubility in aqueous solutions and, therefore, are hampered in their transport as single species within body fluids like blood plasma, interstitium, and the cellular cytoplasm. To overcome this limitation and make lipids readily available for utilization by various tissues, lipids either are bound by specific carrier proteins, for example long-chain fatty acids bound to plasma albumin, or are part of larger lipid-protein complexes referred to as lipoproteins. During the last decade a growing amount of evidence has been presented that the various proteins interacting with lipids not only facilitate their transport in aqueous media but, by interacting with specific targets, also modulate their subsequent biological action or metabolism [1]. Therefore, the notion arises that for better understanding the action of lipids as signalling compounds and the role of lipids in intermediary metabolism,

Address for offprints: J.F.C. Glatz, Department of Physiology, CARIM, Maastricht University, P.O. Box 616, NL-6200 MD Maastricht, The Netherlands (E-mail: glatz@fys.unimaas.nl)

it is essential to have detailed knowledge on the interactions between lipids and their cognate binding proteins.

In view of this growing interest in lipid-protein interaction, the 4th International Conference on Lipid Binding Proteins was held in Maastricht, the Netherlands, in June, 2001. The organization of this meeting was prompted by the success of previous conferences held in Maastricht (1989 and 1992) and Minneapolis (1997). The proceedings of these previous three meetings have been published in *Molecular and Cellular Biochemistry*, volume 98 (1990), volume 123 (1993) and volume 192 (1999), respectively. The present focused issue of *Molecular and Cellular Biochemistry* is a collection of selected papers based on the lectures and posters presented during the conference, and thus provides insight into our current understanding of the functioning and physiological significance of cellular lipid binding proteins.

The conference started with a reflection on the various individual steps involved in cellular fatty acid uptake and utilization [2], and then focused on the following topics: (i) mechanisms of ligand binding and transfer, (ii) physiological significance of lipid binding proteins, (iii) lipid binding proteins and the regulation of gene expression, and (iv) pathophysiology of lipid binding proteins. The major novel findings presented and remaining issues discussed during the conference are summarized below.

Mechanisms of ligand binding and transfer

The precise mechanism by which long-chain fatty acids are released from or taken up by cells is still controversial. Studies not only with model membrane systems, but also with intact cells (adipocytes), and using sophisticated techniques such as dual fluorescence approaches, have provided convincing evidence that fatty acids diffuse very rapidly across the plasma membrane [3]. Passage of the lipid bilayer occurs by flip-flop of the fatty acid in the un-ionized (i.e. protonated) form, with the direction and driving force for net movement being the transmembrane gradient of fatty acids [3, 4]. The diffusional mechanism is seemingly at variance with the identification of a number of membrane-associated fatty acid-binding proteins (FABPs), which each have been linked to function in the transmembrane transport process [1]. This conundrum is not easily being solved. Two of the discussed alternative explanations are that (i) passive diffusional uptake and protein-mediated transmembrane translocation of fatty acids co-exist as separate routes, with the contribution of each to the overall rate of fatty acid transport being dependent on a.o. cell type, fatty acid availability, and hormonal milieu, and (ii) the membrane proteins merely function as fatty acid acceptors (or donors) while the actual movement of fatty acids through the phospholipid bilayer occurs by passive diffusion of the un-ionized species. Because our knowledge on the membrane topology of the membrane-associated FABPs, including the site of interaction with the fatty acid ligand, is still in its infancy, at present a distinction between these alternatives cannot be made.

Inside cells, cytoplasmic FABP ($FABP_c$) functions as acceptor and transcytoplasmic carrier of fatty acids [5]. Interestingly, most of the $FABP_c$ types appear to be not strictly cytosolic, but are 'membrane active', that is, these proteins form a transient complex with the donor or acceptor membrane, upon which the fatty acid ligand is transferred. The helix-turn-helix domain of the $FABP_c$ appears to be involved in this interaction. In contrast, at least two members of the protein family of cellular lipid binding proteins, i.e. L(iver-type)-$FABP_c$ and cellular retinol binding protein (CRBP)-II, were found to be 'membrane-inert', that is, transfer occurs by ligand diffusion through the aqueous phase [5]. These mechanisms of fatty acid transfer were first established for rodent proteins, and now fully confirmed for their human counterparts [6]. Importantly, model studies by Weisiger *et al.* [7] further indicate that membrane-active FABP types catalyze the dissociation of the fatty acid from the donor membrane and binding to the acceptor membrane, while membrane-inactive FABP types catalyze the diffusion of fatty acids across the aqueous cytoplasm. This suggests that different FABP types may have evolved to match the specific transport requirements of the cell type within which they are found.

In recent years much progress has been made with respect to defining the ligand binding properties of the various intracellular lipid binding proteins [5, 8]. The availability of recombinant (mutant) proteins and their crystallographic and solution structures has been instrumental for this, and has revealed remarkable differences among of these proteins for their ligand specificity and affinity, further underscoring the functional uniqueness of each member of the family of intracellular lipid binding proteins. For instance, I(ntestinal-type)-$FABP_c$ binds fatty acids in a bent conformation, and L(iver-type)-$FABP_c$ and intestinal bile acid-binding protein (I-BABP) bind bulky ligands or two fatty acids simultaneously [5, 8]. H(eart-type)- and B(rain-type)-$FABP_c$ bind their ligand in a U-shaped conformation, while H-$FABP_c$ preferentially binds n-6 polyunsaturated fatty acids and B-$FABP_c$ preferentially n-3 polyunsaturated fatty acids [8].

Physiological significance of lipid binding proteins

Both membrane-associated and intracellular lipid binding proteins are increasingly being linked with specific aspects of cellular lipid metabolism, not only in a facilatory manner but, importantly, also as regulators of lipid metabolism. For instance, an intriguing observation is that, at least in muscle, cellular fatty acid uptake is acutely regulated by translocation of fatty acid translocase (FAT/CD36) from intracellular stores to the plasma membrane, which is analogous to the regulation of glucose uptake by the membrane glucose transporter GLUT4 [9]. Translocation of FAT/CD36 to the membrane is triggered by muscle contraction and by insulin [10]. Another membrane protein, fatty acid transport protein-1 (FATP1), facilitates the import of long-chain and very long-chain fatty acids, and may also possess esterification activity toward very long-chain substrates, leading to the hypothesis that FATP1 may be a bi-functional protein with separable transport and catalytic activities [11].

The simultaneous occurrence in the small intestine of several members of the family of intracellular lipid binding proteins has interested many researchers to unveil the specific functions of these proteins. I-FABP$_c$ is mainly contained in the medial portion of the small intestine, while L-FABP$_c$ occurs predominantly in the proximal and I-LBP in the distal part. Targeted gene disruption of I-FABP$_c$ in murine intestine, however, does not compromise dietary fat absorption *in vivo*, but, on the contrary, increased the assimilation of dietary fat [12, 13]. This finding demonstrates that I-FABP$_c$ is not specifically required for intestinal fat absorption, and suggests that the expression of multiple types of intracellular lipid binding proteins in the small intestine represents a strategy for functional redundancy to ensure fat absorption [13]. On the other hand, I-FABP$_c$ null mice were found to develop insulin resistance [13]. Thus, while, for instance, L-FABP$_c$ may compensate for the loss of I-FABP$_c$ so as to apparently make up for the reduced fatty acid uptake, I-FABP$_c$ likely may serve other functions which, in its absence, are not or only poorly compensated for and eventually elicit insulin resistance.

Lipid binding proteins and the regulation of gene expression

It has clearly been demonstrated that long-chain fatty acids directly modulate gene expression by binding to ligand-activated transcription factors, such as the peroxisome proliferator-activated receptor (PPAR), of which three subtypes occur [14]. In this manner, PPARs can also be regarded as cellular lipid binding proteins. While fatty acids are agonists to PPAR, coenzyme A esters of fatty acids were shown *in vitro* to behave as antagonists [15]. Recently, it was shown that L-FABP$_c$ interacts with PPARs, suggesting that L-FABP$_c$ serves to directly traffic its ligand to the nucleus, and thereby directly functions in the regulation of gene expression [16]. Because PPARs have been implicated in many genes expressing proteins involved in lipid metabolism, this latter finding places the intracellular lipid binding proteins in a pivotal position in lipid metabolism.

Interestingly, evidence was presented that in the distal small intestine bile acids regulate gene expression in a similar fashion, which involves the binding of a (ligand-activated) nuclear receptor (farnesoid-X-receptor) to a bile acid-responsive element. Furthermore, it has been hypothesized that intestinal bile acid-binding protein (I-BABP) participates in the carrying of the bile acid ligand to the nuclear receptor, similar to that described for L-FABP$_c$ [17].

Pathophysiology of cellular lipid binding proteins

Various pathophysiological conditions have been associated with alterations in the expression level or subcellular localization of the cellular lipid binding proteins, or with the occurrence of mutant proteins. Conversely, mouse models that are null for certain lipid binding proteins have been created by targeted gene disruption, and are providing support for some of the proposed functions for these proteins, while they also suggest additional possible functions, especially with respect to the etiology of disease. Evidence was presented that in obese Zucker rats, an established model of obesity and insulin resistance (type-2 diabetes), increased rates of fatty acid uptake by heart, skeletal muscle and adipose tissue could be associated with increased plasmalemmal contents of plasma membrane FABP (FABP$_{pm}$; heart and adipose tissue) or of FAT/CD36 (heart, muscle and adipose tissue), but without changes in the expression of these proteins in either tissue [18]. These observations suggest that in obesity and insulin resistance some of the membrane transporters are permanently relocated from an intracellular pool to the plasma membrane, thereby eliciting increased fatty acid uptake rates.

Human FAT/CD36 deficiency has been well documented to result in decreased or absent uptake of long-chain fatty acids by the heart [19, 20]. Studies with FAT/CD36 null mice have confirmed these findings and further support a role for FAT/CD36 in the regulation of cellular fatty acid transport. On the other hand, despite the occurrence of FAT/CD36 in healthy enterocytes of the jejunal villi, intestinal lipid absorption was not affected in FAT/CD36 deficient mice [21], suggesting that this protein is not essential in intestinal lipid absorption or that the deficiency is fully compensated by counter regulation of other fatty acid transporters.

Insight into the functioning of E(pidermal-type)-FABP$_c$ was obtained from the creation of E-FABP null mice [22]. In functional analysis of the skin, the basal transepidermal water loss of the adult homozygous mice showed lower levels compared with the wild type mice, suggesting the functioning of E-FABP$_c$ in establishing the water permeability barrier of the skin. Finally, extracellular fatty acid-binding protein (Ex-FABP), a 21-kD lipocalin that was previously reported to be expressed in cartilage, muscle fibers and granulocytes during chicken embryo development, is also found in adult cartilage but only in pathological conditions. Although the function of this protein is not yet clear, the data suggest that Ex-FABP is a stress protein expressed in tissues were active remodelling is taking place [23]. The latter findings underline the multifunctionality of lipid binding proteins.

Concluding remarks

Thirty years after the discovery of cytoplasmic FABP [24, 25], a group of 13–15 kD intracellular lipid binding proteins has evolved which show a highly conserved tertiary structure, but reveal distinct binding characteristics, rendering similar but also unique functions to these proteins within the cell. Thus, besides the more general function of solubilizing lipid ligands (long-chain fatty acids, retinoids, bile acids), some of these proteins appear to display specific interactions with membranes while others do not. In addition, direct interactions between H-FABP$_c$ and FAT/CD36 [26] and between L-FABP$_c$ and PPARs [16] suggest that ligand transfer may be dependent on protein-protein interaction and suggest that the cytoplasmic lipid binding proteins are involved in targeting of ligands to specific receptors or metabolic pathways. In this way, the properties of the lipid-protein complex, rather than those of the lipid ligand itself, determine the fate of the ligand in the cell.

In the last decade, membrane-associated FABPs have emerged as another group of cellular proteins affecting lipid metabolism [1, 27]. Evidence for their involvement in the cellular uptake (or release) of fatty acids is compelling, although the exact nature of action, especially in relation to passive diffusional uptake of fatty acids, is not yet elucidated. A striking newly discovered feature is the short-term regulation of fatty acid uptake into muscle by cellular redistribution of FAT/CD36, which adds another level of regulation of cellular fatty acid uptake and indicates a pivotal role for FAT/CD36 in this process. The increasing number of reports that deficiencies or altered functioning of both membrane-associated and cytoplasmic lipid binding proteins can be associated with disease states suggests that the cellular lipid binding proteins may be suitable targets for therapeutic intervention.

Acknowledgements

The Maastricht Conference on Cellular Lipid Binding Proteins was made possible by generous donations from the Royal Netherlands Academy of Sciences (KNAW), the Netherlands Heart Foundation, the Cardiovascular Research Institute Maastricht (CARIM), Bristol-Myers Squibb, Laxdale Ltd., Roche Diagnostics GmbH, HyCult biotechnology b.v., and Packard Bioscience b.v.

References

1. Glatz JFC, Storch J: Unravelling the significance of cellular fatty acid-binding proteins. Curr Opin Lipidol 12: 267–274, 2001
2. Van der Vusse GJ, van Bilsen M, Glatz JFC, Hasselbaink DM, Luiken JJFP: Critical steps in cellular fatty acid uptake and utilization. Mol Cell Biochem 239: 9–15, 2002
3. Hamilton JA, Kamp F: Mechanism of cellular uptake of long-chain fatty acids: do we need cellular proteins? Mol Cell Biochem 239: 17–23, 2002
4. Hamilton JA, Kamp F: How are free fatty acids transported in membranes? Is it by proteins or by free diffusion through the lipids? Diabetes 48: 2255–2269, 1999
5. Storch J, Thumser EA: The fatty acid transport function of fatty acid-binding proteins. Biochim Biophys Acta 1486: 28–44, 2000
6. Storch J, Veerkamp JH, Hsu KT: Similar mechanisms of fatty acid transfer from human anal rodent fatty acid-binding proteins to membranes: Liver, intestine, heart muscle, and adipose tissue FABPs. Mol Cell Biochem 239: 25–33, 2002
7. Weisiger RA: Cytosolic fatty acid binding proteins catalyze two distinct steps in intracellular transport of their ligands. Mol Cell Biochem 239: 35–43, 2002
8. Hanhoff T, Lücke C, Spener F: Insights into binding of fatty acids by fatty acid binding proteins. Mol Cell Biochem 239: 45–54, 2002
9. Bonen A, Luiken JJFP, Arumugam Y, Glatz JFC, Tandon NN: Acute regulation of fatty acid uptake involves the cellular redistribution of fatty acid translocase. J Biol Chem 275: 14501–14508, 2000
10. Bonen A, Luiken JJFP, Glatz JFC: Regulation of cellular fatty acid transport and membrane transporters in health and disease. Mol Cell Biochem 239: 181–192, 2002
11. Lewis SE, Listenberger LL, Ory DS, Schaffer JE: Membrane topology of the murine fatty acid transport protein 1. J Biol Chem 276: 37042–37050, 2001
12. Vassileva G, Huwyler L, Poirier K, Agellon LB, Toth MJ: The intestinal fatty acid binding protein is not essential for dietary fat absorption in mice. Faseb J 14: 2040–2046, 2000
13. Agellon LB, Toth MJ, Thomson ABR: Intracellular lipid binding proteins of the small intestine. Mol Cell Biochem 239: 79–82, 2002
14. Van Bilsen M, van der Vusse GJ, Gilde AJ, Lindhout M, van der Lee KAJM: Peroxisome proliferator-activated receptors: Lipid binding proteins controlling gene expression. Mol Cell Biochem 239: 131–138, 2002
15. Helledie T, Jørgensen C, Antonius M, Krogsdam AM, Kratchmarova I, Kristiansen K, Mandrup S: Role of adipocyte lipid-binding protein (ALBP) and acyl-CoA binding protein (ACBP) in PPAR-mediated transactivation. Mol Cell Biochem 239: 157–164, 2002
16. Wolfrum C, Borrmann CM, Borchers T, Spener F: Fatty acids and hypolipidemic drugs regulate peroxisome proliferator-activated receptors alpha- and gamma-mediated gene expression via liver fatty acid binding protein: A signaling path to the nucleus. Proc Natl Acad Sci USA 98: 2323–2328, 2001
17. Landrier JF, Grober J, Zaghini I, Besnard P: Regulation of the ileal bile acid-binding protein gene: An approach to determine its physiological function(s). Mol Cell Biochem 239: 149–155, 2002

18. Luiken JJFP, Arumugam Y, Dyck DJ, Bell RC, Pelsers MMAL, Turcotte LP, Tandon NN, Glatz JFC, Bonen A: Increased rates of fatty acid uptake and plasmalemmal fatty acid transporters in obese Zucker rats. J Biol Chem 276: 40567–40573, 2001
19. Tanaka T, Nakata T, Oka T, Ogawa T, Okamoto F, Kusaka Y, Sohmiya K, Shimamoto K, Itakura K: Defect in human myocardial long-chain fatty acid uptake is caused by FAT/CD36 mutations. J Lipid Res 42: 751–759, 2001
20. Febbraio M, Guy E, Coburn C, Knapp FF, Beets AL, Abumrad NA, Silverstein RL: The impact of overexpression and deficiency of fatty acid translocase (FAT)/CD36. Mol Cell Biochem 239: 193–197, 2002
21. Goudriaan JR, Dahlmans VEH, Febbraio M, Teusink B, Romijn JA, Havekes LM, Voshol PJ: Intestinal lipid absorption is not affected in CD36 deficient mice. Mol Cell Biochem 239: 199–202, 2002
22. Owada Y, Suzuki I, Noda T, Kondo H: Analysis on the phenotype of E-FABP-gene knock-out mice. Mol Cell Biochem 239: 83–86, 2002
23. Descalzi Cancedda F, Dozin B, Zerega B, Cermelli S, Gentili C, Cancedda R: Ex-FABP, extracellular fatty acid binding protein, is a stress lipocalin expressed during chicken embryo development. Mol Cell Biochem 239: 221–225, 2002
24. Ockner RK, Manning JA, Poppenhausen RB, Ho WK: A binding protein for fatty acids in cytosol of intestinal mucosa, liver, myocardium, and other tissues. Science 177: 56–58, 1972
25. Ockner RK. Historic overview of studies on fatty acid-binding proteins. Mol Cell Biochem 98: 3-9, 1990
26. Spitsberg VL, Matitashvili E, Gorewit RC: Association and co-expression of fatty-acid-binding protein and glycoprotein CD36 in the bovine mammary gland. Eur J Biochem 230: 872–878, 1995
27. Abumrad N, Coburn C, Ibrahimi A: Membrane proteins implicated in long-chain fatty acid uptake by mammalian cells: CD36, FATP and FABPpm. Biochim Biophys Acta 1441: 4–13, 1999

Molecular and Cellular Biochemistry **239**: 9–15, 2002.

Critical steps in cellular fatty acid uptake and utilization

Ger J. van der Vusse, Marc van Bilsen, Jan F.C. Glatz, Danny M. Hasselbaink and Joost J.F.P. Luiken

Department of Physiology, Cardiovascular Research Institute Maastricht, Maastricht University, Maastricht, The Netherlands

Abstract

Despite decades of extensive research, the transport routes, mechanisms of uptake and points of flux control of long-chain fatty acids (FA) in mammalian organs are still incompletely understood. In non-fenestratred organs such as heart and skeletal muscle, membrane barriers for blood-borne FA are the luminal and abluminal membranes of endothelial cells, the sarcolemma and the mitochondrial membranes. Transport of FA through the phospholipid bilayer of the cellular membrane is most likely accomplished by diffusion of protonated FA. Evidence is accumulating that membrane-associated proteins, such as plasmalemmal fatty acid-binding protein ($FABP_{pm}$) and fatty acid translocase (FAT/CD36), either alone or in conjunction with albumin binding protein (ABP), are instrumental in enhancing the delivery of FA to the cellular membrane.

Inside the cell, cytoplasmic fatty acid-binding proteins ($FABP_c$) are involved in diffusion of FA from the plasmalemma to the intracellular sites of conversion, such as the mitochondrial outer membrane. After conversion of FA to FACoA, the fatty acyl chain is transported across the mitochondrial inner membrane in a carnitine-mediated fashion.

Uptake and utilization of FA by muscle cells are finely tuned, most likely to avoid the intracellular accumulation of FA, as these are cytotoxic at high concentrations. On a short-term basis, net uptake is, among others, regulated by intracellular translocation of FAT from intracellular stores to the sarcolemma and by the concentration gradient of FA across the sarcolemma. The latter implies that, among others, the rate of FA utilization determines the rate of uptake. The rate of utilization is governed by a variety of factors, including malonylCoA, the ratio acetylCoA/CoA and the availability of competing substrates such as glucose, lactate, and ketone bodies. Long-term regulation of uptake and utilization is accomplished by alterations in the rate of expression of genes, encoding for FA-handling proteins. Circumstantial evidence indicates that FA themselves are able to modulate the expression of FA-handling genes via nuclear transcription factors such as peroxisome proliferator-activated receptors (PPARs). (Mol Cell Biochem **239**: 9–15, 2002)

Key words: fatty acid-binding proteins, endothelium, sarcolemmal transport

Introduction

Long-chain fatty acids (FA) are serving a variety of roles in cell homeostasis. In addition to constituents of membrane phospholipids they are substrates for mitochondrial energy conversion and act as energy store after esterification into the triacylglycerol pool. Moreover, FA are involved in co- and post-translational modification of proteins, i.e. protein acylation, and modulate gene expression via interaction with members of the PPAR family [1–4]. Since most organs, including cardiac and skeletal muscle, are not capable of *de novo* synthesis FA and possess limited amounts of FA stored in the cytoplasmic compartment, they heavily rely on exogenous FA. The mechanisms underlying the uptake and transport of FA are incompletely understood. In this brief review the potentially critical steps in the transport route of blood-borne FA will be discussed with special reference to heart and skeletal muscle. For more extensive reviews on this subject matter see [5, 6].

Address for offprints: G.J. van der Vusse, Cardiovascular Research Institute Maastricht (CARIM), Department of Physiology, Maastricht University, P.O. Box 616, 6200 MD Maastricht, The Netherlands (E-mail: vandervusse@fys.unimaas.nl)

Supply of long-chain fatty acids to organs

Circulating FA are supplied to the organ either complexed to plasma albumin or esterified into glycerol, forming the triacylglycerol core of circulating chylomicrons and very low density lipoproteins [5]. Prior to uptake by the parenchymal cells, FA are liberated from plasma triacylglycerols by the catalytic action of lipoprotein lipase, attached to the luminal side of the endothelium [2]. Both cellular membranes composed of a phospholipid bilayer, and aqueous spaces such as the interstitial and intracellular compartment represent constraints for FA on their route from blood to the mitochondrial outer membrane due to the amphiphilic nature of the FA molecule. The mitochondrial outer membrane is the site where the first committed step in overall FA metabolism takes place, i.e. the activation of FA to fatty acyl CoA catalyzed by acyl CoA synthetase (ACS, Fig. 1).

The endothelium and interstitial compartment

In liver, the undissociated FA–albumin complex diffuses through the fenestrae of the endothelium into the spaces of Disse and the interstitial compartment [7]. However, in organs with non-fenestrated endothelium such as cardiac and skeletal muscle the endothelial cells lining the microvascular compartment form a serious transport barrier for blood-borne FA. Accordingly, FAs need to be dissociated from their plasma albumin carrier prior to transport through the endothelium [8].

Circumstantial evidence indicates that in muscle tissue diffusion of FA through the endothelial clefts is inconsequential [8]. The bulk of FA most likely diffuses through the luminal membrane of the endothelial cell, the endothelial cytoplasm and subsequently through the abluminal membrane to be transmitted to albumin present in the interstitial compartment. Detailed information on the mechanisms underlying trans-endothelial transport of FA in heart and skeletal muscle is, however, lacking. Earlier studies suggest that specific interaction of the albumin–FA complex with the outer leaflet of the luminal membrane or with proteins associated with the membrane facilitates the dissociation of the FA–albumin complex and, hence, the transfer of FA from the vascular to the interstitial compartment [9]. Endothelial cytoplasmic fatty acid-binding protein, $FABP_c$, may play a role in the transport of FA from the luminal to the abluminal endothelial membrane. The diffusion of FA from the abluminal membrane to the plasmalemma of the parenchymal cell is facilitated by interstitial albumin (Fig. 1).

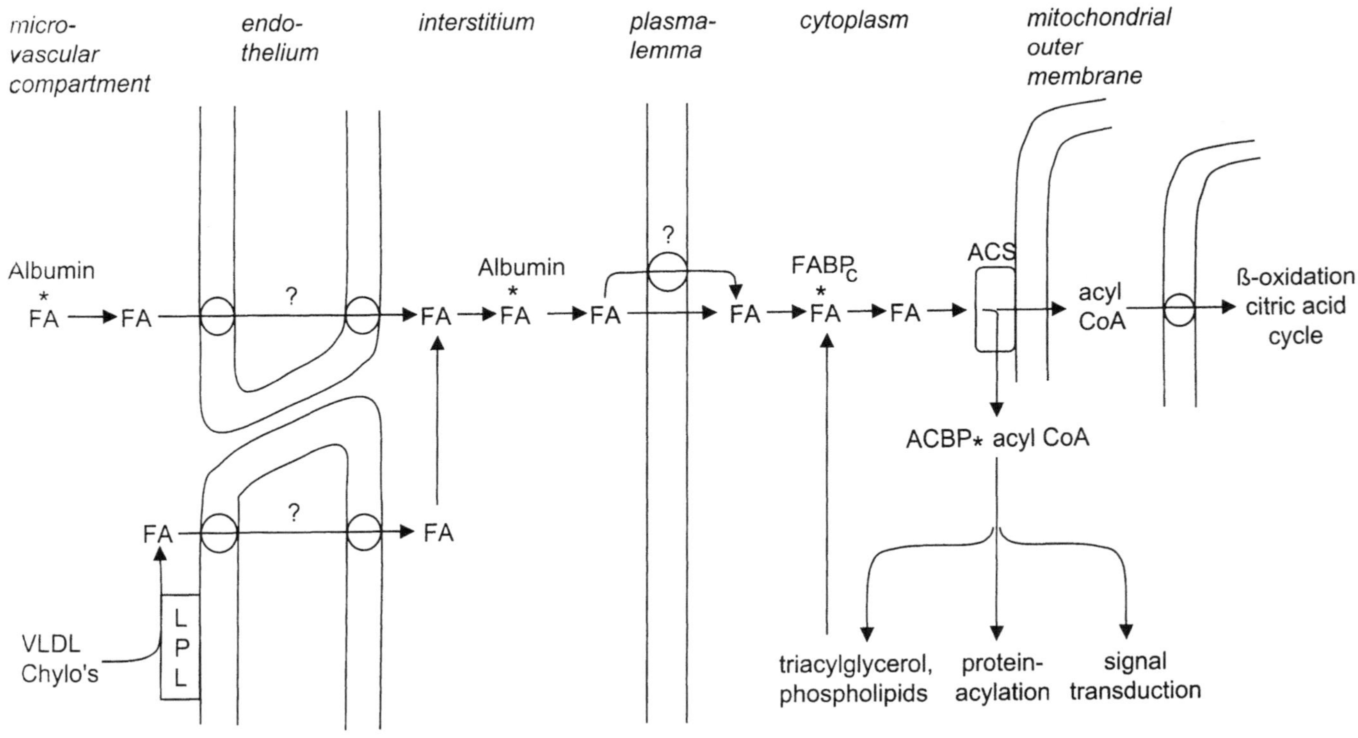

Fig. 1. Schematic representation of long-chain fatty acid uptake and transport in organs with non-fenestrated endothelium. FA – long-chain fatty acid; $FABP_c$ – cytoplasmic fatty acid-binding protein; ACS – acyl CoA synthetase; ACBP – acyl CoA-binding protein; VLDL – very low density lipoprotein; chylo's – chylomicrons; LPL – lipoprotein lipase.

The sarcolemma

Recent findings indicate that the plasmalemma (sarcolemma) of cardiac and skeletal muscle cells represent a major constraint for FA, bound to albumin in the interstitial compartment [10]. The traditional view of transport of FA across biological membranes, including the sarcolemma, composed of a phospholipid bilayer is by passive diffusion of protonated FA driven by the concentration difference of FA in the two compartments separated by the membrane [11–13]. Studies on experimental membranes revealed that diffusion of FA through the phospholipid bilayer is fast, leading the investigators to conclude that auxiliary proteins are not required for transmembrane movement of FA [13]. Evidence, however, is accumulating that proteins are involved in the overall transport of FA from the interstitial compartment to the cytoplasm of the parenchymal cell. At present, at least three different membrane-associated proteins have been identified putatively involved in FA trafficking across the membrane. Stremmel *et al.* [14] identified a 40–43 kDa protein in the plasmalemma of hepatocytes, displaying high affinity for FA. This protein, designated plasmalemmal fatty acid-binding protein, $FABP_{pm}$, is a peripherally bound protein and expressed in a variety of tissues, including heart and skeletal muscle. Antibodies against $FABP_{pm}$ significantly inhibit the cellular uptake of FA [15]. Later studies by Abumrad *et al.* [16] established the existence of an 88 kDa integral membrane protein, called fatty acid translocase, FAT, which turned out to be homologous to the human leukocyte differentiation antigen CD36. This protein is expressed in a host of tissues, but not in liver [16, 17]. Transfection of this membrane protein into fibroblast induced a saturable, high-affinity component of oleate uptake [18]. Thirdly, Schaffer and Lodish [19] discovered a 63 kDa protein via expression cloning of an adipocyte cDNA library in COS7 cells. More recent findings [20] indicate that this integral membrane protein, designated fatty acid transport protein, FATP, possesses catalytic properties, promoting the conversion of FA into fatty acyl CoA, and may therefore modulate the uptake of FA by creating an intracellular sink for FA. Finally, a series of membrane-associated proteins has been discovered expressing a relatively high affinity for albumin [21–23].

In addition to the earlier mentioned inhibitory effect of antibodies of $FABP_{pm}$ on cellular FA uptake, the outcome of recently performed experiments points towards a quantitatively important role of membrane-associated proteins in transmembrane FA transport. Mild treatment of isolated cardiomyocytes with trypsin significantly reduced the uptake of FA [24]. Phloretin, an general inhibitor of membrane-associated transport proteins, severely hampered cardiomyocyte FA uptake. Of interest is the fact that also sulfo-*N*-succinimidyl oleate (SSO), a specific ligand of FAT/CD36, inhibited FA uptake by isolated cardiomyocytes. More recently it was shown that FA uptake by sarcolemmal right-side out giant vesicles prepared from rat heart is also blocked by trypsin, phloretin, SSO, and antibodies against $FABP_{pm}$ [25]. Since these vesicles lack FA converting enzymes an indirect effect of these inhibitory agents on membrane FA transport is rather unlikely. Interestingly, the inhibitory actions of anti-$FABP_{pm}$ and SSO were found to be non-additive, which indicates that $FABP_{pm}$ and FAT/CD36 cooperate in facilitating the translocation of FA across the plasmalemma.

In order to unify the notion that the transport of FA across the phospholipid bilayer of biological membranes is accomplished by passive diffusion with the recent findings that membrane-associated proteins are involved in cellular FA uptake, we propose the following paradigm. Proteins such as albumin-binding proteins (ABP) are trapping the albumin–FA complex at close vicinity of the outer leaflet of the phospholipid bilayer (Fig. 2). The interaction of the albumin–FA complex with ABPs facilitates the dissociation of the ligand from the complex. FA accumulating in the stagnant water layer of the extracellular compartment bind to the peripheral membrane protein $FABP_{pm}$ and are subsequently transmitted to the integral membrane protein, FAT/CD36. Alternatively, FA is directly transferred from albumin to $FABP_{pm}$. This process substantially increases the amount of FA available for transport through the membrane. The latter is accomplished by passive diffusion of protonated FA, most likely by a flip-flop mechanism of the fatty acyl moieties through the outer and inner leaflet of the phospholipid bilayer. It can, however, not be excluded that the integral protein is physically involved in translocating non-protonated long-chain fatty acids through the phospholipid bilayer by facilitated diffusion. Moreover, membrane-associated proteins may also play a crucial role in the release of FA from the inner leaflet of the plasmalemma into the aqueous cytoplasm of the parenchymcal cell.

The cytoplasm

Pioneering studies of Ockner *et al.* [26] have disclosed the presence of cytoplasmic proteins with a high affinity for FA. Mathematical models and *in vitro* studies have later shown that these proteins, designated fatty acid-binding proteins, $FABP_c$, are indispensable for bulk transport of FA from the plasmalemma to the main intracellular site of conversion, i.e. the mitochondrial outer membrane, home of fatty acyl CoA synthetase [27, 28]. $FABP_c$, a 14–15 kDa protein, consists of a family of at least 13 members, slightly differing in amino acid sequence. The physiological role of $FABP_c$ in cardiac myocytes has recently been firmly established by Binas *et al.* [29] and Schaap *et al.* [30]. They investigated the FA uptake in the heart of intact mice carrying a targeted disruption of the gene coding for heart-type $FABP_c$, and FA utilization by

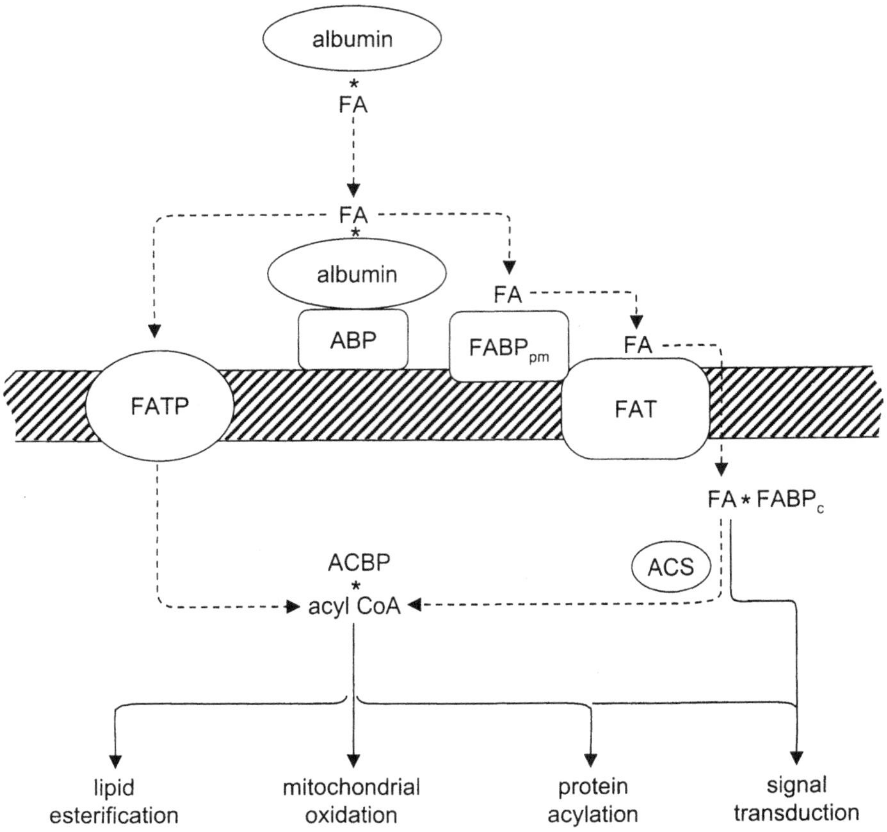

Fig. 2. Hypothetical transport of long-chain fatty acids across the plasmalemma of a muscle cell. FA – long-chain fatty acid; ABP – albumin-binding protein; $FABP_{pm}$ – plasmalemmal fatty acid-binding protein; FAT – fatty acid translocase; FATP – fatty acid transport protein; ACBP – acyl CoA-binding protein; ACS – acyl CoA synthetase.

cardiomyocytes isolated from H-$FABP_c$ deficient mouse heart, respectively. Other studies on immortalized cell lines and other types of parenchymal cells underline the notion that $FABP_c$ is indeed involved in cytoplasmic FA trafficking [31].

FA may be transmitted from the inner leaflet of the phospholipid bilayer to $FABP_c$ (1) by collisional interaction of $FABP_c$ with the membrane or (2) after release of FA into the water layer adjacent to the membrane prior to binding to $FABP_c$. Evidence provided by Storch *et al.* [32] indicates that the former mechanism is true for muscle $FABP_c$ and the latter for liver $FABP_c$. It is of interest to note that Spitsberg *et al.* observed in the bovine mammary gland physical interaction between soluble FABP and membrane-associated FAT/CD36 [33].

It remains to be established whether $FABP_c$ physically interacts with fatty acyl CoA synthetase or with phospholipid domains in close vicinity of the enzyme in the mitochondrial outer membrane. This kind of interaction may facilitate and regulate the transmission of cytoplasmic FA to the enzyme, which catalyses the first committed step in intracellular FA metabolism. The product, fatty acyl CoA, serves as substrate for reacylation of phospholipids, formation of the intracellular lipid store, triacylglycerol, and/or intramitochondrial oxidation to CO_2 and H_2O after transport through the mitochondrial inner membrane in a carnitine-mediated process (Fig. 1). Moreover, acyl CoA is also substrate for acylation of proteins. This type of acylation regulates the function and localization of a variety of cellular proteins via co-translational (myristoylation) and post-translational (palmitoylation) modification [34]. Another protein involved in cellular FA handling, namely the cytoplasmic acyl CoA-binding protein (ACBP), is thought to be involved in the vectorial delivery of the fatty acyl moieties to the distinct metabolic pathways [35].

Regulation of cellular FA utilization

In tissues, such as heart and skeletal muscle, FA are important oxidizable substrates. In cardiac muscle over 70% of energy consumed for electro-mechanical activity is covered by mitochondrial oxidation of FA [2]. The accompanying FA flux from the vascular compartment to the mitochondrial matrix is on the order of 100–150 nmol fatty acyl moieties/g/min. In resting skeletal muscle blood-borne FA utilization is minor, but during prolonged muscular contractile activity FA uptake readily increases to 200 nmol/g/min [36].

Under physiological conditions, uptake is finely tuned to mitochondrial utilization, to prevent intracellular accumulation of FA, being highly toxic at elevated intracellular concentrations [2]. The endogenous triacylglycerol pool acts most likely as an intracellular capacitor to absorb temporary imbalances in uptake and utilization of exogenous FAs. The intracellular FA concentration is also kept as low as possible to create a concentration difference between interstitial and cytoplasmic FA, which is required as driving force for the transition of FA from the extracellular to the intracellular compartment. Noteworthy is the observation of Kiens *et al.* [37] that during skeletal muscle activity at moderate intensity the cellular FA concentration decreases compared to the resting state, promoting the diffusion of blood-borne FA to the muscle cell interior.

Recent findings of Bonen *et al.* [38] indicate that the capacity to extract FA from extracellular sources is also regulated during acute changes in muscle energy requirements. They observed that sarcolemmal giant vesicles prepared from muscles that were electrically stimulated for 25 min display a substantially higher rate of FA uptake than vesicles from non-stimulated muscles. The increase was paralleled by a higher sarcolemmal content of FAT/CD36, most likely due to translocation of the FA transporting protein from an intracellular pool to the cell membrane. Of interest is that insulin is also able to promote the translocation of FAT/CD36 from the intracellular storage site to the sarcolemma (Luiken *et al.*, in press). In addition to translocation, the biological activity of FAT/CD36 may be acutely modulated by phosphorylation of the membrane-associated protein (Luiken *et al.*, manuscript in preparation).

Mitochondrial long-chain acyl CoA formation and/or subsequent oxidation of fatty acyl chains via β-oxidation and citric acid cycle activity offers (an) additional site(s) of control of the rate of muscular FA utilization. The actual regulation of mitochondrial FA oxidation is a multifactorial process in which malonyl CoA, the ratio acetyl CoA/CoA and NADH/NAD^+ play prominent roles [2]. Moreover, the availability of competing substrates such as lactate has profound effects on the rate of muscular FA oxidation. The collected findings indicate that both the rate of mitochondrial FA utilization and the rate of transport of FA from the interstitial compartment into the myocyte interior contribute to the overall rate of muscle cell FA utilization.

Long-term adaptation of cellular FA uptake and transport

Various lines of evidence suggest that chronic alterations in the cellular utilization of FA can be attributed to adjustment of the expression of genes encoding for proteins involved in FA uptake, transport and metabolism. Experiments performed on adipocytes and hepatocytes indicated that substrates for energy conversion themselves are capable of modulating the transcriptional activity of metabolic genes [39–42]. We observed recently in cultured neonatal cardiac muscle cells [4, 43] that exogenous FA also control cardiac transcriptional activity. Chronic exposure of the neonatal cardiomyocytes to physiological concentrations of oleate and palmitate in the incubation medium for 48 h resulted in a 3-fold increase in the mRNA levels of FAT/CD36 and FABP [43]. In addition, the transcription of three mitochondrial FA-handling enzymes, acyl CoA synthetase, carnitine palmitoyl transferase-1 and long-chain acyl-CoA dehydrogenase increased substantially. In this respect it is of interest to note that in the healthy rat, deprivation of food for 48 h increased the plasma FA concentration almost 2-fold, associated with an increase in the cardiac tissue levels of FAT/CD36 and $FABP_c$ mRNA [44]. Moreover, the maturational increase in rat cardiac FA oxidation, starting a couple of days after birth [45, 46], was found to be accompanied by an increased content of FAT/CD36 and of cytoplasmic FABP in the myocardium [17, 47].

The finding that chronic exposure of cardiomyocytes to exogenous FA results in the transcriptional upregulation of a variety of FA-handling genes indicates that the cells are capable of sensing changes in the supply of FA and/or FA derivatives. Recent studies [3, 48] revealed that the promoter region of FAT/CD36 contains *cis*-regulatory elements with binding sites for transcription factors such as COUP/TF, PPAR, RXR, and HFN-4 [48, 49]. PPAR is a likely candidate for cellular FA sensing, since FA may act as natural ligand of the PPARs [50–52]. In particular, the PPARα and PPARβ isoforms may be involved in modulating transcriptional activity in cardiomyocytes by FA.

Future directions

Despite many attempts to elucidate in full detail the molecular mechanism of cellular fatty acid uptake, transport and utilization, the complete picture of transport routes and regulatory processes is still missing. It is for instance unknown whether FA released from the triacylglycerol core of circulating lipoproteins mix with FA supplied by albumin in the microvascular compartment prior to transport through the endothelium and utilization in the parenchymal cells. The route of FA transport across the endothelium and the mechanisms involved are still enigmatic. The precise mechanism of transport of FA across a biological membrane, consisting of a phospholipid bilayer, is incompletely understood. The question as to whether membrane-associated proteins promote the FA membrane trafficking *per se* or enhance the availability of FA at close vicinity of the membrane remains to be answered. Moreover, it is unclear whether the putative membrane-associated FA binding proteins are acting in concert. The role of albumin-binding proteins attached the cellular membranes in cellular FA uptake and transport remains to be established.

Since evidence is accumulating that proteins involved in cellular FA uptake, such as FAT/CD36, can be mobilized from intracellular sites, future studies should be performed to elucidate the nature of the signalling pathways resulting in translocation of membrane FA-binding proteins.

The fine-tuning between mitochondrial FA oxidation and cellular FA uptake and transport is only partly understood, but it is becoming evident that factors increasing mitochondrial activity also act on signalling pathways involved in FAT/CD36 translocation (Luiken *et al.*, unpublished results). Finally, the investigation on the transcriptional regulation of genes encoding for proteins involved in tissue FA handling state is in its infancy. Detailed information is required to elucidate both the short-term and long-term regulation of FA-handling under physiological conditions and the impact of alterations in FA-handling in the diseased state.

Acknowledgments

The authors are greatly indebted to Mrs. Claire Bollen and Mr. Th. Roemen for their help in preparing the manuscript. Work in the authors' laboratories was supported by the Netherlands Heart Foundation, grants D98.012 and D98.015. Marc van Bilsen is an Established Investigator, and Joost Luiken a Dekker post-doctoral fellow of the Netherlands Heart Foundation.

References

1. Resh MD: Fatty acylation of proteins: New insights into membrane targeting of myristoylated and palmitoylated proteins. Biochim Biophys Acta 1451: 1–16, 1991
2. Van der Vusse GJ, Glatz JFC, Stam HCG, Reneman RS: Fatty acid homeostasis in the normoxic and ischemic heart. Physiol Rev 72: 881–940, 1992
3. Van Bilsen M, Van der Vusse GJ, Reneman RS: Transcriptional regulation of metabolic processes: Implications for cardiac metabolism. Eur J Physiol 437: 2–14, 1998
4. Van Bilsen M, De Vries JE, Van der Vusse GJ: Long-term effects of fatty acids on cell viability and gene expression of neonatal cardiac myocytes. Prostagl Leukotr Essent Fatty Acids 57: 39–45, 1997
5. Van der Vusse GJ, Van Bilsen M, Glatz JFC: Cardiac fatty acid uptake and transport in health and disease. Cardiovasc Res 45: 279–293, 2000
6. Glatz JFC, van Nieuwenhoven FA, Luiken JJ, Schaap FG, Van der Vusse GJ: Role of membrane-associated and cytoplasmic fatty acid-binding proteins in cellular fatty acid metabolisms. Prostagl Leukotr Essent Fatty Acids 57: 373–378, 1997
7. Weisiger RA, Pond S, Bass L: Hepatic uptake of protein-bound ligands: Extended sinusoidal perfusion model. Am J Physiol 261: G872–G884, 1991
8. Bassingthwaighte JB, Noodleman L, Van der Vusse GJ, Glatz JFC: Modeling of palmitate transport in the heart. Mol Cell Biochem 88: 51–58, 1989
9. Van der Vusse GJ, Glatz JFC, Van Nieuwenhoven FA, Reneman RS, Bassingthwaighte JB: Transport of long-chain fatty acids across the muscular endothelium. Adv Exp Med Biol 441: 181–191, 1998
10. Luiken JJ, Glatz JFC, Bonen A: Fatty acid transport proteins facilitate fatty acid uptake in skeletal muscle. Can J Appl Physiol 25: 333–352, 2000
11. Zakim D: Thermodynamics of fatty acid transfer. J Membr Biol 176: 101–109, 2000
12. Zakim D: Fatty acids enter cells by simple diffusion. Proc Soc Exp Biol Med 212: 5–14, 1996
13. Hamilton JA, Kamp F: How are free fatty acids transported in membranes? Is it by proteins or by free diffusion through the lipids? Diabetes 48: 2255–2269, 1999
14. Stremmel W, Strohmeyer G, Borchard F, Kochwa S, Berk PD: Isolation and partial characterization of a fatty acid binding protein in rat liver plasma membranes. Proc Natl Acad Sci USA 82: 4–8, 1985
15. Sorrentino D, Stump DD, Potter BJ, Robinson RB, White R, Kiang CL, Berk PD: Oleate uptake by cardiac myocytes is carrier mediated and involves a 40-kD plasma membrane fatty acid binding protein similar to that in liver, adipose tissue and gut. J Clin Invest 82: 928–935, 1988
16. Abumrad NA, El-Maghrabi MR, Amri E-Z, Lopez E, Grimaldi PA: Cloning of a rat adipocyte membrane protein implicated in binding or transport of long-chain fatty acids that is induced during preadipocyte differentiation. J Biol Chem 268: 17665–17668, 1993
17. Van Nieuwenhoven FA, Willemsen PM, Van der Vusse GJ, Glatz JFC: Coexpression in rat heart and skeletal muscle of four genes coding for proteins implicated in long-chain fatty acid uptake. Int J Biochem Cell Biol 31: 489–498, 1999
18. Ibrahimi A, Sfeir Z, Magharaie H, Amri EZ, Grimaldi PA, Abumrad NA: Expression of the CD36 homolog FAT in fibroblast cells: Effects on fatty acid transport. Proc Natl Acad Sci USA 93: 2646–2651, 1996
19. Schaffer JE, Lodish HF: Expression, cloning and characterization of a novel adipocyte long chain fatty acid transport protein. Cell 79: 427–436, 1994
20. Hermann T, Buchkremer F, Gosch I, Hall AM, Bernlohr DA, Stremmel W: Mouse fatty acid transport protein 4 (FATP4): Characterization of the gene and functional assessment as a very long chain acyl-CoA synthetase. Gene 270: 31–40, 2001
21. Schnitzer JE, Sung A, Horvat R, Bravo J: Preferential interaction of albumin-binding proteins, gp30 and gp18, with conformationally modified albumins. Presence in many cells and tissues with a possible role in catabolism. J Biol Chem 34: 24544–24553, 1992
22. Tiruppathi C, Finnegan A, Malik AB: Isolation and characterization of a cell surface albumin-binding protein from vascular endothelial cells. Proc Natl Acad Sci USA 93: 250–254, 1996
23. Popov D, Hasu M, Ghinea N, Simionescu N, Simionescu M: Cardiomyocytes express albumin binding proteins. J Mol Cell Cardiol 24: 989–1002, 1992
24. Luiken JJFP, Van Nieuwenhoven FA, America G, Van der Vusse GJ, Glatz JFC: Uptake and metabolism of palmitate by isolated cardiac myocytes from adult rats: Involvement of sarcolemmal proteins. J Lipid Res 38: 745–758, 1997
25. Luiken JJFP, Turcotte LP, Bonen A: Protein-mediated palmitate uptake and expression of fatty acid transport proteins in heart giant vesicles. J Lipid Res 40: 1007–1016, 1999

26. Ockner RK, Manning JA, Poppenhausen RB, Ho WK: A binding protein for fatty acids in cytosol of intestinal mucosa, liver, myocardium, and other tissues. Science 177: 56–58, 1972
27. Vork MM, Glatz JFC, Van der Vusse GJ: Modeling intracellular fatty acid transport: Possible mechanistic role of cytoplasmic fatty acid-binding protein. Prostagl Leukotr Essent Fatty Acids 57: 11–16, 1997
28. Peeters RA, Veerkamp JH: Does fatty acid-binding protein play a role in fatty acid transport? Mol Cell Biochem 88: 45–49, 1989
29. Binas B, Danneberg H, McWhir J, Mullins L, Clark AJ: Requirement for the heart-type fatty acid-binding protein in cardiac fatty acid utilization. FASEB J 13: 805–812, 1999
30. Schaap FG, Danneberg H, Binas B, Van der Vusse GJ, Glatz JFC: Impaired uptake and oxidation of long-chain fatty acids by cardiac myocytes of mice lacking heart-type fatty acid-binding protein (H-FABP). Circ Res 85: 329–337, 1999
31. Glatz JFC, Storch J: Unravelling the significance of cellular fatty acid-binding proteins. Curr Opin Lipidol 12: 267–274, 2001
32. Storch J, Thumser AE: The fatty acid transport function of fatty acid-binding proteins. Biochim Biophys Acta 1486: 28–44, 2000
33. Spitsberg VL, Matiatashvili E, Gorewit RC: Association and coexpression of fatty-acid protein and glycoprotein CD36 in the bovine mammary gland. Eur J Biochem 230: 872–878, 1995
34. Dunphy JT, Linder ME: Signalling functions of protein palmitoylation. Biochim Biophys Acta 1436: 245–261, 1998
35. Færgeman NJ, Knudsen J: Role of long-chain fatty acyl-CoA esters in the regulation of metabolism and in cell signalling. Biochem J 323: 1–12, 1997
36. Van der Vusse GJ, Reneman RS: Lipid Metabolism in muscle. In: L.B. Rowell, J.T. Shepherd (eds). Handbook of Physiology. Integration of Motor, Circulatory, Respiratory and Metabolic Control During Exercise. Am Phys Soc 1996, pp 952–994
37. Kiens B, Roemen THM, Van der Vusse GJ: Muscular long-chain fatty acid content during graded exercise in humans. J Appl Physiol 278: E352–E357, 1999
38. Bonen A, Luiken JJFP, Arumugam Y, Glatz JFC, Tandon NN: Acute regulation of fatty acid uptake involves the cellular redistribution of fatty acid translocase. J Biol Chem 12: 14501–14508, 2000
39. Distel RJ, Robinson GS, Spiegelman BM: Fatty acid regulation of gene expression. Transcriptional and post-transcriptional mechanisms. J Biol Chem 267: 5937–5941, 1992
40. Grimaldi PA, Knobel SM, Whitesell RR, Abumrad NA: Induction of aP2 gene expression by nonmetabolized long-chain fatty acids. Proc Natl Acad Sci USA 89: 10930–10934, 1992
41. Jump BD, Clarke SD, Thelen A, Liimatta M: Coordinate regulation of glycolytic and lipogenic expression by polyunsaturated fatty acids. J Lipid Res 35: 1076–1084, 1994
42. Meunier-Durmort C, Poirier H, Niot I, Forest C: Upregulation of the expression of the gene for liver fatty acid-binding protein by long-chain fatty acids. Biochem J 319: 483–487, 1996
43. Van der Lee KAJM, Vork MM, De Vries JE, Willemsen PHM, Glatz JFC, Reneman RS, Van der Vusse GJ, Van Bilsen M: Long-chain fatty acid-induced changes in gene expression in neonatal cardiac myocytes. J Lipid Res 41: 279–293, 2000
44. Van der Lee KAJM, Willemsen PHM, Samec S, Seydoux J, Dullo AG, Pelsers MAAL, Glatz JFC, Van der Vusse GJ, Van Bilsen M: Fasting-induced changes in the expression of genes controlling substrate metabolism in the rat heart. J Lipid Res 42: 1752–1758, 2001
45. Glatz JFC, Veerkamp JH: Postnatal development of palmitate oxidation and mitochondrial enzyme activities in rat cardiac and skeletal muscle. Biochim Biophys Acta 711: 327–335, 1982
46. Lopaschuk GD, Collins-Nakai RL, Itoi T: Developmental changes in energy substrate use by the heart. Cardiovasc Res 26: 1172–1180, 1992
47. Van Nieuwenhoven FA, Verstijnen CPHJ, Abumrad NA, Willemsen PH, Van Eys GJ, Van der Vusse GJ, Glatz JFC: Putative membrane fatty acid translocase and cytoplasmic fatty acid-binding protein are co-expressed in rat heart and skeletal muscles. Biochem Biophys Res Commun 207: 747–752, 1995
48. Sack MN, Disch DL, Rockman HA, Kelly DP: A role for Sp_1 and nuclear receptor transcription factors in a cardiac hypertrophic growth program. Proc Natl Acad Sci USA 94: 6438–6443, 1997
49. Gulick T, Cresci S, Caira T, Moore DD, Kelly DP: The peroxisome proliferator-activated receptor regulates mitochondrial fatty acid oxidative enzyme gene expression. Proc Natl Acad Sci USA 91: 11012–11016, 1994
50. Gottlicher M, Widmark E, Li Q, Gustafsson JA: Fatty acids activate a chimera of the clofibric acid-activated receptor and the glucocorticoid receptor. Proc Natl Acad Sci USA 89: 4653–4657, 1992
51. Forman BM, Chen J, Evans RM: Hypolipidemic drugs, polyunsaturated fatty acids, and eicosanoids are ligands for peroxisome proliferator-activated receptors α and δ. Proc Natl Acad Sci USA 94: 4312–4317, 1997
52. Kliewer SA, Sundseth SS, Jones SA, Brown PJ, Wisely GB, Koble CS, Devchand P, Wahli W, Willson TM, Lenhard JM, Lehmann JM: Fatty acids and eicosanoids regulate gene expression through direct interactions with proxisome proliferator-activated receptors α and γ. Proc Natl Acad Sci USA 94: 4318–4323, 1997

Molecular and Cellular Biochemistry **239**: 17–23, 2002.

Mechanism of cellular uptake of long-chain fatty acids: Do we need cellular proteins?

James A. Hamilton, Wen Guo and Frits Kamp
Department of Physiology and Biophysics, Boston University School of Medicine, Boston, MA, USA

Abstract

Defining the mechanism(s) of long-chain fatty acid movement through membranes is vital to understanding whether or not entry of fatty acids into cells can be controlled at the plasma membrane of a typical cell. Is there a protein that acts as gatekeeper, regulating the amount, and possibly the type, of fatty acid that can enter the cell for metabolism? Is the lipid bilayer of the membrane highly permeable to fatty acids, and is the rate of simple diffusion on the time scale of metabolism? We will briefly review efforts to study diffusion in model lipid membranes that are devoid of proteins. We also present new results using dual fluorescence approaches showing that fatty acids diffuse very rapidly across the plasma membrane of the adipocyte. (Mol Cell Biochem **239**: 17–23, 2002)

Key words: fatty acid-binding protein, FAT/CD36, passive diffusion

Introduction

There is no question or disagreement that proteins are vital to the overall 'uptake' of fatty acids into cells. Many intracellular enzymes, not only those directly involved in lipid metabolism, enhance the uptake of fatty acids. In ongoing investigations of the mechanisms of the influx and efflux of fatty acids through membranes, it is important to maintain a focus on the individual steps involved in transport. If we equate fatty acid transport into cells with uptake and do not separate metabolism from the nonenzymatic processes of uptake, we will learn little about the latter.

Membrane transport minimally involves adsorption, transmembrane movement, and desorption of fatty acids. The most controversial aspect of membrane transport is the translocation step. Can translocation occur by simple diffusion, or is a membrane-associated protein required (e.g. fatty acid translocase [FAT/CD36])? If the protein does not act at this step, does it enhance the rate of adsorption or the partitioning of fatty acids into the plasma membrane? These are not issues of semantics but of function, with very different ramifications with regard to how the importation of fatty acids into cells can be modulated or controlled. Here we discuss briefly some of our studies of model membranes and adipocytes and present our perspective on the membrane transport of fatty acids.

Materials and methods

We obtained *cis*-parinaric acid from Molecular Probes and isolated adipocytes from rats as described previously [1]. Other materials and methods have been described in detail elsewhere [2].

Results and discussion

Model membranes

Our interest in the translocation of fatty acids across membranes arose from the interesting controversy concerning the rapidity with which fatty acids could cross a protein-free phospholipid bilayer. Early physical studies with pyrene-

Present address: F. Kamp, Institute of Physiological Chemistry, University of Munich, Schillerstrasse 44, D-80336 Munich, Germany

Address for offprints: J.A. Hamilton, Department of Physiology and Biophysics, Boston University School of Medicine, 715 Albany Street, Boston, MA 02118, USA (E-mail: hamilton@biophysics.bumc.bu.edu)

labeled fatty acids suggested that transbilayer movement is rapid [3], whereas subsequent studies with anthroyloxy fatty acids suggested that it is slow [4, 5]. The latter results were used to support the hypothesis that a membrane protein is required to translocate the fatty acids from one side of the membrane to the other [6]. However, another series of fluorescence studies with natural fatty acids provided indirect evidence for the rapid transbilayer movement of fatty acids [7].

The kinetics and mechanism of fatty acid translocation across a simple bilayer need to be clear from a biophysical perspective so that the mechanism(s) in a cell membrane can be unraveled. We searched for a more direct method of measuring movement of fatty acids across model membranes. First, we attempted NMR spectroscopy, which we had used successfully to detect bile acids in the inner and outer monolayer of small unilamellar vesicles and to observe the exchange (flip-flop) between these two environments [8]. The flip-flop of unconjugated bile acids is rapid (half times of seconds or less) and dependent on the hydrophobicity of the bile acid: the more hydrophobic, the faster the flip-flop [8]. In contrast, we observed a single homogenous signal for fatty acids (data not shown) , even at the low temperatures that allowed observation of two signals for bile acids. The explanation for the lack of two distinct environments could be either that fatty acids flip-flop very rapidly (milliseconds) or that the environments of fatty acids in the two leaflets are not distinct enough to produce different NMR signals. Thus, the NMR experiments resolved nothing about the fatty acid flip-flop. However, in pH-dependent experiments with bile acids in model membranes, we clearly observed that the un-ionized bile acid flip-flopped more rapidly than the ionized bile acid.

We returned to the method of fluorescence, but with a different idea of how to monitor transmembrane movement. If the un-ionized fatty acid crosses a phospholipid bilayer more rapidly than the ionized form, a pH dye that is trapped inside a phospholipid vesicle should detect the delivery of protons to the aqueous compartment inside the vesicle. The time dependence of transmembrane movement of fatty acids could be monitored by the addition of fatty acids to the external buffer of vesicles that contain a fluorescent pH dye trapped in the inner aqueous volume. This approach arguably would provide the most direct assay of fatty acid flip-flop, since binding of fatty acids to vesicles and the ionization reactions occur extremely rapidly. Furthermore, it is not necessary to add a probe to the fatty acid molecule or to perform separation procedures.

Table 1. Summary of flip-flop rates of fatty acids in model membranes

FA	Short formula	Vesicle type	Flip-flop ($t_{1/2}$ upper limit)	Reference
Fatty acids				
Octanoic acid	8:0	SUV	1 sec	10
Lauric acid	12:0	SUV	10 msec	11
		LUV	25 msec	11
Myristic acid	14:0	SUV	10 msec	11
		LUV	25 msec	11
Palmitic acid	16:0	SUV	10 msec	11
		LUV	35 msec	11
Margaric acid	17:0	SUV	350 msec	12
Stearic acid	18:0	SUV	10 msec	11
		LUV	25 msec	11
Oleic acid	18:1 (cis)	SUV	10 msec	11
		LUV	35 msec	11
Elaidic acid	18:1 (trans)	SUV	1 sec	10
Linoleic acid	18:2 (cis, cis)	SUV, LUV	1 sec	34
Linolelaidic acid	18:2 (trans, trans)	SUV	1 sec	10
Parinaric acid	18:4 (cis, trans, trans, cis)	SUV, (LUV)*	1 sec (50 msec)*	2
Eicosanoic acid	20:0	SUV	1 sec	12
Arachidonic acid	20:4 (all cis)	SUV	1 sec	10
		LUV	1 sec	34
Pentaenoic acid	20:5	SUV, LUV	1 sec	34
Docosanoic acid	22:0	SUV	1 sec	12
Docosapentaenoic acid	22:5	SUV, LUV	1 sec	34
Docosahexaenoic acid	22:6	SUV, LUV	1 sec	34
Tetracosanoic acid	24:0	SUV	1 sec	12
Hexacosanoic acid	26:0	SUV	1 sec	12
Monohydroxy FA				
13-Hydroxyoctadecdienoic acid (HODE)	18:2	SUV, LUV	1 sec	34
15-Hydroxytetraenoic acid (HETE)	20:4	SUV, LUV	1 sec	34

*Unpublished results.

This simple fluorescence method has been used in our laboratory [9, 10] to monitor the flip-flop of numerous fatty acids (Table 1). We adapted this method to measure the flip-flop of very long chain fatty acids that are very insoluble in water by first incorporating the fatty acids into the vesicle bilayer and then adjusting the external pH [11] and to quantitate the kinetics of desorption of fatty acids from vesicles [12]. More recently, we used dual fluorescence approaches to distinguish the adsorption and transmembrane steps of the transport of fatty acids in vesicles [2].

Our results with protein-free model membranes can be summarized briefly, with reference to the three essential steps of fatty acid transport, as schematized in Fig. 1.

Adsorption

- Very fast, probably diffusion-limited, and is not strongly dependent on chain length of the fatty acid.
- Can increase the concentration of fatty acid in the membrane to μM and mM.

Transbilayer movement

- Occurs by preferential translocation of the un-ionized fatty acid
- Fast ($t_{1/2}$ < 1 sec) for all fatty acids studied to date (Table 1).

Desorption

- Strongly dependent on chain length and unsaturation of the fatty acid
- Fast ($t_{1/2}$< 1 sec) for common dietary fatty acids
- Rate-limiting step for the overall transport of fatty acids in phospholipid membranes.

We concluded from such model-system studies that none of the steps in membrane transport of common fatty acids has a universal requirement for a protein transporter. However, on the basis of these studies, we could not conclude or state that passive (or 'simple') diffusion is the only possible mechanism in cell membranes. The simple phospholipid bilayers we have studied (phosphatidylcholine and mixed soy phospholipids with and without cholesterol) are not models for all types of membranes, although they are well accepted and extensively used models for typical mammalian cell membranes. Nevertheless, the physical properties of fatty acids, including their interactions with other lipids and with known fatty acid binding proteins, must be well understood as a foundation for interpreting the membrane transport of fatty acids in cells.

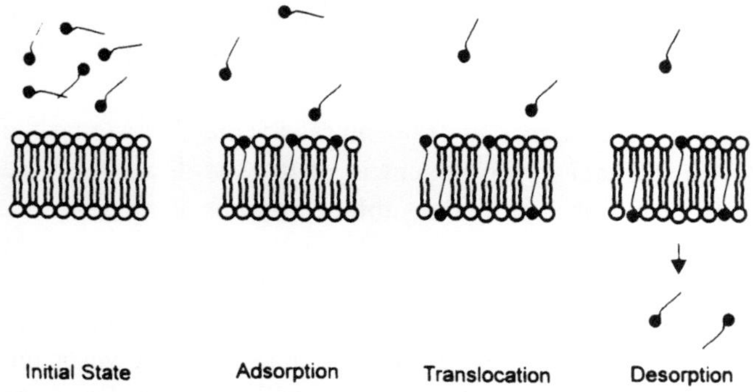

Fig. 1. Simplified scheme of fatty acid transport in a phospholipid bilayer [2].

Adipocytes

The adipocyte has been at the center of the controversy about mechanisms of entry of fatty acids into cells. Some studies concluded that long-chain fatty acids enter adipocytes solely or primarily via protein transporter(s) in the plasma membrane. For instance, one study, using anthroyloxy-labeled stearic acid, concluded that this molecule, but not a medium-chain fatty acid with the same label, required a transporter [13]. The rate of transport, as measured by the rate of accumulation of the probe molecule in triglyceride droplets, was much slower than the rate of metabolism of natural fatty acids. Although the complications of metabolism were eliminated, additional transport steps were encompassed in this study, including desorption from the bilayer and diffusion or transport to the fat droplet. Furthermore, the rate of flip-flop of anthroyloxy-labeled stearic acid may be several orders of magnitude slower than that of stearic acid [14]. In any case, results with covalently labeled fatty acids must be validated by results with natural fatty acids.

Another study of uptake of a fluorescent fatty acid (parinaric acid) into adipocytes showed apparently complete inhibition by phloretin [15]. However, this non-specific 'inhibitor' has numerous effects in membranes, interacting both with membrane lipids and with proteins [16]. In recent studies of uptake of *cis*-parinaric acid into HepG2 cells, we duplicated the results reported for adipocytes. Uptake appeared to be inhibited, but our measurements of intracellular pH showed that transbilayer movement was not diminished by phloretin (Guo and Hamilton, unpublished). We concluded that the fluorescence of parinaric acid could be quenched completely by phloretin but that binding and translocation were unaffected.

A more prevalent view is that both diffusion and protein-mediated transport contribute to the uptake of fatty acids into adipocytes [2, 6]. Estimates of the contributions of the different mechanisms have been made by comparison of uptake with and without inhibitors and by deconvolution of uptake curves. The usefulness of the former analysis can be questioned because the inhibitors generally have low specificity and sometimes inhibit metabolism of fatty acids. In addition, the approaches in these studies required separation procedures, and sometimes phloretin was added in the separation procedure [17]. The analysis of uptake curves often leads to the conclusion that protein-mediated uptake is important at low concentrations of fatty acid and that simple diffusion is important at high concentrations [6]. There is a long-standing disagreement about the application of such an analysis to fatty acids and the possibility of deriving any conclusions about mechanisms from fatty acid uptake curves [18, 19]. Under some of the conditions used for studying and analyzing fatty acid uptake into cells, similar protocols will also yield 'saturation' curves in model phospholipid membranes without any membrane proteins [19]. Although it is not appropriate to conclude that analysis of all fatty acid uptake curves is incorrect, this type of evidence is not conclusive for any mechanism. Furthermore, newer evidence indicates that fatty acids enter adipocytes by diffusion at low ratios of fatty acid to albumin by the flip-flop diffusion mechanism [20].

More recently, it has been argued that studies of genetically engineered animals that lack a putative fatty acid transporter provide a conclusive demonstration of the role of membrane transporters [21, 22]. Do experiments with knockout mice provide final proof of the transporter hypothesis? Not necessarily. Depending on how 'uptake' is measured, it could be diminished if the protein has a direct function in fatty acid metabolism. Moreover, all of the putative fatty acid membrane transporters have other established functions, including enzymatic activity directly related to fatty acid metabolism (fatty acid transport protein, FATP) and uptake of plasma lipoproteins (FAT/CD36). Although CD36 has been described as responsible for most fatty acid uptake into adipocytes and several other cell types [23], the CD36 null mouse is viable [22] and the knockout has not provided insight into the mechanism(s) by which CD36 acts as a transporter. A comparison of adipocytes isolated from wild-type and CD36 null mice found almost no difference in fatty acid uptake at a ratio of fatty acid to albumin of 1.5; the only significant difference reported occurred at a ratio of 0.6 [22]. This result could support the hypothesis that transporters are more important at low ratios of fatty acid to albumin in the serum or interstitial compartment. However, even with this interpretation, there is either a significant contribution of diffusion at the lower fatty acid ratio, or other protein transporters are contributing to fatty acid uptake. Another caution about interpreting these results is that the endogenous levels of unesterified fatty acid in each cell population before addition of the radiolabeled fatty acid were not reported. If a low concentration of radiolabeled fatty acid is added to different cell populations with differing levels of endogenous fatty acid and without pre-equilibration of the label, comparison of uptake is confounded by exchange of labeled fatty acids with cellular fatty acids.

Many investigators have reservations about the methods used to measure fatty acid uptake into cells [13, 24, 25]. The most commonly used methods employ separation procedures, usually at low temperatures and often with the addition of phloretin [17, 26]. These methods measure the fatty acids that remain bound to the cells during the separation procedures but do not provide information about whether any of the fatty acids have translocated across the membrane. Other approaches that monitor the formation of products of intracellular fatty acid metabolism show translocation of fatty acids through the membrane but encompass additional steps of metabolism and intracellular transport.

To address such issues, we have used new assays for cellular uptake of fatty acids that discriminate the events of membrane transport from those of metabolism and that further discriminate among the three key steps of membrane transport of fatty acids: adsorption, transmembrane movement, and desorption. Other goals of our approaches are:

- to provide new approaches based on a specific hypothesis with testable corollaries;
- to eliminate the need to separate fatty acid donors and cells;
- to use natural fatty acids instead of probe molecules.

In applying our hypothesis-based approach to the study of cells, we first demonstrated that externally added fatty acids caused an immediate decrease in intracellular pH in pancreatic β-cells [27] and adipocytes [1]. We interpreted this result as evidence for the mechanism of diffusion by flip-flop. If correct, this argues against some previous conclusions that long-chain fatty acids cannot enter adipocytes or other cells by simple diffusion. The diffusion hypothesis was further supported with data for an alkylamine and a fatty acid dimer with the two fatty acid chains linked in the middle. Both these molecules showed the predicted pH effect, and neither is likely to bind to a fatty acid transporter. These studies [1, 27] used high concentrations of fatty acids (with and without albumin) and were not designed to study the kinetics of fatty acid entry into cells, although others have performed a kinetic analysis of some of the data (inappropriately) [26]. The high levels of fatty acids could result in the precipitation of some of the fatty acid (or other artifacts). New data from our laboratory (see below) show very rapid diffusion of fatty acids into adipocytes and refute the prediction that low curvature of a membrane will greatly retard the rate of flip-flop [26].

We did not conclude from our results that diffusion is the *only* mechanism for movement of fatty acids across these cell membranes or that membrane proteins do not play any role in fatty acid transport [1, 27]. Some investigators might conclude that our new approach merely validates previous conclusions that fatty acids diffuse into cells at high concentrations when the transporter(s) is saturated [26]. However, our studies provided a mechanism of diffusion (flip-flop) and negated the idea that long-chain fatty acids could enter cells only by transporters.

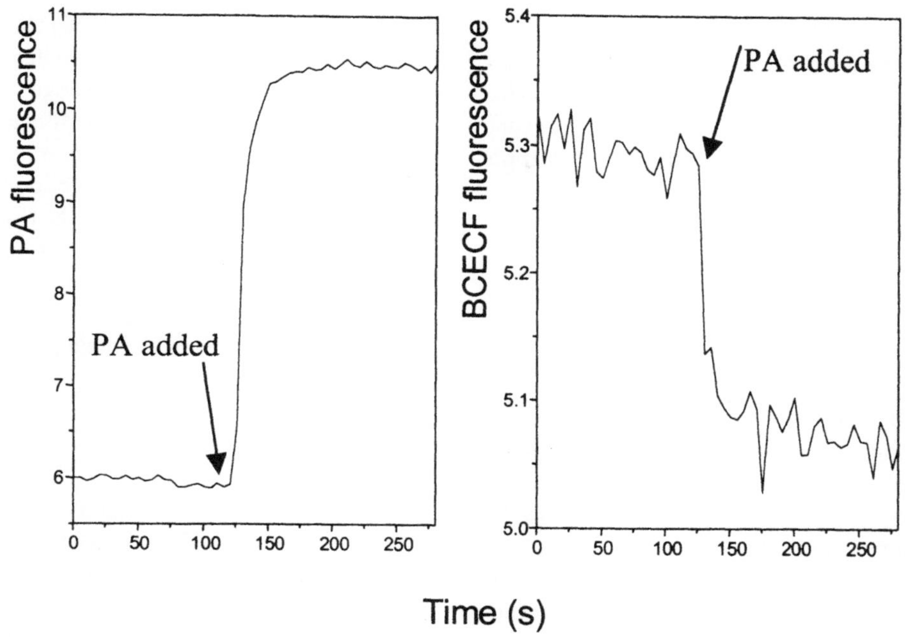

Fig. 2. Parinaric acid dissolved in DMSO was added to freshly isolated rat epididymal adipocytes (10^6 cells/ml) at 37°C in KRB buffer (2.0 ml; pH 7.4) at a final concentration of PA = 200 μM. The fluorescence intensities of PA (λ_{ex} = 303 nm; left panel) and BCECF, a probe for intracellular pH (λ_{ex} = 436 and 505 nm; right panel), were measured alternatively at 5 sec intervals.

The approach of measuring intracellular pH permits monitoring of the arrival of fatty acid in the inner leaflet of the plasma membrane but encompasses the combined steps of adsorption and transmembrane movement. In studies with cells, one cannot assume that adsorption is extremely rapid and does not contribute to the overall measured rate, as in the case of phospholipid vesicles. In addition, *in vivo*, fatty acids dissociate from albumin or from lipoproteins following lipase reactions. In *in vitro* experiments in which fatty acids are added without a natural carrier , artifacts such as the precipitation of fatty acids could decrease the measured rate of binding of fatty acids to the outer leaflet of the plasma membrane.

Insights from newer studies with adipocytes

We recently used two approaches to monitor adsorption independently from translocation, one of which is illustrated in Fig. 2. The fluorescence of the natural fatty acid *cis*-parinaric acid (PA) was monitored simultaneously with the intracellular pH in adipocytes. Following the addition of PA to the external buffer, its fluorescence increased rapidly, reaching an equilibrium level within 10 sec, indicating rapid immersion of the acyl chain into a hydrophobic environment.

Thus, the acyl chain is intercalated in the bilayer and not loosely adsorbed to the polar interface. No information is gleaned about transmembrane movement of the PA because the acyl chain environment is the same, or nearly so, on both sides of the bilayer. However, the change in intracellular pH reflects arrival of the head group (and its ionization) in the inner leaflet. The results presented in Fig. 2 support our hypothesis that the decrease in pH seen inside cells after the addition of fatty acid to the external buffer is the result of the diffusion processes defined in the model membrane studies: adsorption of the fatty acid into the membrane followed by flip-flop. The decrease in pH is simultaneous with binding and demonstrates rapid translocation (within 10 sec) of PA across the plasma membrane.

The dual fluorescence approach described for parinaric acid is not applicable to typical dietary fatty acids, which lack intrinsic fluorescence. Another dual fluorescence approach separates the adsorption step from subsequent steps and is applicable to different types of fatty acids. This approach uses the fluorescent indicator ADIFAB in the external medium, making it possible to determine how much fatty acid binds to the plasma membrane and the rate of binding. When combined with the measurement of intracellular pH, this approach also has demonstrated rapid binding and diffusion of oleic acid into adipocytes [2]. In addition, new studies (Kamp and Hamilton, unpublished) demonstrate the flip-flop mechanism at concentrations of fatty acid much lower (2.0–6.0 μM; without added albumin) than those we used previously. We found linear relationships between the quantity of added fatty acid, the amount of fatty acid that binds to the plasma membrane, and the decrease in intracellular pH. The correspondence of the intracellular changes in pH with the partitioning of fatty acids into the plasma membrane indicates that either the mechanism is diffusion alone or that a parallel mechanism makes a constant contribution. Either interpretation contradicts the hypotheses that diffusion is not a viable mechanism at low concentrations of fatty acid and that the diffusion mechanism becomes progressively more important with increasing concentrations of fatty acid.

Conclusions

From our perspective, diffusion by the flip-flop mechanism is sufficient for supplying fatty acids to cells. To develop this hypothesis further, we will aim future work at quantitation of the contribution of diffusion to the overall flux of fatty acids into cells under a wide variety of conditions. We will direct additional efforts at differentiating the adsorption step from the translocation step of transport of fatty acids in membranes. Such approaches will also be useful in investigating the mechanisms of the putative fatty acid transporters.

Our understanding of the role of membrane proteins in fatty acid transport will be enhanced by a better understanding of their binding properties and by uptake studies that separate metabolism from transport. Experiments with giant vesicles comprising membranes from muscles have supported a role of CD36 in fatty acid transport in the absence of metabolism [30]. In the more complex environment of the cell, new studies have described enrichment of the surface pool of CD36 and depletion of the intracellular pool under conditions where fatty acid uptake is increased in muscle cells [28]. The mechanism by which CD36 might enhance fatty acid uptake is unresolved, and there is no evidence yet that the protein possesses enzymatic activity. On the other hand, most of the members of the FATP family have been shown to possess enzymatic activity [28, 29].

The putative fatty acid transporters exist in a lipid environment that is favorable to the binding and translocation of fatty acids. To understand how these proteins might compete with the phospholipid bilayer, it is vital to know the number of copies of the protein in the membrane, the number of binding sites for fatty acids, the affinity of these sites for fatty acids, and the location of the binding sites. If the fatty acid binding site(s) are in the cytosolic region of the protein, it is less likely that the protein is a translocase. One might then hypothesize that the protein interacts with an intracellular FABP to release or channel the fatty acid [31], although the rapid desorption of fatty acids from a simple phospholipid bilayer [12] has to taken into account in such a model. To facilitate the adsorption step, a membrane protein could attract albumin and enhance the concentration of fatty acid near the membrane surface, which could help overcome the barriers of the unstirred water layer. We do not suggest that this interaction occurs by a specific receptor mechanism, which has been largely disproved [18, 24]. It is also possible that membrane proteins sequester fatty acid molecules to a membrane-bound enzyme or change the conformation of the enzyme to enhance substrate binding.

Preliminary models of the topology of FATP1 [32] and CD36 [33] in a membrane bilayer do not reveal a mode for binding or translocating fatty acids. However, these are very-low-resolution structures, and efforts to define the membrane topology of the putative transport proteins are a step in the right direction. A more definitive experiment would be the reconstitution of purified protein into phospholipid vesicles.

Acknowledgements

This research was supported by NIH grants HL26335 and HL67188. We thank Jeff Simard for preparation of Table 1 and Charu Rewal for help with preparation of the text.

References

1. Civelek VN, Hamilton JA, Tornheim K, Kelly KL, Corkey BE: Intracellular pH in adipocytes: Effects of free fatty acid diffusion across the plasma membrane, lipolytic agonists, and insulin. Proc Natl Acad Sci USA 93: 10139–10144, 1996
2. Hamilton JA, Johnson RA, Corkey B, Kamp F: Fatty acid transport: the diffusion mechanism in model and biological membranes. J Mol Neurosci 16: 99–108, 2001
3. Doody MC, Pownall HJ, Kao YJ, Smith LC: Mechanism and kinetics of transfer of a fluorescent fatty acid between single-walled phosphatidylcholine vesicles. Biochemistry 19: 108–116, 1980
4. Storch J, Kleinfeld AM: Transfer of long-chain fluorescent free fatty acids between unilamellar vesicles. Biochemistry 25: 1717–1726, 1986
5. Kleinfeld AM, Storch J: Transfer of long-chain fluorescent fatty acids between small and large unilamellar vesicles. Biochemistry 32: 2053–2061, 1993
6. Abumrad N, Harmon C, Ibrahimi A: Membrane transport of long-chain fatty acids: Evidence for a facilitated process. J Lipid Res 39: 2309–2318, 1998
7. Daniels C, Noy N, Zakim D: Rates of hydration of fatty acids bound to unilamellar vesicles of phosphatidylcholine or to albumin. Biochemistry 24: 3286–3292, 1985
8. Cabral DJ, Small DM, Lilly HS, Hamilton JA: Transbilayer movement of bile acids in model membranes. Biochemistry 26: 1801–1804, 1987
9. Kamp F, Hamilton JA: pH gradients across phospholipid membranes caused by fast flip-flop of un-ionized fatty acids. Proc Natl Acad Sci USA 89: 11367–11370, 1992
10. Kamp F, Hamilton JA: Movement of fatty acids, fatty acid analogues, and bile acids across phospholipid bilayers. Biochemistry 32: 11074–11086, 1993
11. Kamp F, Zakim D, Zhang F, Noy N, Hamilton JA: Fatty acid flip-flop in phospholipid bilayers is extremely fast. Biochemistry 34: 11928–11937, 1995
12. Zhang F, Kamp F, Hamilton JA: Dissociation of long and very long chain fatty acids from phospholipid bilayers. Biochemistry 35: 16055–16060, 1995

13. Storch J, Lechene C, Kleinfeld AM: Direct determination of free fatty acid transport across the adipocyte plasma membrane using quantitative fluorescence microscopy. J Biol Chem 266: 13473–13476, 1991
14. Kleinfeld AM, Chu P, Romero C: Transport of long-chain native fatty acids across lipid bilayer membranes indicates that transbilayer flip-flop is rate limiting. Biochemistry 36: 14146–14158, 1997
15. Fraser H, Coles SM, Woodford JK, Frolov AA, Murphy EJ, Schroeder F, Bernlohr DA, Grund V: Fatty acid uptake in diabetic rat adipocytes. Mol Cell Biochem 167: 51–60, 1997
16. Pohl P, Rokitskaya TI, Pohl EE, Saparov SM: Permeation of phloretin across bilayer lipid membranes monitored by dipole potential and microelectrode measurements. Biochim Biophys Acta 1323: 163–172, 1997
17. Abumrad NA, Perkins RC, Park JH, Park CR: Mechanism of long chain fatty acid permeation in the isolated adipocyte. J Biol Chem 256: 9183–9191, 1981
18. Zakim D: Fatty acids enter cells by simple diffusion. Proc Soc Exp Biol Med 212: 5–14, 1996
19. Zakim D: Thermodynamics of fatty acid transfer. J Membr Biol 176: 101–109, 2000
20. Trigatti BL, Gerber GE: The effect of intracellular pH on long-chain fatty acid uptake in 3T3-L1 adipocytes: Evidence that uptake involves the passive diffusion of protonated long-chain fatty acids across the plasma membrane. Biochem J 313: 487–494, 1996
21. Coburn CT, Knapp FF Jr, Febbraio M, Beets AL, Silverstein RL, Abumrad NA: Defective uptake and utilization of long chain fatty acids in muscle and adipose tissues of CD36 knockout mice. J Biol Chem 275: 32523–32529, 2000
22. Febbraio M, Abumrad NA, Hajjar DP, Sharma K, Cheng W, Pearce SF, Silverstein RL: A null mutation in murine CD36 reveals an important role in fatty acid and lipoprotein metabolism. J Biol Chem 274: 19055–19062, 1999
23. Coburn CT, Hajri T, Ibrahimi A, Abumrad NA: Role of CD36 in membrane transport and utilization of long-chain fatty acids by different tissues. J Mol Neurosci 16: 117–121, 2001
24. Hamilton JA: Fatty acid transport: difficult or easy? J Lipid Res 39: 467–481, 1998
25. Glatz JFC, Luiken JJFP, Bonen A: Involvement of membrane-associated proteins in the acute regulation of cellular fatty acid uptake. J Mol Neurosci 16: 123–132, 2001
26. Stump DD, Fan X, Berk PD: Oleic acid uptake and binding by rat adipocytes define dual pathways for cellular fatty acid uptake. J Lipid Res 42: 509–520, 2001
27. Hamilton JA, Civelek VN, Kamp F, Tornheim K, Corkey BE: Changes in internal pH caused by movement of fatty acids into and out of clonal pancreatic beta-cells (HIT). J Biol Chem 269: 20852–20856, 1994
28. Watkins PA, Lu JF, Steinberg SJ, Gould SJ, Smith KD, Braiterman LT: Disruption of the *Saccharomyces cerevisiae* FAT1 gene decreases very long-chain acyl-CoA synthetase activity and elevates intracellular very long-chain fatty acid concentrations. J Biol Chem 273: 18210–18219, 1998
29. Watkins PA, Pevsner J, Steinberg SJ: Human very long-chain acyl-CoA synthetase and two human analoges: Initial characterization and relationship to fatty acid transport. Prostagland Leuckot Essent Fatty Acids 60: 323–328, 1999
30. Bonen A, Luiken J, Arumugam Y, Glatz JFC, Tandon NN: Acute regulation of fatty acid uptake involves the redistribution of fatty acid translocase. J Biol Chem 275: 14501–14508, 2000
31. Glatz JFC, Storch J: Unraveling the significance of cellular fatty acid-binding proteins. Curr Opin Lipid 12: 267–274, 2001
32. Lewis SE, Listenberger LL, Ory DS, Schaffer JE: Membrane topology of the murine fatty acid transport protein 1. J Biol Chem 276: 37042–37050, 2001
33. Abumrad N, Coburn C, Ibrahimi A: Membrane proteins implicated in long-chain fatty acid uptake by mammalian cells: CD36, FATP and FABPpm. Biochim Biophys Acta 1441: 4–13, 1999
34. Ek-von Mentzer B, Zhang F, Hamilton JA: Binding of 13-HODE and 15-HETE to phospholipid bilayers, albumin, and intracellular fatty acid binding proteins. J Biol Chem 276: 15575–15580, 2001

Molecular and Cellular Biochemistry **239**: 25–33, 2002.

Similar mechanisms of fatty acid transfer from human anal rodent fatty acid-binding proteins to membranes: Liver, intestine, heart muscle, and adipose tissue FABPs

Judith Storch[1], Jacques H. Veerkamp[2] and Kuo-Tung Hsu[1]

[1]*Department of Nutritional Sciences, Rutgers University, New Brunswick, NJ, USA;* [2]*Department of Biochemistry, University of Nijmegen, Trigon Bldg., Nijmegen, The Netherlands*

Abstract

The mammalian fatty acid-binding proteins (FABPs) are thought to be important for the transport and metabolism of fatty acids in numerous cell types. The transfer of FA from different members of the FABP family to membranes has been shown to occur by two distinct mechanisms, an aqueous diffusion-based mechanism and a collisional mechanism, wherein the FABP interacts directly with membrane acceptors. Much of the work that underlies this concept comes from efforts using rodent FABPs. Given the increasing awareness of links between FABPs and several chronic diseases in humans, it was important to establish the mechanisms of FA transfer for human FABPs. In the present studies, we examined the rate and mechanism of fatty acid transfer from four pairs of human and rodent (rat or mouse, as specified) FABPs: hLFABP and rLFABP, hIFABP and rIFABP, hHFABP and rHFABP, and hAFABP and mAFABP. In the case of human IFABP, both the Ala^{54} and Thr^{54} forms were examined. The results show clearly that for all FABPs examined, the mechanisms of ligand transfer observed for rodent proteins hold true for their human counterparts. Moreover, it appears that the Ala to Thr substitution at residue 54 of the human IFABP does not alter the fundamental mechanism of ligand transfer to membranes, but nevertheless causes a consistent decrease in the rate of transfer. (Mol Cell Biochem **239**: 25–33, 2002)

Key words: fatty acid-binding protein, fatty acid, lipid transport, phospholipid membranes, gene polymorphism

Abbreviations: FA – fatty acid; FABP – fatty acid-binding protein; LFABP – liver FABP; IFABP – intestinal FABP; AFABP – adipose tissue FABP; HFABP – heart muscle FABP; rFABP – rat FABP; mFABP – mouse FABP; hFABP – human FABP; AOFA – anthroyloxy-labeled FA; 12AO – 12-(9-anthroyloxy)-oleic acid; 2AP – 2-(9-anthroyloxy)-palmitic acid; SUV – small unilamellar vesicles; EPC – egg phosphatidylcholine, PS – brain phosphatidylserine; CL – bovine heart cardiolipin; NBC-PC – 1-palmitoyl-2-(lauryl-*N*-(7-nitro2,1,3-benzoxadiazol-4-yl))-phosphatidylcholine

Introduction

A Western diet contains abundant triacylglycerol, and the gastrointestinal tract efficiently digests this lipid and absorbs the high levels of fatty acids (FA) produced, making fatty acid flux of great importance in the absorptive enterocyte. Also notable for high levels of fatty acid metabolism and transport are the hepatocyte, myocyte, and adipocyte. Thus, fatty acid transport into cells, out of cells, and within cells, are critical processes that contribute to the utilization and function of this important nutrient. All of the aforementioned cell types contain high levels of intracellular fatty acid-binding proteins (FABPs), and these have been proposed as transporters of FA through the cytosol, serving to increase the solubility and mobility of FA and, thus, their overall rate of utilization. There is abundant indirect and

Address for offprints: J. Storch, Department of Nutritional Sciences, Rutgers University, 96 Lipman Drive, New Brunswick, NJ 08901, USA (E-mail: storch@aesop.rutgers.edu)

direct support for this concept, as recently reviewed [1, 2].

In previous studies we found that the mechanisms by which the FABPs appear to transport their ligands to acceptor membranes, as well as receive fatty acids from donor membranes, appear to fall into two categories. For the liver FABP (LFABP) (and the related protein, cellular retinol binding protein II), transfer occurs by ligand diffusion through the aqueous phase; the protein releases the ligand into the surrounding medium and transfer to the phospholipid vesicles occurs by aqueous diffusion [3, 4]. In support of such a mechanism, it was found that the FA transfer rate from LFABP was independent of the FABP/phospholipid ratio and vesicle characteristics, but was dependent on the aqueous-phase solubility of the ligand [3–5]. In contrast, we demonstrated that intestinal FABP (IFABP), adipocyte FABP (AFABP) and heart muscle FABP (HFABP) interact directly with phospholipid vesicles through a transient collisional mechanism during the process of fatty acid transfer [3–9]. In support of this mechanism, it was shown that the rate of fatty acid transfer was dependent on the acceptor vesicle concentration and was markedly enhanced by the presence of anionic phospholipids that were hypothesized to interact with positively-charged residues on FABPs. Chemical modification approaches, residue-specific mutagenesis, and the analysis of chimeric FABPs, have identified the FABP helix-turn-helix domain as critical for determining whether diffusional or collision-mediated ligand transfer will occur [10–12].

Initial work was carried out using L-, I-, H-, and AFABPs purified from tissue, hence rat and mouse proteins were employed. Subsequent availability of cloned cDNAs for these rodent proteins showed that recombinant and tissue-purified proteins displayed virtually identical characteristics. Nevertheless, it remains to be shown whether the human orthologues of these FABPs also demonstrate differential transfer mechanisms.

In the human I-FABP gene but not the rodent gene, a potentially functionally important polymorphism has been identified [13]. This polymorphism, resulting in an Ala to Thr substitution at position 54, results in a 2-fold greater long chain FA binding affinity, as determined by calorimetry [13]. The effects of this single amino acid substitution on hIFABP fatty acid transport characteristics are not known. In the present studies, we have examined the rate and mechanism of fatty acid transfer from four pairs of human and rodent (rat or mouse, as specified) FABPs: hLFABP and rLFABP, hIFABP and rIFABP, hHFABP and rHFABP, and hAFABP and mAFABP. In the case of human IFABP, both the Ala^{54} and Thr^{54} forms were examined. The results show clearly that for all FABPs examined, the mechanisms of ligand transfer observed for rodent proteins hold true for their human counterparts. Moreover, it appears that the Ala to Thr substitution at residue 54 of the human IFABP does not alter the fundamental mechanism of ligand transfer to membranes, but nevertheless causes a consistent decrease in the rate of transfer.

Materials and methods

Materials

All human and rodent FABPs were expressed in *E. coli* and purified as previously described [4, 9, 14–16]. Expression vectors for hIFABP-Ala^{54} and hIFABPThr^{54} were generously provided by Dr. James Sacchettini (University of Texas, College Station, TX, USA). Proteins were kept at 4°C or stored frozen at –20°C, and no differences were noted between these preparations. The anthroyloxy-labeled fatty acids (AOFA) 2-(9-anthroyloxy)-palmitic acid (2AP) and 12-(9-anthroyloxy)-oleic acid (12-AO) were obtained from Molecular Probes (Eugene, OR, USA). Phospholipids were from Avanti Polar Lipids (Birmingham, AL, USA): egg phosphatidylcholine (EPC), brain phosphatidylserine (PS), bovine heart cardiolipin (CL), and 1-palmitoyl-2-(lauryl-*N*-(7-nitro2,1,3-benzoxadiazol-4-yl))-phosphatidylcholine (NBD-PC). All other chemicals were reagent grade or higher.

Membrane vesicle preparation

Small unilamellar vesicles (SUV) were prepared with 90 mol% POPC and 10% NBD-PC by sonication and centrifugation in 40 mM Tris.HCl buffer, 150 mM NaCl, 0.1 mM EDTA, pH 7.4 (TBS) except for SUV containing cardiolipin which were prepared in TBS with 1 mM EDTA [17]. In cases where anionic phospholipids were included in SUV (PS, CL) their concentration was 25 mol%, with 65 mol% EPC and 10 mol% NBD-PC. Vesicle phospholipid concentration was determined by analysis of inorganic phosphate [18].

Transfer of 2AP from AFABP to model membranes

The rate of 2AP transfer from FABPs to acceptor vesicles was determined using a stopped-flow fluorescence resonance energy transfer assay as detailed previously [19]. FA transfer was monitored at 25°C in TBS buffer. For experiments requiring increased ionic strength, NaCl was added to the vesicles and the protein samples prior to mixing. Conditions were developed prior to each experiment to ensure that no photobleaching of AOFA was observed. Final concentrations in a typical transfer assay were 10 μM FABP, 1 μM 2AP or 12AO, and 100–500 μM vesicle PL; for the IFABP experiments, 30 μM protein and 3 μM AOFA were used, to compensate for the lower fluorescence quantum yields for AOFA binding, as previously discussed [4]. The fluorescent FA-FABP complex was the fluorescence donor. Upon mixing of FA-FABP complex with acceptor membranes, transfer of FA from protein to acceptor SUVs is directly monitored by the time-dependent decrease in fluorescence, using an SX-18MV stopped-flow spectrofluorimeter (Applied Photophysics, UK) interfaced with an Acorn A5000 computer. The data were analyzed as previously described by fitting to a single exponential with steady state function [4, 14]. Each experiment was repeated 3 or more times, and results are presented as mean ± S.D.

Equilibrium binding of FA to IFABPs

The ADIFAB method was used to analyze the binding of oleic acid to rIFABP, hIFABP-Ala^{54} and hIFABP-Thr^{54}, as detailed previously [19, 20]. Experiments used 2 μM IFABP, 0.2 μM ADIFAB, and increasing concentrations of oleate.

Thermodynamic parameters of AOFA transfer from IFABPs

12AO transfer from rat and human IFABPs to SUV was determined as a function of temperature. The activation energy (Ea) was obtained from the slope of an Arrhenius plot of the data, and the Eyring rate theory was used to determine the thermodynamic parameters of the FA transfer process, as previously described [4].

Results

Human and rat LFABPs

To distinguish between FA transfer occurring by aqueous diffusion from that occurring by collisional interactions, 12AO transfer from rLFABP and hLFABP was examined as a function of increasing acceptor vesicle concentration. For the former mechanism, little or no change in transfer rate is expected; in contrast, for a collisional mechanism the rate of ligand transfer will increase in proportion to the number of collisions, which in turn is dependent on the concentration of acceptor membranes. As shown in Fig. 1, the 12AO transfer rate from neither rLFABP nor hLFABP was found to vary with [SUV]. This suggests that, in agreement with previous results for rLFABP [3, 4], FA transfer from hLFABP occurs by diffusion of the ligand through the aqueous phase.

Further support for this hypothesis comes from experiments that examined the 12AO transfer rate as a function of buffer ionic strength. The rate of transfer of a hydrophobic molecule from a binding protein through an aqueous space to an acceptor membrane will be modulated by factors that alter its solubility, such as ionic strength. Indeed, as shown in Fig. 2, increasing [NaCl] resulted in a logarithmic decrease in the rate of 12AO transfer from both rodent and human LFABPs, consistent with the expected decrease in aqueous solubility of a hydrophobic molecule [21].

If FA transfer from LFABP is occurring via aqueous diffusion, the composition of the acceptor membrane should not affect the rate of transfer, since the rate-determining step in the transfer process is the dissociation of the FA from the FABP. Figure 3 shows that the addition of 25 mol% of phosphatidylserine to acceptor SUV did not alter the 12AO transfer rate from rat or human LFABP. For vesicles containing CL, a relatively small (approximately 50%) but reproducible increase in transfer rate of was observed; this too is in agreement with our previous findings for rLFABP [4]. Although the meaning of this consistent effect of CL-SUV is uncertain, it

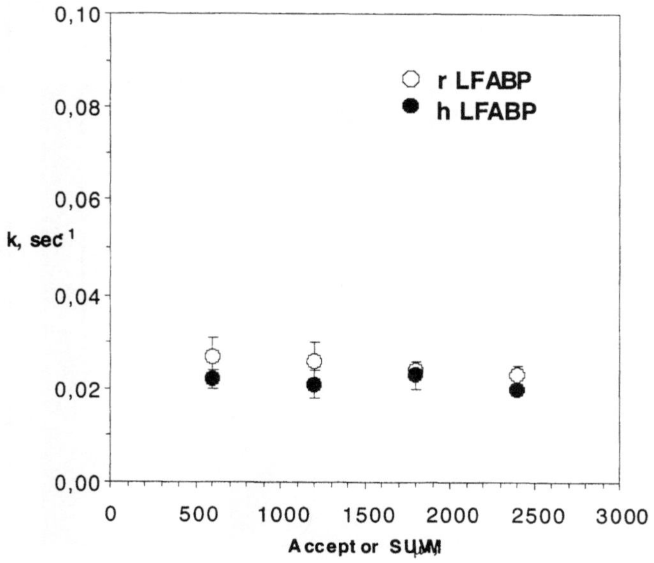

Fig. 1. Effect of acceptor SUV concentration on AOFA transfer from rLFABP and hLFABP to membranes. 10 μM FABP was equilibrated with 1 μM 12AO. Transfer to increasing concentrations of EPC SUV (90 mol% EPC, 10 mol% NBD-PC) was monitored at 25°C, as described in Materials and methods.

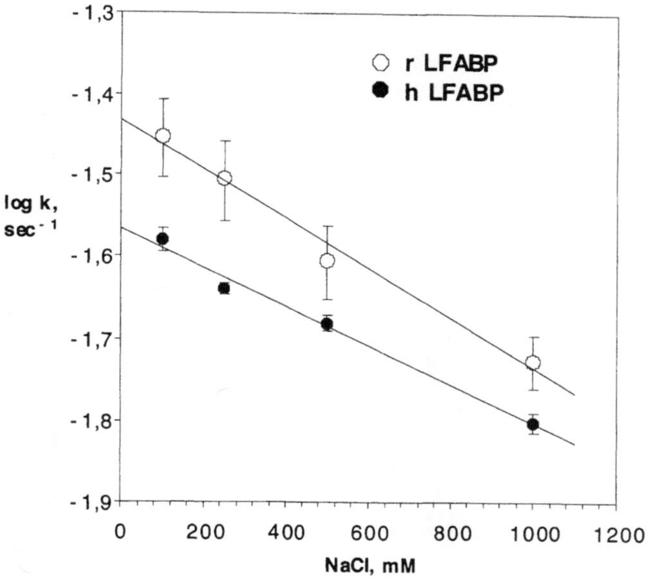

Fig. 2. Effect of ionic strength on AOFA transfer from rLFABP and hLFABP to membranes. Transfer of 12AO from LFABPs to 600 μM SUV was examined as a function of increasing concentrations of buffer NaCl, as described in Materials and methods.

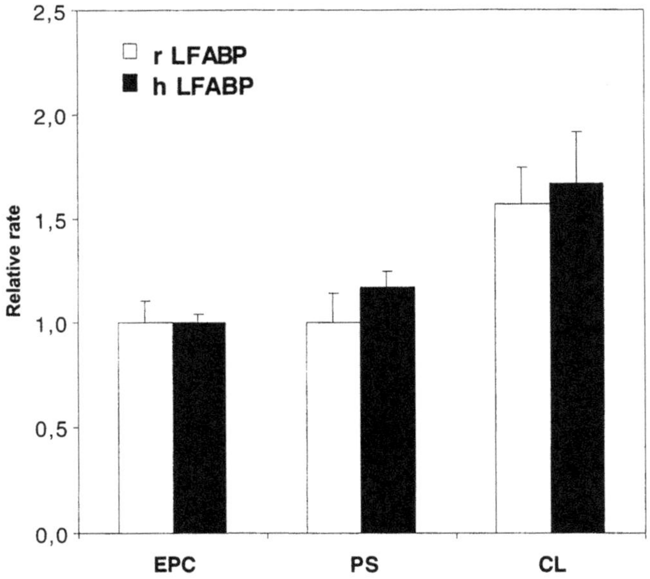

Fig. 3. Effect of acceptor SUV composition on AOFA transfer from human and rat LFABP to membranes. Transfer of 12AO from LFABPs to 600 μM SUVs containing 25 mol% PS or CL. Results are normalized to the rate of transfer from each protein to EPC SUV.

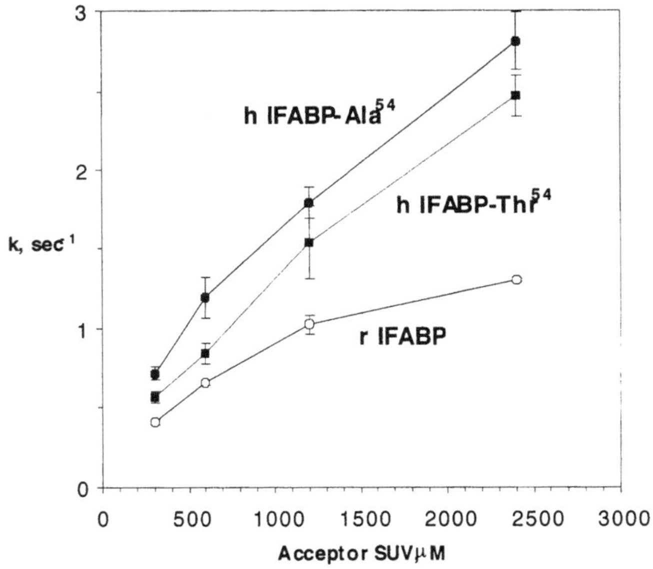

Fig. 4. Effect of acceptor SUV concentration on AOFA transfer from rIFABP, hIFABP-Ala54, and hIFABP-Thr54 to membranes. 30 μM FABP was equilibrated with 3 μM 12AO. Transfer to increasing concentrations of EPC SUV (90 mol% EPC, 10 mol% NBD-PC) was monitored at 25°C, as described in Materials and methods.

may, perhaps, reveal the potential for LFABP to interact with membranes under certain limited circumstances. It is noteworthy that in all the aforementioned studies of the LFABPs, not only were the relative effects observed to be similar, the absolute rates of FA transfer were also quite similar.

Human and rat IFABPs

The equilibrium binding affinities of IFABP and the two forms of hIFABP for oleic acid were determined using the ADIFAB method. The results in Table 1 show that the hIFABP-Thr^{54} has an affinity for 18:1 of 38 μM, approximately 2-fold higher relative to hIFABP-Ala^{54}. The oleate affinity of IFABP is similar to that of the Thr^{54} form.

The mechanism of fatty acid transfer from the IFABPs to membranes was explored by examining effects of acceptor vesicle concentration, acceptor vesicle net charge, and buffer ionic strength on the rate of 12AO transfer, as discussed above. Figure 4 shows that for both rodent and human FABPs, the rate of 12AO transfer to membranes increases as a function of SUV concentration. Moreover, addition of 25 mol% PS or CL to acceptor vesicles resulted in approximately 4-fold and 24-fold increases is relative transfer rates, respectively, relative to EPC vesicles, for transfer of 12AO from rIFABP. For the two human IFABPs, the results were qualitatively similar but even more dramatic – SUVs containing PS or CL resulted in approximately 10-fold and >40-fold higher 12AO transfer rates, respectively, from both hIFABP-Ala^{54} and hIFABP-Thr^{54} (Fig. 5). That membrane concentration and composition affect the rate of ligand transfer from the IFABPs provides strong support for a collisional mechanism of FA transfer: In an aqueous diffusion mechanism, the nature and abundance of the acceptor is irrelevant whereas in

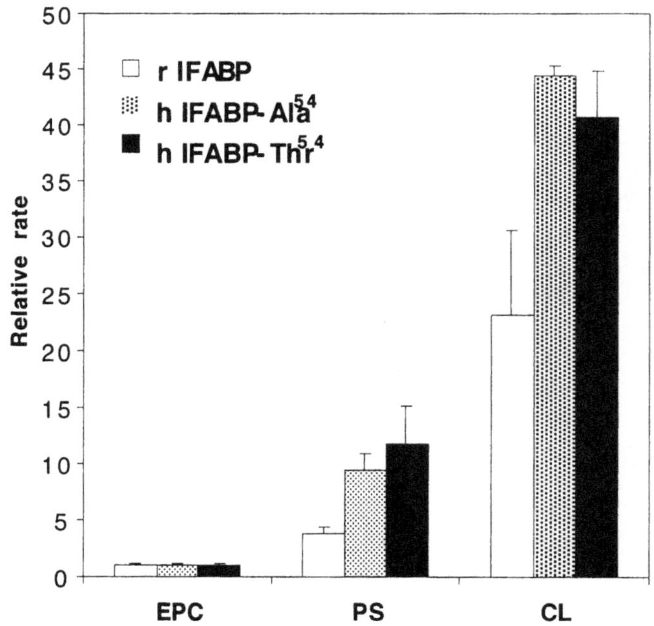

Fig. 5. Effect of acceptor SUV composition on AOFA transfer from rat and human IFABP to membranes. Transfer of 12AO from IFABPs to 300 μM SUVs containing 25 mol% PS or CL. Results are normalized to the rate of transfer from each protein to EPC SUV.

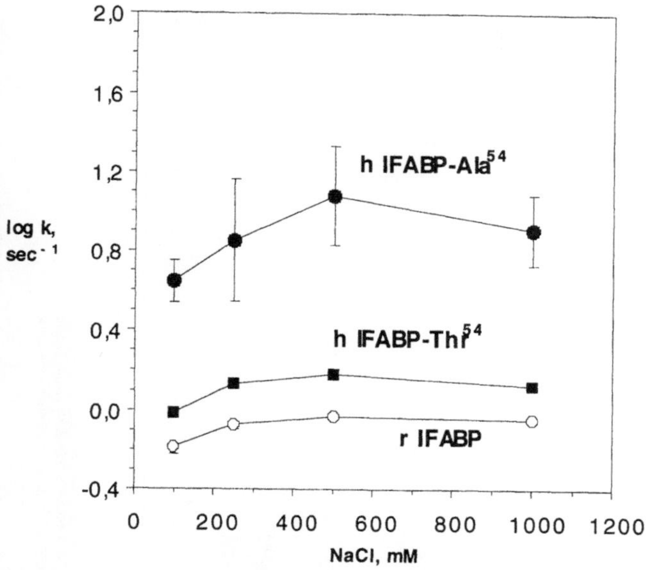

Fig. 6. Effect of ionic strength on AOFA transfer from rIFABP, hIFABP-Ala[54], and hIFABP-Thr[54] to membranes. Transfer of 12AO from IFABPs to 300 μM SUV was examined as a function of increasing concentrations of buffer NaCl, as described in Materials and methods.

the case of direct protein-acceptor interactions, the acceptor properties would, indeed, be expected to modulate the rate of transfer. An examination of the effects of increasing ionic strength on 12AO transfer from the IFABPs reinforces the absence of an aqueous diffusion mechanism, in that the 12AO transfer rates did not vary with 12AO solubility. In fact, and in distinct contrast to the results with the LFABPs, Fig. 6 shows that if anything, transfer rates from the IFABPs increased slightly with increasing [NaCl].

The effect of temperature on 12AO transfer from human and rat IFABPs was examined in order to determine the thermodynamic parameters of the transfer process. The results obtained for rIFABP (Table 1) are similar to those we reported previously for transfer of the stearate derivative 12AS [4], with the calculated free energy of 12AO transfer from rIFABP to SUV arising primarily from an enthalpic process ($\Delta H^{\ddagger}$ = 14.9 kcal/mol), with a relatively small entropic contribution

Table 1. Comparative properties of rat and human IFABPs

	kD	ΔEa	ΔH	TΔS	ΔG
	μM	kcal/mol			
Rat IFABP	33.3 ± 3.4	15.4 ± 0.8	14.9 ± 0.8	–2.9 ± 0.8	17.8 ± 0.1
Human IFABP-Ala[54]	77.5 ± 7.7	13.9 ± 0.4	13.3 ± 0.4	–4.1 ± 0.4	17.4 ± 0.1
Human IFABP-Thr[54]	38.4 ± 6.2	14.8 ± 0.4	14.2 ± 0.4	–3.4 ± 0.4	17.6 ± 0.1

Values for kD determinations represent average ± S.D. for n = 6–8 separate determinations. Values for thermodynamic parameters are average ± S.D. for n = 3 separate determinations.

($T\Delta S^{\ddagger}$ = –2.9 kcal/mol). The results for hIFABP-Ala[54] and hIFABP-Thr[54] are quite similar (Table 1). For the human IFABPs, too, the FA transfer process is seen to be predominantly enthalpic. Consistent with our mechanistic interpretations, the low entropic input may be due to the lack of FA traversing and 'ordering' the surrounding aqueous phase, and the greater enthalpic input the result of charge–charge interactions between IFABPs and SUV.

Under all experimental conditions examined, the absolute rates of 12AO transfer to membranes varied as hIFABP-Ala[54] > hIFABP-Thr[54] > rIFABP. Transfer from the hIFABP-Ala[54] form was on average ≈30% faster than transfer from hIFABP-Thr[54]. Transfer rates from hIFABP-Thr[54], in turn, were ≈64% faster relative to rates from rIFABP.

Human and mouse AFABPs

For AFABP and HFABP, the bulk of experiments were done using the palmitate analogue 2AP, so as to establish similar conditions to those previously used [11, 19]. Results for the oleate analogue 12AO were qualitatively similar (although transfer of 12AO is consistently faster than that of 2AP), always providing the same information regarding the mechanism of AOFA transfer. The results in Fig. 7A show that for mouse and human AFABPs, the rate of 2AP transfer increased in proportion to the concentration of acceptor vesicle phospholipid, consistent with a collisional transfer mechanism for both mAFABP and hAFABP. Furthermore, the two proteins displayed similar increases in transfer rates, of approximately 2–3-fold and 15–20-fold, to vesicles containing 25 mol% PS or CL, respectively (Fig. 8A). Finally, as found previously for mAFABP [7], increasing buffer ionic strength had no effect on the AOFA transfer rate (data not shown). Absolute rates of 2AP transfer were approximately 2-fold greater from human AFABP than from mouse AFABP under all experimental conditions. Taken together, the results indicate that transfer from hAFABP, like that from mAFABP, occurs via collisional interaction of the protein with acceptor membranes.

Human and rat HFABPs

The results for the HFABPs essentially recapitulate those of the AFABPs, and are entirely consistent with a collisional mechanism of FA transfer from HFABPs to membranes. Here, too, a proportional increase in 2AP transfer rate is observed with SUV concentration (Fig. 7B), and increases in AOFA transfer rates secondary to incorporation of acceptor membrane phospholipid negative charge are found for rodent and human proteins (Fig. 8B). Increasing buffer ionic strength resulted in a small increase in 2AP transfer rate (data not shown), similar to our previous observations for rHFABP [8]. The

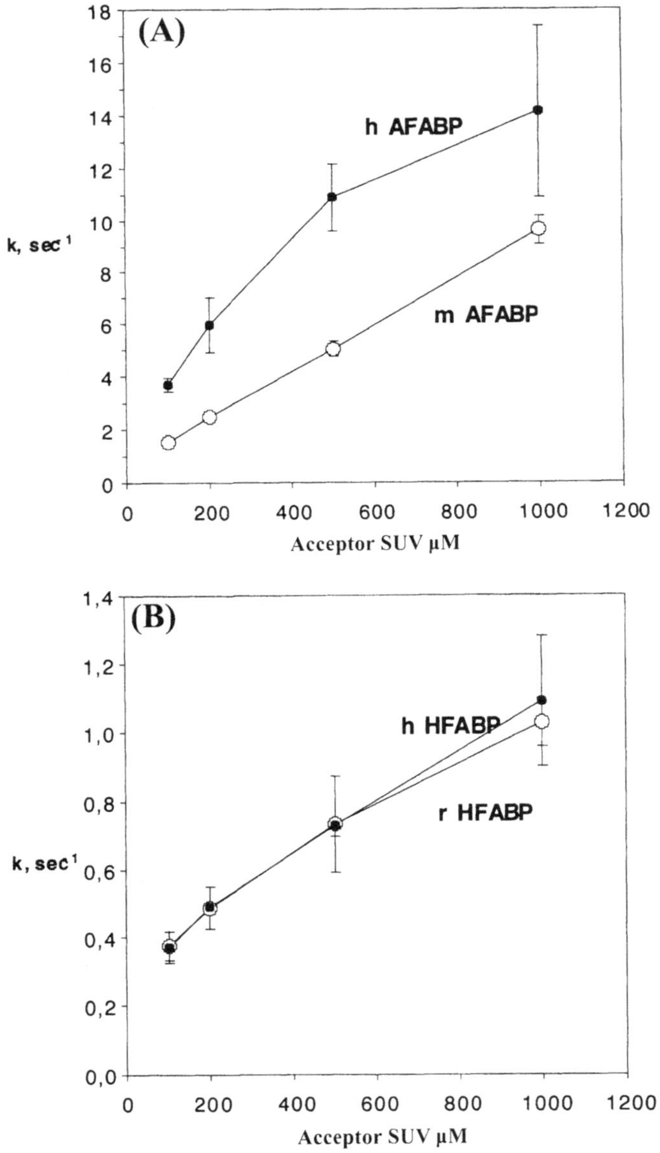

Fig. 7. Effect of SUV concentration on AOFA transfer from (A) mouse and human AFABP, and (B) rat and human HFABP. Ten μM FABP was equilibrated with 1 μM 2AP. Transfer to increasing concentrations of EPC SUV (90 mol% EPC, 10 mol% NBD-PC) was monitored at 25°C, as described in Materials and methods.

absolute rates of 2AP transfer from rat HFABP and human HFABP to zwitterionic membranes were virtually the same; transfer rates to PS- and CL-containing SUV were somewhat greater for hHFABP relative to rHFABP.

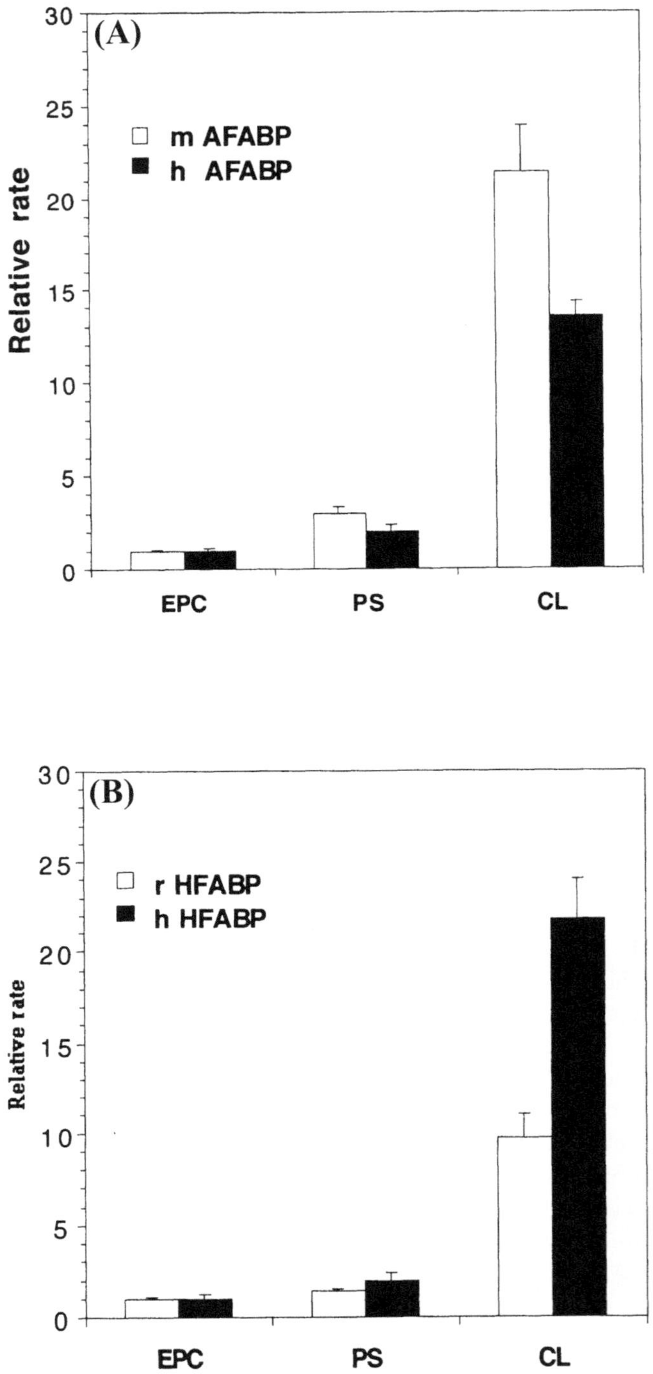

Fig. 8. Effect of acceptor SUV composition on AOFA transfer from (A) mouse and human AFABP, and (B) rat and human HFABP, to membranes. Transfer of 2AP from FABPs to 100 μM SUVs containing 25 mol% PS or CL. Results are normalized to the rate of transfer from each protein to EPC SUV.

Discussion

The FABPs comprise a growing family of small molecular weight mammalian and non-mammalian proteins that bind long chain FA with relatively high affinity. Work from our and others' laboratories has shown that in model systems, the mechanism of FA transfer from some members of the FABP family proceeds by aqueous diffusion, while from others it appears to occur via direct collisional interactions between the FABP and an acceptor membrane [1, 22, 23]. Further studies have begun to define the structural determinants that underlie these fundamentally different ligand transfer mechanisms. Thus far, however, all the FA transport work has been done using rodent proteins. Given the increasing interest in fatty acid transport in human disease states such as obesity, diabetes, and atherosclerosis, it was necessary to determine whether the functional differences found in rat and mouse proteins were present in their human counterparts.

Direct comparisons between rodent and human proteins were made for the FABPs from liver, intestine, adipose tissue, and heart muscle. In all four cases, the results clearly demonstrate that the fatty acid transfer mechanisms for the human FABPs are identical to those of the rat and mouse FABPs. Thus, transfer of fluorescent FA from human LFABP to membranes is entirely consistent with an aqueous diffusion mechanism, whereas from human IFABP, AFABP, and HFABP to SUV occurs during direct protein–membrane interactions. In addition, a comparison of the absolute rates of AOFA transfer shows the identical order for human and rodent FABPs, i.e. AFABP > HFABP > IFABP > LFABP. Taken together, the results provide a strong demonstration that the rodent FABPs are faithful reporters of the FA transport properties of the human FABPs. The high degree of similarity in lipid transfer mechanisms is no doubt a result of the large degree of amino acid sequence identity between these FABP orthologues. The percent identity between rat (or mouse) and human proteins (ExPASy software, Baylor College of Medicine, http://searchlauncher.bcm.tmc.edu), is quite high: 83% for LFABPs, 78% for IFABPs, 92% for AFABPs, and 90% for HFABPs.

Previous work from this laboratory has used mutagenesis and protein engineering to begin to identify the structural elements that underlie the differences in FA transfer mechanisms between different members of the FABP family, as well as the differences in absolute rates of AOFA transfer to membranes. These studies have uniformly pointed to the small helix-turn-helix domain of the FABPs as the key structural domain that determines whether the basic mechanism of FA transfer will be diffusional or collisional. For example, we showed that the absence of the helical region in rIFABP converted the FA transfer mechanism from collisional to diffusional [10]. Moreover, neutralization of positively charged lysine residues in the helical domain, but not in the β-barrel, resulted in large alterations in the rate of FA transfer to membranes and in the sensitivity of transfer to membrane negative charge [11, 19]. Interestingly, the percent identity in the α-I and α-II helixes of the rodent and human FABP pairs is even greater than the overall percent identity: 88% of the α-I and α-II helix residues are identical in the LFABPs, 82% in the IFABPs, and fully 100% for both AFABP and HFABP helixes. Thus, the exceptional conservation of structural elements in the helix-turn-helix domain, which as mentioned above appears to be crucial for determining AOFA transfer properties, suggests the conservation of an important functional characteristic in this protein family.

Several years ago, the presence of a polymorphism in the human IFABP gene that results in an Ala to Thr substitution at amino acid 54 was identified [13]. Numerous population-based studies have now shown that the Thr^{54} allele is associated with insulin resistance and dyslipidemias in several [24–26] although not all [27, 28] populations examined. It was recently found that plasma FA concentrations were higher following a test meal in Pima Indians homozygous for the Thr^{54} form relative to the more prevalent Ala^{54} form [24]. However, since we and others found that the Thr^{54}-I-FABP has a higher binding affinity for FA ([13] and Table 1), the apparently greater level of lipid assimilation in the Thr^{54} homozygotes implies that the effect cannot be simply due to greater binding of FA, since that would likely result in decreased FA export from the enterocyte, not increased. This could, rather, support a more specific role for I-FABP in cellular trafficking, and indeed the present results have shown that AOFA transfer is faster from hIFABP-Ala^{54} than from hIFABP-Thr^{54}. Notably, increased fat absorption was recently reported in human fetal intestinal explants from IFABP-Thr^{54} heterozygotes relative to Ala^{54} homozygotes [29]. On the other hand, a complete absence of I-FABP in a knockout mouse model was also correlated with higher levels of serum lipids [30], implying an inhibitory action of I-FABP on lipid assimilation. In other studies, wt rat IFABP and the two human IFABP forms have been transfected into Caco-2 cells, although inconsistent effects on lipid assimilation were found [31, 32]. Clearly, it will be necessary to dissect the level at which I-FABP effects are being exerted, i.e. FA uptake, intracellular processing, lipoprotein secretion, etc., in order to fully define its role in lipid assimilation and, further, in whole-body energy homeostasis.

The FABPs are becoming increasingly recognized for their potential roles in several chronic diseases. As mentioned above, the hIFABP polymorphism points to an as yet ill-understood role in insulin resistance and the regulation of plasma lipid levels. The IFABP null mouse also displayed alterations in serum lipids, underscoring this possibility [30]. The HFABP null mouse was shown to develop cardiac hypertrophy, thus linking HFABP to another common chronic disease [33]. Another mouse model, null for AFABP, has also been engineered; although results have not been entirely consistent, it is possible that these animals have somewhat altered plasma glucose and insulin levels under certain conditions [34–38]. It is likely that the impressive upregulation of keratinocyte-type FABP expression in the adipocytes of $AFABP^{-/-}$ animals has led to the inconsistent results, and perhaps obscured the relationship of AFABP to the development of insulin resistance and diabetes. Other studies, however, clearly point to an important involvement for AFABP in the development of diet-induced atherosclerosis. When $AFABP^{-/-}$ mice were crossed with the atherosclerosis-prone apolipoprotein E$^{-/-}$ mouse to create a 'double knockout' animal, these ApoE$^{-/-}$/AFABP$^{-/-}$ mice were remarkably resistant to the development of atherosclerosis, strongly indicating a role for AFABP in the accumulation of lipid-rich foam cells in the arterial intima [39]. Further, bone marrow transplantation studies showed that AFABP expressed specifically in the macrophage, rather than adipocyte AFABP, was likely involved in the development of dietary atherosclerosis [40, 41]. In summary, these studies underscore the important roles that human FABPs may play in normal and pathophysiologic states. Further investigation of these FABPs, as well as their rodent orthologues, will continue to shed light on the functions of the FABPs in health and disease.

Acknowledgements

The authors gratefully acknowledge Dr. James Sacchettini (University of Texas, College Station, TX, USA) for providing the expression plasmids for human IFABPs.

References

1. Storch J, Thumser AEA: The fatty acid transport function of fatty acid-binding proteins. Biochim Biophys Acta 1486: 28–44, 2000
2. Glatz JFC, van der Vusse G: Cellular fatty acid-binding proteins: Their function and physiological significance. Prog Lipid Res 35: 243–282, 1996
3. Kim HK, Storch J: Free fatty acid transfer from rat liver fatty acid-binding protein to phospholipid vesicles: Effect of ligand and solution properties. J Biol Chem 267: 77–82, 1992
4. Hsu KT, Storch J: Fatty acid transfer from liver and intestinal fatty acid binding-proteins to membranes occurs by different mechanisms. J Biol Chem 271: 13317–13323, 1996
5. Storch J, Bass NM: Transfer of fluorescent fatty acids from liver and heart fatty acid-binding proteins to model membranes. J Biol Chem 265: 7827–7831, 1990
6. Wootan MG, Bernlohr DA, Storch J: Mechanism of fluorescent fatty acid transfer from adipocyte fatty acid binding protein to membrane. Biochemistry 32: 8622–8627, 1993
7. Wootan MG, Storch J: Regulation of fluorescent fatty acid transfer from adipocyte and heart fatty acid binding proteins by acceptor membrane lipid composition and structure. J Biol Chem 269: 10517–10523, 1994
8. Kim HK, Storch J: Mechanism of free fatty acid transfer from rat heart fatty acid binding protein to phospholipid membranes: Evidence for a collisional process. J Biol Chem 267: 20051–20056, 1992
9. Liou H, Storch J: The helical cap domain is important for fatty acid transfer from adipocyte and heart fatty acid-binding proteins to membranes. FASEB J 12: A514, 1998
10. Corsico B, Cistola DP, Frieden C, Storch J: The helical domain of intestinal fatty acid binding protein is critical for collisional transfer of fatty acids to phospholipid membranes. Proc Natl Acad Sci USA 95: 12174–12178, 1998
11. Liou H-L, Storch J: Role of surface lysine residues of adipocyte fatty acid-binding protein in fatty acid transfer to phospholipid vesicles. Biochemistry 40: 6475–6485, 2001
12. Richieri GV, Low PJ, Ogata RT, Kleinfeld AM: Thermodynamics of fatty acid binding to engineered mutants of the adipocyte and intestinal fatty acid-binding proteins. J Biol Chem 273: 7397–7405, 1998
13. Baier LJ, Sacchettini JC, Knowler WC, Eads J, Paolisso G, Tataranni PA, Mochizuki H, Bennett PH, Bogardus C, Prochazka M: An amino acid substitution in the human intestinal fatty acid binding protein is associated with increased fatty acid binding, increased fat oxidation, and insulin resistance. J Clin Invest 95: 1281–1287, 1995
14. Herr FM, Matarese V, Bernlohr DA, Storch J: Surface lysine residues modulate the collisional transfer of fatty acid from adipocyte fatty acid binding protein to membranes. Biochemistry 34: 11840–11845, 1995
15. Zimmerman AW, van Moerkerk HTB, Veerkamp JH: Ligand specificity and conformational stability of human fatty acid-binding proteins. Int J Biochem Cell Biol 33: 865–876, 2001
16. Veerkamp JH, van Moerkerk HTB, Prinsen CFM, van Kuppevelt TH: Structural and functional studies on different human FABP types. Mol Cell Biochem 192: 137–142, 1999
17. Smith ER, Storch J: The adipocyte fatty acid-binding protein binds to membranes by electrostatic interactions. J Biol Chem 274: 35325–35330, 1999
18. Ames BN: Assay of inorganic phosphate, total phosphate and phosphatases. Meth Enzymol 8: 115–118, 1966
19. Herr FM, Aronson J, Storch J: Role of portal region lysine residues in electrostatic interactions between heart fatty acid binding protein and phospholipid membranes. Biochemistry 35: 1296–1303, 1996
20. Richieri GV, Ogata RT, Kleinfeld AM: Equilibrium constants for the binding of fatty acids with fatty acid-binding proteins from adipocyte, intestine, heart, and liver measured with the fluorescent probe ADIFAB. J Biol Chem 269: 23918–23930, 1994
21. Charlton SC, Smith LC: Kinetics of transfer of pyrene and *rac*-1-oleyl-2-[4-(3-pyrenyl)butanoyl]glycerol between human plasma lipoproteins. Biochemistry 21: 4023–4030, 1982
22. Zucker SD: Kinetic model of protein-mediated ligand transport: Influence of soluble binding proteins on the intermembrane diffusion of a fluorescent fatty acid. Biochemistry 40: 977–986, 2001
23. Weisiger RW, Zucker SD: Transfer of fatty acids between intracellular membranes: Roles of soluble binding proteins, distance, and time. Am J Physiol 282: G105–G115, 2002
24. Pratley RE, Baier L, Pan DA, Salbe AD, Storlien L, Ravussin E, Bogardus C: Effects of an Ala^{54}Thr polymorphism in the intestinal fatty acid-binding protein on responses to dietary fat in humans. J Lipid Res 41: 2002–2008, 2000
25. Agren JJ, Vidgren HM, Valve RS, Laakso M, Uusitupa MI: Postprandial responses of individual fatty acids in subjects homozygous for the threonine- or alanine-encoding allele in codon 54 of the intestinal fatty acid binding protein 2 gene. Am J Clin Nutr 73: 31–35, 2001
26. Georgopoulos A, Aras O, Tsai MY: Codon-54 polymorphism of the fatty acid-binding protein 2 gene is associated with elevation of fasting and postprandial triglyceride in type 2 diabetes. J Clin Endocrinol Metab 85: 3155–3160, 2000
27. Lei HH, Coresh J, Shuldiner AR, Boerwinkle E, Brancati FL: Variants of the insulin receptor substrate-1 and fatty acid binding protein 2 genes and the risk of type 2 diabetes, obesity, and hyperinsulinemia in african-americans: The atherosclerosis risk in communities study. Diabetes 48: 1868–1872, 1999
28. Tahvanainen E, Molin M, Vainio S, Tiret L, Nicaud V, Farinaro E, Masana L, Ehnholm C: Intestinal fatty acid binding protein polymorphism at codon 54 is not associated with postprandial responses to fat and glucose tolerance tests in healthy young europeans. Results from EARS II participants. Atherosclerosis 152: 317–325, 2000
29. Levy E, Menard D, Delvin E, Stan S, Mitchell G, Lambert M, Ziv E, Feoli-Fonseca JC, Seidman E: The polymorphism at codon 54 of the FABP2 gene increases fat absorption in human intestinal explants. J Biol Chem 276: 39679–39684, 2001

30. Vassileva G, Huwyler L, Poirier K, Agellon LBTMJ: The intestinal fatty acid binding protein is not essential for dietary fat absorption in mice. FASEB J 14: 2040–2046, 2000
31. Baier LJ, Bogardus C, Sacchettini JC: A polymorphism in the human intestinal fatty acid binding protein alters fatty acid transport across Caco-2 cells. J Biol Chem 271: 10892–10896, 1996
32. Darimont C, Gradoux N, Persohn E, Cumin F, De Pover A: Effects of intestinal fatty acid-binding protein overexpression on fatty acid metabolism in Caco-2 cells. J Lipid Res 41: 84–92, 2000
33. Binas B, Danneberg H, McWhir J, Mullins L, Clark AJ: Requirement for the heart-type fatty acid binding protein in cardiac fatty acid utilization. FASEB J 13: 805–812, 1999
34. Scheja L, Makowski L, Uysal KT, Wiesbrock SM, Shimshek DR, Meyers DS, Morgan M, Parker RA, Hotamisligil GS: Altered insulin secretion associated with reduced lipolytic efficiency in $aP2^{-/-}$ mice. Diabetes 48: 1987–1993, 1999
35. Uysal KT, Scheja L, Wiesbrock SM, Bonner-Weir S, Hotamisligil GS: Improved glucose and lipid metabolism in genetically obese mice lacking aP2. Endocrinology 141: 3388–3396, 2000
36. Shaughnessy S, Smith ER, Kodukula S, Storch J, Fried SK: Adipocyte metabolism in adipocyte fatty acid binding protein knockout ($aP2^{-/-}$) mice after short-term high-fat feeding. Diabetes 49: 904–911, 2000
37. Hotamisligil GS, Johnson RS, Distel RJ, Ellis R, Papaioannou VE, Spiegelman BM: Uncoupling of obesity from insulin resistance through a targeted mutation in aP2, the adipocyte fatty acid binding protein. Science 274: 1377–1379, 1996
38. Coe NR, Simpson MA, Bernlohr D: Targeted disruption of the adipocyte lipid-binding protein (aP2 protein) gene impairs fat cell lipolysis and increases cellular fatty acid levels. J Lipid Res 40: 967–972, 1999
39. Perrella MA, Pellacani A, Layne MD, Patel A, Zhao D, Schreiber BM, Storch J, Feinberg MW, Hsieh CM, Haber E, Lee ME: Absence of adipocyte fatty acid binding protein prevents the development of accelerated atherosclerosis in hypercholesterolemic mice. FASEB J. 15: 1774–1776, 2001
40. Layne MD, Patel A, Chen YH, Rebel VI, Carvajal IM, Pellacani A, Ith B, Zhao D, Schreiber BM, Yet SF, Lee ME, Storch J, Perrella MA: Role of macrophage-expressed adipocyte fatty acid-binding protein in the development of accelerated atherolsclerosis in hypercholesterolemic mice. FASEB J 15: 2733–3735, 2001
41. Makowski L, Boord JB, Maeda K, Babaev VR, Uysal KT, Morgan MA, Parker RA, Suttles J, Fazio S, Hotamisligil GS, Linton MF: Lack of macrophage fatty-acid-binding protein aP2 protects mice deficient in apolipoprotein E against atherosclerosis. Nature Med 7: 699–705, 2001

Molecular and Cellular Biochemistry **239**: 35–43, 2002.

Cytosolic fatty acid binding proteins catalyze two distinct steps in intracellular transport of their ligands

Richard A. Weisiger

Department of Medicine and the Liver Center, University of California, San Francisco, CA, USA

Abstract

Cytosolic long-chain fatty acid binding proteins (FABPs) are found in tissues that metabolize fatty acids. Like most lipid binding proteins, their specific functions remain unclear. Two classes have been described. Membrane-active FABPs interact directly with membranes during exchange of fatty acids between the protein binding site and the membrane, while membrane-inactive FABPs bind only to fatty acids that are already in aqueous solution. Despite these binding proteins, most fatty acids in cell cytoplasm appear to be bound to membranes. This paper reviews data suggesting that FABPs catalyze transfer of fatty acids between intracellular membranes, often across considerable intracellular distances. This process occurs in three distinct steps: dissociation of the fatty acid from a 'donor' membrane, diffusion of the fatty acid across the intervening water layer, and binding to an 'acceptor' membrane. Membrane-active FABPs catalyze dissociation of the fatty acid from the donor membrane and binding to the acceptor membrane, while membrane-inactive FABPs catalyze diffusion of fatty acids across the aqueous cytosol. Thus, FABPs catalyze all three steps in intracellular transport. A simple quantitative model has been developed that predicts the rate of intracellular transport as a function of the concentration, affinity and diffusional mobility of the binding protein. Different FABPs may have evolved to match the specific transport requirements of the cell type within which they are found. (Mol Cell Biochem **239**: 35–43, 2002)

Key words: fatty acid binding proteins, FABP, diffusion, dissociation, rate constants, kinetic models

Introduction

Soluble fatty acid binding proteins (FABPs) are among the most abundant proteins in cells. However, no clear understanding of their function has emerged [1–4], apart from suggestions that they are somehow involved in intracellular fatty acid transport [2, 5–8]. Recent data indicate that they may play an important role in regulating fatty acid metabolism. For example, disrupting the gene for cardiac FABP in mice greatly reduces fatty acid uptake and metabolism in heart and skeletal muscle tissue [9], while a natural mutant of intestinal FABP in humans causes increased fatty acid absorption [10]. Rates of fatty acid uptake and metabolism typically rise when cytosolic FABP levels are increased by transfection or induction [11–20] and fall when the concentration or binding affinity of FABP is reduced [21–23]. These data suggest that FABPs are intimately involved in cellular fatty acid metabolism, but do not tell us how they work.

One possible explanation for these observations is that FABPs are required for timely exchange of fatty acids between the plasma membrane and intracellular sites of metabolism, most of which are also on membranes. In other words, FABPs may regulate transfer of fatty acids between membranes within cells. This hypothesis is supported by considerable literature data. Thus, the rate of cytoplasmic fatty acid transport is proportional to the concentration of FABP in liver and intestinal mucosa [6, 24]. Moreover, inhibition or removal of FABP blocks the cytoplasmic transport of fatty acids almost completely, indicating that alternative pathways cannot replace the function of FABP [25, 26]. If intracellular transport of fatty acids is sufficiently slow, it could limit access of the fatty acid to the cytoplasmic enzymes responsible

Address for offprints: R.A. Weisiger, Division of Gastroenterology and the Liver Center, 357 Sciences Building, University of California, San Francisco, CA 94143-0538, USA (E-mail: dickw@itsa.ucsf.edu)

for fatty acid metabolism [27]. Thus, FABP seems well-positioned to regulate the rate of fatty acid uptake, efflux and metabolism in cells.

This review will summarize recent data suggesting that FABPs regulate the rate of intracellular transport of long-chain fatty acids in two distinct and complementary ways: First, by catalyzing the rate of dissociation of fatty acids from membranes into the cell water. Second, by catalyzing diffusion of the fatty acids across cytosolic water layers that they could not otherwise penetrate efficiently because of their low aqueous solubility. These simple concepts lead to new model for understanding the transport functions of FABPs that makes a number of experimentally testable predictions.

The crucial importance of membranes

Long chain fatty acids have an extremely limited solubility in water, estimated to be as low as 10^{-10} M for palmitate [28], and bind avidly to membranes [29]. Most long chain fatty acids in cell cytoplasm are bound to membranes. Thus, measurements using a fluorescent fatty acid analog (NBD-stearate) found that 61–95% of the fatty acid was bound to cytoplasmic membranes in liver [30], intestinal mucosa [24], and HepG2 cells [25] despite the presence of substantial concentrations of FABP. Palmitate appears to exhibit a similar high degree of membrane binding in liver cytoplasm as evidenced by a comparably slow rate of intracellular transport [27]. Removal of FABP increases membrane binding to nearly 100% [25]. Membranes not only bind fatty acids, but process them as well. Fatty acids enter and exit cells through the plasma membrane and are metabolized within membranes [31]. Thus, intracellular fatty acid transport may be viewed as the process by which fatty acids are transferred from one membrane to another, i.e. from a 'donor' membrane (to which they avidly bind) through the cell water (where they are virtually insoluble) to an 'acceptor' membrane (where they may be metabolized or secreted). This process may occur directly, or may involve binding to intermediate membranes along the way. This review considers how FABPs participate in this process.

Classification of fatty acid binding proteins

To date, 14 members of the FABP gene family have been identified [8]. Major types include FABP from liver (L-FABP), heart (H-FABP), intestine (I-FABP), and adipocytes (A-FABP or aP2). These FABPs may be divided into two general classes based on their ability to interact with membranes. Most forms of FABP (e.g. from heart, adipocyte and intestine) transfer fatty acids to and from membranes during brief collisional encounters [32–36]. This process appears to involve a transient interaction between a positively charged region of the binding protein and negative charges on the membrane surface, resulting in a conformational change that permits direct transfer of fatty acid between the protein binding site and the membrane [8, 32, 33, 37, 38]. We will refer to this FABP class as being *membrane-active* [39, 40]. The second class of FABP molecules lacks the ability to interact with membranes, and is only able to bind fatty acid monomers that are already in aqueous solution. The best studied example of this class is L-FABP [34, 36]. We will refer to this class as being *membrane-inactive* [39, 40].

Mechanisms of cytoplasmic transport

Transfer of fatty acids between membranes requires a minimum of three steps, dissociation from a donor membrane, diffusion across the intervening water layer, and binding to an acceptor membrane. Although this process is reversible, fatty acids will tend to move from membranes with higher fatty acid concentrations (e.g. the plasma membrane) to membranes with lower concentrations (e.g. the outer mitochondrial membrane). The overall transfer rate at steady-state reflects the rates of all component steps. If one step is much slower than the others, then it will determine the overall transfer rate. Such a step is referred to as being 'rate-limiting'. Available data suggest that dissociation [40] and aqueous diffusion [27] are slower than rebinding, and are thus most likely to be rate-limiting. Three distinct transfer mechanisms are possible (Fig. 1).

Spontaneous transfer

The simplest mechanism is for the fatty acid to spontaneously dissociate from the donor membrane and to diffuse across the cytosol to the acceptor membrane without ever binding to FABP (Fig. 1, top). However, this mechanism is limited by the extremely small solubility of the fatty acid in water and its strong affinity for membranes, both of which greatly limit the concentration that can exist in the soluble (diffusible) phase. According to Fick's law of diffusion [41], the diffusional flux is proportional to the concentration difference across the water layer (ΔC) divided by the diffusion distance. However, ΔC cannot be greater than the solubility of the fatty acid in water. Thus, this mechanism is very slow unless the distance separating the membranes is very short [39].

Transfer catalyzed by membrane-inactive FABP

Membrane-inactive binding proteins such as L-FABP increase the diffusional flux by increasing the concentration of the fatty acid in the aqueous phase, often by many orders of magnitude [42]. The greater solubility of the fatty acid allows

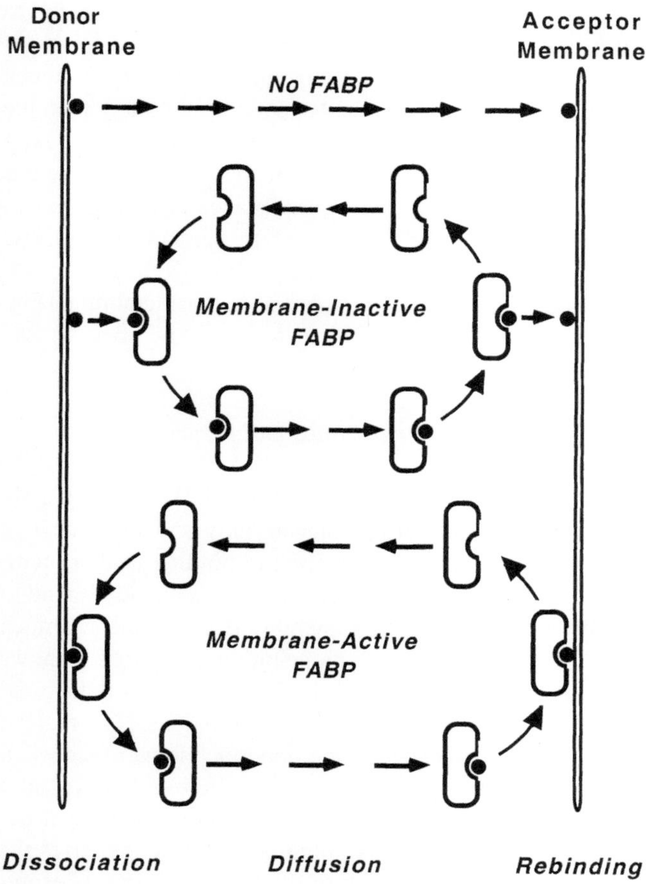

Fig. 1. Three mechanisms for transfer of fatty acid from a donor to an acceptor membrane. Transfer of a fatty acid (solid circle) from a donor membrane to an acceptor membrane may occur spontaneously (top) or with the assistance of a membrane-inactive (middle) or membrane-active (bottom) binding protein. In each case, transfer requires three steps: dissociation from the donor membrane, diffusion across the water layer, and binding to the acceptor membrane. Spontaneous transfer is very slow unless the membrane separation is small (less than d_m, see text). Membrane-inactive binding proteins catalyze the diffusional step by increasing the concentration of the fatty acid in the soluble (diffusible) pool. Membrane-active FABPs also catalyze dissociation and rebinding during collisional interactions with the membrane. Thus, FABPs catalyze all three transport steps. These diagrams depict the catalytic cycle only. In fact, each step is reversible and transfer occurs in both directions with a net rate proportional to the difference in the fatty acid concentration between the two membranes.

ΔC to be much larger, thus supporting much greater diffusional fluxes despite the somewhat lower diffusional mobility of the protein-bound form (Fig. 1, middle). The process of augmenting transport by adding a second diffusional flux of protein-bound fatty acid is known as *co-diffusion*. In this process, FABP serves as a true carrier system to transport the fatty acid across the water layer in a manner that is completely analogous membrane carriers [6]. The only difference is that membrane carriers transport hydrophilic molecules across a hydrophobic barrier, while FABP transports hydrophobic molecules across a hydrophilic barrier [6].

Transfer catalyzed by membrane-active FABP

Membrane-active forms of FABP can catalyze diffusion across water layers just like membrane-inactive forms. However, they can also catalyze dissociation of the fatty acid from the donor membrane [36] and rebinding of the fatty acid to the acceptor membrane (Fig. 1, bottom) [8]. Thus, membrane-active FABPs catalyze all three steps in fatty acid transfer. This may explain why most forms of FABP studied so far are membrane-active.

Alternative mechanisms

Fatty acids could also reach the acceptor membrane without ever entering the water phase by lateral diffusion within membranes. Because the endoplasmic reticulum and plasma membrane are connected by membrane bridges, it has been suggested that fatty acid uptake by cells may occur solely by this process [43]. Fatty acids could also move through the cytoplasm while bound to mobile membrane vesicles using an energy-dependent process [44]. However, both of these mechanisms are too slow to support observed rates of transport. The most convincing evidence for this is that removing FABP from cells or blocking binding to FABP using a competitive inhibitor almost completely blocks the intracellular movement of fatty acids [25, 26].

To better understand the functions of fatty acid binding proteins, Stephen Zucker and I modeled the transport of a fatty acid between two planar membranes in the presence of FABP. The derivation of this model has already been published [39]. In this paper I will present our results in intuitive terms and discuss their implications for understanding the regulation of fatty acid metabolism.

Possible outcomes

Consider what can happen after an individual fatty acid dissociates from a donor membrane into the cytosol (either spontaneously or catalyzed by a membrane-active FABP). Because diffusion is a random process, the fatty acid will move unpredictably through the aqueous cytosol until it eventually rebinds to a membrane. The diffusional path traced through the cytosol is referred to as an 'excursion'. Four types of excursion are possible (Fig. 2). First, the fatty acid may reach and bind to the acceptor membrane, thus resulting in a 'successful' excursion (Fig. 2A). Second, the fatty acid may rebind to the donor membrane before it reaches the acceptor membrane, resulting in an 'unsuccessful' excursion (Fig. 2B). Excursions are always more likely to be unsuccessful than successful because the fatty acid molecules start out closer to the donor membrane and are thus more likely to randomly

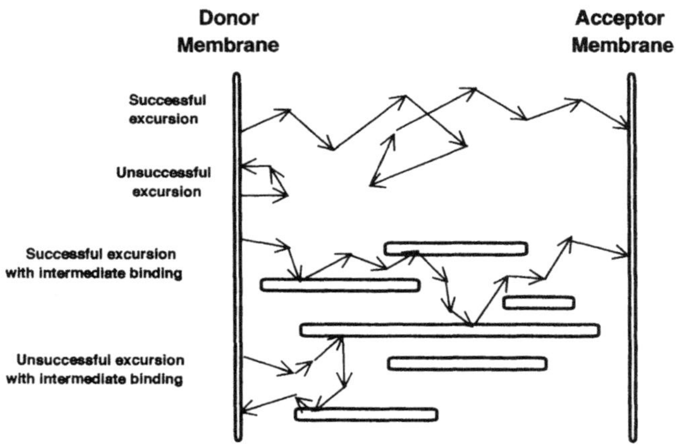

Fig. 2. Possible outcomes following dissociation of a fatty acid molecule from a donor membrane. (A) Diffusion to an acceptor membrane resulting in successful transfer. (B) Rebinding to the same membrane resulting in unsuccessful transfer. (C and D) Binding to an intermediate membrane before ultimately binding to either the donor or acceptor membrane. FABPs increase the probability of successful transfer (see text).

collide with it than with the acceptor membrane. Third, the fatty acid may transiently bind to one or more intermediate membranes before finally binding the donor or acceptor membranes (Figs 2C and 2D). Intermediate membranes are defined as membranes from which fatty acid molecule dissociates before it undergoes metabolism.

Mean diffusional excursion

What determines the probability of a successful excursion? Inspection of some simulated diffusional paths is instructive. Figure 3A shows the diffusional paths following dissociation of 20 consecutive fatty acid molecules in the absence of FABP. These random paths were generated using a computer program available from the author. Notice that only one of these excursions was successful. Most of the fatty acids diffused only a short distance into the water before they rebound the to donor membrane, and nearly half rebound immediately with no visible diffusional path. The average distance that the fatty acid diffuses into the water before rebinding to the donor membrane is referred to as the *mean diffusional excursion*, and is shown by the broken vertical line in Fig. 3A. Abbreviated d_m, the mean diffusional excursion is a critical determinant of the rate of transfer of amphipathic molecules between membranes. As shown in Fig. 3B, binding proteins increase d_m. Before we consider how they do this, we must address the importance of membrane separation.

The distance separating the membranes (d) greatly affects the probability of a successful excursion. If the membrane separation is much greater than the mean diffusional excursion ($d >> d_m$), then the probability of success will be very low (because most of the fatty acid molecules will rebind to the donor membrane before they encounter the acceptor membrane). On the other hand, if the membrane separation is much less than the mean diffusional excursion ($d << d_m$), then the probability of success approaches 50% (because each fatty acid is equally likely to encounter the donor and acceptor membranes). If follows that as the distance separating the two membranes is reduced, the rate of fatty acid transfer will rise until it approaches a maximum equal to half the rate of dissociation of the fatty acid from the donor membrane (Fig. 4). We refer to this maximum as J_{max}.

FABP increases the mean diffusional excursion

We now consider the affect of adding FABP. In Fig. 3A, we assumed that collision of an aqueous fatty acid molecule with either membrane always resulted in binding to that membrane. If most of the aqueous fatty acid is already bound to FABP, however, then these membrane collisions are much less likely to result in membrane binding. Figure 3B shows what happens when enough FABP is added to reduce the probability of rebinding to 1% of all collisions. Note that the fatty acid stays in aqueous solution much longer than when FABP is absent despite many membrane collisions, typically allowing it to diffuse much further into the water layer before rebinding. In this way binding to FABP increases the mean diffusional excursion of the fatty acid. In this example, half of the excursions (2 of 4) were successful, suggesting that the transfer rate is now near J_{max}. These simulations show that FABP can increase the transfer rate to J_{max} by increasing the mean diffusional excursion of the fatty acid into the water phase (d_m) until it is greater than the distance separating the donor and acceptor membranes (d).

With this intuitive background, we now examine the simple equation that describes the effects of FABP on the steady-state rate of fatty acid transport between the two membranes as previously derived [39].

$$J = \frac{J_{max}}{1 + d/d_m} \tag{1}$$

Here, J is the steady rate of fatty acid transfer between the membranes and J_{max} is half the rate of dissociation of the fatty acid from the donor membrane. Note that when d is much smaller than d_m, Eq. 1 simplifies to $J = J_{max}$. This can happen for one of two reasons. First, the membranes may be so close together so that the mean diffusional excursion of the fatty acid is greater than the membrane separation even when no FABP is present. In this case, adding a membrane-inactive FABP does not further increase the flux despite an increase in d_m. If the membrane separation is initially greater than d_m, however, then few excursions will be successful and the trans-

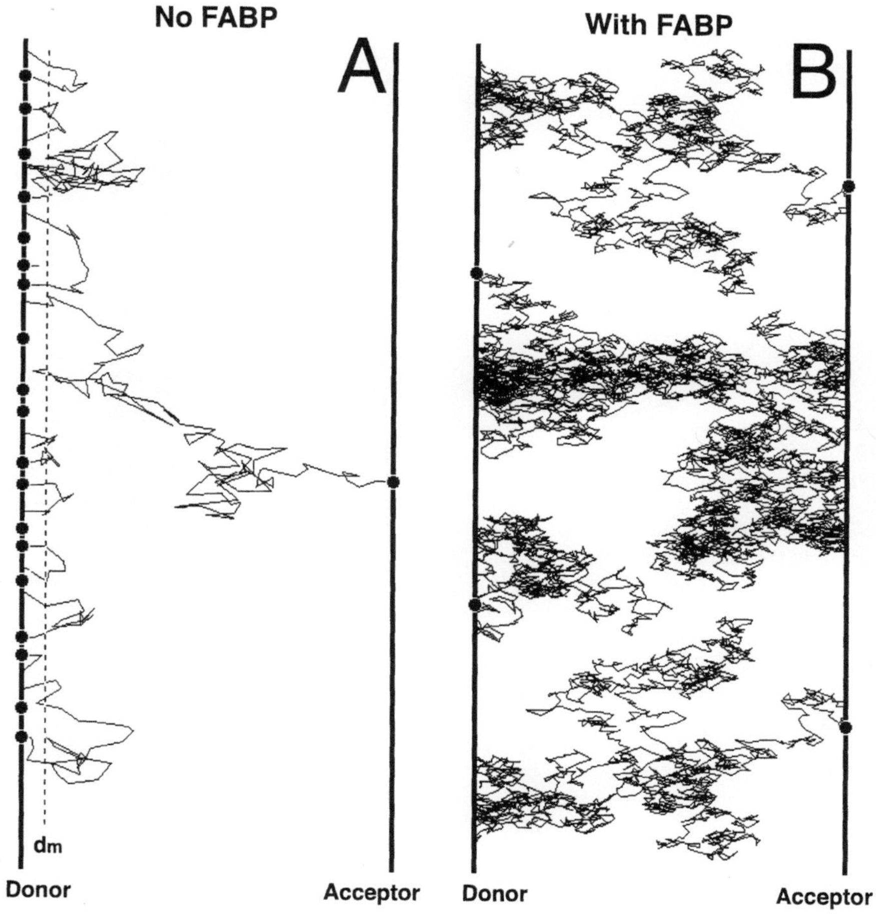

Fig. 3. Effect of FABP on sample diffusional paths. All fatty acid molecules start on the donor membrane and diffuse randomly until they rebind to either the donor or the acceptor membrane. Lines trace the random paths for each fatty acid molecule through the aqueous phase and solid circles show the final binding location. The membrane separation (d) is the same for both panels. (Panel A) In the absence of FABP, d is much greater than the mean diffusional excursion of the fatty acid (d_m, broken line). As a result, most of the fatty acids (19 of 20 in this example) rebind to the donor membrane before they reach the acceptor membrane, half without leaving any visible trace. (Panel B) Enough FABP has been added to reduce by 99% the probability that collision of the fatty acid with a membrane will result in binding to that membrane. Slower rebinding allows the fatty acid to remain in solution much longer, increasing the mean diffusional excursion of the fatty acid into the water phase (by a factor of 40 in this example). Four sample diffusional paths are shown, of which 50% resulted in transfer to the acceptor membrane.

port rate will be much less than J_{max}. In this case, adding a membrane-inactive FABP can increase the rate by increasing d_m, but only until J equals J_{max}. In summary, by increasing d_m, FABP can increase transport between the membranes until diffusion is no longer rate-limiting and the rate is equal to J_{max}.

In contrast, membrane-active FABPs not only increase d_m, but also increase J_{max} by catalyzing the rate of dissociation of the fatty acid from the donor membrane [8]. As a result, membrane-active FABPs can catalyze transport even when the membrane separation is very small and the flux is already at J_{max}. This suggests that membrane-active binding proteins may be specialized to catalyze transfer between closely spaced membranes, such as between the sarcolemma and the outer mitochondrial membrane in the heart myocyte.

If membrane-active FABP's can do everything membrane-inactive FABPs can do and they can also catalyze dissociation, then why do cells need membrane-inactive FABPs at all? The answer may be that the ability to catalyze dissociation comes at the cost of a reduced d_m. Membrane-active FABPs may be less effective in preventing rebinding of fatty acids to the donor membrane. With membrane-inactive FABPs, only the tiny amount of unbound fatty acid in the cytosol can bind to the membrane. For membrane-active FABPs however, collisional encounters of the protein with the membrane offer an additional rebinding mechanism, resulting in a smaller mean diffusional excursion for a given amount of FABP. Membrane-active FABPs may also bind to membranes so frequently that their diffusional mobility is reduced. Figure 5 compares the binding of a membrane-active FABP (H-FABP) to negatively-

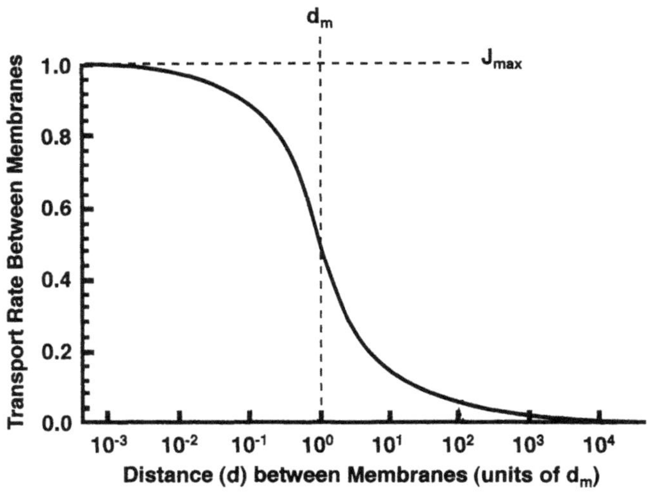

Fig. 4. Effect of membrane separation on transfer rate. When the membrane separation (d) is much larger than the mean diffusional excursion (d_m), most excursions are unsuccessful and diffusion is rate-limiting (right side of plot). Decreasing d until it is less than d_m causes the rate to increase to a maximum, J_{max}, that is equal to half the rate of dissociation of the fatty acid from the donor membrane. Under these conditions, dissociation of the fatty acid from the membrane is rate-limiting (left side of plot). All FABPs increase d_m by increasing the mean diffusional excursion. Membrane-inactive FABPs also increase J_{max} by increasing the rate of dissociation from the donor membrane (see text). This curve is valid whether or not FABP is present. If a membrane-active FABP is present, J_{max} is half the catalyzed rate of dissociation.

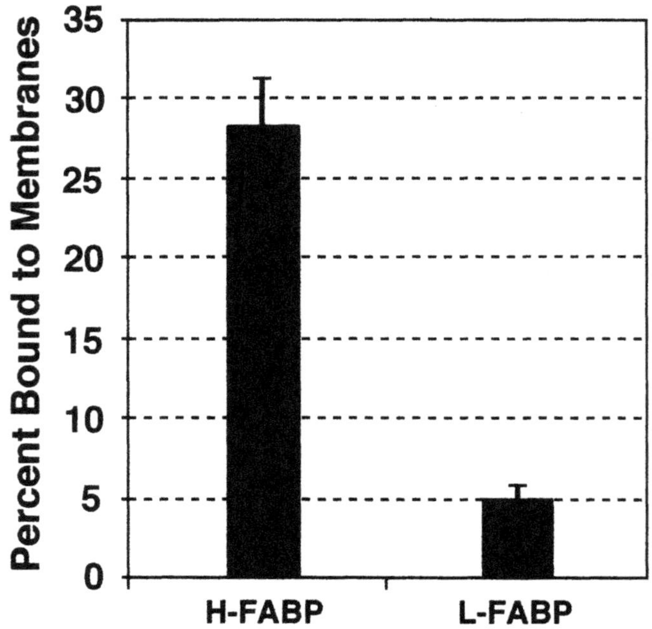

Fig. 5. Binding of FABPs to membrane vesicles. FABPs (labeled with ^{125}I) were incubated with negatively-charged liposomes (labeled with ^{3}H) and then separated on a Sephacryl S-100 column using methods described in [46]. A much greater percentage of the membrane-active H-FABP co-eluted with the membrane vesicles than for the membrane-inactive L-FABP, indicating more avid membrane binding. If similar binding occurs to cytoplasmic membranes, it would slow the movement of membrane-active FABPs through membrane-rich cell cytoplasm. H-FABP was kindly provided by Nathan Bass, while I-FABP was from Judith Storch.

charged membranes with that for a membrane-inactive form (L-FABP). The greater binding of H-FABP to membranes would be expected to reduce its diffusional mobility in a membrane-rich cytoplasm. If similar binding to biological membranes occurs in cell cytoplasm, then this would help to explain why certain cells need a membrane-inactive FABP. Cells that would benefit the most would have cytoplasm that is densely packed with membranes and the need to transport fatty acids efficiently over longer intracellular distances. The two cell types known to have membrane-inactive FABPs, liver and intestinal mucosa, meet both of these criteria.

The mean diffusional excursion can be defined in terms of the affinity and concentration of the binding proteins as shown below [39].

$$d_m = \frac{2D_l}{P_{wm}} + \frac{2P\,D_p\,K_a}{P_{wm}} \tag{2}$$

where D_l and D_p are the diffusion constants of the unbound and protein-bound fatty acids in cytosol, P and K_a are the concentration and binding affinity of the FABP, and P_{wm} is the permeability of the membrane-water interface that determines the rate of rebinding. Notice that when the binding protein concentration is zero, this equation simplifies to the diffusional excursion in the absence of binding protein, $d_m = 2D_l/P_{wm}$. The rightmost term in Eq. (3) gives the additional diffusional excursion due to protein binding. From inspection, it is clear that the largest values of d_m will occur for binding proteins with high concentrations, high binding affinities and high diffusion constants.

As noted earlier, J_{max} is simply half the rate of dissociation of the fatty acid from the donor membrane,

$$J_{max} = \frac{C_d\,P_{mw}}{2} \tag{3}$$

where C_d is the concentration of the fatty acid in the donor membrane and P_{mw} is the permeability of the membrane-water interface that determines the rate of dissociation. If the concentration of the fatty acid in the acceptor membrane is not zero, then bidirectional transfer will occur. In this case, J_{max} is determined by the difference in the fatty acid concentrations between the two membranes, ΔC_m.

$$J_{max} = \frac{\Delta C_m\,P_{mw}}{2} \tag{4}$$

These equations were used to generate the curve shown in Fig. 4. Addition of a membrane-inactive binding protein in-

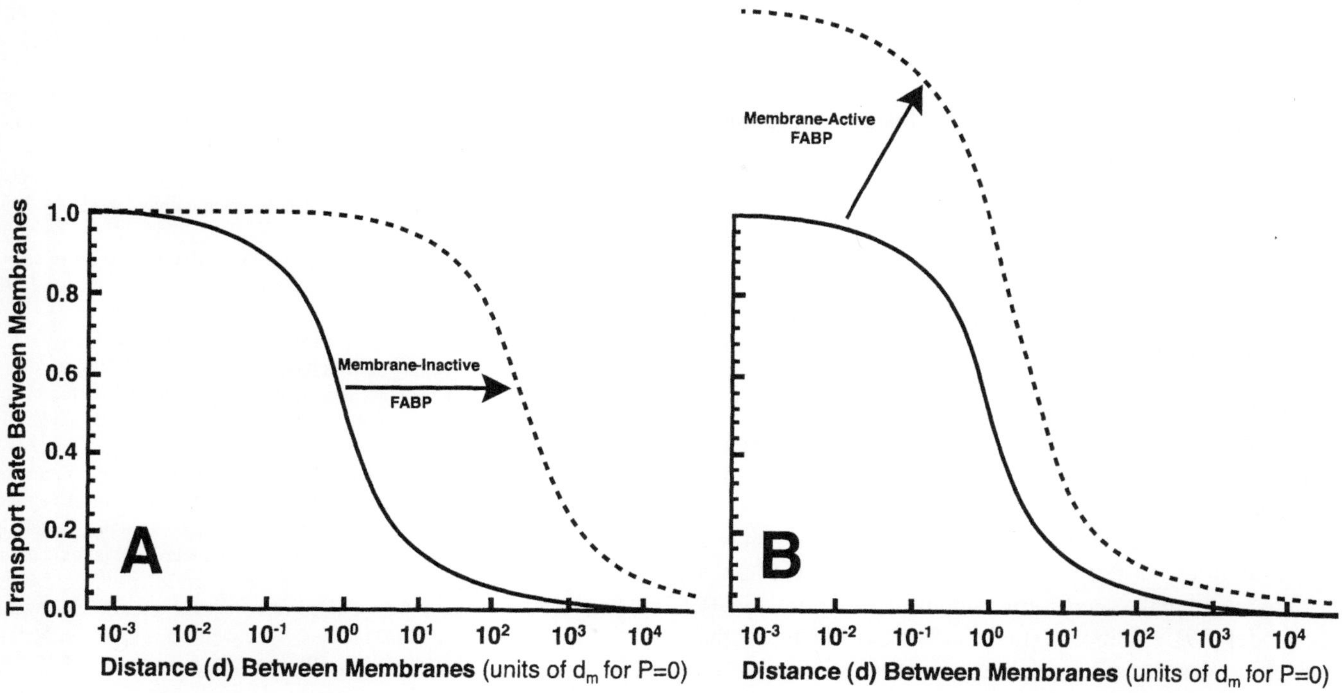

Fig. 6. Effect of added FABP on transport flux. Membrane-inactive binding proteins shift the curve to the right (panel A), allowing the transport flux to remain at J_{max} for larger membrane separations. Membrane-active binding proteins shift the curve upwards (panel B) by catalyzing J_{max}, but may shift it to the right as well. Solid line is identical to curve in Fig. 4. Distance between membranes is given in units of d_m in the absence of binding proteins.

creases the transport rate by shifting this curve to the right, allowing the rate to remain a J_{max} for larger membrane separations (Fig. 6A). Addition of a membrane-active binding protein not only increases d_m, but also increases J_{max} (Fig. 6B).

Future directions

The synthesis presented here raises a number of questions that will need to be addressed before the functional importance of FABPs can be fully assessed. Some of the most important are:

Magnitude of intracellular fatty acid concentration gradients

A primary function of FABPs may be to minimize formation of intracellular concentration gradients that could limit the rates of fatty acid uptake, efflux or metabolism. For example, recent data suggest that fatty acids are metabolized in liver cells faster than they can be replenished by diffusion from the cell surface [27, 45]. If so, hepatic fatty acid metabolism may be limited (and thus regulated) by formation of concentration gradients within the cell. In a like manner, the decrease in myocardial fatty acid metabolism when H-FABP is deleted [9] could be readily explained by formation of a steep fatty acid concentration gradient between the plasma membrane and outer mitochondrial membrane in the myocyte. Studies are needed to determine whether these predicted gradients are actually present and to assess their magnitude.

Characterizing intracellular transport kinetics

Membrane carriers have traditionally been characterized by analyzing their kinetics using transport models. In a like manner, intracellular carriers such as the FABPs can be characterized by analyzing intracellular transport kinetics using Eqs 1–4. Comparing the transport kinetics for different types of FABP should help define how each form is matched to the function it serves in its specific cell type. Experiments should include determining the value of d_m in living cells and comparing the diffusional mobilities of different forms of FABP within membrane-rich cytoplasm.

Interaction of membrane-active FABPs with biological membranes

Most experimental data supporting a role of FABP in catalyzing dissociation of fatty acids from membranes were obtained using artificial membrane vesicles. These findings need to be confirmed using biological membranes, ideally within living cells. The possibility that membrane-active FABPs may selectively direct fatty acids to certain intracellular membranes (such as the outer mitochondrial membrane) also needs to be tested.

Application to more physiologic settings

Like all models, Eqs 1–4 are based on certain simplifying assumptions including planar membranes, a steady transport rate and equilibrium between aqueous and protein-bound FABP. These assumptions may not always apply *in vivo*. Fortunately, a numerical model has been developed that can be used in these cases [39]. This model makes the same predictions as Eqs 1–4, under comparable conditions, but is not constrained by the same assumptions. We have already shown that binding of the fatty acids to intermediate membranes prior to reaching the acceptor membrane does not invalidate our model [39]. Additional studies are needed to assess the impact of other factors on model predictions.

Summary

The transport functions of FABPs and other soluble lipid binding proteins are now becoming clear. Like membrane carriers, these cytosolic carriers catalyze a crucial transport process. The availability of new models will permit quantitative analysis of intracellular transport kinetics, and should lead to rapid advances in our understanding of how these ubiquitous proteins control the distribution and metabolism of their ligands.

Acknowledgements

This study was supported by National Institutes of Health Grants DK 32898 (RAW) and by Liver Center Grant DK 26743.

References

1. Bernlohr DA, Coe NR, LiCata VJ: Fatty acid trafficking in the adipocyte. Semin Cell Dev Biol 10: 43–49, 1999
2. Glatz JFC, van der Vusse GJ: Cellular fatty acid-binding proteins: Their function and physiological significance. Prog Lipid Res 35: 243–282, 1996
3. Bass NM, Kaikaus RM, Ockner RK: Physiology and molecular biology of hepatic cytosolic fatty acid-binding protein. In: N. Tavoloni, P. D. Berk (eds). Hepatic Transport and Bile Secretion: Physiology and Pathophysiology. Raven Press, New York, 1993, pp 421–446
4. Ockner RK, Kaikaus RM, Bass NM: Fatty acid binding proteins: Recent concepts of regulation and function. Prog Clin Biol Res 375: 189–204, 1992
5. Tipping E, Ketterer B: The influence of soluble binding proteins on lipophile transport and metabolism in hepatocytes. Biochem J 195: 441–452, 1981
6. Weisiger RA: When is a carrier not a membrane carrier? The cytoplasmic transport of amphipathic molecules. Hepatology 24: 1288–1295, 1996
7. Zucker SD, Goessling W, Gollan JL: Intracellular transport of small hydrophobic compounds by the hepatocyte. Semin Liver Dis 16: 159–167, 1996
8. Storch J, Thumser AE: The fatty acid transport function of fatty acid-binding proteins. Biochim Biophys Acta 55624: 1–17, 2000
9. Binas B, Danneberg H, McWhir J, Mullins L, Clark AJ: Requirement for the heart-type fatty acid binding protein in cardiac fatty acid utilization. FASEB J 13: 805–812, 1999
10. Baier LJ, Sacchettini JC, Knowler WC, Eads J, Paolisso G, Tataranni PA, Mochizuki H, Bennett PH, Bogardus C, Prochazka M: An amino acid substitution in the human intestinal fatty acid binding protein is associated with increased fatty acid binding, increased fat oxidation, and insulin resistance. J Clin Invest 95: 1281–1287, 1995
11. Renaud G, Foliot A, Infante R: Increased uptake of fatty acids by the isolated rat liver after raising the fatty acid binding protein concentration with clofibrate. Biochem Biophys Res Commun 80: 327–334, 1978
12. Capuzzi DM, Lackman RD, Uberti MO, Reed MA: Stimulation of hepatic triglyceride synthesis and secretion by clofibrate. Biochem Biophys Res Commun 60: 1499–1508, 1974
13. Murphy EJ: L-FABP and I-FABP expression increase NBD-stearate uptake and cytoplasmic diffusion in L cells. Am J Physiol 275: G244–G249, 1998
14. Burczynski FJ, Fandrey S, Wang G, Pavletic PA, Gong Y: Cytosolic fatty acid binding protein enhances rat hepatocyte [^{3}H]palmitate uptake. Can J Physiol Pharmacol 77: 896–901, 1999
15. Baier LJ, Bogardus C, Sacchettini JC: A polymorphism in the human intestinal fatty acid binding protein alters fatty acid transport across Caco-2 cells. J Biol Chem 271: 10892–10896, 1996
16. Prows DR, Murphy EJ, Moncecchi D, Schroeder F: Intestinal fatty acid-binding protein expression stimulates fibroblast fatty acid esterification. Chem Phys Lipids 84: 47–56, 1996
17. Murphy EJ, Prows DR, Jefferson JR, Schroeder F: Liver fatty acid-binding protein expression in transfected fibroblasts stimulates fatty acid uptake and metabolism. Biochim Biophys Acta 1301: 191–198, 1996
18. Atshaves BP, Foxworth WB, Frolov A, Roths JB, Kier AB, Oetama BK, Piedrahita JA, Schroeder F: Cellular differentiation and I-FABP protein expression modulate fatty acid uptake and diffusion. Am J Physiol 274: C633–C644, 1998
19. Wolfrum C, Buhlmann C, Rolf B, Borchers T, Spener F: Variation of liver-type fatty acid binding protein content in the human hepatoma cell line HepG2 by peroxisome proliferators and antisense RNA affects the rate of fatty acid uptake. Biochim Biophys Acta 1437: 194–201, 1999
20. Darimont C, Gradoux N, Persohn E, Cumin F, De Pover A: Effects of intestinal fatty acid-binding protein overexpression on fatty acid metabolism in Caco-2 cells. J Lipid Res 41: 84–92, 2000
21. Ockner RK, Manning JA: Fatty acid binding protein. Role in esterification of absorbed long chain fatty acid in rat intestine. J Clin Invest 58: 632–641, 1976
22. Wu-Rideout MYC, Elson C, Shrago E: The role of fatty acid binding protein on the metabolism of fatty acids in isolated rat hepatocytes. Biochem Biophys Res Commun 71: 809–816, 1976

23. Memon RA, Bass NM, Moser AH, Fuller J, Appel R, Grunfeld C, Feingold KR: Down-regulation of liver and heart specific fatty acid binding proteins by endotoxin and cytokines *in vivo*. Biochim Biophys Acta 1440: 118–126, 1999
24. Luxon BA, Milliano MT: Cytoplasmic transport of fatty acids in rat enterocytes: Role of binding to fatty acid-binding protein. Am J Physiol 277: G361–G366, 1999
25. Luxon BA, Milliano MT: Facilitation of cytoplasmic transport of fatty acids via co-diffusion is not specific for fatty acid binding protein. Am J Physiol 273: C859–C867, 1997
26. Luxon BA: Inhibition of binding to fatty acid binding protein reduces the intracellular transport of a fatty acids. Am J Physiol 271: G113–G120, 1996
27. Luxon BA, Holly DC, Milliano MC, Weisiger RA: Sex differences in multiple steps in the hepatic transport of palmitate support a balanced uptake mechanism. Am J Physiol 274: G52–G61, 1998
28. Vorum H, Brodersen R, Kragh-Hansen U, Pedersen AO: Solubility of long-chain fatty acids in phosphate buffer at pH 7.4. Biochim Biophys Acta Lipids Lipid Metab 1126: 135–142, 1992
29. Richieri GV, Ogata RT, Kleinfeld AM: Thermodynamics of fatty acid binding to fatty acid-binding proteins and fatty acid partition between water and membranes measured using the fluorescent probe ADIFAB. J Biol Chem 270: 15076–15084, 1995
30. Luxon BA, Weisiger RA: Sex differences in intracellular fatty acid transport: Role of cytoplasmic binding proteins. Am J Physiol 265: G831–G841, 1993
31. Katz N: Metabolism of lipids. In: R.G. Thurman, F.C. Kauffman, K. Jungermann (eds). Regulation of Hepatic Metabolism: Intra- and Intercellular Compartmentation. Plenum Press, New York, 1986, pp 237–253
32. Gericke A, Smith ER, Moore DJ, Mendelsohn R, Storch J: Adipocyte fatty acid-binding protein: Interaction with phospholipid membranes and thermal stability studied by FTIR spectroscopy. Biochemistry 36: 8311–8317, 1997
33. Herr FM, Aronson J, Storch J: Role of portal region lysine residues in electrostatic interactions between heart fatty acid binding protein and phospholipid membranes. Biochemistry 35: 1296–1303, 1996
34. Hsu KT, Storch J: Fatty acid transfer from liver and intestinal fatty acid-binding proteins to membranes occurs by different mechanisms. J Biol Chem 271: 13317–13323, 1996
35. Kim HK, Storch J: Mechanism of free fatty acid transfer from rat heart fatty acid-binding protein to phospholipid membranes. Evidence for a collisional process. J Biol Chem 267: 20051–20056, 1992
36. Thumser AE, Storch J: Liver and intestinal fatty acid-binding proteins obtain fatty acids from phospholipid membranes by different mechanisms. J Lipid Res 41: 647–656, 2000
37. Herr FM, Matarese V, Bernlohr DA, Storch J: Surface lysine residues modulate the collisional transfer of fatty acid from adipocyte fatty acid binding protein to membranes. Biochemistry 34: 11840–11845, 1995
38. Corsico B, Cistola DP, Frieden C, Storch J: The helical domain of intestinal fatty acid binding protein is critical for collisional transfer of fatty acids to phospholipid membranes. Proc Natl Acad Sci USA 95: 12174–12178, 1998
39. Weisiger RA, Zucker SD: Transfer of small molecules between intracellular membranes: Roles of soluble binding proteins, distance and time. Am J Physiol 2002 (in press)
40. Kleinfeld AM, Chu P, Romero C: Transport of long-chain native fatty acids across lipid bilayer membranes indicates that transbilayer flip-flop is rate limiting. Biochemistry 36: 14146–14158, 1997
41. Fick A: Über diffusion. Ann Physik 94: 59–86, 1855
42. Weisiger RA: Cytoplasmic transport of lipids: Role of binding proteins. Comp Biochem Physiol 115B: 319–331, 1996
43. Scow RO, Blanchette-Mackie EJ, Smith LC: Role of capillary endothelium in the clearance of chylomicrons: A model for lipid transport. Circ Res 49: 149–162, 1976
44. Crawford JM, Gollan JL: Transcellular transport of organic anions in hepatocytes: Still a long way to go. Hepatology 14: 192–197, 1991
45. Luxon BA, Milliano MT, Weisiger RA: Induction of hepatic cytosolic fatty acid binding protein with clofibrate accelerates both membrane and cytoplasmic transport of palmitate. Biochim Biophys Acta 1487: 309–318, 2000
46. Weisiger RA: Saturable stimulation of fatty acid transport through model cytoplasm by soluble binding protein. Am J Physiol 277: G109–G119, 1999

Molecular and Cellular Biochemistry **239**: 45–54, 2002.

Insights into binding of fatty acids by fatty acid binding proteins

Thorsten Hanhoff,[1] Christian Lücke[2] and Friedrich Spener[1]

[1]*Department of Biochemistry, University of Münster, Münster;* [2]*Institute of Biophysical Chemistry, University of Frankfurt, Frankfurt, Germany*

Abstract

Members of the phylogenetically related intracellular lipid binding protein (iLBP) are characterized by a highly conserved tertiary structure, but reveal distinct binding preferences with regard to ligand structure and conformation, when binding is assessed by the Lipidex method (removal of unbound ligand by hydrophobic polymer) or by isothermal titration calorimetry, a true equilibrium method. Subfamily proteins bind retinoids, subfamily II proteins bind bulky ligands, examples are intestinal bile acid binding protein (I-BABP) and liver fatty acid binding protein (L-FABP) which binds 2 ligand molecules, preferably monounsaturated and n-3 fatty acids. Subfamily III intestinal fatty acid binding protein (I-FABP) binds fatty acid in a bent conformation. The fatty acid bound by subfamily IV FABPs has a U-shaped conformation; here heart (H-) FABP preferably binds n-6, brain (B-) FABP n-3 fatty acids. The ADIFAB-method is a fluorescent test for fatty acid in equilibrium with iLBP and reveals some correlation of binding affinity to fatty acid solubility in the aqueous phase; these data are often at variance with those obtained by the other methods. Thus, in this review published binding data are critically discussed, taking into account on the one hand binding increments calculated for fatty acid double bonds on the basis of the 'solubility' hypothesis, on the other hand the interpretation of calorimetric data on the basis of crystallographic and solution structures of iLBPs. (Mol Cell Biochem **239**: 45–54, 2002)

Key words: FABP, tertiary structure, fatty acid, binding, ADIFAB, isothermal titration calorimetry, Lipidex

Introduction

Since 1972, when Ockner *et al.* reported the first fatty acid binding protein (FABP) [1], a multitude of different types of intracellular lipid binding proteins (iLBPs) have been identified in various organisms and tissues. Over the years it was realized that iLBPs evolved from a common ancestor gene more than 900 million years ago [Schaap et al., this issue]. Despite the high degree of diversification in amino acid sequence and ligand binding specificity, a common 3-dimensional fold has been conserved throughout the protein family. All iLBPs consist of a 10-stranded antiparallel β-sheet forming a clam shell-like structure ('β-clam') [2]. A helix-turn-helix domain with two short, nearly parallel α-helices covers the opening of the β-clam structure (Fig. 1). The interior of the protein contains a large cavity, partially filled with ordered water molecules, that serves as binding pocket for respective ligand molecules. Thus, protein side-chains and bound water molecules are involved in hydrogen-bonding networks that coordinate the ligand at the binding site. A so-called portal region for the entry and exit of fatty acid ligands has been postulated to be located between α-helix II and the two turns connecting β-strands βC and βD as well as βE and βF [3].

Here we will look at structural differences within the family of iLBPs and the inherent consequences for their binding interactions with fatty acids. We shall discuss the three main methods currently used for the determination of binding specificities of orthologous and paralogous iLBPs. We critically examine the binding data and question on how they can be reconciled with structural differences in the binding cavities of iLBPs.

Address for offprints: F. Spener, Department of Biochemistry, University of Münster, Wilhelm-Klemm-Str. 2, 48149 Münster, Germany (E-mail: spener@uni-muenster.de)

Fig. 1. Ribbon representation showing the X-ray structure of human E-FABP [14]. The 10 antiparallel β-strands (βA–βJ) and 2 short helices (αI and αII) form a closed protein structure with a large cavity inside. As member of the iLBP subfamily *IV*, E-FABP displays the following characteristic features: (i) The bound palmitate (ball-and-stick model) adopts a U-shaped conformation near the portal region and (ii) the N-terminal residues form a helical loop. This figure was produced with MOLSCRIPT [64] and Raster3D [65].

The structural basis of binding

Four major iLBP subfamilies have been categorized based on sequence homology and ligand binding characteristics [4]:

(i) The intracellular retinoid binding proteins can be further subdivided into the cellular retinoic acid binding proteins (CRABPs) and the cellular retinol binding proteins (CRBPs). The functional group of the retinoid ligand is always deeply immersed into the binding cavity, the isoprenoid tail extended and the β-ionone ring located close to the helix-turn-helix domain. In CRABP types I and II, which bind solely all-*trans* retinoic acid, residues R111, R132 and Y134 are involved in direct ligand binding [5], similar to the scenario in iLBP subfamily *IV*, where the corresponding residues R106, R126 and Y128 interact with the carboxylate group of the bound fatty acid. In CRBP types I and II, which can bind either all-*trans* retinol or retinal, the functional group of the ligand interacts with Q108 [6, 7]. Recently, two different CRBP type III molecules have been reported [8, 9], which bind retinol only – possibly because of the replacement of Q108 by a histidine residue.

(ii) The liver-type fatty acid binding protein (L-FABP) and intestinal bile acid binding protein (I-BABP, also known as ileal lipid binding protein) are closely related based on sequence homology and both stand out because of their unusual ligand binding specificities. L-FABP, which has been reported to bind a broad range of ligand molecules including heme, bilirubin and certain eicosanoids, is the only FABP that forms a complex with two fatty acid molecules at the same time [10, 11]. One ligand molecule, located at the bottom of the protein cavity in a bent conformation, is coordinated *via* an extensive hydrogen-bonding network to S39, R122 and S124, a similar binding scenario as in iLBP subfamily *IV* where the corresponding residues T40 (only in heart-type FABP), R126 and Y128 interact with the carboxylate group of the bound fatty acid. The second ligand molecule in L-FABP adopts a rather linear shape, with the solvent-exposed carboxylate end sticking out of the fatty acid portal and the acyl chain extending down the cavity towards the center of the other fatty acid molecule. Even though I-BABP contains the respective residues S38, R121 and S123 as well, fatty acid binding is very weak. Instead, I-BABP preferentially complexes bile acid molecules based on mostly hydrophobic interactions of the steroid moiety with protein side-chains at the bottom of the cavity. Thereby, the steroid backbone is located in a similar position as the hydrocarbon tail (C8–C18) of the first fatty acid ligand in L-FABP, while the carboxylate tail is extended towards the protein:solvent interface [12].

(iii) The intestinal fatty acid binding protein (I-FABP) is rather singular in both sequence and ligand binding characteristics. The bound fatty acid adopts a slightly bent conformation, reverse in direction to the second fatty acid in L-FABP. Thus, the solvent-inaccessible carboxylate group is located deep inside the protein cavity directly coordinated to the side-chain of R106, which corresponds to R111 in the CRABPs, residue 108 in the CRBPs and R106 in iLBP subfamily *IV*. The ω-tail extends towards the F55 ring, homologous to F57 in iLBP subfamily *IV* (except for epidermal-type FABP) and in most CRBPs (except for murine CRBP III), which may act as a lid at the fatty acid portal [3]. While the convex portion of the fatty acid chain is lined with mostly hydrophobic protein side-chains, the concave face apparently interacts with a cluster of well-ordered bound water molecules [13].

(iv) This iLBP subfamily comprises the largest number of different types of fatty acid binding proteins, i.e. adipocyte- (A-), brain- (B-), epidermal- (E-), heart- (H-) and myelin-type (M-)FABP. These proteins are characterized by an additional structural element, a 3_{10}-helical loop, comprising four residues at the N-terminus (see Fig. 1). In general, they all bind only a single fatty acid molecule usually in a U-shaped conformation. The carboxylate group is coordinated via an extensive hydrogen-bonding network to R106, R126 and Y128, while the ω-tail is located close to F57 (L60 in E-FABP) at the fatty acid portal. The pseudo-*si* face of the bent fatty acid chain is lined by mostly hydrophobic protein side-chains, while the pseudo-*re* face interacts with a cluster of well-ordered bound water molecules [13], a scenario comparable to the ligand binding arrangement in I-FABP. Several unique features in this iLBP subfamily have been reported only recently. First, human E-FABP contains 6 cysteine residues, of which C120 and C127 form a disulfide bridge inside the protein cavity [14]. Secondly, human B-FABP, which binds also long-chain polyunsaturated fatty acids, complexes docosahexaenoic acid (DHA) in a helical conformation [15].

Most of the above-described structural details are based on X-ray crystallographic results, even though in more recent years also several high-resolution NMR solution structures have been reported. Early NMR studies on iLBPs were restricted to monitoring local structural aspects, for example by focussing on isolated resonances of methyl groups and aromatic rings to detect conformational changes upon ligand binding [16] or by introducing protein modifications like fluoro-derivatives to probe ligand transfer processes [17, 18]. Furthermore, ^{13}C-labelled ligands were employed to study protein:lipid interactions [19–21]. With the event of high-resolution NMR spectroscopy, including heteronuclear techniques for data collection with ^{13}C- and/or ^{15}N-labelled protein samples, well-resolved solution structures of iLBPs began to appear mid of the 1990's [12, 22–28, Rademacher *et al.*, this issue]. Subsequently, NMR studies have been extended to include the investigation of protein dynamics [29–36]. Such structural studies in solution have the advantage that dynamic properties of the protein structure, e.g. flexibility of loop structures [24, 28], different backbone conformational states [25, 32] or side-chain mobility [30], can be identified more easily. In particular, ^{15}N relaxation studies in combination with hydrogen/deuterium exchange experiments are currently applied extensively by NMR to investigate differences in the backbone dynamics behaviour of various iLBP types, since local conformational disorder and differences in protein stabilities have hinted at a direct correlation between protein backbone dynamics and ligand binding behaviour [23–26, 29, 31–36]. In short summary, these data on protein dynamics in solution have led to the following results: (1) The protein structures are stabilized by ligand binding, showing reduced backbone flexibility in the holo forms. (2) Differences in the backbone dynamics of various iLBPs can be correlated to preferences for specific ligands and their relative binding affinities. (3) In the portal region, the backbone structures generally display an increased conformational variability, which is reflected by either decreased order parameters (derived from ^{15}N relaxation data), increased hydrogen/deuterium exchange, the appearance of multiple spin-systems representing separate conformational states for single residues, or even a complete lack of observable resonances due to dynamic processes. In the future, NMR data may provide a handle to investigate time-dependent effects – such as protein dynamics, folding and stability, conformational transitions, or kinetics of ligand binding – more intensely, in order to gain a better understanding of the molecular processes underlying binding affinity and specificity within the iLBP family.

Thermodynamics of binding

A detailed knowledge of the tertiary structures of different iLBP types provides useful insights into the overall structural features, conformational changes upon ligand binding and binding stoichiometry. Within the functional scope assumed for FABPs, however, the question always arises, which ligand binds best to which FABP type. Therefore, different methods for determining fatty acid binding affinities have been established.

The Lipidex assay

First methods were based on the charcoal model, which involved separation of unbound fatty acids from those bound by FABP. The so-called 'Lipidex assay', established by Glatz and Veerkamp [37], replaced charcoal by Lipidex 1000 material – a hydroxyalkoxypropyl-derivative of Sephadex G-25 with hydrophobic properties, which at 0°C binds unbound fatty acids only. Hereby, delipidated FABP is equilibrated in buffer with radiolabeled fatty acid at 37°C, before the solution is cooled to 0°C and Lipidex is added to bind unbound fatty acids; the Lipidex/fatty acid complex is sedimented and radioactivity in the supernatant, representative for fatty acid bound to FABP, is determined. With knowledge of the total protein concentration and the ratio of unbound to bound fatty acid, the dissociation constant can be derived from a Scatchard plot. The dissociation constants determined for various fatty acid/FABP complexes by this method [14, 16, 38–44] are generally between 0.2–4 μM. As an example, dissociation constants for oleic acid bound to FABPs from different tissues or species are shown in Table 1. Orthologous FABPs, e.g. H-FABPs, show very similar values around 0.2–0.4 μM, while those for paralogous human FABPs measured in the same laboratory can differ more than 5-fold, e.g. from 0.3–1.6 μM in the case of M-FABP and A-FABP, respectively. One problem of this assay is the physical separation of fatty acids bound to Lipidex material or FABP by centrifugation at 0°C, as they may move from the latter to the former and therefore the value for fatty acids bound to FABP becomes too low. Kinetic measurements showed that fatty acids dissociated within seconds from FABP in the presence of albumin acting as sink [45]. A second uncertainty is the cooling to 0°C, because dissociation constants change with temperature. Because of these problems, K_d values determined by the Lipidex method are considered as generally too high.

Table 1. Dissociation constants for oleic acid bound to orthologous and paralogous FABPs determined with the Lipidex assay

FABP type		Human	Bovine	Porcine	Murine	Rat
				K_d [μM]		
A-FABP		1.56 ± 0.25[a]				0.80 ± 0.02[e]
B-FABP					0.44 ± 0.027[j]	
E-FABP		1.60 ± 0.2[b]				
H-FABP		0.43 ± 0.04[a]		0.43 ± 0.09[c]		0.38 ± 0.11[c]
		0.2 ± 0.05[c]				0.2[g]
I-FABP		0.57 ± 0.04[a]				2.87 ± 0.16[f]
						2.07 ± 0.27[h]
L-FABP	1st	0.89 ± 0.03[a]	0.24[i]			1.77 ± 0.08[f]
			0.7[d]			1.45 ± 0.23[h]
	2nd	n.d.[a]	2.15[i]			n.d.[f]
			n.d.[d]			n.d.[h]
M-FABP		0.31 ± 0.03[a]				

Data taken from [a][44], [b][14], [c][39], [d][43], [e][42], [f][38], [g][40], [h][41], [i][16], [j][60]. Data for the 2nd binding site of L-FABPs that were not determined are given as n.d.

The ADIFAB assay

In order to avoid these problems, the group of Alan Kleinfeld in 1992 introduced the so-called 'ADIFAB assay'. ADIFAB is rat intestinal fatty acid binding protein modified with fluorescent acrylodan at Lys^{27} (in α-helix I) [46]. Upon excitation at 386 nm the emission maximum of the apo form is at 432 nm, consequently shifting to 505 nm in a concentration dependent manner with the binding of fatty acid. Since ADIFAB-fluorescence is very sensitive to even small amounts of organic solvents, consequently ethanol, etc have to be strictly avoided in this assay.

The fluorescence emission ratio $R = I_{505}/I_{432}$ is a sensitive marker for fatty acid binding. To determine the dissociation constant (K_d^*) for a fatty acid bound to ADIFAB, a known amount of ADIFAB is titrated with fatty acid and the ratio R is calculated after each addition. $[FA]_{unbound}$ can be calculated according to equation 1 and $[ADIFAB]_{bound}$ (i.e. fatty acid bound to ADIFAB) by equation 2.

$$[FA]_{unbound} = K_d^* \cdot Q \cdot \frac{(R - R_0)}{(R_{max} - R)} \tag{1}$$

$$[ADIFAB]_{bound} = \frac{[ADIFAB]_{total} \cdot Q \cdot (R - R_0)}{R_{max} - R + Q \cdot (R - R_0)} \tag{2}$$

In these equations Q equals $I_{432}(\text{free})/I_{432}(\text{bound})$, i.e. the ratio of fluorescence intensity at 432 nm of ADIFAB both without and saturated with fatty acid. R_0 is the ratio $I_{505}(\text{free})/I_{432}(\text{free})$ of ADIFAB without fatty acid, while R_{max} is the same ratio of ADIFAB saturated with fatty acid ($I_{505}(\text{bound})/I_{432}(\text{bound})$). Q and R_{max} can be determined by fitting a nonlinear expression for R in dependence of Q, R_{max} and K_d^*. By restriction of one parameter and floating of the other two, the ranges of Q and R_{max} were estimated to be $17 < Q < 22$ and $10 < R_{max} < 13$, respectively. For the calculation of the K_d^*, Q and R_{max} were set to their average values of 19.5 and 11.5, respectively. The authors calculated that the maximum absolute error in K_d^* determined with this method is about ± 20% due to this averaging. Errors due to fatty acids bound to the glass of the cuvette, a general phenomenon affecting all fatty acid binding assays, are minimized here by control experiments that determine the amount of glass-bound fatty acids. A definite advantage of the ADIFAB assay is its ability to provide data on binding kinetics as well, i.e. rate constants for the transfer of fatty acid from the fatty acid/FABP complex to ADIFAB [45, 47].

Once the K_d^* values are known for every fatty acid in equilibrium with ADIFAB, addition of any other FABP type allows the determination of its dissociation constant (K_d) for the respective fatty acids. First, exact amounts of ADIFAB and FABP in buffer are placed into a cuvette to measure R_0. Next, a titration with fatty acid is carried out in 10 min steps to attain equilibrium before R is measured. In each step, the amount of fatty acid bound to FABP, $[FA]_{bound}$, is calculated according to equation 3:

$$[FA]_{bound} = [FA]_{total} - [FA]_{unbound} - [ADIFAB]_{bound} \tag{3}$$

For a single-affinity binding site the data can be analyzed with a Scatchard plot according to equation 4, yielding a straight line with the slope $-1/K_d$ and the y-intercept being directly proportional to binding stoichiometry n.

$$\frac{\dfrac{[FA]_{bound}}{[FABP]_{total}}}{[FA]_{unbound}} = -\frac{1}{K_d}\frac{[FA]_{bound}}{[FABP]_{total}} + \frac{n}{K_d} \tag{4}$$

Table 2. Dissociation constants for fatty acids bound to orthologous and paralogous FABPs determined by the ADIFAB assay

FABP type	SA	PA	OA	DHA	LA	AA	LNA
	K_d [nM]						
mA-FABP	76	77	61	107	101	139	196
hA-FABP	80	83	57	84	92	182	167
hB-FABP	2.3	7	7	13	11	18	21
bE-FABP		27	36	39	63	66	97
hM-FABP	2.4	6	5	10	7	15	29
rH-FABP	4	9	6	10	22	45	38
bH-FABP	2	2.4	4.4		22	34	30
hH-FABP	2.3	7	4	9	11	27	55
rL-FABP	9	23	9	23	27	48	69
hL-FABP	23	60	15	19	57	100	240
rI-FABP	6	20	27	213	89	197	367
hI-FABP	6	23	35	309	101	125	317
mCRABP-I	500	223	360		380	670	540
mCRBP-II	750	2600	2200		3600	2900	10000

Data taken from [48, 49, 63]. For L-FABP, values are shown for the high-affinity site only. Abbreviations of fatty acids are: SA – stearic acid; PA – palmitic acid; OA – oleic acid; DHA – docosahexaenoic acid; LA – linoleic acid; AA – arachidonic acid; LNA – linolenic acid.

By repeating this experiment at different temperatures the thermodynamic parameters ΔH and –TΔS can be calculated via van't Hoff analysis by plotting $\ln(K_d)$ against 1/T.

Using this method, over the last years several laboratories published elaborate studies on fatty acid binding to either wild-type or mutant proteins of different FABP types [45–53]. A sampling of the data obtained for wild-type FABPs is presented in Table 2. Comparing these data with Lipidex data (Table 1) reveals that dissociation constants determined by the ADIFAB assay are at least one order of magnitude lower. In particular affinities of orthologous H-FABPs are high for saturated and monounsaturated fatty acids, with K_d values below 9 nM, while Lipidex-derived dissociation constants are up to 100 times higher. A comparison of only ADIFAB data shows that dissociation constants for fatty acids with the same number of carbon atoms, e.g. stearic, oleic, linoleic and α-linolenic acids, increase exponentially with decreasing levels of saturation. This led to the formulation of the so-called 'solubility hypothesis' by Kleinfeld *et al.* [46, 48], which states that the affinity of a given fatty acid for any FABP is driven in a first approximation by the solubility of the fatty acid in the bulk aqueous phase. This solubility can be expressed by the partition coefficient (K_p) of the FA between a membrane and an aqueous phase. This coefficient, defined as

$$K_p = \frac{[FA]_{membrane}}{[FA]_{unbound}}$$

can be determined by the ADIFAB method as

$$K_p = \frac{[FA]_{total} - [FA]_{unbound}}{[FA]_{unbound}} \frac{V_a}{V_m}$$

where V_a and V_m are the volumes of the aqueous and the membrane phase, respectively, as described by Anel *et al.* [54].

Richieri *et al.* [48] stated that the binding site of I-FABP acts similar to a nonpolar solvent. This means that affinity for different fatty acids is mainly determined by the entropic contribution of the hydrophobic effect [55]. In Table 2 and Fig. 2, fatty acids are presented in the order of their relative solubility defined by the aqueous/membrane partition coefficient [54]. The graph in Fig. 2 reveals that fatty acid binding to ADIFAB and I-FABP follows the solubility rule, with the exception of DHA. However, this exponential relationship is less obvious when fatty acid binding to other FABPs, such as L-FABP (subfamily *II*) or the proteins of iLBP subfamily *IV* (H-, M-, B- and A-FABP), is considered. In recognition of these differences of fatty acid binding to the different FABP types, Kleinfeld's group calculated additional free energy contributions for each double bond in a fatty acid and derived increments for a given fatty acid bound by a given FABP type by relating them to stearic acid binding and standardizing it to I-FABP (DHA binding not considered) [49]. The graphs of the data obtained from this thermodynamic treatment of K_d values measured (Fig. 3) indicate a deviation from the solubility hypothesis. In other words, the increments shown in Fig. 3 are thermodynamic parameters that reflect the specific physico-chemical properties of the binding site in a given FABP type, i.e. the molecular architecture of the residues and ordered water in the binding site that interact with a given fatty acid bound. In comparison to I-FABP – the almost perfect 'organic solvent' – an additional release of free energy upon ligand binding was observed for

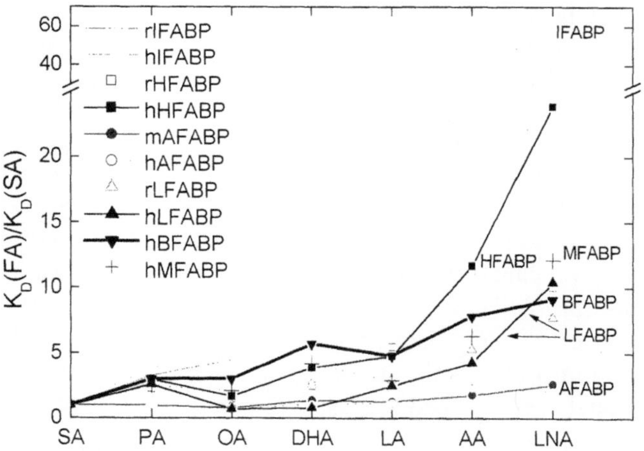

Fig. 2. Variation of FABP binding affinities with relative solubility of the fatty acid at 37°C. K_d values were normalized for each protein to the value for stearate binding [K_d(SA)] and plotted according to increasing fatty acid solubility. Reprinted with permission from [49]. Copyright 2000 American Chemical Society.

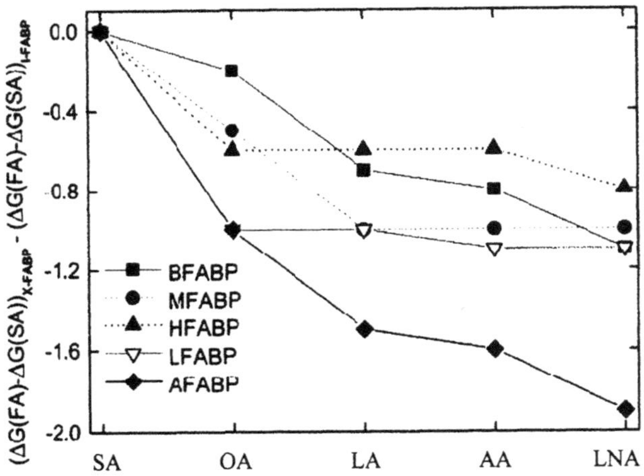

Fig. 3. The double bond-mediated increase in the free energy of binding. The difference in binding free energy, at 37°C, between each FABP and fatty acid and the value for stearate was computed. The value for this difference for the I-FABP was subtracted from the corresponding values for each of the other FABPs shown in the figure. Reprinted with persmission from [49]. Copyright 2000 American Chemical Society.

all FABP types shown in Fig. 3. Consequently, aside from the hydrophobic effect there must be other attractive forces that favor FA ligands with increased numbers of double bonds, in spite of their higher solubility in the aqueous phase relative to their saturated counterparts.

Isothermal titration calorimetry

A third method for collecting data on ligand/iLBP interactions, the isothermal titration calorimetry (ITC), was first applied by Miller and Cistola in 1993 [56]. ITC allows the measurement of the heat of interaction when a FABP interacts with a ligand in equilibrium. In an ITC experiment, one binding partner (usually the ligand) is titrated in small aliquots into a solution containing the other binding partner (the FABP). The heat required to maintain the temperature inside the sample cell constant is measured and provides the raw data of the ITC experiment (Fig. 4, upper panel). The peaks correspond to the heat released or absorbed in each titration step. Integration of these peaks obtained after each ligand addition yields the apparent binding heat Q_{app} (Fig. 4, lower panel). This heat is the sum of the binding heat Q, the heat of dilution Q_{dil} and the heat by nonspecific effects Q_{ns} (equation 5). Q_{app} is proportional to the volume inside the calorimetric cell, V_0, the change in concentration of bound ligand $\Delta[FA]_{bound}$ and the association enthalpy ΔH

$$Q_{app} = Q + Q_{dil} + Q_{ns} = \Delta[FA]_{bound} \times V_0 \times \Delta H \tag{5}$$

V_0 is known. Q_{dil} and Q_{ns} can be determined by a blank titration of ligand into buffer. ΔH, K_d and n can be calculated by nonlinear regression from

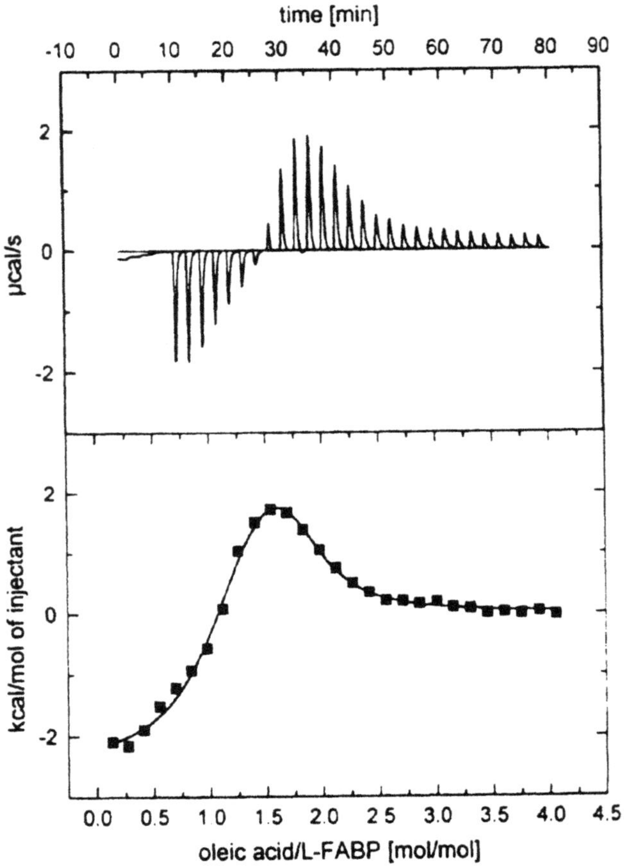

Fig. 4. Binding of oleic acid to bovine L-FABP measured by ITC. The upper panel shows the differential power signal that was recorded during the experiment. In the lower panel, the binding isotherm obtained after integration and normalization per mol of added ligand is displayed. Graph taken from [43].

$$Q = \frac{nM_t \Delta H V_0}{2}\left[1 + \frac{X_t}{nM_t} + \frac{1}{nK_d M_t} - \sqrt{\left(1 + \frac{X_t}{nM_t} + \frac{1}{nK_d M_t}\right)^2 - \frac{4X_t}{nM_t}}\right] \tag{6}$$

where M_t is the concentration of FABP in the cell, n the number of binding sites and X_t the bulk concentration of ligand. The isotherms can be fitted to a one or two binding-site model and allows to calculate the dissociation constant K_d [57]. Using simple thermodynamics, the free enthalpy and entropy can be calculated according to

$$\Delta G = -RT \ln K = \Delta H - T\Delta S$$

Up to date, a number of studies have appeared where interactions of fatty acids with FABP types have been examined by ITC [15, 41, 54, 56, 57]. These data can now be compared with those resulting from ADIFAB measurements. In Table 3, K_d values for fatty acid binding to orthologous L-FABPs are shown, arranged in the order of fatty acid solubility in the bulk aqueous phase. Data from ADIFAB measurements do not completely agree with the solubility hypothesis, as the saturated stearic and palmitic acids show slightly higher dissociation constants than oleic acid even though both are less soluble. Higher affinity for oleic acid than for the saturated fatty acids is also seen in ITC data, and interestingly Lipidex assay data also indicate a higher affinity of L-FABP for oleic acid than for palmitic acid [44]. K_d values obtained by ADIFAB assay and ITC for unsaturated fatty acids in general agree quite well to be in the low nanomolar range within the margin of error, while data for stearic acid binding differ by almost two orders of magnitude.

ITC measurements with orthologous L-FABPs allowed to trace differences in affinities and thermodynamic parameters back to protein structure: In rodent (mouse and rat) L-FABP, Tyr54 together with an ordered water molecule was recognized to participate in binding of the carboxylate headgroup and the hydrocarbon chain of the 2nd oleic acid molecule bound [10]; consequently this binding to the 2nd site resulted in a high negative enthalpic contribution. In contrast, the substitution of Tyr54 by an alanine residue in bovine and human L-FABP led to positive binding enthaplies that can be seen as a result of destabilization of the ordered water molecule. Yet, as dissociation constants for oleic acid binding to all L-FABPs are within the same range, differences in enthalpies must be compensated by respective entropies [59].

Another case in point are the data for fatty acid binding to human B-FABP (Table 4). Dissociation constants measured by ITC are 4–11 times higher than by the ADIFAB assay. Whereas ADIFAB data follow the solubility rule, though with exceptions, ITC clearly indicates high affinity for n-3 α-linolenic and docosahexaenoic acids, even though α-linolenic acid is the most soluble fatty acid. Moreover, dissociation constants for saturated palmitic and stearic acids were too high to be measured by ITC and thus determined by Lipidex assay; indeed only weak binding of saturated acids to B-FABP was observed by the latter method, in contrast to ADIFAB data [49]. Weak binding of palmitic acid was already reported for murine B-FABP [60] together with high affinity for long-chain polyunsaturated fatty acids, e.g. α-linolenic acid and DHA. The preference of B-FABP for polyunsaturated fatty acids, espe-

Table 4. Comparison of binding data for human B-FABP obtained by ADIFAB and ITC

Fatty acid	ADIFAB, 37°C[a] K_d [nM]	ITC, 30°C[b] K_d [nM]
SA	2.3	13500[c]
PA	7	7100[c]
OA	7	46.7 ± 1.4
DHA	13	53.4 ± 4.1
LA	11	115 ± 19
AA	18	207 ± 19
LNA	21	27.5 ± 1.3

[a][49], [b][15] [c]By Lipidex assay and referenced to K_d = 47 nM for OA as obtained by ITC.

Table 3. Comparison of binding data for orthologous L-FABP obtained by either the ADIFAB assay or ITC

Fatty acid	Rat ADIFAB [a,b]	Rat ADIFAB [c]	Mouse ITC [d]	Human ADIFAB [a]	Human ITC [d]	Bovine ITC [d]	Bovine ITC [e]
				K_d[μM]			
SA	9 ± 1		265 ± 4 328 ± 4	23	1900 ± 20 13000 ± 6780	95 ± 8 566 ± 215	
PA	23 ± 2	7.2 ± 1 28 ± 1		60			
OA	9 ± 2	2.6 ± 0.1	21 ± 8 312 ± 73	15	6 ± 4 940 ± 390	47 ± 13 1310 ± 120	260 ± 10 4900 ± 300
DHA	23		13 ± 1 514 ± 17	19	3 ± 1 200 ± 30	52 ± 35 430 ± 190	
LA	29 ± 3	10.1 ± 0.1 250 ± 11		57			
AA	48 ± 9	12 ± 0.1 211 ± 7		100			540 ± 70 3000 ± 1600
LNA	69 ± 12	27.8 ± 0.3 350 ± 12		240			

K_d values for the low-affinity site are given in each 2nd line when determined. [a][49], [b][48], [c][47], [d][59], [e][43].

Table 5. Fatty acid binding to E-FABP obtained by the ADIFAB and Lipidex assays

Fatty acid	Human[a] K_d [nM]	Bovine[b] K_d [nM]
SA	290 ± 60	
PA		27 ± 3
OA	1600 ± 200	36 ± 4
DHA		39 ± 5
LA		63 ± 8
AA	1730 ± 250	66 ± 8
LNA		91 ± 12

[a]Binding data obtained by Lipidex [14]; [b]binding data obtained by ADIFAB [63].

cially for DHA, can be explained on the basis of the crystal structures which are available for these complexes with human B-FABP. The DHA is buried inside the cavity in a helical, low energy conformation where the double bonds of the fatty acid show π–π interactions between themselves and with Phe[104] and Met[115], while oleic acid is bound in the typical U-shaped conformation that allows for interaction of the double bond with Phe[16] and for a large number of van der Waals interactions with amino acid residues within the binding cavity [15].

Prendergast *et al.* [53] measured fatty acid binding to I-FABP by ITC and compared the values obtained to those determined by the ADIFAB assay as reported by Richieri *et al.* [61]. The former argued that a point to point substraction of the heat of dilution of the ligand in 'protein-free' buffer solution from the values obtained in the presence of protein, as commonly done, is incorrect; rather they suggest that only the heat of dilution of free ligand in the titration cell must be substracted. They applied this modified procedure of background subtraction and obtained now lower K_d values (0.2 ± 0.06 μM for oleic acid, 0.06 ± 0.016 μM for arachidonic acid), which were considered comparable to those determined by the ADIFAB method (0.027 and 0.197 μM, respectively) [61]. But these values do not conform with the solubility hypothesis.

Conclusions

Up to date the Lipidex assay, the ADIFAB assay and ITC are commonly used to elaborate isotherms representative for fatty acid binding to FABPs. The Lipidex assay can be easily carried out and affords a simple technical setup only. As this method does not operate in real equilibrium, K_d values derived are considered too high; however, it is applicable to ligands of low solubility. The ADIFAB assay, on the other hand, operates within real equilibria, and furnishes not only binding constants, but also thermodynamic parameters and binding kinetics as well. With ADIFAB readily available the essential technical setup is a fluorescence spectrophotometer. Disadvantages are narrow assay conditions and the choice of ligands restricted to mostly fatty acids (which is also the case for the Lipidex assay). Binding of a broad range of natural as well as xenobiotic ligands to FABPs can be examined, however, by ITC. It allows to measure directly the heat change during the binding process and therefore provides direct access to thermodynamic parameters and K_d. Disadvantages are the need of an expensive microtitration calorimeter and the high amounts of protein used up in each experiment.

The affinity of a fatty acid for a FABP can be classified into three primary components: (i) the energetic strain imposed on the ligand in the bound state, (ii) the specific interactions of the protein with the bound ligand, and (iii) the desolvation of the binding cavity and the ligand upon binding [62]. In contrast, the solubility hypothesis in its original form does not account for any of these components, but defines the affinity on the basis of the partition coefficient of the fatty acid between aqueous phase and membrane phase ('organic solvent'). In this respect the binding cavity of I-FABP was considered as organic solvent-like. However, the study of the thermodynamics of fatty acid binding to FABPs by the ADIFAB method revealed that the free energy of binding is almost entirely enthalpic and solubility is reflected in the small entropy contribution [61]. With the introduction of increments of the binding free energy for each double bond (Fig. 3) [49] the solubility hypothesis takes now into account the specific interaction of the protein with the bound ligand, i.e. point (ii). A particular example for this kind of specific interaction is highlighted by π–π interactions of DHA double bonds with Phe[104] and Met[115] side chains of human B-FABP as seen by x-ray crystallography described before, but not seen in the complex with human H-FABP where both residues are replaced by leucin; the latter resulted in a 100-fold higher dissociation constant [15]. Further evidence for specific ligand–protein interactions can also be provided by NMR, as backbone dynamics can be correlated to ligand preferences (see section 'The structural basis of binding', p. 46).

In summary, Kleinfeld *et al.* published over the years a series of studies carried out under identical conditions, which furnish a treasure of binding data, the basis of the solubility hypothesis. K_d values for mono- and polyunsaturated fatty acids obtained by ADIFAB and ITC do agree to some extent. Remarkable differences are seen for C16 and C18 saturated fatty acids. According to ADIFAB measurement, these fatty acids have always the highest affinities for the various FABP types, yet ITC provides more differentiated data. Thus agreement was found for their binding to E- and I-FABP, but not to the other types. Moreover, ITC furnished more nuanced binding data for unsaturated fatty acids, as L-FABP has highest affinity for monounsaturated and n-3 polyunsaturated fatty acids, while H-FABP has preferences for n-6 polyunsaturated fatty acids. In the opinion of the authors of this review, differences in binding affinity of paralogous FABPs for a distinct fatty acid structure is proof for its functional uniqueness in the cell where it is expressed.

References

1. Ockner RK, Manning JA, Poppenhausen RB, Ho WK: A binding protein for fatty acids in cytosol of intestinal mucosa, liver, myocardium, and other tissues. Science 177: 56–58, 1972
2. Sacchettini JC, Gordon JI, Banaszak LJ: The structure of crystalline *Escherichia coli*-derived rat intestinal fatty acid-binding protein at 2.5-Å resolution. J Biol Chem 263: 5815–5819, 1988
3. Sacchettini JC, Scapin G, Gopaul D, Gordon JI: Refinement of the structure of *Escherichia coli*-derived rat intestinal fatty acid binding protein with bound oleate to 1.75-Å resolution. Correlation with the structures of the apoprotein and the protein with bound palmitate. J Biol Chem 267: 23534–23545, 1992
4. Hohoff C, Spener F: Fatty acid binding proteins and mammary-derived growth inhibitor. Fett Lipid 100: 252–263, 1998
5. Kleywegt GJ, Bergfors T, Senn H, Le Motte P, Gsell B, Shudo K, Jones TA: Crystal structures of cellular retinoic acid binding proteins I and II in complex with all-trans-retinoic acid and a synthetic retinoid. Structure 2: 1241–1258, 1994
6. Cowan SW, Newcomer ME, Jones TA: Crystallographic studies on a family of cellular lipophilic transport proteins. Refinement of P2 myelin protein and the structure determination and refinement of cellular retinol-binding protein in complex with all-trans-retinol. J Mol Biol 230: 1225–1246, 1993
7. Winter NS, Bratt JM, Banaszak LJ: Crystal structures of holo- and apo-cellular retinol-binding protein II. J Mol Biol 230: 1247–1259, 1993
8. Folli C, Calderone V, Ottonello S, Bolchi A, Zanotti G, Stoppini M, Berni R: Identification, retinoid binding, and x-ray analysis of a human retinol-binding protein. Proc Natl Acad Sci USA 98: 3710–3715, 2001
9. Vogel S, Mendelsohn CL, Mertz JR, Piantedosi R, Waldburger C, Gottesman ME, Blaner WS: Characterization of a new member of the fatty acid-binding protein family that binds all-trans-retinol. J Biol Chem 276: 1353–1360, 2001
10. Thompson J, Winter N, Terwey D, Bratt J, Banaszak L: The crystal structure of the liver fatty acid-binding protein. A complex with two bound oleates. J Biol Chem 272: 7140–7150, 1997
11. Haunerland N, Jagschies G, Schulenberg H, Spener F: Fatty-acid-binding proteins. Occurrence of two fatty-acid-binding proteins in bovine liver cytosol and their binding of fatty acids, cholesterol, and other lipophilic ligands. Hoppe Seyler's Z Physiol Chem 365: 365–376, 1984
12. Lücke C, Zhang F, Hamilton JA, Sacchettini JC, Rüterjans H: Solution structure of ileal lipid binding protein in complex with glycocholate. Eur J Biochem 267: 2929–2938, 2000
13. Scapin G, Young AC, Kromminga A, Veerkamp JH, Gordon JI, Sacchettini JC: High resolution X-ray studies of mammalian intestinal and muscle fatty acid-binding proteins provide an opportunity for defining the chemical nature of fatty acid:protein interactions. Mol Cell Biochem 123: 3–13, 1993
14. Hohoff C, Börchers T, Rüstow B, Spener F, van Tilbeurgh H: Expression, purification, and crystal structure determination of recombinant human epidermal-type fatty acid binding protein. Biochemistry 38: 12229–12239, 1999
15. Balendiran GK, Schnütgen F, Scapin G, Börchers T, Xhong N, Lim K, Godbout R, Spener F, Sacchettini JC: Crystal structure and thermodynamic analysis of human brain fatty acid-binding protein. J Biol Chem 275: 27045–27054, 2000
16. Schulenberg-Schell H, Schäfer P, Keuper HJ, Stanislawski B, Hoffmann E, Rüterjans H, Spener F: Interactions of fatty acids with neutral fatty-acid-binding protein from bovine liver. Eur J Biochem 170: 565–574, 1988
17. Li E, Qian SJ, Yang NC, d'Avignon A, Gordon JI: 19F nuclear magnetic resonance studies of 6-fluorotryptophan-substituted rat cellular retinol binding protein II produced in *Escherichia coli*. An analysis of four tryptophan substitution mutants and their interactions with all-trans-retinol. J Biol Chem 265: 11549–11554, 1990
18. Li E, Qian SJ, Winter NS, d'Avignon A, Levin MS, Gordon JI: Fluorine nuclear magnetic resonance analysis of the ligand binding properties of two homologous rat cellular retinol-binding proteins expressed in *Escherichia coli*. J Biol Chem 266: 3622–3629, 1991
19. Sacchettini JC, Hauft SM, Van Camp SL, Cistola DP, Gordon JI: Developmental and structural studies of an intracellular lipid binding protein expressed in the ileal epithelium. J Biol Chem 265: 19199–19207, 1990
20. Cistola DP, Walsh MT, Corey RP, Hamilton JA, Brecher P: Interactions of oleic acid with liver fatty acid binding protein: A carbon-13 NMR study. Biochemistry 27: 711–717, 1988
21. Cistola DP, Sacchettini JC, Banaszak LJ, Walsh MT, Gordon JI: Fatty acid interactions with rat intestinal and liver fatty acid-binding proteins expressed in *Escherichia coli*. A comparative 13C NMR study. J Biol Chem 264: 2700–2710, 1989
22. Zhang F, Lücke C, Baier LJ, Sacchettini JC, Hamilton JA: Solution structure of human intestinal fatty acid binding protein: Implications for ligand entry and exit. J Biomol NMR 9: 213–228, 1997
23. Wang L, Li Y, Abildgaard F, Markley JL, Yan H: NMR solution structure of type II human cellular retinoic acid binding protein: Implications for ligand binding. Biochemistry 37: 12727–12736, 1998
24. Lücke C, Zhang F, Rüterjans H, Hamilton JA, Sacchettini JC: Flexibility is a likely determinant of binding specificity in the case of ileal lipid binding protein. Structure 4: 785–800, 1996
25. Lücke C, Rademacher M, Zimmerman AW, van Moerkerk HT, Veerkamp JH, Rüterjans H: Spin-system heterogeneities indicate a selected-fit mechanism in fatty acid binding to heart-type fatty acid-binding protein (H-FABP). Biochem J 354: 259–266, 2001
26. Hodsdon ME, Cistola DP: Discrete backbone disorder in the nuclear magnetic resonance structure of apo intestinal fatty acid-binding protein: Implications for the mechanism of ligand entry. Biochemistry 36: 1450–1460, 1997
27. Hodsdon ME, Ponder JW, Cistola DP: The NMR solution structure of intestinal fatty acid-binding protein complexed with palmitate: Application of a novel distance geometry algorithm. J Mol Biol 264: 585–602, 1996
28. Lassen D, Lücke C, Kveder M, Mesgarzadeh A, Schmidt JM, Specht B, Lezius A, Spener F, Rüterjans H: Three-dimensional structure of bovine heart fatty-acid-binding protein with bound palmitic acid, determined by multidimensional NMR spectroscopy. Eur J Biochem 230: 266–280, 1995
29. Hodsdon ME, Cistola DP: Ligand binding alters the backbone mobility of intestinal fatty acid-binding protein as monitored by 15N NMR relaxation and 1H exchange. Biochemistry 36: 2278–2290, 1997
30. Constantine KL, Friedrichs MS, Wittekind M, Jamil H, Chu CH, Parker RA, Goldfarb V, Mueller L, Farmer II BT: Backbone and side chain dynamics of uncomplexed human adipocyte and muscle fatty acid-binding proteins. Biochemistry 37: 7965–7980, 1998

31. Lücke C, Fushman D, Ludwig C, Hamilton JA, Sacchettini JC, Rüterjans H: A comparative study of the backbone dynamics of two closely related lipid binding proteins: Bovine heart fatty acid binding protein and porcine ileal lipid binding protein. Mol Cell Biochem 192: 109–121, 1999
32. Lu J, Lin CL, Tang C, Ponder JW, Kao JL, Cistola DP, Li E: The structure and dynamics of rat apo-cellular retinol-binding protein II in solution: Comparison with the X-ray structure. J Mol Biol 286: 1179–1195, 1999
33. Lu J, Lin CL, Tang C, Ponder JW, Kao JL, Cistola DP, Li E: Binding of retinol induces changes in rat cellular retinol-binding protein II conformation and backbone dynamics. J Mol Biol 300: 619–632, 2000
34. Krishnan VV, Sukumar M, Gierasch LM, Cosman M: Dynamics of cellular retinoic acid binding protein I on multiple time scales with implications for ligand binding. Biochemistry 39: 9119–9129, 2000
35. Franzoni L, Lücke C, Pérez C, Cavazzini D, Rademacher M, Ludwig C, Spisni A, Rossi GL, Rüterjans H: Structure and backbone dynamics of apo- and holo-cellular retinol-binding protein in solution. J Biol Chem 277: 21983–21997, 2002
36. Guitérréz-González LH, Ludwig C, Hohoff C, Rademacher M, Hanhoff T, Rüterjans H, Spener F, Lücke C: Solution structure and backbone dynamics of human epidermal-type fatty acid-binding protein (E-FABP). Biochem J 364: 725–737, 2002
37. Glatz JF, Veerkamp JH: A radiochemical procedure for the assay of fatty acid binding by proteins. Anal Biochem 132: 89–95, 1983
38. Lowe JB, Sacchettini JC, Laposata M, McQuillan JJ, Gordon JI: Expression of rat intestinal fatty acid-binding protein in *Escherichia coli*. Purification and comparison of ligand binding characteristics with that of *Escherichia coli*-derived rat liver fatty acid-binding protein. J Biol Chem 262: 5931–5937, 1987
39. Paulussen RJ, van der Logt CP, Veerkamp JH: Characterization and binding properties of fatty acid-binding proteins from human, pig, and rat heart. Arch Biochem Biophys 264: 533–545, 1988
40. Vork MM, Glatz JF, Surtel DA, van der Vusse GJ: Assay of the binding of fatty acids by proteins: Evaluation of the Lipidex 1000 procedure. Mol Cell Biochem 98: 111–117, 1990
41. Nemecz G, Hubbell T, Jefferson JR, Lowe JB, Schroeder F: Interaction of fatty acids with recombinant rat intestinal and liver fatty acid-binding proteins. Arch Biochem Biophys 286: 300–309, 1991
42. Dutta-Roy AK, Huang Y, Dunbar B, Trayhurn P: Purification and characterization of fatty acid-binding proteins from brown adipose tissue of the rat. Biochim Biophys Acta 1169: 73–79, 1993
43. Rolf B, Oudenampsen-Krüger E, Börchers T, Faergeman NJ, Knudsen J, Lezius A, Spener F: Analysis of the ligand binding properties of recombinant bovine liver-type fatty acid binding protein. Biochim Biophys Acta 1259: 245–253, 1995
44. Veerkamp JH, van Moerkerk HT, Prinsen CF, van Kuppevelt TH: Structural and functional studies on different human FABP types. Mol Cell Biochem 192: 137–142, 1999
45. Richieri GV, Ogata RT, Kleinfeld AM: Kinetics of fatty acid interactions with fatty acid binding proteins from adipocyte, heart, and intestine. J Biol Chem 271: 11291–11300, 1996
46. Richieri GV, Ogata RT, Kleinfeld AM: A fluorescently labeled intestinal fatty acid binding protein. Interactions with fatty acids and its use in monitoring free fatty acids. J Biol Chem 267: 23495–23501, 1992
47. Richieri GV, Ogata RT, Kleinfeld AM: Thermodynamic and kinetic properties of fatty acid interactions with rat liver fatty acid-binding protein. J Biol Chem 271: 31068–31074, 1996
48. Richieri GV, Ogata RT, Kleinfeld AM: Equilibrium constants for the binding of fatty acids with fatty acid-binding proteins from adipocyte, intestine, heart, and liver measured with the fluorescent probe ADIFAB. J Biol Chem 269: 23918–23930, 1994
49. Richieri GV, Ogata RT, Zimmerman AW, Veerkamp JH, Kleinfeld AM: Fatty acid binding proteins from different tissues show distinct patterns of fatty acid interactions. Biochemistry 39: 7197–7204, 2000
50. Richieri GV, Low PJ, Ogata RT, Kleinfeld AM: Mutants of rat intestinal fatty acid-binding protein illustrate the critical role played by enthalpy-entropy compensation in ligand binding. J Biol Chem 272: 16737–16740, 1997
51. Richieri GV, Ogata RT, Kleinfeld AM: Fatty acid interactions with native and mutant fatty acid binding proteins. Mol Cell Biochem 192: 77–85, 1999
52. Richieri GV, Low PJ, Ogata RT, Kleinfeld AM: Thermodynamics of fatty acid binding to engineered mutants of the adipocyte and intestinal fatty acid-binding proteins. J Biol Chem 273: 7397–7405, 1998
53. Kurian E, Kirk WR, Prendergast FG: Affinity of fatty acid for rat intestinal fatty acid binding protein: Further examination. Biochemistry 35: 3865–3874, 1996
54. Anel A, Richieri GV, Kleinfeld AM: Membrane partition of fatty acids and inhibition of T cell function. Biochemistry 32: 530–536, 1993
55. Tanford C: The hydrophobic effect and the organization of living matter. Science 200: 1012–1018, 1978
56. Miller KR, Cistola DP: Titration calorimetry as a binding assay for lipid-binding proteins. Mol Cell Biochem 123: 29–37, 1993
57. Wiseman T, Williston S, Brandts JF, Lin LN: Rapid measurement of binding constants and heats of binding using a new titration calorimeter. Anal Biochem 179: 131–137, 1989
58. LaLonde JM, Levenson MA, Roe JJ, Bernlohr DA, Banaszak LJ: Adipocyte lipid-binding protein complexed with arachidonic acid. Titration calorimetry and X-ray crystallographic studies. J Biol Chem 269: 25339–25347, 1994
59. Wolfrum C, Börchers T, Sacchettini JC, Spener F: Binding of fatty acids and peroxisome proliferators to orthologous fatty acid binding proteins from human, murine, and bovine liver. Biochemistry 39: 1469–1474, 2000
60. Xu LZ, Sanchez R, Sali A, Heintz N: Ligand specificity of brain lipid-binding protein. J Biol Chem 271: 24711–24719, 1996
61. Richieri GV, Ogata RT, Kleinfeld AM: Thermodynamics of fatty acid binding to fatty acid-binding proteins and fatty acid partition between water and membranes measured using the fluorescent probe ADIFAB. J Biol Chem 270: 15076–15084, 1995
62. Young AC, Scapin G, Kromminga A, Patel SB, Veerkamp JH, Sacchettini JC: Structural studies on human muscle fatty acid binding protein at 1.4 Å resolution: Binding interactions with three C18 fatty acids. Structure 2: 523–534, 1994
63. Kingma PB, Bok D, Ong DE: Bovine epidermal fatty acid-binding protein: Determination of ligand specificity and cellular localization in retina and testis. Biochemistry 37: 3250–3257, 1998
64. Kraulis PJ: MOLSCRIPT: A program to produce both detailed and schematic plots of protein structures. J Appl Crystallogr 24: 946–950, 1991
65. Merritt EA, Bacon DJ: Raster3D: Photorealistic molecular graphics. Methods Enzymol 277: 505–524, 1997

Molecular and Cellular Biochemistry **239**: 55–60, 2002.

The effect of charge reversal mutations in the α-helical region of liver fatty acid binding protein on the binding of fatty-acyl CoAs, lysophospholipids and bile acids

Robert M. Hagan, Joanna K. Davies and David C. Wilton
Division of Biochemistry and Molecular Biology, School of Biological Sciences, University of Southampton, Bassett Crescent East, Southampton, UK

Abstract

Liver fatty acid binding protein (LFABP) is unique among the various types of FABPs in that it can bind a variety of ligands in addition to fatty acids. LFABP is able to bind long chain fatty acids with a 2:1 stoichiometry and the crystal structure has identified two fatty acid binding sites in the binding cavity. The presumed primary site (site 1) involves the fatty acid binding with the carboxylate group buried in the cavity whereas the fatty acid at site 2 has the carboxylate group solvent-exposed within the ligand portal region and in the vicinity of α-helix II. The α-helical region contains three cationic residues, K20, K31, K33 and modelling studies suggest that K31 on α-helix II could make an electrostatic contribution to anionic ligands binding to site 2. The preparation of three charge reversal mutants of LFABP, K20E, K31E and K33E has allowed an investigation of the role of site 2 in ligand binding, particularly those ligands with a bulky anionic head group. The binding of oleoyl CoA, lysophosphatidic acid, lysophosphatidylcholine, lithocholic acid and taurolithocholate 3-sulphate to LFABP has been studied using the α-helical mutants. The results support the concept that such ligands bind at site 2 of LFABP where solvent exposure allows the accommodation of their bulky anionic group. (Mol Cell Biochem **239**: 55–60, 2002)

Key words: liver FABP, lysophospholipid, bile acid, acyl CoA, DAUDA, charge reversal mutagenesis

Introduction

Among the family of fatty acid binding proteins, liver FABP (LFABP) is presently unique in binding two molecules of fatty acid. This binding is directly demonstrated in the crystal structure of rat liver FABP that contains two oleic acid molecules bound in the central cavity [1]. The large size of this cavity [2] also provides an explanation for the ability of LFABP to bind a variety of more bulky non-polar anionic ligands, be it at lower affinities than long chain fatty acids. Two of these ligands, oleoyl CoA and haem, have been modelled into the crystal structure [3]. In the case of oleoyl CoA, modelling placed the bulky and poly-anionic ADP derivative of the oleoyl CoA in the ligand portal adjacent to the α-helical region of the protein that allows solvent exposure of much of this anionic group. It would be predicted that ligands such as lysophospholipids or bile acids would bind with a similar orientation, allowing bulky and poly-anionic residues to be accommodated in the portal region.

In a previous publication [4], we described how LFABP was able to bind to anionic vesicles under conditions of low ionic strength and that such binding was monitored by the conformational change associated with binding that resulted in the release of the fluorescent ligand, DAUDA. It was the loss of fluorescence intensity associated with DAUDA release that provided a convenient method of measuring this interfacial binding [4]. In order to determine which cationic residues on the protein surface might be involved in binding

Address for offprints: D.C. Wilton, Division of Biochemistry and Molecular Biology, School of Biological Sciences, University of Southampton, Bassett Crescent East, Southampton SO16 7PX, UK (E-mail: D.C.Wilton@soton.ac.uk)

to the anionic interface, we have in the first instance mutated those cationic residues that are within the α-helical region.

The mutants, K20E, K31E and K33E [5] together with a double mutant (K31E, K33E) have been produced as a result of charge reversal mutagenesis. The ability of these mutants to bind to anionic vesicles has been investigated [5]. In addition, it has been observed that the charge reversal mutations had a minimal effect on the binding of oleic acid and DAUDA [5]. In the case of oleic acid, the primary binding site involves the carboxylate buried in the protein cavity while it is the second, lower affinity, binding site that involves solvent exposure of the carboxyl group in the portal region [1]. DAUDA binds with 1:1 stoichiometry [6] and mutations had no affect on its binding affinity [4]. We therefore conclude that DAUDA binding is in an orientation involving the primary fatty acid-binding site where the carboxyl group is located within the binding cavity of the protein [4].

In this paper we have investigated the binding of other ligands to the charge reversal mutants of LFABP. We are able to show that where a ligand has a bulky anionic group, that should not be accommodated in the binding cavity, binding is affected by charge reversal mutations in the α-helical region. In particular, the mutation of K31 in the α-helix II had the greatest effect on the binding of these bulky ligands.

Materials and methods

Materials

11-((5-dimethylaminonaphthalene-1-sulphonyl)amino)-undecanoic acid (DAUDA) was obtained from Molecular Probes (Junction City, OR, USA). All other chemicals were obtained from Sigma Chemicals (Poole, Dorset, UK).

Molecular biology

A synthetic gene of rat LFABP [7] was cloned into the pET-11a vector and transformed into competent BL21 (DE3) cells [4]. The preparation of the mutants, K20E, K31E and K33E has been described [5]. The double mutant (K31E,K33E) used the forward primer 5′-GACCTGATCCAGGAAGGT-GAAGATATCAAAGGTGTT-3′ (mutant sequences are underlined).

Protein expression and purification

The expression and purification of wild-type and mutant LFABPs has been described [4, 5]. All proteins gave a single band after SDS polyacrylamide gel electrophoresis.

Binding assays

The binding of oleoyl CoA, lysophospholipids and bile salts to LFABP was determined by performing fluorescence displacement assays using the fluorescent fatty acid probe, DAUDA. Assays (1 ml) were performed in 10 mM Hepes buffer pH 7.5 containing 1 μM DAUDA and 8 μg of LFABP. The fluorescence was recorded at 500 nm with excitation at 350 nm using a Hitachi F2000 fluorimeter. Displacing ligands were added in 1 μL aliquots as a 0.5 mM solution in methanol and the decrease in fluorescence intensity was recorded. All assays were performed in triplicate. Titrations were corrected for blank titrations involving methanol alone. The final methanol concentration did not exceed 1% by volume. The corrected displacement curves were plotted as % of starting fluorescence intensity in the absence of ligand. K_i apparent values for the displacing ligand were determined using the method of Kane and Bernlohr [8]. In the case of some ligands, notably lysophospholipids, the % displacement did not reach 50% due to the low affinity and low solubility of the displacing ligand and no K_i apparent value could be determined.

Results

The binding of oleoyl CoA to LFABP

Fatty acyl CoAs have long been recognised as ligands for LFABP *in vitro* and confirmed using recombinant proteins [9, 10]. However their role *in vivo* as potential ligands has been compromised by the discovery of high affinity binding proteins for acyl CoAs, called acyl CoA binding proteins (ACBPs) [11]. The binding affinity of acyl CoAs for LFABP is in the μM range with the possibility of a second low affinity binding site [10] and this compares with the very high affinity binding to ACBP [11]. It is possible that LFABP only has a role in acyl CoA metabolism when concentration of the acyl CoA exceeds that of ACBP such as in fasting when the acyl CoA could bind to the more abundant LFABP [10]. An oleoyl CoA molecule has been modelled into the binding cavity of LFABP that places the bulky anionic portion of the ligand in a solvent exposed position within the portal region. The anionic pyrophosphate in this model interacts with K31, K57 and perhaps K36 [3].

The binding of oleoyl CoA to the charge reversal mutants was studied indirectly measuring displacement of DAUDA by this ligand. The results of DAUDA displacement assays under standard assay conditions (10 mM Hepes buffer) are shown in Fig. 1 and highlights a significant reduction in the ability of oleoyl CoA to displace DAUDA with the two helix

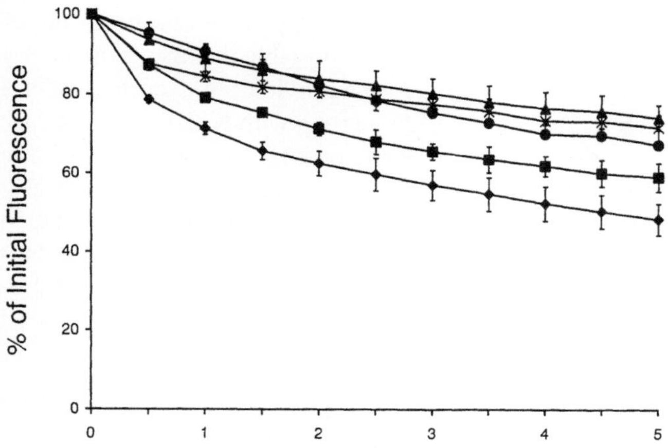

Fig. 1. The displacement of DAUDA from LFABP by oleoyl CoA. To a mixture of LFABP (8 μg) and DAUDA (1 nmol) in 10 mM Hepes buffer pH 7.5 (1 ml) was added by titration up to 5 nmoles of oleoyl CoA in methanol. The percentage loss of fluorescence was recorded. Values are the mean of three titrations ± S.D. ◆ = wild-type LFABP; ■ = K20E LFABP; ▲ = K31E LFABP; ● = K33E LFABP; X = K31E, K33E.

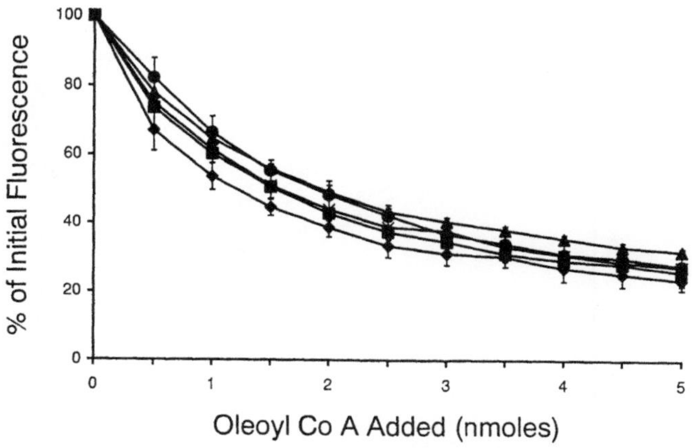

Fig. 2. The displacement of DAUDA from LFABP by oleoyl CoA. As for Fig. 1 except that the titration was performed in the presence of 0.1 M NaCl. Values are the mean of three titrations ± S.D. ◆ = wild-type LFABP; ■ = K20E LFABP; ▲ = K31E LFABP; ● = K33E LFABP; X = K31E, K33E.

II mutants, K31E and K33E together with the double mutant (K31E, K33E). However the percentage displacement for all curves was low compared with previous data (in 50 mM phosphate buffer) [12] and did not permit the determination of K_i apparent values for most ligands. When the displacement curves were repeated in the presence of 0.1 M NaCl in order to enhance hydrophobic interactions, the assay became significantly more sensitive to DAUDA displacement (Fig. 2). The K_i apparent values are shown in Table 1 where it can be seen that the affinity of wild-type for oleoyl CoA increases about 5-fold in the presence of the NaCl. This result suggests that hydrophobic interactions play an important role in the binding of this ligand to LFABP under these assay conditions. The binding of oleoyl CoA under conditions of both low and high ionic strength was least affected by the K20E mutation in helix I.

These results suggest that there are only small specific electrostatic contributions involving cationic residues within the helical region with least involvement of helix I. Overall, the size of the oleoyl CoA with its extensive polar pyrophosphate and phosphate containing headgroup must mean that individual electrostatic contributions from the protein to ligand are minimised and much of the head group is solvent exposed. This conclusion is entirely consistent with the modelled structure of oleoyl CoA bound to liver FABP [3] in which the substantial non-polar oleoylpantetheine residue is buried in the binding cavity of the protein.

The binding of bile acids to LFABP

Two bile acids have been demonstrated to bind to liver FABP with reasonable affinity, these are lithocholic acid [13] and taurolithocholate 3-sulphate [13, 14]. These two ligands were evaluated in terms of their ability to displace DAUDA from wild type and charge reversal mutant liver FABPs. The results for lithocholic acid are seen in Fig. 3. It can be clearly seen that the K31E and the double mutant have significantly reduced ability to bind lithocholic acid and the K_i apparent values are shown in Table 1. A more dramatic situation is seen

Table 1. The binding of oleoyl CoA and bile salts to wild-type and charge reversal mutants of LFABP

	Wild-type LFABP K_i apparent (μM)	K20E LFABP K_i apparent (μM)	K31E LFABP K_i apparent (μM)	K33E LFABP K_i apparent (μM)	K31E, K33E LFABP Ki apparent (μM)
Oleoyl CoA	0.81	N/A	N/A	N/A	N/A
Oleoyl CoA (0.1M NaCl)	0.21 ± 0.04	0.34 ± 0.02	0.49 ± 0.03	0.49	0.42 ± 0.06
Lithocholic acid	0.31 ± 0.04	0.45 ± 0.02	0.80 ± 0.04	0.45 ± 0.04	0.78 ± 0.07
Taurolithocholic acid 3-sulphate	0.29 ± 0.03	0.42 ± 0.08	N/A	1.25 ± 0.07	N/A

The K_i apparent values for ligands were determined from the DAUDA displacement curves shown in Figs 1–7 using the method of Kane and Bernlohr [8]. N/A: if 50% displacement is not achieved by the added ligand it is not possible to calculate the K_i apparent value for that ligand and no value is shown. K_i apparent values for oleoyl CoA are also determined from titrations in the presence of 0.1M NaCl. All values are calculated from the mean of three titrations ± S.D. except where only two of the three titrations gave 50% displacement.

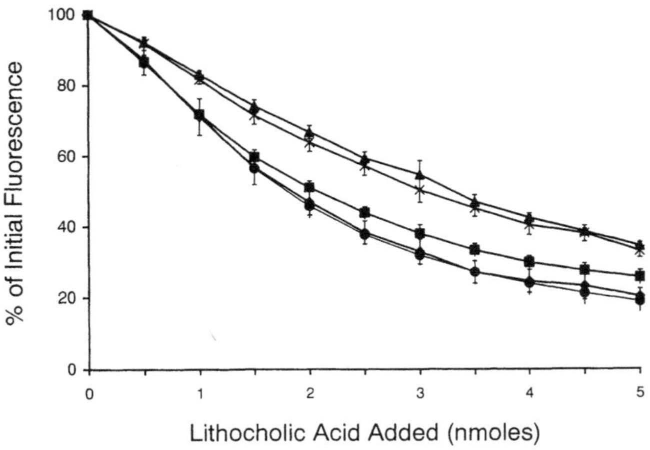

Fig. 3. The displacement of DAUDA from LFABP by lithocholic acid. To a mixture of LFABP (8 μg) and DAUDA (1 nmol) in 10 mM Hepes buffer pH 7.5 (1 ml) was added by titration up to 5 nmoles of lithocholic acid in methanol. The percentage loss of fluorescence was recorded. Values are the mean of three titrations ± S.D. ◆ = wild-type LFABP; ■ = K20E LFABP; ▲ = K31E LFABP; ● = K33E LFABP; X = K31E, K33E.

in the case of taurolithocholate 3-sulphate (Fig. 4), where the ability of this ligand to displace DAUDA is greatly reduced in the case of the K31E and the double mutant. It was not possible to achieve 50% displacement of the DAUDA (required for K_i calculations) under these assay conditions with the K31E and the double mutant.

We would argue that in each case the anionic side chain is located in the portal region while stabilisation of this anionic charge by cationic residues, particularly K31 makes a larger contribution to the binding energy in the case of the taurolithocholate 3-sulphate. It is possible that the sp^3 geometry of the sulphonate in taurine provides an opportunity for a greater interaction of this group with K31 than is the case of the sp^2 geometry of the carboxylate of the lithocholic acid. This binding topology is consistent with that of glycocholate bound to ileal lipid binding protein where structural analysis by NMR [15] has demonstrated that the glycine-conjugated carboxylate tail of the ligand protrudes from the proposed bile acid portal into the aqueous environment.

The binding of lysophospholipids to LFABP

Liver FABP binds lysophospholipids with low affinity however the most effective ligand is lysophosphatidic acid [12] and this ligand has been implicated in LFABP function [16]. It would be anticipated that the bulky glycerophosphate anion of this ligand would reside in the ligand portal adjacent to the α-helix II region and binding of these ligands would be sensitive to charge reversal mutations in this region. That this is indeed the case demonstrated by the DAUDA displacement curves (Fig. 5). It can be clearly seen that as a result of a lack of displacement, binding must be minimal for the K31E and K33E while no binding could be detected for the double mutant. It was not possible to determine K_i apparent values for this ligand because 50% displacement could not be achieved. The considerable di-anionic nature of the phosphate group together with its geometry could be the reason for the dramatic effect of charge reversal (K31E

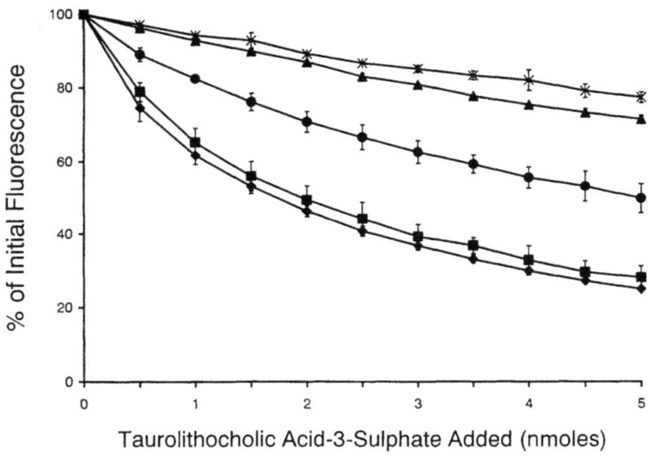

Fig. 4. The displacement of DAUDA from LFABP by taurolithocholate 3-sulphate. To a mixture of LFABP (8 μg) and DAUDA (1 nmol) in 10 mM Hepes buffer pH 7.5 (1 ml) was added by titration up to 5 nmoles of taurolithocholate 3-sulphate in methanol. The percentage loss of fluorescence was recorded. Values are the mean of three titrations ± S.D. ◆ = wild-type LFABP; ■ = K20E LFABP; ▲ = K31E LFABP; ● = K33E LFABP; X = K31E, K33E.

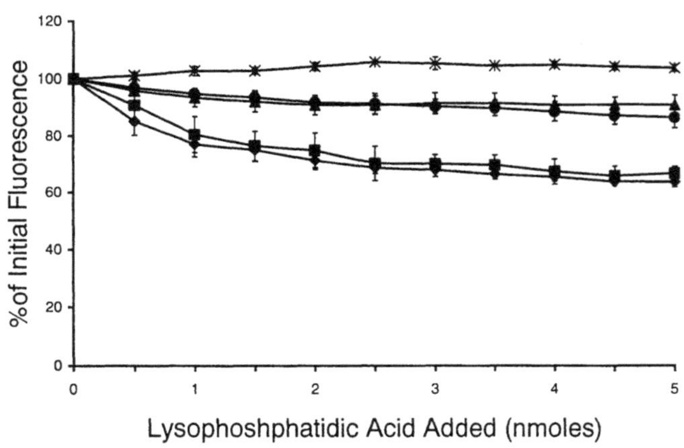

Fig. 5. The displacement of DAUDA from LFABP by lysophosphatidic acid. To a mixture of LFABP (8 μg) and DAUDA (1 nmol) in 10 mM Hepes buffer pH 7.5 (1 ml) was added by titration up to 5 nmoles of lysophosphatidic acid in methanol. The percentage loss of fluorescence was recorded. Values are the mean of three titrations ± S.D. ◆ = wild-type LFABP; ■ = K20E LFABP; ▲ = K31E LFABP; ● = K33E LFABP; X = K31E, K33E.

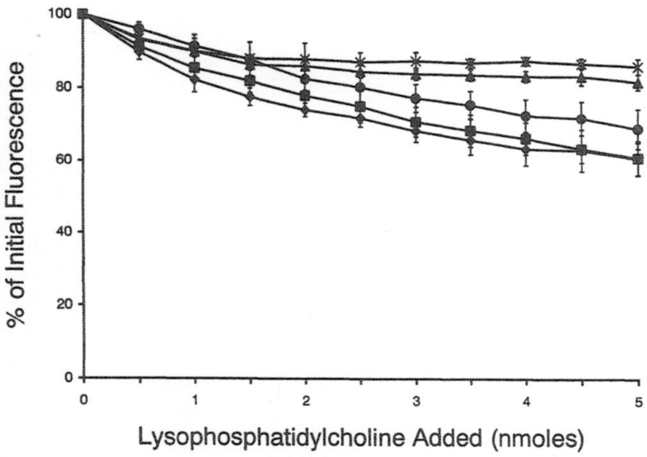

Fig. 6. The displacement of DAUDA from LFABP by lysophosphatidylcholine. To a mixture of LFABP (8 μg) and DAUDA (1 nmol) in 10 mM Hepes buffer pH 7.5 (1 ml) was added by titration up to 5 nmoles of lysophosphatidylcholine in methanol. The percentage loss of fluorescence was recorded. Values are the mean of three titrations ± S.D. ◆ = wild-type LFABP; ■ = K20E LFABP; ▲ = K31E LFABP; ● = K33E LFABP; X = K31E, K33E.

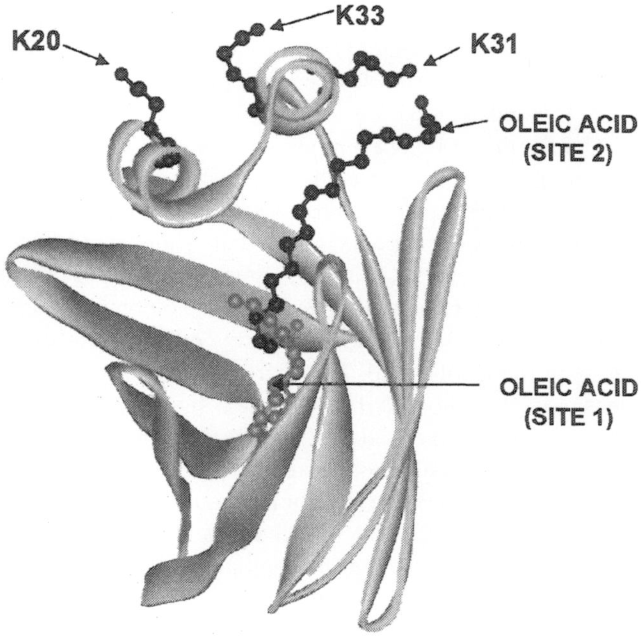

Fig. 7. Ribbon diagram of rat liver FABP showing the positions of bound oleic acids and residues K20, K31 and K33. Derived from the crystal structure of rat liver FABP with two bound oleic acid ligands [1]. The protein molecule is orientated to most clearly show the positions of the three lysine residues in the α-helical region. The orientation of oleic acid at site 2 shows the exposed carboxyl group adjacent to K31.

and also the double mutant) on the binding of lysophosphatidic acid.

The binding of lysophosphatidylcholine was also investigated and the results of DAUDA displacement are shown in Fig. 6. The effects of charge reversal are similar to those seen with lysophosphatidic acid but much less dramatic. Presumably the loss of di-anionic character of the phosphate together with the presence of the compensating positive charge on the choline, which produces a zwitterionic molecule, greatly reduces the electrostatic contribution of K31 to ligand binding.

Discussion

The binding cavity of LFABP is significantly larger than that of other members of the FABP family, having a an internal volume 610 Å^3 compared with values of 450, 391 and 353 Å^3 for ALBP, HFABP and IFABP respectively [2]. The most apparent consequence of this very large cavity is the ability of LFABP to bind two molecules of fatty acid (oleic acid) [1]. However, LFABP is also able to bind a variety of other ligands including the physiological ligands such as fatty-acyl CoAs, bile salts, lysophospholipids and haem. The affinity of these ligands (with the exception of haem [17]) is lower than that for fatty acids with K_d values in the μM range. The bulky anionic headgroup associated with many of these ligands strongly suggests that such ligands bind in an orientation consistent with them using the second oleic site. In this orientation the anionic head group is solvent exposed in the portal region of the protein (Fig. 7). In the case of oleoyl CoA, this has been modelled into LFABP with the large poly-anionic headgroup containing the pyrophosphate and ribose 3′-phosphate extending from the portal region into the solvent [3]. In this orientation interactions are predicted with the α-helical region and specifically K31 [3]. Lysophospholipids and bile acids have not been modelled into LFABP however in the case of ileal lipid binding protein, the NMR structure of the holo-protein places the conjugated bile acid side chain of glycocholate in a portal region and solvent exposed. Although haem has been modelled into LFABP [3] this ligand has not been investigated in the present study.

We have produced charge reversal mutants of the lysine residues present in the α-helical region of LFABP (K20E; K31E and K33E) together with the double mutant (K31E, K33E) [5] in order to investigate the ability of LFABP to bind to anionic vesicles [4]. The ability of DAUDA to bind to LFABP was not significantly affected by charge reversal mutagenesis suggesting that this fluorescent fatty acid analogue uses the primary fatty acid binding site of the LFABP with the carboxyl group buried in the protein cavity and not located in the portal region [5]. Similarly, these mutations had only a minimal effect on oleic acid binding based on DAUDA displacement [5]. However, it would be predicted the ligands for LFABP containing bulky anionic head would utilise a site 2 ligand orientation with the headgroup located in the portal region allowing a high degree of solvent exposure. The binding of such ligands should be adversely affected by charge reversal mutagenesis of lysine residues in the α-helical region, particularly K31 [3].

The results of ligand binding studies with these charge reversal mutants clearly establish that the binding of bile acids and lysophospholipid is adversely affected. In fact the binding of lysophosphatidic acid is essentially eliminated with the double mutant (K31E, K33E). A dramatic effect of charge reversal mutagenesis is also seen with the binding of taurolithocholic acid 3-sulphate whereas the effect is less apparent with lithocholic acid. The charge geometry and charge density of the anionic group may be a significant factor in determining the contribution of electrostatic interactions with K31. It is not possible to predict the precise orientation of these bile acids in the cavity but the results suggest that the side chain carboxyl of the lithocholic acid is located in the portal region, as is the taurine conjugated derivative found in taurolithocholic acid 3-sulphate. Such a prediction would require that the sulphate was located within the binding cavity. In the case of ALBP, the crystal structure containing the ligand, hexadecane sulphonate, revealed that the sulphonate occupied the same site within the cavity as the carboxyl of palmitic acid [18].

The binding data for oleoyl CoA was more ambiguous in that the charge reversal mutants appeared to have a minimal affect on the binding of this ligand, as judged by DAUDA displacement. In contrast, the modelling of this ligand into the crystal structure [3] places the pyrophosphate and ribose 3′-phosphate CoA within the portal region and extending into the solvent. The most likely explanation is that the large size, polyanionic nature and solvent exposure of this headgroup reduce the contribution from individual electrostatic interactions. Moreover, the effect of salt concentration on binding suggests a major contribution from non-polar interactions as part of the energetics of binding.

Although the discussion of ligand binding to LFABP has focused on the role of electrostatics in interpreting the binding data, it must be appreciated that lysine is a potential hydrogen bond donor whereas a solvent-exposed aspartate will be a hydrogen bond acceptor under normal physiological conditions. Thus hydrogen bonding will have a significant role in determining the precise contribution of cationic residues in the α-helical region to ligand binding.

If the interpretation of the mutagenesis and binding data is correct, then it would be predicted that any ligand for LFABP that has a bulky or poly-anionic headgroup should bind at fatty acid binding site 2 [1]. Where such a ligand has a relatively small or flexible hydrophobic group buried in the cavity it may also be possible to also bind a second ligand in the cavity, involving fatty acid binding site 1.

References

1. Thompson J, Winter N, Terwey D, Bratt J, Banaszak L: The crystal structure of the liver fatty acid-binding protein – A complex with two bound oleates. J Biol Chem 272: 7140–7150, 1997
2. Thompson J, Ory J, ReeseWagoner A, Banaszak L: The liver fatty acid binding protein – comparison of cavity properties of intracellular lipid binding proteins. Mol Cell Biochem 192: 9–16, 1999
3. Thompson J, Reese-Wagoner A, Banaszak L: Liver fatty acid binding protein: Species variation and the accommodation of different ligands. Biochim Biophys Acta Mol Cell Biol Lipids 1441: 117–130, 1999
4. Davies JK, Thumser AA, Wilton DC: Binding of recombinant rat liver fatty acid-binding protein to small anionic phospholipid vesicles results in ligand release: A model for interfacial binding and fatty acid targeting. Biochemistry 38: 16932–16940, 1999
5. Davies JK, Wilton DC: Effect of charge reversal mutations within the α-helical region of liver fatty acid binding protein on the ligand and membrane binding properties of the protein. (submitted)
6. Wilkinson TI, Wilton DC: Studies on fatty acid-binding proteins – the binding-properties of rat-liver fatty acid-binding protein. Biochem J 247: 485–488, 1987
7. Worrall AF, Evans C, Wilton DC: Synthesis of a gene for rat-liver fatty-acid-binding protein and its expression in Escherichia-coli. Biochem J 278: 365–368, 1991
8. Kane CD, Bernlohr DA: A simple assay for intracellular lipid-binding proteins using displacement of 1-anilinonaphthalene 8-sulfonic acid. Anal Biochem 233: 197–204, 1996
9. Hubbell T, Behnke WD, Woodford JK, Schroeder F: Recombinant liver fatty-acid-binding protein interacts with fatty acyl-coenzyme-A. Biochemistry 33: 3327–3334, 1994
10. Rolf B, OudenampsenKruger E, Borchers T, Faergeman NJ, Knudsen J, Lezius A, Spener F: Analysis of the ligand binding properties of recombinant bovine liver-type fatty acid binding protein. Biochim Biophys Acta Lipids Lipid Metab 1259: 245–253, 1995
11. Knudsen J, Jensen MV, Hansen JK, Faergeman NJ, Neergaard TF, Gaigg B: Role of acylCoA binding protein in acylCoA transport, metabolism and cell signaling. Mol Cell Biochem 192: 95–103, 1999
12. Thumser AA, Voysey JE, Wilton DC: The binding of lysophospholipids to rat-liver fatty-acid-binding protein and albumin. Biochem J 301: 801–806, 1994
13. Thumser AA, Wilton DC: The binding of cholesterol and bile salts to recombinant rat liver fatty acid-binding protein. Biochem J 320: 729–733, 1996
14. Dietrich A, Dieminger W, Fuchte K, Stoll GH, Schlitz E, Gerok W, Kurz G: Functional-significance of interaction of h-fabp with sulfated and nonsulfated taurine-conjugated bile-salts in rat-liver. J Lipid Res 36: 1745–1755, 1995
15. Lucke C, Zhang FL, Hamilton JA, Sacchettini JC, Ruterjans H: Solution structure of ileal lipid binding protein in complex with glycocholate. Eur J Biochem 267: 2929–2938, 2000
16. Thompson J, Ory J, ReeseWagoner A, Banaszak L: The liver fatty acid binding protein – comparison of cavity properties of intracellular lipid binding proteins. Mol Cell Biochem 192: 9–16, 1999
17. Stewart JM, Slysz GW, Pritting MA, Mullereberhard U: Ferriheme and ferroheme are isosteric inhibitors of fatty acid binding to rat liver fatty acid binding protein. Biochem Cell Biol Biochim Biol Cell 74: 249–255, 1996
18. Lalonde JM, Bernlohr DA, Banaszak LJ: X-ray crystallographic structures of adipocyte lipid-binding protein complexed with palmitate and hexadecanesulfonic acid – properties of cavity binding-sites. Biochemistry 33: 4885–4895, 1994

Molecular and Cellular Biochemistry **239**: 61–68, 2002.

Solution structure of fatty acid-binding protein from human brain

Martin Rademacher,[1] Aukje W. Zimmerman,[2] Heinz Rüterjans,[1] Jacques H. Veerkamp[2] and Christian Lücke[1]

[1]*Institut für Biophysikalische Chemie, Johann Wolfgang Goethe-Universität, Frankfurt a. M., Germany;* [2]*Department of Biochemistry, University of Nijmegen, Nijmegen, The Netherlands*

Abstract

Human brain-type fatty acid-binding protein (B-FABP) has been recombinantly expressed in *Escherichia coli* both unlabelled and ^{15}N-enriched for structure investigation in solution using high-resolution NMR spectroscopy. The sequential assignments of the ^{1}H and ^{15}N resonances were achieved by applying multidimensional homo- and heteronuclear NMR experiments. The ensemble of the 20 final energy-minimized structures, representing human B-FABP in solution, have been calculated based on a total of 2490 meaningful distance constraints. The overall B-FABP structure exhibits the typical backbone conformation described for other members of the FABP family, consisting of ten antiparallel β-strands (βA to βJ) that form two almost orthogonal β-sheets, a helix-turn-helix motif that closes the β-barrel on one side, and a short N-terminal helical loop. A comparison with the crystal structure of the same protein complexed with docosahexaenoic acid [12] reveals only minor differences in both secondary structure and overall topology. Moreover, the NMR data indicate a close structural relationship between human B-FABP and heart-type FABP with respect to fatty acid binding inside the protein cavity. (Mol Cell Biochem **239**: 61–68, 2002)

Key words: β-barrel, lipid binding protein, fatty acid carrier, ^{15}N isotope enrichment, NMR spectroscopy

Abbreviations: 2D – two-dimensional; 3D – three-dimensional; CRBP – cellular retinol-binding protein; CRABP – cellular retinoic acid-binding protein; DHA – docosahexaenoic acid; FA – fatty acid; FABP – fatty acid-binding protein; A-FABP – adipocyte-type FABP; B-FABP – brain-type FABP; H-FABP – heart-type FABP; I-FABP – intestinal-type FABP; ILBP – ileal lipid-binding protein; L-FABP – liver-type FABP; M-FABP – myelin-type FABP; HSQC – heteronuclear single-quantum correlation; NOE – nuclear Overhauser effect; NOESY – nuclear Overhauser enhancement and exchange spectroscopy; TOCSY – total correlation spectroscopy; RMSD – root-mean-square deviation

Introduction

The main function of fatty acid-binding proteins (FABPs) is the intracellular binding and transport of hydrophobic ligands like fatty acids (FA). By modulating the FA concentration they may influence gene expression, cellular growth, and the function of enzymes, membranes, ion channels and receptors. The brain fatty acid-binding protein (B-FABP) belongs to a family of highly conserved intracellular lipid-binding proteins with low molecular mass (14–16 kDa). Nine different FABP types have been identified up to now, named after the first tissue of isolation [1, 2]. They display 28–70% sequence similarity and are expressed in a broad variety of tissues. According to sequence and binding characteristics, the members of the FABP family can be grouped into four subfamilies: (i) the cellular retinol- and retinoic acid-binding proteins (CRBP and CRABP); (ii) the ileal lipid-binding protein (ILBP; binds bile acids) and liver- (L-)FABP (binds two FA); (iii) intestinal- (I-)FABP (binds the FA in a linear conformation); and (iv) adipocyte- (A-), brain- (B-), heart- (H-), epidermal- (E-) and myelin-type (M-)FABP, which bind the FA in a highly bent or U-shaped conformation [2].

Three different FABP types, B-FABP, E-FABP and H-FABP, are present in the brain [3]. They show a spatio-temporally

Address for offprints: C. Lücke, Institut für Biophysikalische Chemie, Johann Wolfgang Goethe-Universität, Marie-Curie-Str. 9, 60439 Frankfurt a. M., Germany (E-mail: lueck@bpc.uni-frankfurt.de)

differential expression during development and maturity [4]. B-FABP is expressed at high levels in neuroepithelial precursor cells of the developing brain, and later becomes restricted to radial cells and immature astrocytes [5, 6]. At the adult state, B-FABP expression persists only in glial cells of the olfactory bulb. Since B-FABP is produced in large amounts during neurogenesis and/or neuronal migration, it may play an important role during central nervous system development [6]. Human fetal retina and brain as well as malignant glioma also express B-FABP [7].

In contrast to other FABP types, murine B-FABP appears to display a higher affinity for long-chain polyunsaturated fatty acids such as arachidonic acid and docosahexaenoic acid (DHA) than for oleic acid [8]. Richieri *et al.*, on the other hand, reported a weaker affinities of human B-FABP for polyunsaturated FA than for saturated and monounsaturated FA [9]. Considering the binding of several different FA to various FABP types, B-FABP apparently shows the highest binding affinity together with H-FABP and M-FABP [9, 10]. A putative function of B-FABP is the storage and/or delivery of FA that are essential as components of phospho- and glycolipids in membranes [11]. Alternatively, the protein could be involved in transport of fatty acids to specific sites of regulation or may protect polyunsaturated FA from undergoing free radical-catalyzed peroxidation.

Recently two crystal structures of human B-FABP have been solved, in complex with DHA and oleate [12]. The overall structure is typical of FABP family members. Detailed knowledge about the structural and functional properties of B-FABP compared to other FABP types, nevertheless, may shed more light on its physiological significance. In the present study, we have employed high-resolution NMR spectroscopy as an analytical tool to investigate various structural aspects of B-FABP in solution.

Materials and methods

Expression and purification of recombinant human B-FABP

Recombinant human B-FABP complexed with a mixture of bacterial FA was obtained as described elsewhere [10]. Human B-FABP cDNA in pBluescript was a kind gift of Dr. F. Shimizu (Otsuka GEN Research Institute, Tokushima, Japan). Briefly, B-FABP cDNA was cloned into a pET3d expression vector (Novagen, Madison, WI, USA) via *Bam*HI and *Nco*I restriction sites. The protein was expressed in *E. coli* BL21(DE3) cells. Extraction from inclusion bodies and purification were performed as previously described for H-FABP mutants [13], except that the anion-exchange step was omitted. For ^{15}N-enrichment, the cells were grown in M9 minimal medium containing ^{15}N ammonium chloride (Cambridge Isotope Laboratories, Andover, MA, USA). The B-FABP preparations showed the usual FA binding and stability characteristics [10].

NMR spectroscopy

All NMR experiments were performed at pH 7.0 and 298 K. The protein samples were prepared at 2–4 mM concentration in argon-purged 20 mM phosphate buffer (H_2O:D_2O = 90:10 v/v) containing 0.05% NaN_3. NMR data collection has been carried out on a Bruker DMX spectrometer operating at 600.13 MHz ^{1}H resonance frequency and using a 5 mm triple-resonance (^{1}H/^{13}C/^{15}N) probe with XYZ-gradient capability. Homonuclear two-dimensional (2D) spectra (TOCSY and NOESY) as well as ^{15}N-edited multidimensional spectra (HSQC, HTQC, TOCSY-HSQC and NOESY-HSQC) were collected. The TOCSY experiments were performed with spin-lock times of 80 or 6 msec (to obtain COSY-type information). Mixing times of 147 and 200 msec were used for the 3D and 2D NOESY experiments, respectively. The homonuclear spectra were recorded in a phase-sensitive mode with time-proportional phase incrementation of the initial pulse. Quadrature detection was used in both dimensions with the carrier placed in the center of the spectrum on the water resonance. All three-dimensional (3D) experiments made use of pulsed field gradients for coherence selection and artifact suppression, and utilized gradient sensitivity enhancement schemes wherever appropriate [14, 15]. Quadrature detection in the indirectly-detected dimensions was achieved by either the States-TPPI or the echo/antiecho method. ^{1}H chemical shifts were referenced to external sodium 2,2-dimethyl-2-silapentane-5-sulfonate (Cambridge Isotope Laboratories, Andover, MA, USA). Heteronuclear spectra were calibrated in the ^{15}N dimension according to the method of Wishart *et al.* [16].

The spectral data were processed on a Silicon Graphics Indy workstation with the Bruker XWIN-NMR 1.3 software package. Peak-picking and data analysis of the transformed spectra were performed using the AURELIA 2.5.9 program (Bruker).

Constraint generation and structure calculation

The sequence-specific ^{1}H and ^{15}N resonance assignments have been obtained via the classical assignment strategy proposed by Wüthrich [17]. The NOE-derived distance constraints were determined from 2D homonuclear NOESY and 3D ^{15}N-edited NOESY-HSQC spectra. Automated assignments of the NOEs, based only on chemical shifts, were obtained with the program *nmr2st* [18]. The upper distance limits were set by an internal calibration based on the intensities of sequential and medium-range NOE values obtained for residues within well-defined secondary structure elements. The cross-peak intensities were grouped into five different distance categories of

2.5, 3.0, 3.5, 4.5 and 6.0 Å. To compute the protein structure, simulated annealing combined with molecular dynamics in torsion angle space was performed using the DYANA 1.5 program package [19]. Assignments of meaningful NOE cross-peaks were made by applying a structure-aided filtering strategy in repeated rounds of structure calculations. Starting *ab initio*, 100 conformers were calculated in 8000 annealing steps each. A total of 106 stereo-specific assignments of prochiral methylene and isopropyl groups were obtained using the program GLOMSA [20]. Pseudo-atom correction for unassigned stereo partners and magnetically equivalent protons was applied as proposed by Wüthrich *et al.* [21]. No hydrogen bond constraints were used in the structure calculation.

In the final calculation run, a total of 300 structures were computed. Subsequent energy minimization in the presence of the NOE-derived distance restraints was performed on the 20 best DYANA conformers using the DISCOVER module of the INSIGHT II program package (Molecular Simulations Inc., San Diego, CA, USA). The CVFF force field was used [22] with a dielectric constant equal to r (distance in Å). A force constant of 20 kcal mol^{-1} $Å^{-2}$ was applied in the NOE restraint term. The resulting structures were analysed with PROCHECK-NMR [23].

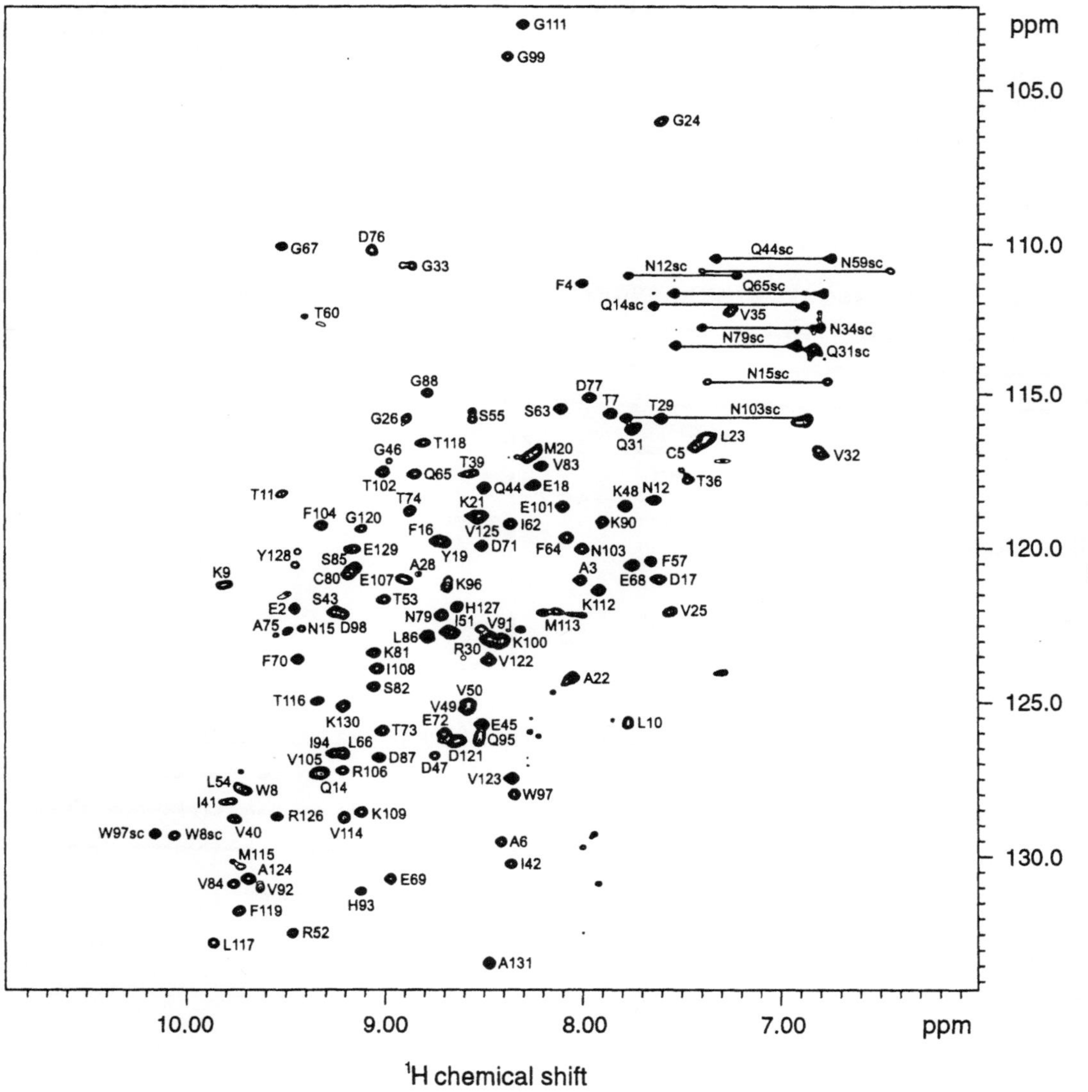

Fig. 1. $^1H/^{15}N$-HSQC spectrum of human B-FABP at pH 7.0 and 298K (^1H frequency of 600.13 MHz). The sequence-specific peak assignments of the backbone and side-chain (sc) amide groups are indicated. Due to spin-system heterogeneities, some residues display multiple signals, which are not additionally labelled.

Results and discussion

Sequential resonance assignments

Nearly complete sequence-specific ^{1}H and ^{15}N backbone assignments have been obtained for all residues except K37. For S13 and T56 the ^{15}N assignment is missing however. Sequential backbone connectivities throughout the entire polypeptide chain were obtained via a combined analysis of the 3D ^{15}N-edited TOCSY-HSQC and NOESY-HSQC spectra. The side-chain proton assignments derived from the 3D data were confirmed and completed using the 2D homonuclear TOCSY spectra. A combination of 2D ^{1}H/^{15}N- HTQC and 3D ^{15}N-edited NOESY-HSQC spectra was employed to assign the side-chain amide resonances of all asparagine and glutamine residues. Remarkably, in the case of the R78 side-chain, also the resonances of one NH_2 group could be identified. All ^{15}N chemical shift values were eventually taken from the 2D ^{1}H/^{15}N-HSQC spectrum (Fig. 1).

Solution structure of human B-FABP

In order to identify the secondary structure elements, short- and medium-range NOEs between backbone protons were assigned based on the 2D and 3D NOESY data (Fig. 2). Consequently, extended β-strand structures, indicated by strong sequential Hα-HN connectivities, were found for the segments spanning residues 6–14, 39–44, 49–54, 59–64, 69–74, 78–86, 91–96, 101–108, 113–118 and 123–130. Helical structures, characterized by strong sequential HN-HN as well as weaker Hα-HN(i,i+1), HN-HN(i,i+2), Hα-HN(i,i+2), Hα-HN(i,i+3), Hα-Hβ(i,i+3), and possibly Hα-HN(i,i+4) NOEs, were present in segments 1–4, 16–23 and 27–35. In addition, NOE connectivity patterns common to antiparallel β-sheet structures were detected between the backbone protons of β-strands βA through βD, βE through βJ, as well as between βA and βJ. Only between β-strands βD and βE no backbone NOE connectivities were observed.

On the basis of 2490 NOE-derived non-redundant distance constraints, 300 structures were generated as described under

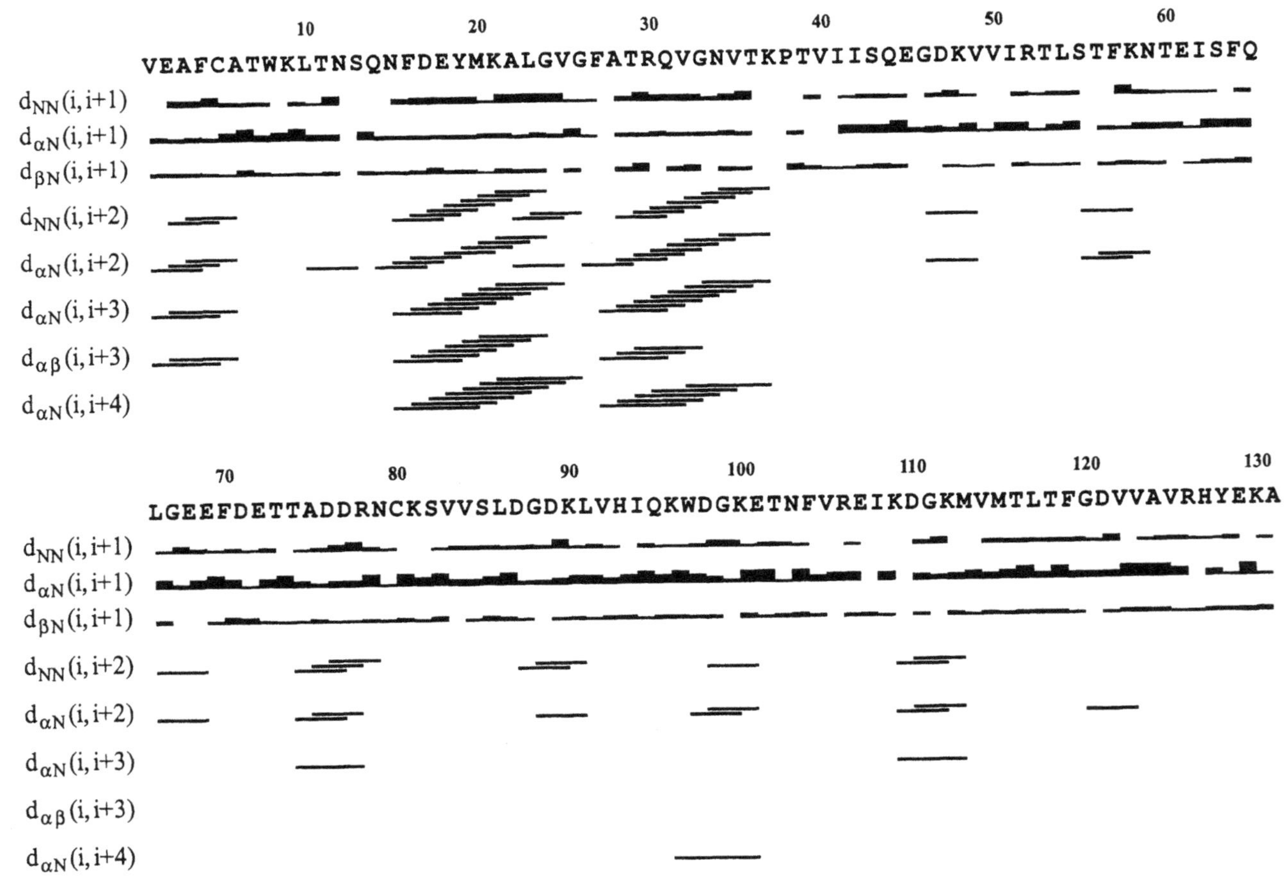

Fig. 2. Survey of short- and medium-range NOEs based on 2D and 3D NOESY data. For sequential connectivities, the thickness of the bars represents the NOE intensities; medium-range NOEs are identified by lines connecting the two coupled residues. Helical structure elements are indicated by NOEs spanning 2–4 residues, while β-strands are characterized by strong $d_{\alpha N}$(i,i+1) connectivities.

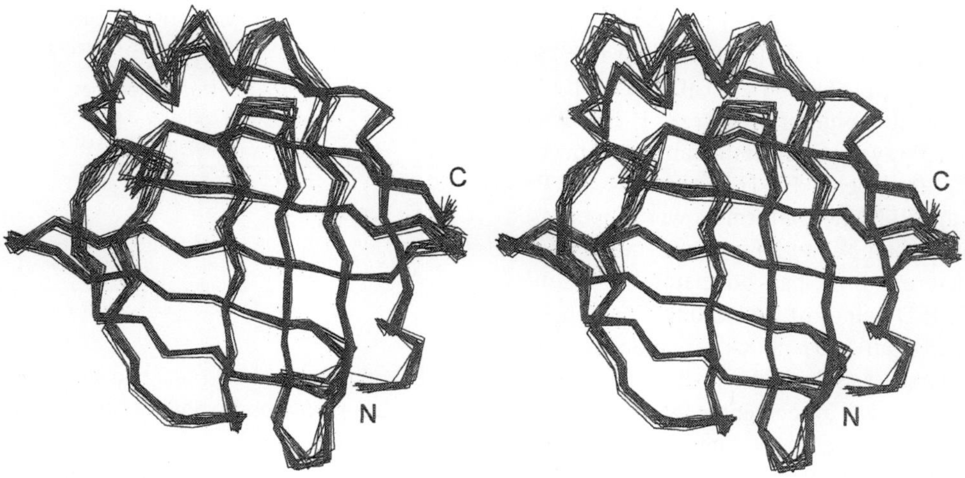

Fig. 3. Stereodiagram showing the Cα traces of the 20 conformers representing the solution structure of B-FABP after restrained energy minimization. An increased conformational dispersion is clearly visible in the portal region encompassing helix II as well as β-turns CD and EF. The N- and C-terminal ends are labelled.

'Materials and methods'. The 20 best structures with the lowest target function values (below 0.1 Å^2) were subsequently subjected to restrained energy minimization. The resulting final structure ensemble is shown in Fig. 3 and the structural statistics are presented in Table 1. While the backbone root-mean-square deviation (RMSD) for the residues 2–130 is 0.85 ± 0.12 Å, it drops down to 0.72 ± 0.10 Å when the residues of the so-called portal region (residues 24–38, 55–58, 74–78) are excluded. This indicates an increased conformational dispersion around the fatty acid portal, whereas the rest of the protein structure is quite well defined.

Table 1. Structural statistics of the 20 selected solution structures of human B-FABP after restrained energy minimization

Restraint statistics	
Total number of meaningful distance restraints	2490
Intraresidual (i = j)	265
Sequential (|i - j| = 1)	627
Medium-range (1 < |i - j| ≤ 4)	420
Long-range (|i - j| > 4)	1178
Restraint violations after energy minimization	
Number of restraint violations > 0.2 Å	5
Number of restraint violations > 0.3 Å	0
Maximal restraint violation	0.25 Å
Structural precision, RMSD (Å)	
Backbone atoms[a] (residues 2–130)	0.85 ± 0.12
All heavy atoms (residues 2–130)	1.41 ± 0.10
Backbone atoms[a] (residues 2–23, 39–54, 59–73, 79–130)	0.72 ± 0.10
All heavy atoms (residues 2–23, 39–54, 59–73, 79–130)	1.30 ± 0.08
Ramachandran plot analysis (%)	
Residues in most favoured regions	83.2
Residues in additionally allowed regions	14.8
Residues in generously allowed regions	1.0
Residues in disallowed regions	1.0

[a]N,C$^{\alpha}$,C′,O.

Overall, the solution structure of human B-FABP (Fig. 4) resembles a so-called β-clam, consisting of a continuous β-sheet that encompasses 10 antiparallel β-strands (βA to βJ) with a gap between β-strands βD and βE. In detail, the N-terminal residues form a helical loop (V1–F4), which leads into β-strand βA (A6–Q14) with a β-bulge at T11–N12. Following the first β-strand, two short α-helices, helix αI (F16–L23) and αII (F27–V35), are arranged in a helix-turn-helix motif. The rest of the structure is defined by a series of antiparallel β-strands

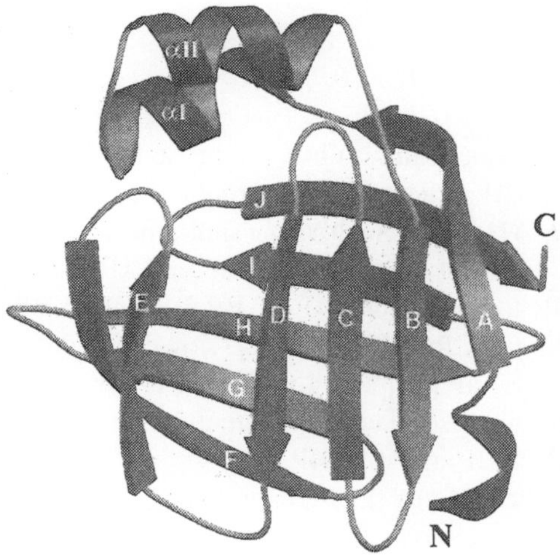

Fig. 4. Ribbon diagram of human B-FABP in solution. The β-strands (βA through βJ) and α-helices (αI and αII) are labelled. An additional helical loop is present at the N-terminus. The ligand binding site is located inside the large cavity defined by the β-barrel structure. (This figure has been produced with MOLSCRIPT [39] and Raster3D [40].)

connected by mostly hairpin turns: βB (T39–Q44), βC (V49–L54), βD (N59–F64), βE (E69–T74), βF (R78–L86), βG (L91–K96), βH (E101–I108), βI (M113–T118), βJ (V123–K130).

In order to compare the solution structure of B-FABP with the 2.1 Å resolution X-ray structure of the same protein complexed with DHA [12] (deposited as 1FDQ in the RCSB Protein Data Bank), the 20 NMR conformers were superposed with each molecule (chain A and B) in the asymmetric unit of the crystal structure. For the backbone atoms of residues 2–130, average RMSD values of 0.81 ± 0.08 Å and 0.82 ± 0.09 Å were obtained relative to both X-ray monomers, which in turn show a backbone RMSD of 0.43 Å between each other. This indicates a good overall agreement between crystal and solutions structures. Moreover, both length and location of the secondary structure elements are comparable.

The helical loop at the N-terminus is a unique feature found in all members of the FABP subfamily IV, which consists of A-, B-, E-, H- and M-FABP. All retinoid-binding proteins, on the other hand, only show a small bulge in an otherwise extended N-terminal structure [24–26], while the other FABPs, i.e. I-FABP, L-FABP and ILBP, have a slightly shorter N-terminus that does not allow the formation of a helical turn. The helical conformation at the N-terminus (V1–F4) could play an additional role, besides the hydrophobic cluster, in the stabilization of the bottom part of the β-barrel. Both V1 and F4 show NOE contacts to residues in the GH turn region. While the highly conserved side-chain of F4 (replaced by L7 in E-FABP) is a central part of the hydrophobic cluster, V1 interacts with the side-chains of both L66 and L86 to form a nonpolar barrier for the external solvent. This might explain the high protein stability found for most members of the FABP subfamily IV [10].

Comparison with H-FABP

Several unique features observed in the NMR spectra of human B-FABP indicate a close structural and functional relationship to H-FABP. Such strong similarities have not become evident for most other members of the FABP family studied by NMR to date, such as E-FABP, I-FABP, ILBP, CRBP I, CRBP II and CRABP II [26–32]. However, additional studies on A-FABP [31] and possibly M-FABP (no NMR data available at present) could eventually give results comparable to B-FABP and H-FABP.

First, because of the high sequence homology between brain- and heart-type FABP (67% for human B-FABP compared to both human and bovine H-FABP), the resonance assignments of many amino acid spin-systems in B-FABP are nearly identical to H-FABP [34, 35]. In fact, more than 50% of all amino acid residues and more than 80% of the conserved residues show maximal chemical shift deviations ≤ 0.30 ppm for the backbone and side-chain protons (Table 2), which emphasizes the high structural similarity between these two proteins [36].

Table 2. Maximal ^{1}H chemical shift deviations found for the residues of human B-FABP relative to human and bovine H-FABP

Δppm	No. of residues (relative to bovine H-FABP)	No. of residues (relative to human H-FABP)
< 0.10	35	28
0.11–0.30	40	42
0.31–0.50	13	14
> 0.50	44	48

Secondly, similar to bovine and human H-FABP [35], several residues that are located mainly around the so-called portal region show spin-system heterogeneities in human B-FABP. More precisely, 12 out of 14 residues that display multiple proton spin-systems are located in this particular region, such as residues G26, Q31, V32, G33, F57, A75 with two, residues L23, V35, T36, K58 and T60 with three, and residue N59 with even four separate spin-systems. These data indicate that up to 4 different stable conformational states occur in the portal region of human B-FABP due to the mixture of endogenous FA present in the sample, like previously reported in the case of bovine H-FABP [35]. The two non-portal residues that display proton spin-system heterogeneities are F104 and V105. This may be due to the contact between the bound ligand and the F104 phenyl ring, which is absent in H-FABP. Thus, B-FABP appears to bind FA via a 'selected-fit' backbone arrangement in the portal region – analogous to H-FABP.

Finally, the labile ^{1}H resonances S82 OγH (5.89 ppm) and H93 Nε2H (11.93 ppm), which usually would not be detected in NMR spectra, have been observed in B-FABP due to very slow exchange with the solvent. The same two resonances have recently been found in H-FABP [37] and were attributed to an extremely intricate hydrogen-bonding network involving several (13) bound water molecules as well as hydrophilic side-chains – in particular E72, S82, H93 and R106 – inside the protein cavity [38]. This electrostatic network of water molecules appears to be an integral part of the H-FABP structure, which prevents the imidazole ring of H93 inside the binding cavity to titrate within the stability limits of the protein (pH 4–9) both in the apo and the holo form [37]. Thus, the side-chains of H93 and S82 apparently are inaccessible to the external solvent, even when the ligand is not present in the binding cavity. This very stable water layer structure inside the protein interior may contribute to the high conformational stability and FA affinity found for both H-FABP and B-FABP. A nearly identical binding scenario can therefore be assumed for B-FABP, since residues E72, S82, H93 and R106 are all conserved, and since a U-shaped ligand with a few (5) highly ordered water molecules is found inside the binding cavity of the oleate complex [12].

The major difference between human B-FABP and H-FABP appears to be the presence of a phenyalanine ring at position 104 inside the protein cavity. In the B-FABP:DHA complex, F104 seems to form a π–π interaction with the C4–C5 double bond of the ligand, possibly providing a rationale for the binding of long-chain polyunsaturated FA [12]. Such a π–π interaction may also occur in mouse and rat B-FABP, where F104 is replaced by a cysteine, whereas in bovine B-FABP the side-chain of residue L104 is unable to form any π–π interactions. Certainly, just as for heart-type FABPs, a hydrophobic residue at position 104 is generally favourable for FA binding to brain-type FABPs.

While the B-FABP:oleate complex shows the ligand in a U-shaped conformation like in H-FABP, the bound DHA adopts a helical shape [12]. The latter arrangement is supposed to be facilitated by π–π interactions of the 6 *cis* double bonds present in DHA with several protein side-chains. The ω-tail of DHA protrudes into the central part of the highly ordered water layer, possibly disrupting the stabilizing effect of the hydrogen-bonding network. As a consequence, only 3 (out of 6) non-linked water positions inside the cavity of the B-FABP:DHA complex are found to be conserved (within 0.1 Å) relative to H-FABP. This disturbance of the internal water layer might be the reason for the weaker binding affinities of B-FABP and H-FABP for long-chain polyunsaturated FA, compared to saturated and monounsaturated FA, as reported by Richieri *et al.* [9].

Conclusions

Based on nearly complete sequential ^{1}H and ^{15}N resonance assignments, the large number of subsequently derived distance constraints has led to a well-resolved solution structure of human B-FABP. The overall fold is highly analogous to other FABPs, and a comparison with the crystal structure of holo B-FABP [12] revealed no significant differences in the backbone conformation (average backbone RMSD of 0.81 ± 0.08 Å). Several features, including high sequence identity (67%), similar proton resonance assignments, spin-system heterogeneities in the portal region, a U-shaped conformation of the bound ligand and an electrostatic network of highly ordered water molecules, indicate a significant similarity in ligand binding between B-FABP and H-FABP, except for the presence of F104 that possibly creates a preference for long-chain polyunsaturated FA in the brain protein. The present study provides the basis for future ^{15}N relaxation studies to obtain additional information on B-FABP backbone dynamics.

Data deposition

The atom coordinates of the 20 conformers representing the solution structure of human B-FABP have been deposited at the RCSB Protein Data Bank under the PDB accession code 1JJX. The ^{1}H and ^{15}N resonance assignments for recombinant human B-FABP have been deposited at the BioMagResBank (http://www.bmrb.wisc.edu) under the accession number BMRB-5320.

Acknowledgements

The European Large Scale Facility for Biomolecular NMR at the University of Frankfurt is kindly acknowledged for the use of its equipment.

References

1. Veerkamp JH, Maatman RGHJ: Cytoplasmic fatty acid-binding proteins: Their structure and genes. Prog Lipid Res 34: 17–52, 1995
2. Hohoff C, Spener F: Fatty acid binding proteins and mammary-derived growth inhibitor. Fett Lipid 100: 252–263, 1998
3. Veerkamp JH, Zimmerman AW: Fatty acid-binding proteins of nervous tissue. J Mol Neurosci 16: 133–142, 2001
4. Owada Y, Yoshimoto T, Kondo H: Spatio-temporally differential expression of genes for three members of fatty acid binding proteins in developing and mature rat brains. J Chem Neuroanat 12: 113–122, 1996
5. Feng L, Hatten ME, Heintz N: Brain lipid-binding protein (BLBP): A novel signaling system in the developing mammalian CNS. Neuron 12: 895–908, 1994
6. Kurtz A, Zimmer A, Schnütgen F, Brüning G, Spener F, Müller T: The expression pattern of a novel gene encoding brain fatty-acid binding protein correlates with neuronal and glial cell development. Development 120: 2637–2649, 1994
7. Godbout R, Bisgrove DA, Shkolny D, Day RS: Correlation of B-FABP and GFAP expression in malignant glioma. Oncogene 16: 1955–1963, 1998
8. Xu LZ, Sánchez R, Sali A, Heintz N: Ligand specificity of brain lipid-binding protein. J Biol Chem 271: 24711–24719, 1996
9. Richieri GV, Ogata RT, Zimmerman AW, Veerkamp JH, Kleinfeld AM: Fatty acid binding proteins from different tissues show distinct patterns of fatty acid interactions. Biochemistry 39: 7197–7204, 2000
10. Zimmerman AW, van Moerkerk HTB, Veerkamp JH: Ligand specificity and conformational stability of human fatty acid-binding proteins. Int J Biochem Cell Biol 33: 865–876, 2001
11. Green P, Glozman S, Kamensky B, Yavin E: Developmental changes in rat brain membrane lipids and fatty acids: The preferential prenatal accumulation of docosahexaenoic acid. J Lipid Res 40: 960–966, 1999
12. Balendiran GK, Schnütgen F, Scapin G, Börchers T, Xhong N, Lim K, Godbout R, Spener F, Sacchettini JC: Crystal structure and thermodynamic analysis of human brain fatty acid-binding protein. J Biol Chem 275: 27045–27054, 2000

13. Zimmerman AW, Rademacher M, Rüterjans H, Lücke C, Veerkamp JH: Functional and conformational characterization of new mutants of heart fatty acid-binding protein. Biochem J 344: 495–501, 1999
14. Kay LE, Keifer P, Saarinen T: Pure absorption gradient enhanced heteronuclear single quantum correlation spectroscopy with improved sensitivity. J Am Chem Soc 114: 10663–10665, 1992
15. Schleucher J, Sattler M, Griesinger C: Coherence selection via gradients without loss of sensitivity. The 3D-HNCO experiment. Angew Chem Int Ed Eng 32: 1489–1491, 1993
16. Wishart DS, Bigam CG, Yao J, Abildgaard F, Dyson HJ, Oldfield E, Markley JL, Sykes BD: ^{1}H, ^{13}C and ^{15}N chemical shift referencing in biomolecular NMR. J Biomol NMR 6: 135–140, 1995
17. Wüthrich K: NMR of Proteins and Nucleic Acids. Wiley, New York, 1986
18. Pristovšek P, Lücke C, Reincke B, Ludwig B, Rüterjans H: Solution structure of the functional domain of *Paracoccus denitrificans* cytochrome c_{552} in the reduced state. Eur J Biochem 267: 4205–4212, 2000
19. Güntert P, Mumenthaler C, Wüthrich K: Torsion angle dynamics for NMR structure calculation with the new program DYANA. J Mol Biol 273: 283–298, 1997
20. Güntert P, Braun W, Wüthrich K: Efficient computation of three-dimensional protein structures in solution from nuclear magnetic resonance data using the program DIANA and the supporting programs CALIBA, HABAS and GLOMSA. J Mol Biol 217: 517–530, 1991
21. Wüthrich K, Billeter M, Braun W: Pseudo-structures for the 20 common amino acids for use in studies of protein conformations by measurements of intramolecular proton–proton distance constraints with nuclear magnetic resonance. J Mol Biol 169: 949–961, 1983
22. Dauber-Osguthorpe P, Roberts VA, Osguthorpe DJ, Wolff DJ, Genest M, Hagler AT: Structure and energetics of ligand binding to proteins: *E. coli* dihydrofolate reductase trimethoprin, a drug-receptor system. Proteins 4: 31–47, 1988
23. Laskowski RA, MacArthur MW, Moss DS, Thornton JM: AQUA and PROCHECK-NMR: Programs for checking the quality of protein structures solved by NMR. J Appl Crystallogr 26: 283–291, 1993
24. Noy N: Retinoid-binding proteins: Mediators of retinoid action. Biochem J 348: 481–495, 2000
25. Folli C, Calderone V, Ottonello S, Bolchi A, Zanotti G, Stoppini M, Rudolfo B: Identification, retinoid binding, and X-ray analysis of a human retinol-binding protein. Proc Natl Acad Sci USA 98: 3710–3715, 2001
26. Lücke C, Pérez C, Cavazzini D, Rademacher M, Ludwig C, Spisni A, Rossi GL, Rüterjans H: Structure and backbone dynamics of apo- and holo-cellular retinol-binding protein in solution. J Biol Chem 277: 21983–21997, 2002
27. Gutiérrez-González LH, Ludwig C, Hohoff C, Rademacher M, Hanhoff T, Rüterjans H, Spener F, Lücke C: Solution structure and backbone dynamics of human epidermal-type fatty acid-binding protein (E-FABP). Biochem J 364: 725–737, 2002
28. Hodsdon ME, Cistola DP: Discrete backbone disorder in the nuclear magnetic resonance structure of apo intestinal fatty acid-binding protein: Implications for the mechanism of ligand entry. Biochemistry 36: 1450–1460, 1997
29. Zhang F, Lücke C, Baier LJ, Sacchettini JC, Hamilton JA: Solution structure of human intestinal fatty acid-binding protein: Implications for ligand entry and exit. J Biomol NMR 9: 213–228, 1997
30. Lücke C, Zhang F, Rüterjans H, Hamilton JH, Sacchettini JC: Flexibility is a likely determinant of binding specificity in the case of ileal lipid binding protein. Structure 4: 785–800, 1996
31. Lu J, Lin C-L, Tang C, Ponder JW, Kao JLF, Cistola DP, Li E: The structure and dynamics of rat apo-cellular retinol-binding protein II in solution: Comparison with the X-ray structure. J Mol Biol 286: 1179–1195, 1999
32. Wang L, Li Y, Abildgaard F, Markley JL, Yan H: NMR solution structure of type II human cellular retinoic acid binding protein: Implications for ligand binding. Biochemistry 37: 12727–12736, 1998
33. Constantine KL, Friedrichs MS, Wittekind M, Jamil H, Chu C-H, Parker RA, Goldfarb V, Mueller L, Farmer BT: Backbone and side chain dynamics of uncomplexed human adipocyte and muscle fatty acid-binding proteins. Biochemistry 37:7965–7980, 1998
34. Lassen D, Lücke C, Kveder M, Mesgarzadeh A, Schmidt JM, Specht B, Lezius A, Spener F, Rüterjans H: Three-dimensional structure of bovine heart fatty-acid-binding protein with bound palmitic acid, determined by multidimensional NMR spectroscopy. Eur J Biochem 230: 266–280, 1995
35. Lücke C, Rademacher M, Zimmerman AW, van Moerkerk HTB, Veerkamp JH, Rüterjans H: Spin-system heterogeneities indicate a selected-fit mechanism in fatty acid binding to heart-type fatty acid-binding protein (H-FABP). Biochem J 354: 259–266, 2001
36. Wishart DS, Sykes BD, Richards FM: Relationship between nuclear magnetic resonance chemical shift and protein secondary structure. J Mol Biol 222: 311–333, 1991
37. Lücke C, Huang S, Rademacher M, Rüterjans H: New insights into intracellular lipid binding proteins: The role of buried water. Prot Sci (in press)
38. Young ACM, Scapin G, Kromminga A, Patel SB, Veerkamp JH, Sacchettini JC: Structural studies on human muscle fatty acid binding protein at 1.4 Å resolution: Binding interactions with three C18 fatty acids. Structure 2: 523–534, 1994
39. Kraulis PJ: MOLSCRIPT: A program to produce both detailed and schematic plots of protein structures. J Appl Crystallogr 24: 946–950, 1991
40. Merritt EA, Bacon DJ: Raster3D: Photorealistic molecular graphics. Meth Enzymol 277: 505–524, 1997

Molecular and Cellular Biochemistry **239**: 69–77, 2002.

Evolution of the family of intracellular lipid binding proteins in vertebrates

Frank G. Schaap, Ger J. van der Vusse and Jan F.C. Glatz
Department of Physiology, Cardiovascular Research Institute Maastricht (CARIM), Maastricht University, Maastricht, The Netherlands

Abstract

Members of the family of intracellular lipid binding proteins (iLBPs) have been implicated in cytoplasmic transport of lipophilic ligands, such as long-chain fatty acids and retinoids. iLBPs are low molecular mass proteins (14–16 kDa) sharing a common structural fold. The iLBP family likely arose through duplication and diversification of an ancestral iLBP gene. Phylogenetic analysis undertaken in the present study indicates that the ancestral iLBP gene arose after divergence of animals from fungi and plants. The first gene duplication was dated around 930 millions of years ago, and subsequent duplications in the succeeding 550 millions of years gave rise to the 16 iLBP types currently recognized in vertebrates. Four clusters of proteins, each binding a characteristic range of ligands, are evident from the phylogenetic tree. Evolution of different binding properties probably allowed cytoplasmic trafficking of distinct ligands. It is speculated that recruitment of an iLBP during evolution of animals enabled the mitochondrial oxidation of long-chain fatty acids. (Mol Cell Biochem **239**: 69–77, 2002)

Key words: fatty acid-binding protein, lipid binding protein, evolution, gene duplication

Abbreviations: CRABP – cellular retinoic acid-binding protein; CRBP – cellular retinol binding protein; LCFA – long-chain fatty acid; FABP – fatty acid-binding protein; K_a – nonsynonymous substitution rate; K_s – synonymous substitution rate; iLBP – intracellular lipid binding protein; mya – millions of years ago

Introduction

Efficient transport of hydrophobic molecules in aqueous solutions is supposedly accomplished by specialized binding proteins. The calycins constitute a superfamily of such lipophilic ligand binding proteins and are widely expressed throughout the animal kingdom [1]. Moreover, genomic sequencing revealed the existence of putative homologues in prokaryotes and yeast as well [2]. Calycins are subdivided into three families: the avidins, the lipocalins, and the intracellular lipid binding proteins (iLBPs) [3]. Amino acid homology between these families is low (typically less than 10% identity) yet they share a similar structural fold designated as the β-barrel [3].

The avidins are biotin binding proteins expressed in certain oviparous vertebrates. Biotin is a vitamin cofactor found in, for instance, acetyl-CoA carboxylase and is bound with very high affinity by avidin (M_r 16.4 kDa). A prokaryotic equivalent (streptavidin, M_r 14.5 kDa) is found in certain *Streptomyces* strains [1]. The lipocalins form a large family of proteins (M_r 18–45 kDa) with great structural and functional diversity [3]. They are usually secretable proteins and are recognized by specific receptors on the cell surface. Lipocalins have been implicated in diverse biological processes, such as pheromone transport, olfaction, and cryptic coloration [3]. Expression is apparently not confined to members of the animal kingdom, as bacterial lipocalins have been reported [2].

iLBPs (M_r 14–16 kDa) make up the third family of calycins. These cytoplasmic proteins are thought to be involved in cellular uptake and intracellular transport of various hydrophobic compounds, such as long-chain fatty acids (LCFAs) and retinoids [4, 5]. Thus far, iLBP expression seems confined to species of the animal kingdom. Within the animal kingdom, iLBPs have been identified in both vertebrates and invertebrates. Although invertebrate iLBPs are briefly discussed, this

Address for offprints: F.G. Schaap, Academic Medical Centre, AMC Liver Centre, Meibergdreef 69-71, 1105 BK Amsterdam, The Netherlands (E-mail: f.g.schaap@amc.uva.nl)

study dealt with the evolutionary relationships of vertebrate iLBPs. Note that iLBPs are unrelated to plant lipid transfer proteins (M_r 8–9 kDa) which bind LCFAs and their CoA esters as well, and have been implicated in the transfer of phospholipids between organelles [6].

Vertebrate iLBPs

A cytoplasmic protein hypothesized to be involved in intracellular transport of LCFAs in mammalian tissues, was first described over 30 years ago [7] and was later termed fatty acid-binding protein (FABP) [8]. Initially, a single protein was thought to be present in hepatic, intestinal, adipose, and cardiac tissue. Nowadays it is known that the mentioned tissues each express a distinct FABP type, namely liver-type FABP (L-FABP$_1$), intestinal-type FABP (I-FABP), adipocyte lipid binding protein (ALBP), and heart-type FABP (H-FABP) [4, 9–11]. Together with the cellular retinol- and retinoic acid-binding proteins (CRBPs, CRABPs) [5], these proteins constitute the iLBP family. Note that, two distinct L-FABPs have been identified in liver of among others the axolotl [12]. L-FABP$_1$ refers to the protein originally isolated from rat liver [8], whereas L-FABP$_2$ denotes the basic protein initially identified in chicken liver [13].

iLBPs are low molecular mass proteins (14–16 kDa) and they share a 3-dimensional structure consisting of two orthogonal β-sheets and a helical cap [14]. This structural design provides an internal cavity which serves as binding site for hydrophobic ligands. The various iLBPs have distinct ligand binding properties. For instance, H-FABP shows high-affinity binding of LCFAs in a 1:1 stoichiometry, whereas L-FABP$_1$ has been reported to bind a wider variety of ligands including LCFAs, acyl-CoA esters, and cholesterol. A larger cavity volume presumably enables L-FABP$_1$ to bind these bulkier ligands, and to accommodate up to two LCFA molecules per protein molecule [15].

iLBPs have been identified in fishes, amphibians, reptiles and birds, and mammals [16]. Most of our knowledge on iLBP characteristics, however, derives from the study of mammalian iLBPs. Recently, two novel CRBPs (CRBP$_3$ and CRBP$_4$) were identified expanding the number of distinct mammalian iLBPs to 15 [17, 18]. Note that two independent groups reported cloning of a novel CRBP which they both termed CRBP$_3$. These proteins turned out to be distinct CRBPs. Throughout the paper CRBP$_3$ refers to the protein identified by Folli *et al.* [17] while CRBP$_4$ refers to the protein identified by Vogel *et al.* [18]. Cloning of the murine CRBP$_4$ orthologue was reported by Conforti *et al.* [19]. Some properties of the mammalian iLBPs are listed in Table 1. An orthologue of L-FABP$_2$, (the 16th iLBP type in vertebrates) has not been identified yet in mammals. In mouse, amino acid identity and similarity between the different iLBPs range between 16–72 and 36–85%, respectively.

Table 1. Characteristics of the mammalian intracellular lipid-binding proteins

Protein name	Main occurrence[a]	Ligands	Gene[b]	Gene locus[c]	exon#1[d]	exon#2	exon#3	exon#4	protein length[e]
Heart-FABP (H-FABP)	(cardiac) myocytes	LCFA	FABP3	1p32–p33	73	173	102	54	132
Brain-FABP (B-FABP)	brain astrocytes	LCFA	FABP7	6q22–q23	73	173	102	51	131
Adipocyte LBP (ALBP)	adipose tissue	LCFA, retinoic acid	FABP4	8q21	73	173	102	51	131
Myelin P2 (MLBP)	Schwann cells	LCFA, retinoids	FABP8	8q21–q22	73	173	102	51	131
Testicular LBP (TLBP)	spermatozoa	?	FABP9	(8q21[f])	73	173	102	51	131
Epidermal-FABP (E-FABP)	skin	LCFA, eicosanoids	FABP5	15q12–q24.3[g]	79	173	102	54	134
Cellular RBP$_1$ (CRBP$_1$)	ovary	retinol	RBP1	3q23	73	179	102	54	134
Cellular RBP$_2$ (CRBP$_2$)	intestine	retinol, retinaldehyde	RBP2	3q23	73	179	102	51	133
Cellular RBP$_3$ (CRBP$_3$)	kidney	retinol		12pter–p13.31					134
Cellular RBP$_4$ (CRBP$_4$)	heart	retinol	RBP7						133
Cellular RABP$_1$ (CRABP$_1$)	testis	retinoic acid	RBP5	15q24	70	179	114	51	136
Cellular RABP$_2$ (CRABP$_2$)	skin	retinoic acid	RBP6	1q21.3	70	179	117	51	137
Intestinal-FABP (I-FABP)	intestinal epithelium	LCFA	FABP2	4q28–q31	67	173	108	51	131
Ileal LBP (ILBP)	ileal enterocytes	LCFA, bile salts	FABP6	5q23–q35	67	176	90	54	127
Liver-FABP$_1$ (L-FABP$_1$)	liver	LCFA, cholesterol, heme	FABP1	2p11	67	173	93	51	126

[a]The major tissue or cell type expressing a particular iLBP is indicated. [b]Gene nomenclature according to human genome database conventions. [c]The chromosomal location refers to the locus in the human genome. [d]Exon sizes (excluding 5'/3' UTR sequences) are given in nucleotides. [e]The length of the mature protein is given in amino acids. [f]BLAST analysis of the draft human genome sequence suggests that a human TLBP orthologue is encoded by a contig (NT_007972) also harboring the ALBP, E-FABP, and MLBP genes. [g]Additional genes encoding human E-FABP appear to be present on among others chromosome 8q21 and 11q12. All data was compiled from [4, 5, 9, 10, 11] and public databases. Note the terminology used to distinguish between iLBPs binding a single class of ligands (FABPs, CRBPs, and CRABPs) and those displaying affinity for a wider range of ligands (LBPs). Abbreviations: LCFA – long-chain fatty acid; FABP – fatty acid-binding protein; LBP – lipid binding protein; RA – retinoic acid; RBP – retinol binding protein; RABP – retinoic acid-binding protein; UTR – untranslated region.

All vertebrate iLBP genes characterized thus far consist of four exons with a similar intron phase (Table 1). This common gene organization suggests that duplication and diversification of a common ancestral iLBP gene gave rise to the present-day iLBP types. In humans, most iLBP genes reside on different chromosomes, only the genes encoding ALBP/E-FABP/MLBP/TLBP (see [f] in Table 1) and $CRBP_1/CRBP_2$ seem tightly linked and may have arisen through direct gene duplication. Basic local alignment search tool (BLAST) analysis of the draft sequence indicates that multiple genes encoding E-FABP are present in the human genome (see [g] in Table 1). In addition to transcribed genes, several nonfunctional pseudogenes have been identified in mammalian genomes. For example, an intronless H-FABP pseudogene is found at human chromosomal region 13q13–q14 [20].

Invertebrate iLBPs

An invertebrate iLBP (Muscle-FABP, M-FABP) was first isolated from flight muscle of the desert locust *Schistocerca gregaria* over a decade ago [21]. During migratory flight, fatty acids are mobilized from triacylglycerol stores in fat bodies and are the preferred substrate for muscle energy production. Expression of M-FABP in flight muscle of the migratory locust *Locusta migratora* amounts up to 18% of the cytoplasmic protein pool and has been implicated in sustaining prolonged flight periods during migration [22]. Elucidation of the three dimensional structure of flight muscle FABP revealed conservation of the characteristic β-barrel structure [23]. iLBPs have also been identified in other Arthropoda, such as the tobacco hawkmoth *Manduca sexta* and the mite *Blomia tropicalis*, as well as in other invertebrate phyla, such as Nematoda and Platyhelminthes. In general, invertebrate iLBP genes seem to consist of 2 or 3 exons, e.g. the locust M-FABP gene contains three exons [24].

Despite dissimilar gene organization, invertebrate and vertebrate iLBPs are clearly related as evidenced by sequence homology (25–47% amino acid identity) and conservation of structural features. This suggests that a common ancestral iLBP gene was present before the vertebrate/invertebrate split which occurred approximately 650 millions of years ago (mya).

Aim of the study

A previous study by Schleicher *et al.* established the basic evolutionary pattern of the iLBP family [25]. Their phylogenetic tree, which was constructed using the neighbour-joining method [26] and included invertebrate iLBP species as well, showed several regions where the branching order was unresolved. Hence, the relationships between individual iLBPs were not fully defined. In the present study, maximum-likelihood analysis was used to further study the evolutionary relations between vertebrate iLBPs. Two newly identified iLBPs, i.e. $CRBP_3$ and $CRBP_4$, were included in the analysis as well.

Materials and methods

Sequences of vertebrate iLBPs (see Appendix) were taken from the GenBank (release 125) and SwissProt (release 39) databases. Entries containing partial sequence information were not included in the study. Amino acid sequences were aligned using ClustalW [27] which allowed guiding by an initial protein alignment based on the structural information available for 13 distinct iLBP types (reference 28 and Protein DataBank at http://www.rcsb.org/pdb/). The resulting protein alignment was used to direct the alignment of protein-coding nucleotide sequences. Von Ebner's gland protein (accession number NP_002288), which belongs to the lipocalin family of calycins and has a size (158 amino acids) comparable with iLBPs, was used as an outgroup. No additional gaps were introduced by including the outgroup in the alignment. Positions showing gaps were excluded from the alignment, resulting in an aligned sequence length of 125 amino acids or 375 nucleotides.

Phylogenetic trees were constructed using maximum-likelihood methods implemented in the PUZZLE program [29] and in programs of the PHYLIP package [30]. Maximum-likelihood amino acid distances were calculated for orthologous sequence pairs using the PUZZLE program [29]. For determination of (non)synonymous substitution distances between pairs of nucleotide sequences, the method described by Comeron was employed [31]. Species divergence times used for calculating substitution rates were taken from reference 32 (rodents/other mammals: 112 mya, reptiles/aves: 276 mya).

Results and discussion

Phylogenetic relationships between vertebrate iLBPs

A maximum-likelihood tree depicting the evolution of the iLBP family is presented in Fig. 1. Among phylogenetic methods, maximum-likelihood analysis offers several advantages including the ability to statistically compare different tree topologies using a likelihood ratio test [26, 33]. Especially when evolutionary rates differ among lineages, this methodology most likely infers the correct evolutionary relationships between (gene) species [34]. Nonetheless, the tree topology could not be completely resolved, i.e. some uncertainty existed in the

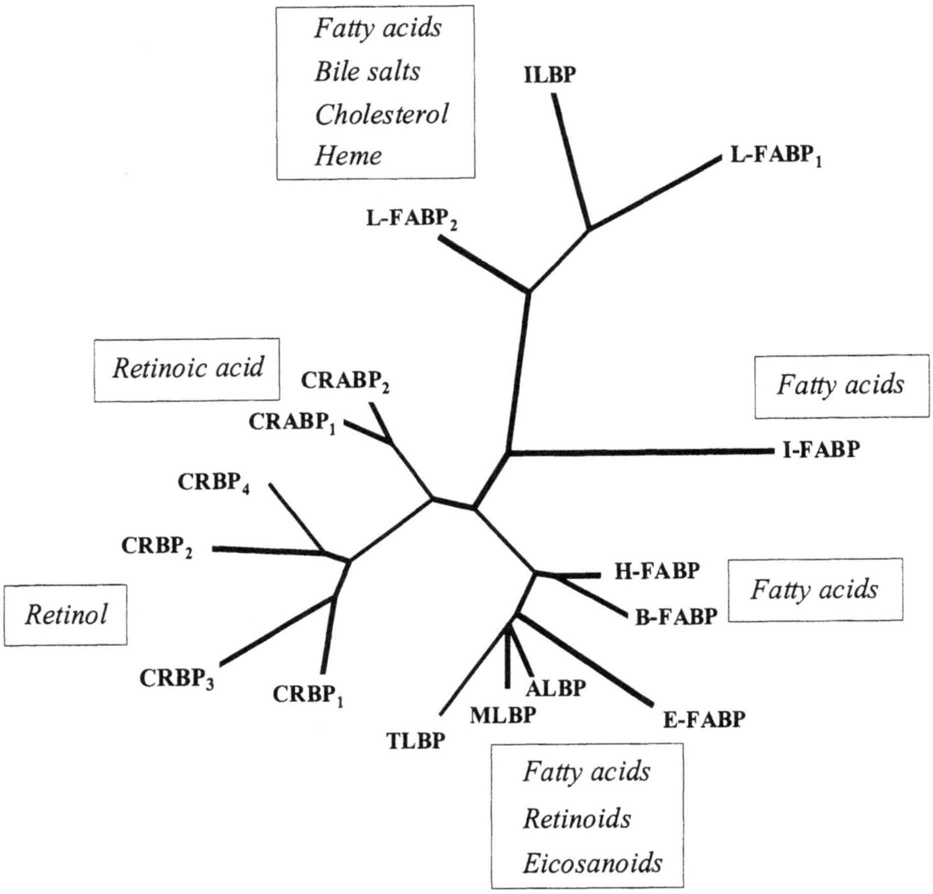

Fig. 1. Unrooted phylogenetic tree showing the evolutionary relationships between vertebrate iLBPs. The ligand-binding specificity of the various iLBP types is indicated. The tree was constructed by maximum-likelihood analysis of a set of 65–72 aligned iLBP sequences. Note that for clarity, the terminal branching pattern representing species evolution has been omitted.

branching order of E-FABP/ALBP/MLBP/TLBP. The topology depicted in Fig. 1 had the highest log likelihood employing either the amino acid-based data set or the nucleotide-based data set. However, it should be noted that some of the 15 possible branching patterns in the E-FABP/ALBP/MLBP/TLBP cluster were not significantly worse than the branching order shown in Fig. 1.

Four different groups of proteins, each apparently binding a distinct range of ligands, are evident from the analysis. Within the ILBP/L-FABP$_1$/L-FABP$_2$ cluster rather bulky ligands, such as heme, bile salts, and cholesterol, are bound. I-FABP binds solely LCFAs, while proteins of the CRBPs/CRABPs cluster bind vitamin A derivatives (retinol and retinoic acid). All proteins within the H-FABP/B-FABP/E-FABP/TLBP/ALBP/MLBP group seem capable of binding LCFAs, and E-FABP, ALBP, MLBP, and presumably TLBP also to bind retinoids and eicosanoids. Importantly, it should be kept in mind that the ligand binding specificity of TLBP has not been determined yet. MLBP and TLBP together form a subgroup and are both presumed to have an inner membrane-associated localization. MLBP is present between the cytoplasmic side of opposing myelin membranes in Schwann cells [35] while TLBP is located between the inner acrosomal and outer nuclear membrane of the sperm head [36].

iLBPs are thought to promote intracellular trafficking of hydrophobic molecules. Diverse iLBPs may have evolved to enable cytoplasmic transport of distinct ligands. For example, simultaneous expression of L-FABP$_1$, ILBP, I-FABP, CRBP$_2$, and CRABP$_1$ found in intestinal tissue may provide soluble carrier proteins for cholesterol, bile salts, LCFAs, retinol, and retinoic acid, respectively. The need for multiple iLBPs with apparently identical ligand binding properties (e.g. H-FABP and B-FABP) seems however less clear.

Evolutionary rates of vertebrate iLBPs

The length of the branches in a phylogenetic tree is a measure for the evolutionary distance (i.e. number of substitutions per site) over the period of time represented by that branch.

If the evolutionary rates (i.e. substitutions per site per year) in the branches leading from an ancestral node are known, then the date of divergence (i.e. gene duplication time) can be estimated. The constancy of evolutionary rate in different lineages, the so-called molecular clock hypothesis, was evaluated using a likelihood ratio test [33]. The analysis indicated that the molecular clock hypothesis had to be rejected (data not shown). Apparently, the iLBP family did not evolve in a simple clock-like manner. Gene duplication times could thus not be estimated by assuming a single substitution rate for all lineages. Note that most iLBPs characterized thus far are of mammalian origin. As this group of animals spans a relatively short period (ca. 110 millions of years) on the evolutionary timescale, the constancy of evolutionary rates of individual iLBPs over longer time periods could not be adequately inspected.

Rate heterogeneity was further assessed by calculating evolutionary rates for individual iLBPs (Table 2). For this, the amino acid/nucleotide distance between orthologous sequence pairs was computed and divided by twice the species divergence time. Substitutions in protein-coding nucleotide sequences were divided in silent changes (synonymous substitutions) and changes that altered the encoded amino acid (nonsynonymous substitutions). This analysis indicated that the amino acid substitution rate differed widely (0.016–1.24 amino acid substitutions per amino acid site per billion of years) between iLBPs, with the slowest ($CRABP_I$) and fastest rate (ILBP) differing by a factor of almost 80. The synonymous substitution rate (K_s) was virtually constant among iLBPs (1.72–3.90 nucleotide substitutions per synonymous nucleotide site per billion of years), while nonsynonymous substitution rates (K_a) showed larger variation (0.007–0.62 nucleotide substitutions per nonsynonymous nucleotide site per billion of years). The average (non)synonymous substitution rate of iLBPs (K_s = 2.37 and K_a = 0.33 (non)synonymous substitutions per (non)synonymous site per billion of years) were comparable with values obtained for other mammalian genes (K_s = 2.07 and K_a = 0.38 (non)synonymous substitutions per (non)synonymous site per billion of years) [37].

When normalized against K_s, the ratio of K_a/K_s provides an index of the evolutionary conservation of a protein. From a large comparative study of rodent and human genes, an average ratio of 0.16 can be calculated [37]. In general, the retinoid binding proteins seemed to be best conserved (e.g. bovine and mouse $CRABP_I$ are fully conserved at the amino acid level). Retinoids play an important role in controlling cellular growth and differentiation [5]. For exerting their biological function, retinoids interact with nuclear receptors. The cellular retinoid binding proteins presumably have a modulatory role in this process by controlling cellular retinoid levels and/or by conveying the retinoid to the nuclear receptor. Strong conservation of retinoid binding proteins may thus be related to their confined ligand binding properties, or because they need to interact with other proteins. For example, holo-$CRBP_I$ is the substrate for the enzyme retinoldehydrogenase [38]. Surprisingly, mice lacking both CRABP genes appeared developmentally normal [39] suggesting that there is redundancy in cellular binding sites for retinoic acid.

Table 2. Substitution rates of vertebrate iLBPs

	Amino acid substitution rate	Synonymous substitution rate (K_s)	Nonsynonymous substitution rate (K_a)	K_a/K_s ratio
H-FABP	0.65	2.22	0.32	0.14
B-FABP	0.60	2.28	0.32	0.14
ALBP	0.64	2.22	0.28	0.13
MLBP	0.72	2.38	0.28	0.12
TLBP	0.57	1.97	0.28	0.14
E-FABP	1.17	3.90	0.53	0.14
$CRBP_1$	0.24	2.13	0.10	0.05
$CRBP_2$	0.50	1.84	0.25	0.13
$CRBP_3$	–	–	–	–
$CRBP_4$	0.49	2.22	0.23	0.10
$CRABP_1$	0.016	1.85	0.007	0.004
$CRABP_2$	0.29	1.92	0.15	0.08
I-FABP	1.14	2.40	0.55	0.23
ILBP	1.24	2.62	0.62	0.24
$LFABP_1$	1.10	1.72	0.46	0.27
$LFABP_2$	0.68	2.82	0.32	0.12

Orthologous distances were estimated as described in Materials and methods. For calculation of substitution rates, the average distance between rodent (rat, mouse, and squirrel) and other mammalian orthologues (man, rabbit, cow, and pig) was divided by twice their divergence time (112 mya). Substitution rates for TLBP and L-$FABP_2$ were based on comparison of sequences from mouse/rat, which diverged 40.7 mya, and lizard/chicken (reptiles/aves split: 276 mya), respectively [32]. Thus far, $CRBP_3$ cDNA has been cloned from a single species only, hence substitution rates could not be determined. All values (except for the K_a/K_s ratio which is dimensionless) are expressed as substitutions per site per billion of years.

Dating of gene duplication times

For determining the time points that gene duplications occurred in the iLBP family, synonymous distances between paralogous sequence pairs would be most suitable as the synonymous substitution rate is fairly constant for all iLBPs (Table 2). However, in most cases synonymous distances could not be reliably calculated. For estimating gene duplication times in the CRBPs/CRABPs cluster, synonymous distances (Table 3, lower part) were used as nonsynonymous substitution rates varied too widely in this group (Table 2). For the other estimates of gene duplication times, nonsynonymous distances (Table 3, upper part) were employed.

For example, to calculate the divergence date of H-FABP and B-FABP, their nonsynonymous distance (0.264) was divided by twice their average nonsynonymous substitution rate

Table 3. (Non)synonymous distances between vertebrate iLBPs

	Nonsynonymous distances														
	ALBP	B-FABP	$CRABP_1$	$CRABP_2$	$CRBP_1$	$CRBP_2$	$CRBP_3$	$CRBP_4$	E-FABP	H-FABP	I-FABP	ILBP	L-$FABP_1$	L-$FABP_2$	MLBP
B-FABP	0.388														
$CRABP_1$	0.649	0.641													
$CRABP_2$	0.715	0.601	0.199												
$CRBP_1$	0.689	0.655	0.613	0.632											
$CRBP_2$	0.660	0.664	0.686	0.643	0.406										
$CRBP_3$	0.869	0.889	0.759	0.681	0.373	0.455									
$CRBP_4$	0.656	0.679	0.621	0.628	0.336	0.352	0.466								
E-FABP	0.377	0.548	0.733	0.761	0.925	0.888	n.a.	n.a.							
H-FABP	0.330	0.264	0.649	0.594	0.698	0.600	0.746	0.645	0.492						
I-FABP	0.734	0.727	0.707	0.755	0.845	0.861	1.052	0.874	0.966	0.722					
ILBP	0.990	1.000	1.058	0.985	0.945	0.936	1.017	1.142	1.054	0.937	0.950				
L-$FABP_1$	1.173	0.838	0.867	0.791	0.884	0.878	0.952	0.972	1.199	0.893	0.938	0.573			
L-$FABP_2$	0.841	0.747	0.977	0.853	0.786	0.939	1.014	0.954	1.037	0.776	0.881	0.577	0.611		
MLBP	0.220	0.362	0.655	0.660	0.674	0.716	0.810	0.649	0.377	0.348	0.783	1.138	1.144	0.865	
TLBP	0.297	0.447	n.a.	n.a.	0.841	0.821	1.059	0.718	0.473	0.422	0.823	1.285	1.177	0.921	0.261

	Synonymous distances				
	$CRBP_1$	$CRBP_2$	$CRBP_3$	$CRBP_4$	CRABP1
$CRBP_2$	1.91				
$CRBP_3$	1.38	1.14			
$CRBP_4$	1.28	1.60	1.21		
$CRABP_1$	1.95	2.17	1.47	1.42	
$CRABP_2$	2.55	2.24	1.42	2.35	1.44

Average (non)synonymous distances between paralogous iLBPs were calculated using the method described by Comeron [31]. Values are expressed as (non)synonymous substitutions per (non)synonymous nucleotide site. All available orthologues (see Appendix) were included in the analysis. N.A. means that Kimura-correction for multiple hits [26] was not applicable and rates could, thus, not be calculated.

(0.32 and 0.32 substitutions per site per billion of years, respectively) yielding a value of 409 mya. Another example, to date the divergence of the ILBP/L-$FABP_1$ and L-$FABP_2$ lineages from their last common ancestor the average synonymous distance between the two groups (average of 0.577 and 0.611) was divided by twice the average synonymous substitution rate in each lineage (0.55 and 0.32 substitutions per site per billion of years) yielding an estimate of 679 mya. Divergence times of the other lineages were calculated in the same way.

The combined results are presented in the tree shown in Fig. 2. The tree was rooted, that is the location of the deepest branch (i.e. duplication of the primordial iLBP gene) was determined by including a lipocalin sequence (outgroup) during tree reconstruction. Note that the divergence time between the CRABPs/CRBPs and H-FABP/B-FABP/E-FABP/TLBP/MLBP/ALBP groups was not estimated because the nonsynonymous substitution rates varied too much between the two lineages. Within the CRBPs/CRABPs cluster, some divergence times have not been calculated because the substitution rates for $CRBP_3$ could not be determined as this iLBP has been cloned from a single species only. Note, the paradoxal dating of the divergence of the ALBP and MLBP/TLBP lineages (457 mya) and the MLBP/TLBP split (467 mya) apparently reflecting the earlier mentioned uncertainty in their branching order. Alternatively, the substitution rates for TLBP may have been underestimated as these values were based on comparison of mouse and rat TLBP.

A time range is given for divergence of the deepest branch. The upper value (1101 mya) was based on the average distance and substitution rate between the ILBP/L-$FABP_1$/L-$FABP_2$ and I-FABP/H-FABP/B-FABP/E-FABP/ALBP/MLBP groups. Due to variation in the individual substitution rates this value is likely to overestimate the divergence time. The lower value (927 mya) was calculated by using average distance and substitution rate between ILBP/L-$FABP_1$ and I-FABP/E-FABP as substitution rates of these iLBPs varied less.

The earliest iLBP gene duplication probably occurred around 930 mya. Around the time of the vertebrate/invertebrate split (700–600 mya), 3 or 4 lineages presumably already existed (Fig. 2). Thus far characterized invertebrate iLBPs appear to show the highest homology with vertebrate iLBPs of the large CRABPs/CRBPs/H-FABP/B-FABP/E-FABP/TLBP/ALBP/MLBP group. For example, $FABP_2$ isolated from the midgut of the tobacco hawkmoth *Manduca sexta* shares 37–47% identity with members of this group.

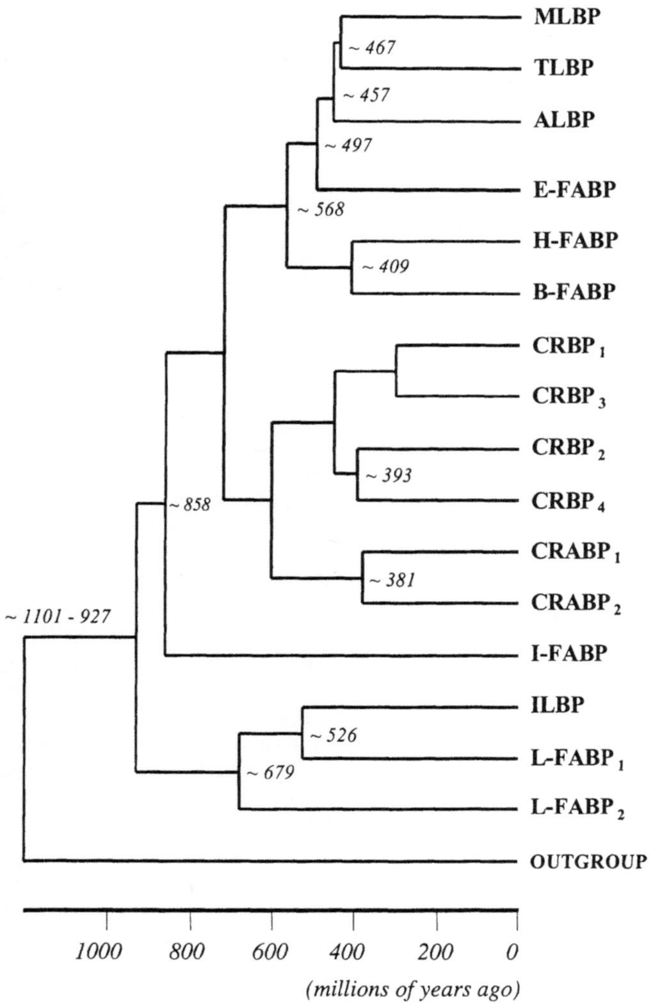

Fig. 2. Scheme for the evolution of the vertebrate iLBP family. The tree was rooted by including an outgroup (Von Ebner's gland protein) in the phylogenetic analysis. Gene duplication times are shown in millions of years ago and were estimated as described in the text.

Table 4. Distribution of iLBPs in the vertebrate classes

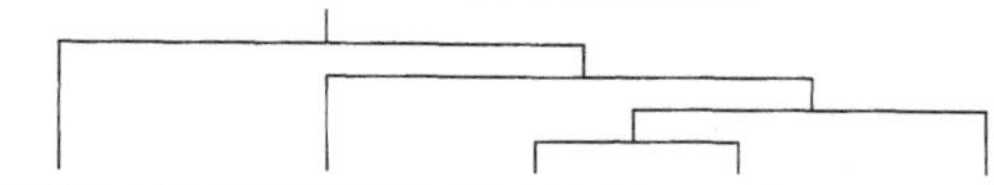

	Fishes (450 mya)	Amphibia (360 mya)	Reptiles (310 mya)	Aves (310 mya)	Mammals
H-FABP	•	•		•	•
B-FABP	•			•	•
ALBP					•
MLBP					•
TLBP					•
E-FABP					•
$CRBP_1$					•
$CRBP_2$					•
$CRBP_3$					•
$CRBP_4$					•
$CRABP_1$	•				•
$CRABP_2$		•		•	•
I-FABP	•	•			•
ILBP					•
$LFABP_1$	•	•		•	•
$LFABP_2$	•	•	•	•	

The presence of the various iLBP types in one of the major vertebrate groups is indicated. The time of divergence of these groups from the mammalian lineage is given in parentheses [32]. Data was compiled from SwissProt, PIR and GenBank databases (see also Appendix).

The phylogenetic distribution of distinct iLBPs in vertebrates (Table 4) seems to correspond well with the estimated times of gene duplications. Note that the time of the $CRABP_1$/$CRABP_2$ split may have been underestimated as a $CRABP_1$ homologue was found to be present in the pufferfish which diverged 450 mya from the mammalian lineage [40]. Likewise, the H-FABP/B-FABP split is apparently underestimated as H-FABP and B-FABP orthologues have both been found in fishes (see Appendix). In liver of the axolotl *Ambystoma mexicanum* two distinct iLBPs have been identified [12]. The first represents the mammalian L-$FABP_1$ homologue, while L-$FABP_2$ is closely related to the basic (pI 9.0) L-FABP initially identified in chicken liver [13]. L-$FABP_2$ diverged from the L-$FABP_1$/ILBP group circa 679 mya and is therefore likely to be present in all vertebrates. Thus far, this iLBP type has been recognized only in fishes, amphibians, reptiles and birds. It will be interesting to learn whether this iLBP type is also expressed in mammals or whether its gene has been silenced. However, homology searches against the mammalian division of the database of expressed sequence tags, or the draft version of the human genome sequence, did not reveal any matches to L-$FABP_2$.

Evolutionary requirement for iLBPs

The ancestral iLBP gene may have encoded a universal hydrophobic ligand binding protein, which acquired specialized binding properties (such as retinoid binding) during evolution. Because most present-day iLBPs are able to bind LCFAs, the ancestral iLBP protein was likely to bind these compounds as well. Homology searches using either reconstructed ancestral iLBP sequences (data not shown) or present-day iLBP sequences, indicate that iLBPs are not present in the yeast (i.e. *Saccharomyces cerevisiae*) or plant (i.e. *Arabidopsis thaliana*) genome. Note that a 15-kDa protein with fatty acid-binding properties but apparently unrelated to vertebrate iLBPs, has been reported for the yeast *Yarrowia lipolytica* [41]. Expression of iLBPs seems thus confined to members of the animal kingdom suggesting that the ancestral iLBP gene arose after divergence of animals from fungi and plants (around 1200–1000 mya).

Most organisms are capable of catabolizing LCFAs to some

extent. For example, the yeast *Saccharomyces cerevisiae* is able to grow on LCFAs as sole carbon source, while during germination fatty acid oxidation occurs in fat-storing plants seeds [42]. In prokaryotes, LCFAs are oxidized in the cytoplasm and ensuing oxidative phosphorylation occurs at the inner cell membrane. In yeast and plants, oxidation of LCFAs is confined to peroxisomes [43], while oxidative phosphorylation occurs in the mitochondria. Peroxisomal supply of LCFAs in the form of acyl-CoA esters is probably accomplished by acyl-CoA binding protein [44]. In animals, LCFA oxidation and oxidative phosphorylation occur in the same compartment, i.e. the mitochondrion, a situation reminiscent to prokaryotes.

During early eukaryotic evolution, mitochondria were acquired as endosymbionts [45, 46]. For LCFAs to reach these newly-gained organelles, the hydrophobic molecules needed to cross the cytoplasmic fluid in an efficient manner. The emergence of an iLBP in higher eukaryotes may have been the evolutionary solution to this problem. Mitochondrial oxidation of LCFAs has the advantage of directly linking the β-oxidation and oxidative phosphorylation pathways, and is more efficient in terms of energy yield than peroxisomal fatty acid oxidation. Both iLBP expression and mitochondrial β-oxidation are apparently restricted to higher eukaryotes. This possible correlation is supported by the observation that certain marine invertebrates (echinoderms, molluscs and crustaceans) seem to lack iLBPs. Although these animals possess functional mitochondria, they appear to oxidize LCFAs in peroxisomes [47]. In these groups of species, mitochondria may be involved in oxidation of chain-shortened fatty acids. From our recent studies on LCFA utilization by H-FABP$^{-/-}$ cardiac myocytes it appears that H-FABP is facilitative – but not obligatory – for mitochondrial oxidation of LCFAs [48]. In the absence of H-FABP, LCFAs are probably oxidized in mitochondria located within short diffusion distance of the sarcolemma. Hence, H-FABP can be seen as an evolutionary adaptation to achieve high rates of LCFA utilization in cells where the diffusion distance between the site of cellular uptake and the site of metabolic conversion is relatively large.

In conclusion, a scheme for the evolution of the vertebrate iLBP family is presented. An ancestral iLBP gene probably arose after divergence of animals from fungi and plants. The first iLBP gene duplication was estimated to have occurred around 930 millions of years ago. Subsequent duplications gave rise to the 16 distinct iLBP types currently known in vertebrates. The appearance of an iLBP during evolution of higher eukaryotes may have substantially enhanced the capacity of mitochondria and other organelles to metabolize LCFAs by facilitating transport of these hydrophobic substances from the cell membrane to the intracellular site of metabolic conversion.

Appendix

The following iLBP sequences were employed in this study. Accession numbers refer to entries in the SwissProt (italicized) or GenBank databases. Species abbreviations are as follows: Am (*Ambystoma mexicanum*, axolotl), Ap (*Anolis pulchellus*, lizard), Bt (*Bos taurus*, cow), Gg (*Gallus gallus*, chicken), Hb (*Halaelurus bivius*, catshark), Hs (*Homo sapiens*, man), Lp (*Lepidosiren paradoxa*, lungfish), Mm (*Mus musculus*, mouse), Oc (*Oryctolagus cuniculus*, rabbit), Om (*Oncorhynchus mykiss*, trout), Rn (*Rattus norvegicus*, rat), Rs (*Rhamdia sapo*, catfish), Ss (*Sus scrofa*, pig), St (*Spermophilus tridecemlineatus*, squirrel), Tr (*Takifugu rubripes*, pufferfish), Xl (*Xenopus laevis*, African clawed frog)

H-FABP: Bt (X12710), Hs (X56549), Mm (U02883), Om (U95296), Rn (J02773), Ss (X98558), St (AF327854), **B-FABP:** Dr (AF237712), Gg (X65459), Hs (AJ002962), Mm (S69799), Rn (U02096), **ALBP:** Bt (X89244), Hs (J02874), Mm (K02109), Rn (U75581), Ss (Y16039), St (AF327855), **MLBP:** Bt (*P02690*), Hs (X62167), Oc (J03744), Mm (S39508), **TLBP:** Mm (AK005745), Rn (U09022), **E-FABP:** Bt (U55188), Hs (M94856), Mm (X70100), Rn (S83247), **CRBP$_1$:** Bt (*P02694*), Hs (M11433), Mm (X60367), Rn (M16459), **CRBP$_2$:** Hs (U13831), Mm (X74154), Rn (M13949), Ss (X77639), **CRBP$_3$:** Hs (NM_031491), **CRBP$_4$:** Hs (AF399927), Mm (NM_022020), **CRABP$_1$:** Bt (X07436), Hs(S74445), Mm (NM_013496), Rn (reference 49), Tr (Y12240), **CRABP$_2$:** Hs (M68867), Mm (M87538), Rn (U23407), Xl (S74933), **I-FABP:** Dr (AF180921), Hs (M18079), Mm (M65034), Rn (M35992), Xl (L19946), **ILBP:** Hs (U19869), Mm (U00938), Oc (Z54345), Rn (S52878), Ss (J05136), **L-FABP$_1$:** Am *(P81399)*, Bt (X86904), Gg (AF380999), Hs (M10617), Mm (Y14660), Rn (M35991), Ss (X77640), **L-FABP$_2$:** Am (*P81400*), Ap (U28756), Dr (AF254642), Gg (AF380998), Hb *(P81653)*, Lp *(P82289)*, Rs (*P80856*)

References

1. Flower DR: Structural relationship of streptavidin to the calycin protein superfamily. FEBS Lett 333: 99–102, 1993
2. Flower DR, Sansom CE, Beck ME, Attwood TK: The first prokaryotic lipocalins. Trends Biochem Sci 20: 498–499, 1995
3. Flower D: The lipocalin protein family: Structure and function. Biochem J 318: 1–14, 1996
4. Glatz JFC, van der Vusse GJ: Cellular fatty acid-binding proteins: Their function and physiological significance. Prog Lipid Res 35: 243–282, 1996
5. Donovan M, Olofsson B, Gustafson AL, Dencker L, Eriksson U: The cellular retinoic acid binding proteins. J Steroid Biochem Mol Biol 53: 459–465, 1995
6. Spener F, Mukherjea M: Nonenzymatic proteins mediating intracellular lipid transport and metabolism. Current status and emerging trends. Subcell Biochem 16: 1–19, 1990
7. Levi AJ, Gatmaitan Z, Arias IM: Two hepatic cytoplasmic protein fractions, Y and Z, and their possible role in the hepatic uptake of bilirubin, sulfobromophthalein, and other anions. J Clin Invest 48: 2156–2167, 1969
8. Ockner RK, Manning JA, Poppenhausen RB, Ho WK: A binding protein for fatty acids in cytosol of intestinal mucosa, liver, myocardium, and other tissues. Science 177: 56–58, 1972
9. Veerkamp JH, Maatman RG: Cytoplasmic fatty acid-binding proteins: Their structure and genes. Prog Lipid Res 34: 17–52, 1995
10. Börchers T, Spener F: Fatty acid binding proteins. Curr Top Membr 40: 261–294, 1994

11. Bernlohr DA, Simpson MA, Hertzel AV, Banaszak LJ: Intracellular lipid-binding proteins and their genes. Annu Rev Nutr 17: 277–303, 1997
12. Di Pietro SM, Veerkamp JH, Santomé JA: Isolation, amino acid sequence determination and binding properties of two fatty-acid-binding proteins from axolotl (*Ambistoma mexicanum*) liver. Eur J Biochem 259: 127–134, 1999
13. Ceciliani F, Monaco HL, Ronchi S, Faotto L, Spadon P: The primary structure of a basic (pI 9.0) fatty acid-binding protein from liver of Gallus domesticus. Comp Biochem Physiol B Biochem Mol Biol 109: 261–271, 1994
14. Banaszak L, Winter N, Xu Z, Bernlohr DA, Cowan S, Jones TA: Lipid-binding proteins: A family of fatty acid and retinoid transport proteins. Adv Protein Chem 45: 89–151, 1994
15. Thompson J, Winter N, Terwey D, Bratt J, Banaszak L: The crystal structure of the liver fatty acid-binding protein. A complex with two bound oleates. J Biol Chem 272: 7140–7150, 1997
16. Londraville RL: Intracellular fatty acid-binding proteins: Putting lower vertebrates in perspective. Braz J Med Biol Res 29: 707–720, 1996
17. Folli C, Calderone V, Ottonello S, Bolchi A, Zanotti G, Stoppini M, Berni R: Identification, retinoid binding, and x-ray analysis of a human retinol-binding protein. Proc Natl Acad Sci USA 98: 3710–3715, 2001
18. Vogel S, Mendelsohn CL, Mertz JR, Piantedosi R, Waldburger C, Gottesman ME, Blaner WS: Characterization of a new member of the fatty acid-binding protein family that binds all-trans-retinol. J Biol Chem 276: 1353–1360, 2001
19. Conforti L, Tarlton A, Mack TG, Mi W, Buckmaster EA, Wagner D, Perry VH, Coleman MP: A Ufd2/D4Cole1e chimeric protein and overexpression of Rbp7 in the slow Wallerian degeneration (WldS) mouse. Proc Natl Acad Sci USA 97: 11377–11382, 2000
20. Prinsen CF, Weghuis DO, Kessel AG, Veerkamp JH: Identification of a human heart FABP pseudogene located on chromosome 13. Gene 193: 245–251, 1997
21. Haunerland NH, Chisholm JM: Fatty acid binding protein in flight muscle of the locust, *Schistocerca gregaria*. Biochim Biophys Acta 1047: 233–238, 1990
22. Haunerland NH: Fatty acid binding protein in locust and mammalian muscle. Comparison of structure, function and regulation. Comp Biochem Physiol B Biochem Mol Biol 109: 199–208, 1994
23. Haunerland NH, Jacobson BL, Wesenberg G, Rayment I, Holden HM: Three-dimensional structure of the muscle fatty-acid-binding protein isolated from the desert locust *Schistocerca gregaria*. Biochemistry 33: 12378–12385, 1994
24. Wu Q, Andolfatto P, Haunerland NH: Cloning and sequence of the gene encoding the muscle fatty acid binding protein from the desert locust, *Schistocerca gregaria*. Insect Biochem Mol Biol 31: 553–562, 2001
25. Schleicher CH, Cordoba OL, Santomé JA, Dell'Angelica EC: Molecular evolution of the multigene family of intracellular lipid-binding proteins. Biochem Mol Biol Int 36: 1117–1125, 1995
26. Goldman N: Phylogenetic estimation. In: M.J. Bishop, C.J. Rawlings (eds). DNA and Protein Sequence Analysis – A Practical Approach. IRL Press, Oxford, 1997, pp 279–312
27. Thompson JD, Higgins DG, Gibson TJ: CLUSTAL W: Improving the sensitivity of progressive multiple sequence alignment through sequence weighting, position-specific gap penalties and weight matrix choice. Nucleic Acids Res 22: 4673–4680, 1994
28. Sander C: The FSSP database: Fold classification based on structure–structure alignment of proteins. Nucleic Acids Res 24: 206–209, 1996
29. Strimmer K, Von Haeseler A: Quartet puzzling: A quartet maximum-likelihood method for reconstructing tree topologies. Mol Biol Evol 13: 964–969, 1996
30. Felsenstein J: PHYLIP. Version 3.57, distributed by the author. University of Washington, Seattle, 1993
31. Comeron JM: A method for estimating the numbers of synonymous and nonsynonymous substitutions per site. J Mol Evol 41: 1152–1159, 1995
32. Kumar S, Hedges SB: A molecular timescale for vertebrate evolution. Nature 392: 917–920, 1998
33. Huelsenbeck JP, Rannala B: Phylogenetic methods come of age: Testing hypotheses in an evolutionary context. Science 276: 227–232, 1997
34. Hasegawa M, Kishino H, Saitou N: On the maximum likelihood method in molecular phylogenetics. J Mol Evol 32: 443–445, 1991
35. Trapp BD, Dubois-Dalcq M, Quarles RH: Ultrastructural localization of P2 protein in actively myelinating rat Schwann cells. J Neurochem 43: 944–948, 1984
36. Oko R, Morales CR: A novel testicular protein, with sequence similarities to a family of lipid binding proteins, is a major component of the rat sperm perinuclear theca. Dev Biol 166: 235–245, 1994
37. Makalowski W, Boguski MS: Evolutionary parameters of the transcribed mammalian genome: An analysis of 2,820 orthologous rodent and human sequences. Proc Natl Acad Sci USA 95: 9407–9412, 1998
38. Ottonello S, Scita G, Mantovani G, Cavazzini D, Rossi GL: Retinol bound to cellular retinol-binding protein is a substrate for cytosolic retinoic acid synthesis. J Biol Chem 268: 27133–27142, 1993
39. Lampron C, Rochette-Egly C, Gorry P, Dollé P, Mark M, Lufkin T, LeMeur M, Chambon P: Mice deficient in cellular retinoic acid binding protein II (CRABPII) or in both CRABPI and CRABPII are essentially normal. Development 121: 539–548, 1995
40. Kleinjan DA, Dekker S, Guy JA, Grosveld FG: Cloning and sequencing of the CRABP-I locus from chicken and pufferfish: Analysis of the promoter regions in transgenic mice. Transgenic Res 7: 85–94, 1998
41. Dell'Angelica EC, Ermacora MR, Santomé JA: Purification and partial characterization of a fatty acid-binding protein from the yeast, *Yarrowia lipolytica*. Biochem Mol Biol Int 39: 439–445, 1996
42. Harwood J: Fatty acid metabolism. Ann Rev Plant Physiol Plant Mol Biol 39: 101–138, 1988
43. Kunau WH, Buhne S, de la Garza M, Kionka C, Mateblowski M, Schultz Borchard U, Thieringer R: Comparative enzymology of beta-oxidation. Biochem Soc Trans 16: 418–420, 1987
44. Knudsen J, Færgeman NJ, Skott H, Hummel R, Borsting C, Rose TM, Andersen JS, Højrup P, Roepstorff P, Kristiansen K: Yeast acyl-CoA-binding protein: Acyl-CoA-binding affinity and effect on intracellular acyl-CoA pool size. Biochem J 302: 479–485, 1994
45. Gray MW: Origin and evolution of organelle genomes. Curr Opin Genet Dev 3: 884–890, 1993
46. Gray MW, Burger G, Lang BF: Mitochondrial evolution. Science 283: 1476–1481, 1999
47. Stewart JM, Carlin RC, MacDonald JA, Van-Iderstine S: Fatty acid binding proteins and fatty acid catabolism in marine invertebrates: Peroxisomal β-oxidation. Invert Reprod Devel 25: 73–82, 1994
48. Schaap FG, Binas B, Danneberg H, van der Vusse GJ, Glatz JFC: Impaired long-chain fatty acid utilization by cardiac myocytes isolated from mice lacking the heart-type fatty acid binding protein gene. Circ Res 85: 329–337, 1999
49. Rajan N, Kidd GL, Talmage DA, Blaner WS, Suhara A, Goodman DS: Cellular retinoic acid-binding protein messenger RNA: Levels in rat tissues and localization in rat testis. J Lipid Res 32: 1195–1204, 1991

Molecular and Cellular Biochemistry **239:** 79–82, 2002.

Intracellular lipid binding proteins of the small intestine

Luis B. Agellon,[1] Matthew J. Toth[2] and Alan B.R. Thomson[3]

[1]*Canadian Institutes of Health Research Group in Molecular and Cell Biology of Lipids and Department of Biochemistry,* [3]*Department of Medicine, University of Alberta, Edmonton, Canada;* [2]*Novartis Institute for Biomedical Research, Summit, NJ, USA*

Abstract

The small intestine contains three distinct proteins belonging to the intracellular lipid binding protein family: the liver-type fatty acid binding protein (L-FABP), the intestinal fatty acid binding protein (I-FABP) and the ileal lipid binding protein (ilbp). The function of these proteins in the small intestine has remained enigmatic. Targeted gene disruption studies may shed insights into the physiological importance of these proteins. In the case of I-FABP, this approach has demonstrated that the complete elimination of this protein in murine intestine does not compromise dietary fat absorption *in vivo* but is associated with the development of insulin resistance. (Mol Cell Biochem **239**: 79–82, 2002)

Key words: fatty acid binding protein, intestinal fat absorption, I-FABP, L-FABP, ILBP

Introduction

The intracellular lipid binding protein family is comprised of small molecular weight proteins that share a similar general structure [1]. These soluble proteins are present in high abundance in a wide variety of tissues and exhibit distinctive preferences for a wide assortment of small hydrophobic molecules. Although these proteins have been studied extensively since their discovery about 30 years ago [2], their exact function and physiologic significance remains poorly understood.

Different nutrients are absorbed along different segments of the small intestine (Fig. 1). The absorption of lipids is facilitated by bile acids, lipases and proteases, which are supplied by bile and pancreatic juices that are released into the proximal region of the small intestine. Fatty acids are mainly absorbed in the proximal half of the small intestine. Bile acids, which are mostly in the conjugated form, are recovered in the distal portion of the small intestine.

Features of intestinal intracellular lipid binding proteins

The small intestine contains three different types of the intracellular lipid binding proteins (Fig. 2). While the distribution of these proteins overlap along the length of the small intestine, each of the different types is maximally represented in different regions. In mice, the mRNA of liver fatty acid binding protein (L-FABP) is found in the proximal two-thirds of the small intestine. The intestinal fatty acid binding protein (I-FABP) mRNA is detectable along the entire length of the small intestine and is maximally represented near the medial segment. The ileal lipid binding protein (ilbp) mRNA is present only in the distal one-third of the small intestine.

The L-FABP is also found in the liver, and the crystal structure of this subtype has been determined both in its apo-form as well as in a complex with oleic acid [3, 4]. The expression of the L-FABP gene appears to be regulated by some of its ligands. In rats, fat feeding stimulated L-FABP gene expression [5]. Moreover, it was recently demonstrated that L-FABP expression in mouse intestine could be stimulated in the distal portion of the small intestine by direct application

Address for offprints: L.B. Agellon, 328 Heritage Medical Research Centre, University of Alberta, Edmonton, AB T6G 2S2, Canada (E-mail: luis.agellon@ualberta.ca)

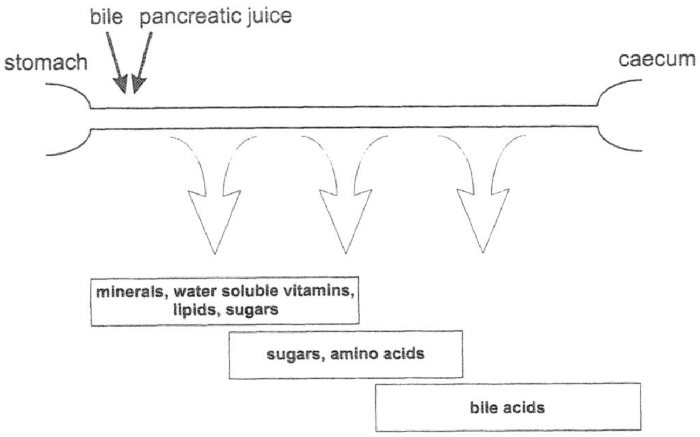

Fig. 1. Absorption of dietary nutrients along the length of the small intestine.

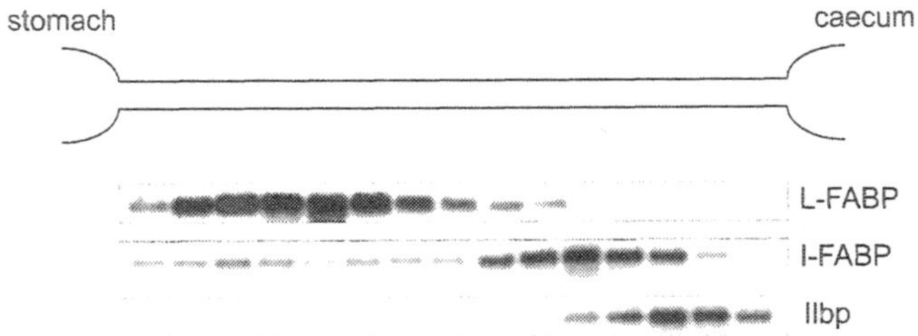

Fig. 2. The intracellular lipid binding proteins of the murine small intestine. The small intestine of a female C57BL/6J mouse was divided into ~15 equal segments from the base of the stomach to the caecum and analyzed by RNA blotting. The figure shows the distribution of mRNA encoding the liver-fatty acid binding protein (L-FABP), the intestinal-fatty acid binding protein (I-FABP) and the ileal lipid binding protein (ilbp).

of PPAR agonists [6]. Thus, the extinction of L-FABP gene expression towards the distal portion of the small intestine may be due to the decreasing concentrations of specific ligands.

In contrast to the L-FABP gene, the I-FABP gene is expressed uniquely in the small intestine. This subtype of the intracellular lipid binding protein is reported to exhibit a preference for saturated fatty acids [7, 8]. The structure of I-FABP is also known, and has been determined both by x-ray crystallography [9] and nuclear magnetic resonance spectroscopy [10]. Unlike L-FABP, the expression of the I-FABP gene is apparently not explained by the availability of its ligand. A change in I-FABP mRNA abundance was not observed after fat feeding of male Swiss mice [5]. One study however demonstrated that feeding male NZB mice with the PPARα agonist WY14643 increased I-FABP mRNA abundance in the small intestine [11]. The discrepancy may be attributable to differences in the genetic composition of the two strains.

The ilbp is found in the distal third of the small intestine, and it is generally thought that ilbp gene expression is restricted to this portion of the organ. Characterization of ilbp ligand binding demonstrated that this protein can bind both bile acids and fatty acids, but bile acids appear to be the preferred ligand [12–14]. There is controversy on the ability of bile acids to regulate ilbp gene expression. Ilbp mRNA abundance is not altered by TCA feeding in female C57BL/6J mice [15], but is increased in male Swiss mice [16]. It was recently discovered that bile acids regulate the transcription of target genes via the nuclear receptor known as farnesoid X receptor (FXR) [17, 18], and ilbp gene expression is lost when mice are made deficient in FXR [19]. In mice possessing the normal ilbp allele, the induction of the ilbp gene expression in response to cholic acid feeding was also demonstrated, but the genetic background of the mice was not controlled in that study. It remains possible that the genetic background of the specific strains of mice is important in determining the responsiveness of the ILBP gene to regulation by bile acids via FXR.

Effect of I-FABP deficiency

In their initial description of the intracellular lipid binding proteins, Ockner *et al.* [2] suggested these proteins may account for differences in the absorption of fatty acids in the small intestines. The *Fabpi* gene represented a candidate for targeted gene disruption because it is expressed along the entire length of the small intestine. Since the *Fabpi* gene is expressed solely in this organ, the elimination of I-FABP was not expected to have a primary effect on the function of other organs.

Mice that are deficient in I-FABP are viable and fertile [20]. The difference in body weight gain is the most overt distinguishing feature between I-FABP-deficient mice and their normal littermates. Male I-FABP-deficient mice exhibit a higher gain in body weight as compared to normal male littermates and this effect appears to be independent of the fat content of the diet. In females, the I-FABP-deficient mice gain less weight than normal littermates but this was only observed when the mice were on a high fat diet. Therefore, there appears to be a gender disparity in the consequence of I-FABP deficiency. Nevertheless, these differences in body weights were correlated with the plasma triacylglycerol concentrations. A consequence of I-FABP-deficiency that is common to both genders is the increased concentration of insulin in the plasma. Despite the hyperinsulinemia, the plasma glucose concentrations were comparable between normal and I-FABP-deficient mice.

On the high fat diet, the livers of male I-FABP-deficient mice became significantly larger than that of normal male mice and this increase in liver mass did not appear to be due to increased lipid content. In contrast, the livers of female I-FABP-deficient mice were slightly smaller compared to that of normal female mice on both the low and high fat diets. The observed difference between the male and female responses may be related to the abundance of hepatic L-FABP. The liver of female rats was previously shown to contain ~2-fold greater abundance of L-FABP than liver of male rats [21]. It has been suggested that this may account for the difference between the sexes in the way fats are metabolized in this organ [22].

The *Fabpi* gene-disrupted mice clearly demonstrate that I-FABP is not specifically required for the assimilation of dietary fat *in vivo*. L-FABP and ilbp are not expressed throughout the entire length of the small intestine, but these proteins can be found in intestinal segments where I-FABP is represented. In addition, it has been shown that ilbp is capable of binding fatty acids [12, 13]. It is possible therefore that the expression of multiple types of intracellular lipid binding proteins in the small intestine represents a strategy for functional redundancy to ensure the absorption of fatty acids in the small intestine.

It is thought that I-FABP transports fatty acids inside the cell via a collisional mediated process in which the acquisition and release of the fatty acid ligand involves direct interaction of the protein with the donor and target membranes [23]. L-FABP, in contrast to I-FABP, does not require interaction with membranes and can facilitate the aqueous diffusion of fatty acids in the cell [24]. This might explain why there is no apparent gross impairment in the assimilation of dietary fat in I-FABP deficient mice. Nevertheless, in the absence of I-FABP, the delivery of fatty acids to specific subcellular compartments may be impaired (Fig. 3). This in turn may lead to the development of pathology in the intact animal. The exact contribution of L-FABP and ilbp in the transcellular transport of fatty acids may become more evident when mice lacking these proteins are intercrossed with the I-FABP-deficient mice.

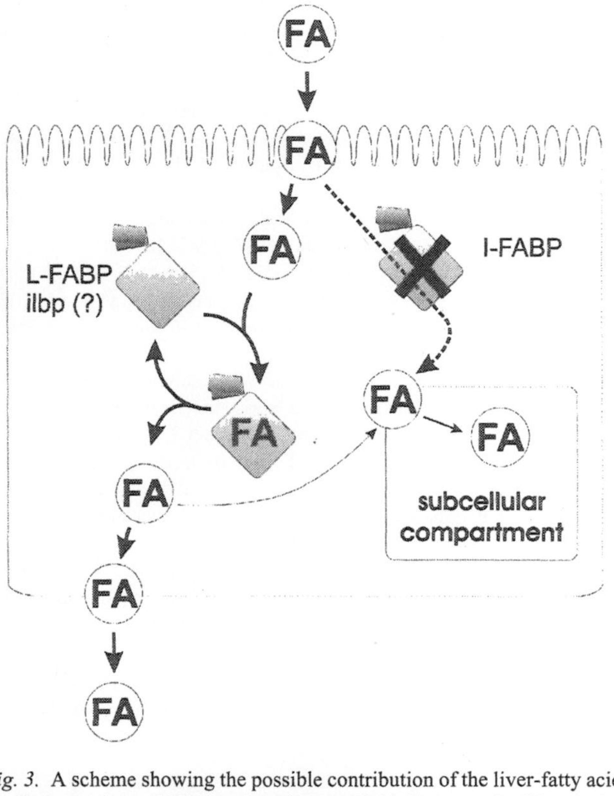

Fig. 3. A scheme showing the possible contribution of the liver-fatty acid binding protein (L-FABP) and the ileal lipid binding protein (ilbp) in the transcellular transport of dietary fatty acids in the intestinal cells of mice deficient in the intestinal-lipid binding protein (I-FABP). The dashed line depicts the less efficient delivery of fatty acids to specific subcellular compartments in the absence of I-FABP.

Significance of I-FABP in the development of pathologic states

Studies have demonstrated an association between the polymorphism at residue 54 of the human I-FABP and a high prevalence of insulin resistance or postprandial lipemia in a number of human populations [25–27]. The precise significance of the polymorphism is not clear and in some populations individuals that possess the A54T allele do not appear to be affected [28]. Increased weight gain and hyperinsulinemia are two clinical features that are common to both human subjects with the A54T polymorphism and male mice deficient in I-FABP. The observed similarity seems paradoxical because it has been reported that the A54T mutation polymorphism increases the affinity of I-FABP to its fatty acid ligand [25], whereas I-FABP is lacking in *Fabpi* gene-disrupted mice. It is possible that the A54T polymorphism interferes with the metabolism of fatty acids in a manner that is not related to its affinity for fatty acids. Non-esterified fatty acid concentration in the plasma of human subjects with A54T polymorphism tended to be higher than in normal subjects after a low fat meal, but is increased significantly after a high fat meal [29]. Additionally, intraportal infusion of a long chain fatty acid in rats has been shown to cause insulin resistance [30]. It is possible that increased intraportal flux of fatty acid into the liver underlies the development of features that are common to both to I-FABP-deficient mice and subjects homozygous to the A54T polymorphism.

Summary

It is now clear that I-FABP-deficiency does not decrease the capacity of mice to absorb of dietary fat. On the contrary, the loss of I-FABP appears to increase the assimilation of dietary fats. The *Fabpi* gene-disrupted mouse has also uncovered the importance of I-FABP in systemic energy homeostasis. The elimination in intact mice of L-FABP or ilbp, by themselves or in combination with the deficiency of other intestinal lipid binding proteins, will be useful in understanding the importance of each of these proteins in the metabolism of lipids by the small intestine.

Acknowledgments

This work was funded by grants from the Canadian Institutes of Health Research and Novartis Pharmaceuticals Canada. LBA is a Senior Scholar of the Alberta Heritage Foundation for Medical Research.

References

1. Thompson J, Ory J, Reese-Wagoner A, Banaszak L: The liver fatty acid binding protein – comparison of cavity properties of intracellular lipid-binding proteins. Mol Cell Biochem 192: 9–16, 1999
2. Ockner RK, Manning JA, Poppenhausen RB, Ho WKL: A binding protein for fatty acids in cytosol of intestinal mucosa, liver, myocardium, and other tissues. Science 177: 56–58, 1972
3. Thompson J, Winter N, Terwey D, Bratt J, Banaszak L: The crystal structure of the liver fatty acid-binding protein. A complex with two bound oleates. J Biol Chem 272: 7140–7150, 1997
4. Thompson J, Reese-Wagoner A, Banaszak L: Liver fatty acid binding protein: species variation and the accommodation of different ligands. Biochim Biophys Acta 1441: 117–130, 1999
5. Poirier H, Niot I, Degrase P, Monnot MC, Bernard A, Besnard P: Fatty acid regulation of fatty acid-binding protein expression in the small intestine. Am J Physiol 273: G289–G295, 1997
6. Poirier H, Niot I, Monnot MC, Braissant O, Meunier-Durmont C, Costet P, Pineau T, Wahli W, Willson TM, Besnard P: Differential involvement of peroxisome-proliferator-activated receptors α and δ in fibrate and fatty-acid-mediated inductions of the gene encoding liver fatty-acid-binding protein in the liver and the small intestine. Biochem J 355: 481–488, 2001
7. Lowe JB, Sacchettini JC, Laposata M, McQuillan JJ, Gordon JI: Expression of rat intestinal fatty acid-binding protein in Escherichia coli. Purification and comparison of ligand binding characteristics with that of *Escherichia coli*-derived rat liver fatty acid-binding protein. J Biol Chem 262: 5931–5937, 1987
8. Richieri GV, Ogata RT, Zimmerman AW, Veerkamp JH, Kleinfeld AM: Fatty acid binding proteins from different tissues show distinct patterns of fatty acid interactions. Biochemistry 39: 7179–7204, 2000
9. Sacchettini JC, Gordon JI, Banaszak LJ: Crystal structure of rat intestinal fatty-acid-binding protein. Refinement and analysis of the *Escherichia coli*-derived protein with bound palmitate. J Mol Biol 208: 327–339, 1989
10. Zhang F, Lucke C, Baier L, Sacchettini JC, Hamilton JA: Solution structure of human intestinal fatty acid binding protein: Implication for ligand entry and exit. J Biomol NMR 9: 213–228, 1997
11. Motojima K: Differential effects of PPARα activators on induction of ectopic expression of tissue-specific fatty acid binding protein genes in the mouse liver. Int J Biochem Cell Biol 32: 1085–1092, 2000
12. Sacchettini JC, Hauft SM, Van Camp SL, Cistola DP, Gordon JI: Developmental and structural studies of an intracellular lipid binding protein expressed in the ileal epithelium. J Biol Chem 265: 19199–19207, 1990
13. Gong YZ, Everett ET, Schwartz DA, Norris JS, Wilson FA: Molecular cloning, tissue distribution, and expression of a 14-kDa bile acid-binding protein from rat ileal cytosol. Proc Natl Acad Sci USA 91: 4741–4745, 1994
14. Fujita M, Fujii H, Kanda T, Sato E, Hatakeyama K, Ono T: Molecular cloning, expression, and characterization of a human intestinal 15-kDa protein. Eur J Biochem 233: 406–413, 1995
15. Torchia EC, Cheema SK, Agellon LB: Coordinate regulation of bile acid biosynthetic and recovery pathways. Biochem Biophys Res Commun 255: 128–133, 1996
16. Grober J, Zaghini E, Fujii H, Jones SA, Kliewer SA, Willson TM, Ono T, Besnard P: Identification of a bile acid-responsive element in the human ileal bile acid-binding protein gene. J Biol Chem 274: 29749–29754, 1999
17. Makishima M, Okamoto AY, Repa JJ, Tu H, Learned RM, Luk A, Hull MV, Lustig KD, Mangelsdorf DJ, Shan B: Identification of a nuclear receptor for bile acids. Science 284: 1362–1365, 1999
18. Wang H, Chen J, Hollister K, Sowers LC, Forman BM: Endogenous bile acids are ligands for the nuclear receptor FXR/BAR. Mol Cell 3: 543–553, 1999
19. Sinal CJ, Tohkin M, Miyata M, Ward JM, Lambert G, Gonzales FJ: Targeted disruption of the nuclear receptor FXR/BAR impairs bile acid and lipid homeostasis. Cell 102: 731–744, 2000
20. Vassileva G, Huwyler L, Poirier K, Agellon LB, Toth MJ: The intestinal fatty acid binding protein is not essential for dietary fat absorption in mice. FASEB J 14: 2040–2046, 2000
21. Bass NM, Manning JA, Ockner RK, Gordon JI, Seetharam S, Alpers DH: Regulation of the biosynthesis of two distinct fatty acid-binding proteins in rat liver and intestine. Influences of sex difference and of clofibrate. J Biol Chem 260: 1432–1436, 1985
22. Luxon BA, Weisiger RA: Sex differences in intracellular fatty acid transport: Role of cytoplasmic binding proteins. Am J Physiol 265: G831–G841, 1993
23. Thumser AE, Strorch J: Liver and intestinal fatty acid-binding proteins obtain fatty acids from phospholipid membranes by different mechanisms. J Lipid Res 41: 647–656, 2000
24. Storch J, Thumser AEA: The fatty acid transport function of fatty acid-binding proteins. Biochim Biophys Acta 1486: 28–44, 2000
25. Baier LJ, Sacchettini JC, Knowler WC, Eads J, Paolisso G, Tataranni PA, Mochizuki H, Bennett PH, Bogardus C, Prochazka M: An amino acid substitution in the human intestinal fatty acid binding protein is associated with increased fatty acid binding, increased fat oxidation, and insulin resistance. J Clin Invest 95: 1281–1287, 1995
26. Agren JJ, Valve R, Vidgren H, Laakso M, Uusitupa M: Postprandial lipemic response is modified by the polymorphism at codon 54 of the fatty acid-binding protein 2 gene. Arterioscler Thromb Vasc Biol 18: 1606–1610, 1998
27. Chiu KC, Chuang LM, Yoon C: The A54T polymorphism at the intestinal fatty acid binding protein 2 is associated with insulin resistance in glucose tolerant Caucasions. BMC Genet 2: 7, 2001
28. Hayakawa T, Nagai Y, Nohara E, Yamashita H, Takamura T, Abe T, Nomura G, Kobayashi K: Variation of the fatty acid binding protein 2 gene is not associated with obesity and insulin resistance in Japanese subjects. Metabolism 48: 655–657, 1999
29. Pratley RE, Baier L, Pan DA, Salbe AD, Storlien L, Ravussin E, Bogardus C: Effects of an Ala54Thr polymorphism in the intestinal fatty acid-binding protein on the responses to dietary fat in humans. J Lipid Res 41: 2002–2008, 2002
30. Benthem L, Keizer K, Wiegman CH, de Boer SF, Strubbe JH, Steffens AB, Kuipers F, Scheurink AJ: Excess portal venous long-chain fatty acids induce syndrome X via HPA axis and sympathetic activation. Am J Physiol 279: E1286–E1293, 2000

Molecular and Cellular Biochemistry **239:** 83–86, 2002.

Analysis on the phenotype of E-FABP-gene knockout mice

Yuji Owada,[1] Ichiro Suzuki,[1] Tetsuo Noda[2] and Hisatake Kondo[1]

Divisions of [1]Histology and [2]Molecular Genetics, Department of Cell Biology, Graduate School of Medicine, Tohoku University, Sendai, Japan

Abstract

The fatty acids are shown to be critical in the maintenance of the water permeability barrier that is ascribed to the lipids in the intracellular milieu of the cornified cell layer in the epidermis. In view of this importance in the skin, we examined the phenotype of epidermal fatty acid binding protein (E-FABP)-deficient mice. In spite of total lack of E-FABP expression in the various tissues of E-FABP deficient mice, these animals appeared normal in gross and histological examination. In Northern blot analysis for other FABPs, the gene expression of heart (H-)-type FABP is specifically elevated in the liver of neonatal heterozygous and homozygous mice, suggesting the functional compensation of H-FABP for E-FABP deficiency during their development. In functional analyses of the skin, the basal transepidermal water loss (TEWL) of the adult homozygous mice showed lower levels compared with the wild-type mice, and the impairment of recovery in TEWL was observed in the homozygous mice when the lipid barrier of the skin was disrupted by acetone. These results demonstrate that E-FABP is responsible for the water permeability barrier of the skin, although the molecular mechanism remains to be further elucidated. (Mol Cell Biochem **239**: 83–86, 2002)

Key words: epidermal FABP, gene knock out, epidermal water barrier

Introduction

Fatty acid binding proteins (FABPs) are soluble non-enzymatic proteins of low molecular mass (about 15 kDa) that are capable of binding long chain fatty acids. Multiple members of the genetically related family have been identified in the rodent, each of which has been originally named according to the tissue of its first isolation [1]. Among them, epidermal FABP (E-FABP) is most widely expressed throughout the body tissues, not only in the epidermis but also in the mammary gland, brain, liver, kidney, adipose tissue and some endothelial cells [2–6].

Because the members of FABP family show relatively similar ligand binding properties *in vitro* [7] and heterogeneous expression patterns in animal tissues, their individual physiological functions are rather difficult to define. This feature, together with most wide occurrence throughout the body tissues, provides a rationale to generate mice with a disruption of the E-FABP gene and analyze its phenotype with primary attention to the epidermal function such as the water barrier.

Materials and methods

The details of the gene targeting technology were described elsewhere [8]. Briefly, embryonic stem cells were electroporated with the linearized targeting vector, which consisted of DT-A, a 1.5 Kb 5′-homologous region, nuclear LacZ-PA, loxP-neo cassette and 9.1 Kb 3′-homologous region. Correctly targeted ES clones were injected into the blastocysts from C57BL/6 mice, which were then transferred into the uteri of pseudo-pregnant ICR recipients. Male chimeras with extensive ES cell contributions to their coats were bred with C57BL/6 female mice. Tail DNA from agouti F1 offspring was genotyped by Southern blotting analysis. F1 heterozygous females and males were interbred and the littermates were genotyped.

The details of Northern and Western blottings were described elsewhere [3, 4, 8]. For Northern blotting, 30 ug of total RNAs from brain, skin and liver of each genotype mouse were extracted and loaded in on the 1% agarose gel. The radiolabeled specific oligonucleotide probes complementary

Address for offprints: H. Kondo, Division of Histology, Graduate School of Medicine, Tohoku University, 2-1 Seiryo-machi, Aoba-ku, Sendai 981, Japan to murine E-, H-, B-, A-, L- and I-FABP cDNAs were used. For Western blotting, polyclonal anti-rat E-FABP antibody was used (a kind gift of Dr. F. Spener, Muenster, Germany).

For β-galactosidase staining, adult E-FABP knockout mice were perfused transcardially with 1% paraformaldehyde/PBS. After preparation of frozen sections (10 um thickness) from the mice, the sections were stained by incubation in solution containing 20 mM $K_3Fe(CN)_6$, 20 mM $K_4Fe(CN)_6$, 1 mM $MgCl_2$ and 1 mg/ml 5-bromo-4-chloro-3 indolyl-β-galactosidase (X-gal).

The assessment of transepidermal water loss (TEWL) was performed in the shaved back skins using an evaporimeter (ServoMed, Stockholm, Sweden). The fatty acid composition in the dorsal skin was analyzed by way of the standard extraction procedure and application to gas chromatography (Hewlett-Packard Model 5890 gas chromatograph, Avondal, PA, USA).

The electron microscopic examination of skins was performed according to the conventional procedure employing 4% paraformaldehyde/5% glutaraldehyde in 0.1 M phosphate buffer (pH 7.4) and 1% buffered osmium tetroxide, and *en bloc* staining with 2% uranyl acetate.

Results

Neither macroscopic nor microscopic abnormalities were detected in E-FABP deficient mice: they were viable and exhibited a normal sex ratio, weight gain, and fertility. The histological organization of skin, brain, kidney, lung and alimentary tract, in which E-FABP expression had originally been observed, appeared normal (data not shown).

In Northern and Western blot analyses, the absence of E-FABP mRNA and protein, respectively, was confirmed in the skin, brain and liver of E-FABP deficient mice on P0 and P7. The expression level of E-FABP mRNA and protein are decreased to 50% in those tissues of the heterozygous mice (Fig. 1).

When the expression of mRNAs for H-, B-, A-, L- and I-FABP was examined in the neonatal brain, skin and liver of all three genotype-groups, no distinct changes in the expression of B-, A-, L- and I-FABP were observed among the three groups. However, the expression of H-FABP mRNA was markedly increased in the liver of heterozygous and homozygous mice at postnatal day 7 (P7), in contrast to undetectable levels in the wild type mouse liver (Fig. 1B). Such elevation in expression of H-FABP mRNA was also observed in the liver on P0 (data not shown). At present, we examine whether or not this increase in mRNA expression is accompanied by its enhanced expression at protein levels and which cells are responsible for this enhanced expression. To examine whether or not E-FABP-deficient cells survive and develop normally in the gene knockout mice, enzyme histochemical analysis for β-galactosidase was performed in the

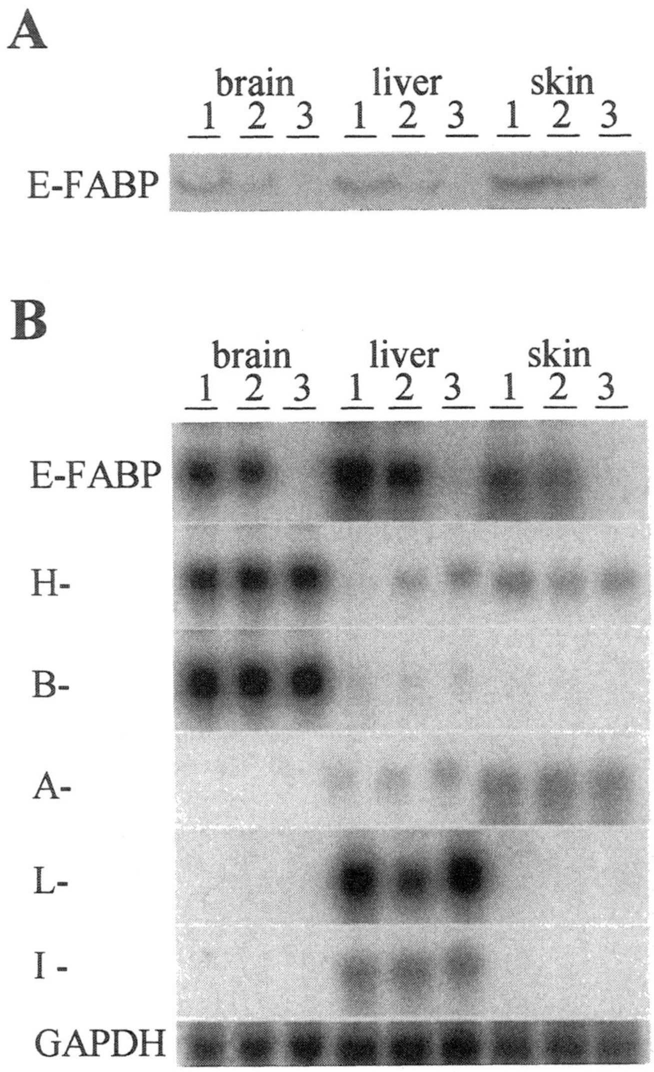

Fig. 1. Western blot (A) and Northern blot (B) analyses of FABPs (E-, H-, B-, A-, L- and I-FABP) in the brain, skin and liver of P7 mice having different genotypes (1: wild, 2: heterozygous, 3: homozygous). Note the total absence of E-FABP protein in the three tissues from the homozygous mice and also note the increased expression of H-FABP in the liver of the heterozygous and homozygous mice. As the internal control, the expression of GAPDH mRNA is shown.

brain of the homozygous mice. The nuclear β-galactosidase activity was observed in many neurons of the cerebral cortex and the hippocampal pyramidal layer (Fig. 2A), and the cerebellar Bergman's glial cells (Fig. 2B). The expression pattern of β-galactosidase-positive cells were consistent with our previous result on the mRNA-expressing cells detected by *in situ* hybridization histochemistry [3].

No obvious differences were seen in the epidermal content of the major C14, C16, C18, C20, and C22 fatty acids between the homozygous mice and the wild counterparts as judged by gas chromatographic analysis (Table 1).

Chronological changes of TEWL following the water bar-

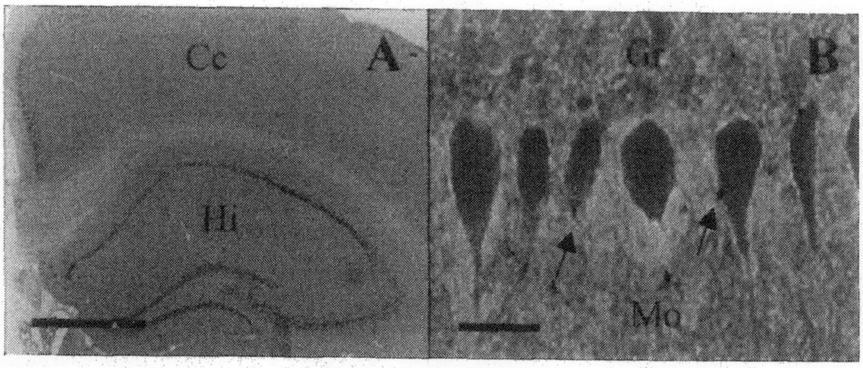

Fig. 2. β-galactosidase histochemistry in the coronal section through hippocampus (A) and cerebellum (B) in the homozygous mouse. Note the positive expression of β-galactosidase activity in the cerebral cortex, hippocampal pyramidal layer and cerebellar Bergman's glial cells (arrows). CC – cerebral cortex; Gr – granule cell layer; Hi – hippocampus; Mo – molecular layer. Bar in A – 1 mm, in B – 20 um.

Table 1. Percent of constituent fatty acids in the skin

	Fatty acid						
	14:0	16:0	18:0	18:1	18:2	20:4	22:6
wild (wt %)	1.9 ±0.15	29.2 ±0.17	4.3 ±0.47	29.7 ±0.71	26.3 ±1.16	0.7 ±0.08	0.5 ±0.06
homozygous (wt %)	1.8 ±0.33	27.9 ±1.14	3.4 ±0.14	29.6 ±0.03	28.2 ±1.91	0.8 ±0.21	0.6 ±0.20

Fatty acid – number of carbone atoms:number of double bonds; wt % – mean values ± S.D. (n = 3).

rier disruption by acetone are shown in Fig. 3. In the pretreatment state, the value of TEWL of the homozygous mice was significantly lower than (almost half of) that of wild-type mice. In the wild-type mice on the other hand, a progressive increase of TEWL was observed during the first 30 min after acetone treatment, and TEWL reached and remained at maximum (about 180% of the pre-treatment level) during the following 3.5 h. Thereafter, TEWL decreased progressively and it reached down to 125% of the pre-treatment level at 36 h and eventually returned almost to the pre-treatment level at 48 h. In the homozygous mice, on the other hand, TEWL increased from the lower basal level, and it reached the maximum level similar to that of the wild type mice after the acetone treatment. Thereafter, TEWL decreased in a time course similar to the wild one and eventually reached the lower basal level at 96 h after the acetone treatment. TEWL was not affected in both the homozygous and wild mice when treated with saline (data not shown). No clear differences in electron microscopy of appearance of the lamellar bodies and their exocytotic features after the acetone treatment between the homozygous mouse skin and the wild one during the time course after the acetone treatment (data not shown), the latter of which were in accord with a previous report [9]: In short, the paucity of the lamellar bodies in the keratinocytes shortly after the acetone

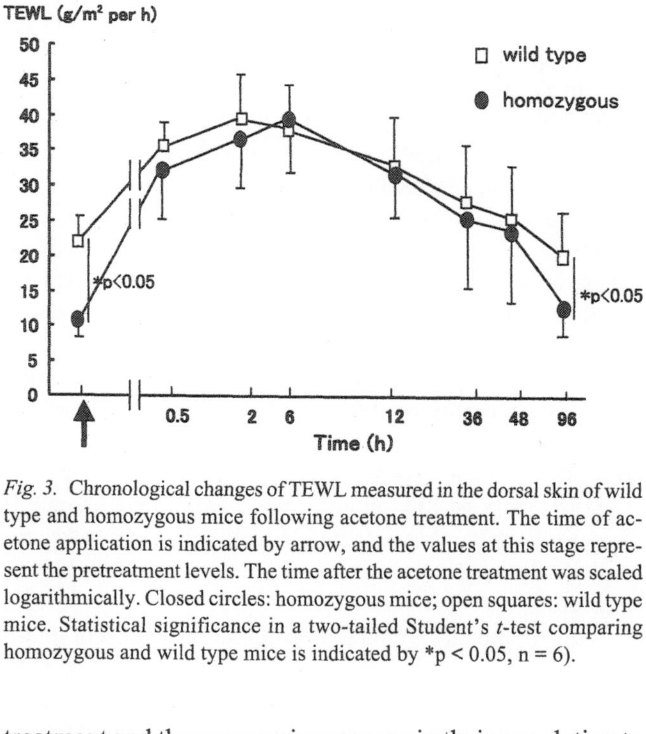

Fig. 3. Chronological changes of TEWL measured in the dorsal skin of wild type and homozygous mice following acetone treatment. The time of acetone application is indicated by arrow, and the values at this stage represent the pretreatment levels. The time after the acetone treatment was scaled logarithmically. Closed circles: homozygous mice; open squares: wild type mice. Statistical significance in a two-tailed Student's *t*-test comparing homozygous and wild type mice is indicated by $*p < 0.05$, n = 6).

treatment and the progressive recover in their population to the pretreatment condition by 24 h after the treatment. No significant differences were noticed among the three genotypic mice in electron microscopic assessment in terms of the general ultrastructure of the keratinocytes including lamellar bodies, which had been well documented in the literatures [9, 10] (data not shown).

Discussion

The superficial lack of obvious phenotypes other than the changes in TEWL in the homozygous mice raises the possibility of compensation by other FABP(s) for E-FABP deficiency during the development. Indeed, among five other

FABPs examined in this study, the expression level of H-FABP mRNA in the liver was markedly elevated in the homozygous mutant on P0 and P7. Considering the fact that H-FABP is one of the most structurally related proteins to E-FABP in the FABP family and its binding property to fatty acids is very similar between the two FABPs [11], it is highly possible that, in the present homozygous mice, H-FABP plays compensatory roles in the tissues in which E-FABP is originally expressed during development. In previous studies of gene-knockout mice for adipocyte FABP (A-FABP), a compensatory increase in the gene expression for E-FABP has been demonstrated to occur in the adipose cells, resulting in normal composition of fatty acids, uptake of fatty acids and morphology of the adipose tissue [12, 13]. In contrast, H-FABP-knockout mice have been shown by another studies to be lacking in such a redundancy in the gene expression for any other FABPs, resulting in impairment of cardiac fatty acid utilization [14, 15].

Three distinct features in the present chronological changes of TEWL after acetone treatment represents the sole and clear phenotype of E-FABP gene knockout mice: the first feature is the lower basal level of TEWL in the mutant skin than the wild before acetone-treatment. The second is a rapid catching-up of TEWL in the mutant skin to the maximally increased level in the wild skin within 1 h after acetone treatment, which is indicative of a more abrupt barrier damage than in the wild. The last one is a much longer recovery time to reach the lower basal level of TEWL in the mutant than that in the wild mice. However, the present study disclosed no clear differences between the mutant and the wild in terms of the percent composition of major fatty acids in the epidermis, and the population density of the lamellar bodies and their exocytotic features in the outermost granule cells, all of which are generally known to be intimately involved in the epidermal water barrier [9].

It should be also noted that, similar to the present mutant skin, the water barrier of aged skin is not impaired or rather enhanced under the undisturbed condition, while the recovery of the aged water barrier is impaired after the treatment by acetone or by tape-stripping [16–18]. Although the basic mechanisms underlying the aged skin and the present mutant mouse skin are different, the similarity in the water barrier function between the two cases may lead us to further elucidation.

Acknowledgements

This work was supported by grants from Ministry of Education, Science and Culture of Japan, No 11470001 and 11694235 (to H.K.) and a grant from the Cosmetology Research Foundation, Japan (to Y.O.)

References

1. Glatz JFC, van der Vusse GJ: Cellular fatty acid-binding proteins: Their function and physiological significance. Prog Lipid Res 35: 243–282, 1996
2. Siegenthaler G, Hotz R, Chatellard-Gruaz D, Didierjean L, Hellman U, Saurat JH: Purification and characterization of the human epidermal fatty acid-binding protein: Localization during epidermal cell differentiation *in vivo* and *in vitro*. Biochem J 302: 363–371, 1994
3. Owada Y, Yoshimoto T, Kondo H: Spatio-temporally differential expression of genes for three members of fatty acid binding proteins in developing and mature rat brain. J Chem Neuroanat 12: 113–122, 1996
4. Owada Y, Yoshimoto T, Kondo H: Increased expression of the mRNA for brain- and skin-type but not heart-type fatty acid binding proteins following kainic acid systemic administration in the hippocampal glia of adult rats. Mol Brain Res 42: 156–160, 1996
5. Masouye I, Saurat JH, Siegenthaler G: Epidermal fatty-acid-binding protein in psoriasis, basal and squamous cell carcinomas: An immunohistochemical study. Dermatology 192: 208–213, 1996
6. O'Shaughnessy RFL, Seery JP, Celis JE, Frischauf A-M, Watt FM: PA-FABP, a novel marker of human epidermal transit amplifying cells revealed by 2D protein gel electrophoresis and cDNA array hybridization. FEBS Lett 486: 149–154, 2000
7. Coe NR, Bernlohr DA: Physiological properties and functions of intracellular fatty acid-binding proteins. Biochim Biophys Acta 1391: 287–306, 1998
8. Owada Y, Takano H, Yamanaka H, Kobayashi H, Sugitani Y, Tomioka Y, Suzuki I, Suzuki R, Terui T, Mizugaki M, Tagami H, Noda T, Kondo H: Altered water barrier function in epidermal fatty acid binding protein (E-FABP)-deficient mice. J Invest Dermatol 118: 430–435, 2002
9. Menon GK, Feingold KR, Elias PM: Lamellar body secretory response to barrier disruption. J Invest Dermatol 98: 279–289, 1992
10. Odland GP, Holbrook K: The lamellar granules of the epidermis. Curr Prob Dermatol 9: 29–49, 1987
11. Kane CD, Coe NR, Vanlandingham B, Krieg P, Bernlohr DA: Expression, purification, and ligand binding analysis of recombinant keratinocyte lipid-binding protein (MAL-1), an intracellular lipid-binding protein found overexpressed in neoplastic skin cells. Biochemistry 35: 2894–2900, 1996
12. Hotamisligil GS, Johnson RS, Distel RJ, Ellis R, Papaioannou VE, Spiegelman BM: Uncoupling of obesity from insulin resistance through a targeted mutation in aP2, the adipocyte fatty acid binding protein. Science 274: 1377–1379, 1996
13. Coe NR, Simpson MA, Bernlohr DA: Targeted disruption of the adipocyte lipid-binding protein (aP2 protein) gene impairs fat cell lipolysis and increases cellular fatty acid levels. FASEB J 40: 967–972, 1999
14. Binas B, Danneberg H, McWhir, J, Mullins L, Clark AJ: Requirement for the heart-type fatty acid binding protein in cardiac fatty acid utilization. FASEB J 13: 805–812, 1999
15. Schaap FG, Binas B, Danneberg H, van der Vusse GJ, Glatz JF: Impaired long-chain fatty acid utilization by cardiac myocytes isolated from mice lacking the heart-type fatty acid binding protein gene. Circ Res 85: 329–337, 1999
16. Kligman AM: Perspectives and problems in cutaneous gerontology. J Invest Dermatol 73: 39–46, 1979
17. Wilhelm KP, Cua AB, Maibach HI: Skin aging. Effect on transepidermal water loss, stratum corneum hydration, skin surface pH, and casual sebum content. Arch Dermatol 127: 1806–1809, 1991
18. Ya-Xian Z, Suetake T, Tagami H: Number of cell layers of the stratum corneum in normal skin-relationship to the anatomical location on the body, age, sex and physical parameters. Arch Dermatol Res 291: 555–559, 1999

Molecular and Cellular Biochemistry **239**: 87–93, 2002.

Sterol carrier protein-2: Not just for cholesterol any more

Eric J. Murphy

Department of Pharmacology, Physiology, and Therapeutics, School of Medicine and Health Sciences, University of North Dakota, Grand Forks, ND, USA

Abstract

Although sterol carrier protein-2 (SCP-2) mediates cholesterol esterification in L-cell fibroblasts and stimulates an accumulation of cholesterol in these cells, a potential role for SCP-2 in fatty acid uptake and trafficking has not been appreciated. Certainly, recent experiments have shown that SCP-2 binds fatty acids *in vitro* with an affinity similar to that observed for fatty acid binding proteins. Because of the ubiquitous tissue distribution of SCP-2, as opposed to the specific distribution of fatty acid binding proteins, as well as the need for fatty acid trafficking in all cells, I have recently proposed that SCP-2 is the universal fatty acid trafficking protein. This supposition is based on a number of observations made with L-cell fibroblasts expressing either the 13.2 kDa SCP-2 or the 15 kDa proSCP-2. In L-cells expressing the 13.2 kDa SCP-2, fluorescent fatty acid uptake was increased by 10–30% depending upon the probe used. In 15 kDa proSCP-2 expressing cells, fluorescent fatty acid uptake was increased 20–40% depending upon the probe used. However, only expression of the 15 kDa pro-SCP-2 increased the cytoplasmic diffusion of the fluorescent fatty acid. Expression of either protein increased the uptake of [^{3}H]-oleic acid 1.9-fold compared to control, with targeting of [^{3}H]-oleic acid for esterification into cholesteryl esters. The 13.2 kDa SCP-2 did target a significant amount of [^{3}H]-oleic acid for esterification into the triacylglycerol pool. Expression of either protein markedly reduced total cellular phospholipid levels, however both proteins increased cholesteryl ester levels. Interestingly, expression of the 15 kDa proSCP-2 decreased ethanolamine plasmalogen levels with a concomitant increase in choline plasmalogen. Expression of both proteins increased PUFA content of the phospholipids, although this effect was greater in 15 kDa proSCP-2 expressing cells. Hence, expression of SCP-2 increased fatty acid uptake and targeted fatty acid to unique lipid pools, suggesting that SCP-2 may effectively serve as universal fatty acid binding and trafficking protein. (Mol Cell Biochem **239**: 87–93, 2002)

Key words: sterol carrier protein, fatty acid binding protein, cholesterol transport, L-cell fibroblasts

Introduction

Sterol carrier protein-2 (SCP-2) is thought to mediate cholesterol trafficking and metabolism [1–3]. SCP-2 is a small ubiquitous protein encoded for by a single gene that contains two initiation sites [4, 5], encoding for the 58 kDa SCP_x and the 15 kDa proSCP-2 [4, 6–8]. The 15 kDa proSCP-2 contains a N-terminal 20 amino acid putative mitochondrial targeting sequence [9, 10] and a C-terminal peroxisomal targeting sequence [7]. Recent evidence suggests that this pre-sequence significantly alters the tertiary structure of the proSCP-2 relative to SCP-2, resulting in the targeting of proSCP-2 to the cytofacial leaflet of the peroxisomal membrane [11]. In cultured cells and most tissues, proSCP-2 undergoes rapid post-translational cleavage resulting in the 13.2 kDa form of SCP-2 [9, 10]. For the 13.2 kDa SCP-2, cellular localization studies using immunogold and immunofluorescence methods have indicated that the protein is associated with peroxisomes [12], endoplasmic reticulum [13, 14], mitochondria [13, 15], and cytosol [13, 16]. However, the 58 kDa SCP_x is exclusively localized in the peroxisomal matrix [12, 17]. Hence, there is a differential targeting of SCP_x, SCP-2, and proSCP-2, although the functional significance of this differential targeting and its impact on lipid metabolism is unclear.

Address for offprints: E.J. Murphy, Department of Pharmacology, Physiology and Therapeutics, University of North Dakota, 501 North Columbia Road, Room 3700, Grand Forks, ND 58203, USA (E-mail: emurphy@medicine.nodak.edu)

Despite the lack of evidence clearly defining a physiological role for SCP-2, a number of studies suggest a function in lipid metabolism [14, 18, 19]. SCP-2 binds sterols [3, 20], cholesterol [20–22], fatty acids [22–25], acylCoAs [22, 25–27], and phospholipids [21], with affinities ranging in the nM to low μM range [25]. The cholesterol binding is consistent with the premise that SCP-2 facilitates cholesterol movement as demonstrated by the enhancement of cholesterol transfer between model membranes [21, 28, 29] and L-cell membranes [16, 30] by SCP-2. Further, proSCP-2 expression increases cholesterol uptake [2] and cholesterol esterification [1]. In contrast, expression of the 13.2 kDa SCP-2 does not enhance cholesterol uptake suggesting the expression of the 15 kDa proSCP-2 containing the N-terminal pre-sequence is essential for proper function in cholesterol uptake. Thus, this sequence may be important in targeting the protein to intracellular locations other than the peroxisome [11]. However, these studies do not demonstrate a potential for SCP-2 to act as an ubiquitous fatty acid binding protein, but rather emphasize its importance as a cholesterol trafficking protein.

Until recently a role for SCP-2 in fatty acid uptake and trafficking was unrealized. Studies *in vitro* show clearly that SCP-2 binds fatty acids with a high affinity similar to that reported for fatty acid binding proteins [22–25]. In L-cells transfected with SCP-2, both 13.2 kDa SCP-2 and 15 kDa proSCP-2 expression increases uptake and intracellular trafficking of fluorescently labeled fatty acid [18], with the pre-sequence having little apparent impact on fatty acid uptake. However, the effect of expressing the proSCP-2 containing the pro-sequence on fatty acid targeting is not fully understood. Further, recent structural studies demonstrate the presence of a hydrophobic tunnel in SCP-2 capable of binding fatty acids and fatty acylCoAs [31, 32]. This further supports a role of SCP-2 in fatty acid binding and in fatty acid targeting. This is an important point, as SCP-2 also binds the activated form of fatty acids, acyl-CoA, with a high affinity [22, 26, 27] and is found associated with lipid processing intracellular organelles [11–14]. Expression of the 58 kDa SCP_x in L-cells stimulates not only an increase in 13.2 kDa SCP-2 levels, but also increases the esterification of oleic acid onto cholesterol, further implicating a role in fatty acid trafficking [33].

While the 13.2 and 15 kDa forms of SCP-2 increase fatty acid uptake into L-cells and bind acylCoA and fatty acids with a high affinity, the role of these proteins in fatty acid targeting to specific lipid pools is unclear. The potential role for SCP-2 in fatty acid uptake and targeting was assessed using cells stably transfected with the murine cDNA encoding for either the 13.2 kDa SCP-2 or 15 kDa proSCP-2. The results presented herein indicate that SCP-2 not only increases fatty acid uptake [18], but also enhances fatty acid targeting to specific lipid pools, suggesting that SCP-2 may function as an ubiquitous fatty acid trafficking protein.

Methods

Cells

Murine L-cells (L $arpt^-tk^-$) previously transfected with the cDNA encoding for the 15 or 13.2 kDa forms of SCP-2 were used in this study [2]. These two cell lines both contain the 13.2 kDa form of SCP-2, consistent with the reported rapid cleavage of the 20 amino acid pre-sequence [9, 10]. The protein levels expressed as a percentage of total cytosolic protein are 0.030 and 0.036% for the 15 and 13.2 kDa SCP-2 expressing cells, respectively [2]. These levels are within the physiological range found in most tissue [34]. Control and transfected L-cell clones were grown to confluency on 100 mm^2 tissue culture plates in Higuchi medium containing 10% fetal bovine serum (Hyclone, Logan, UT, USA) [35].

Radiotracer fatty acid uptake and targeting

Prior to incubation with fatty acids, the medium was removed, the dishes rinsed with an equal volume of phosphate buffered saline (PBS), and then incubated with 30 ml of buffer. To this buffer, 150 μl of either fatty acid (0.1 μg/μl ethanol) was added to the cells to give a final fatty acid concentration of 1.8 μM and incubated at 37°C. The [^{3}H]oleic acid added was diluted from a stock solution with a specific activity of 10 Ci/mmole using cold oleic acid to give a final specific activity of 1 μCi/nmol. Cells were incubated for 30 min in the presence of radiotracer.

Lipid extraction

Cellular lipids were extracted from confluent control and transfected cells using n-hexane:2-propanol (3:2 v/v) [36, 37]. Prior to extraction, cell culture medium was removed and the cells washed twice with ice-cold phosphate buffered saline to remove extracellularly bound [^{3}H]oleic acid. Following removal of the last wash, the cell plate was floated on liquid N_2 to minimize lipid breakdown during cell removal [38]. Following removal from the liquid N_2, 2 ml of 2-propanol was added to the cell plate and the cells removed by scraping. The 2-propanol aliquot containing the cells was transferred to a tube containing 6 ml of n-hexane. The cell plate was washed with another 2 ml aliquot of 2-propanol which was also transferred to the tube containing the n-hexane, resulting in n-hexane:2-propanol (3:2 v/v).

Cell extracts were centrifuged at 2500 rpm to pellet the denatured protein and other cellular debris. The lipid containing organic liquid was decanted and saved until analysis. The residual protein pellet was dried overnight at room temperature to evaporate residual solvent. All lipid extracts were stored under a N_2 atmosphere at –80°C.

Protein assay

Proteins were measured using a modified dye-binding assay [39]. The dried protein residue from the extracts was digested overnight in 0.2 M KOH at 65°C. Following digestion, aliquots were used to measure the protein concentration by converting absorbances to concentrations using a bovine serum albumin standard curve.

Thin layer chromatography

Neutral lipids were separated by thin layer chromatography using Silica gel G plates (Analtech, Newark, DE, USA) that were heat activated at 110°C for 1 h. Samples were streaked onto the plates. The developing solvent was petroleum ether: diethyl ether:acetic acid (75:25:1.3 by volume). This solvent system resolves all the neutral lipids including cholesteryl esters and triacylglycerols. The phospholipids remain at the origin. Bands were visualized using 1 mM 6-p-toluidino-2-napthalene sulfonic acid dissolved in 50 mM Tris (pH 7.4) [40]. Bands corresponding to authentic lipid standards were scraped into liquid scintillation vials and radioactivity measured using liquid scintillation counting. Prior to the addition of 10 ml of Scinitverse (Fisher Scientific, Pittsburgh, PA, USA), 0.5 ml of water was added to the sample to aid in displacing the lipid from the silica.

Statistics

All groups were compared using one-way ANOVA and Tukey–Kramer multiple comparisons post-test using Instat II (GraphPad, San Diego, CA, USA). All values are expressed as mean ± S.D. Statistical significance was defined as $p < 0.05$. The n is defined as the number of cultures used to determine each data point.

Results

Phospholipid and neutral lipid levels

The effects of SCP-2 and proSCP-2 expression on phospholipid and neutral lipid levels has been reported previously [1, 19]. Expression of either form of SCP-2 reduces total cellular phospholipid content by 24%, with a concomitant decrease in the cholesterol to phospholipid ratio [1, 19]. proSCP-2 expression increased choline plasmalogen levels by 30% with a decrease in ethanolamine plasmalogen levels by 25% [19]. This is important as plasmalogens are rapidly turned over in the brain, consistent with a role in signal transduction [41].

proSCP-2 expression increases cholesterol levels [1], while expression of either proteins elevates cholesteryl ester levels [1], suggesting a net increase in esterification (Fig. 1). In contrast, the 13.2 kDa SCP-2 increased triacylglycerol levels, while expression of 15 kDa proSCP-2 decreased triacylglycerol levels (Fig. 1). Hence, SCP-2 expression enhances neutral lipid levels, suggesting a net increase in cellular biosynthetic capacity in the presence of SCP-2.

Fatty acid uptake

Previous data have demonstrated that SCP-2 and proSCP-2 expression enhance both *cis*-parinaric acid and NBD-stearate uptake in L-cells expressing these proteins [18]. However, both of these fatty acids are poorly esterified, hence give a measure of fatty acid uptake in the absence of esterification. Using [^{3}H]18:1 n-9 as a radiotracer, fatty acid uptake was increased by expression of either protein by 1.9-fold (Fig. 2). This increase in fatty acid uptake is similar to that for L-FABP [42] in L-cell fibroblasts expressing this protein at 11–13-fold greater than the expression of SCP-2 [1, 2].

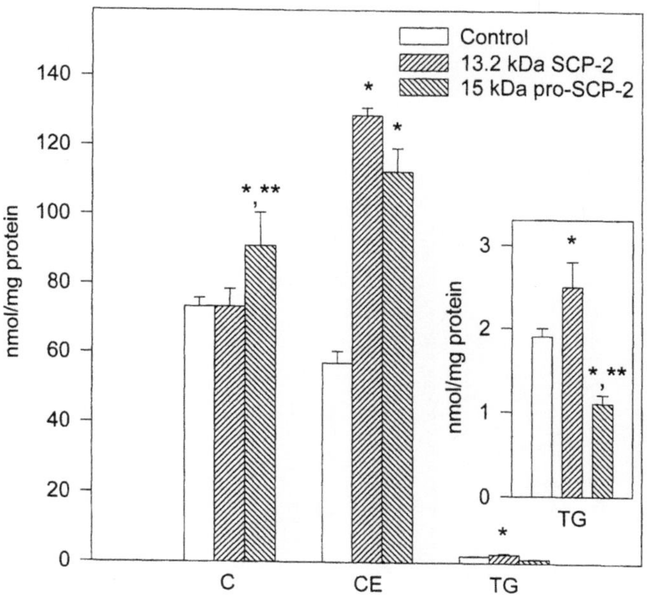

Fig. 1. proSCP-2 and SCP-2 expression alter neutral lipid levels in L-cells relative to control. Values are expressed as nmol/mg protein and represent mean ± S.D., n = 5. *Significant from control, $p < 0.05$. **Significant from 13.2 kDa SCP-2 expressing cells, $p < 0.05$. Adapted from Murphy and Schroeder [1].

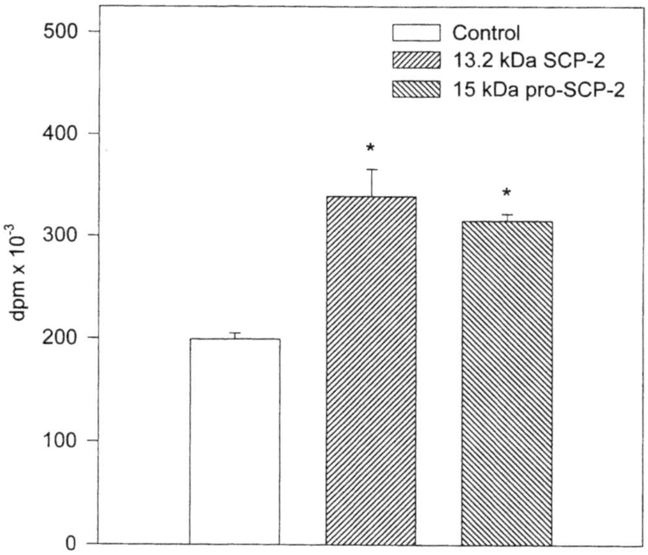

Fig. 2. proSCP-2 and SCP-2 expression increased fatty acid uptake in L-cells relative to control. Values are expressed as dpm × 10^3 and represent mean ± S.D., n = 5. *Significant form control, $p < 0.05$.

Fatty acid targeting

Although SCP-2 and proSCP-2 expression enhanced [^{3}H]18:1 n-9 uptake, it was unknown whether expression of these proteins could alter the targeting of [^{3}H]18:1 n-9. Separating the esterified neutral lipids, phospholipids, and unesterified fatty acid fractions, clearly both of these proteins enhanced the

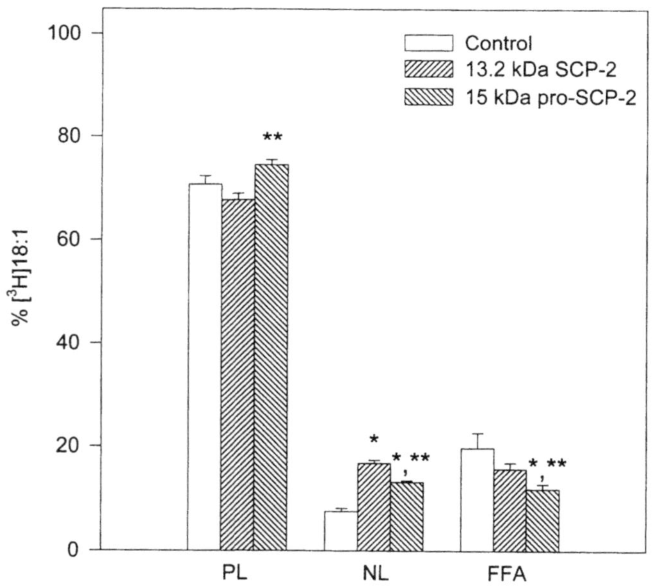

Fig. 3. proSCP-2 and SCP-2 expression in L-cells targets fatty acid predominantly to the neutral lipid for esterification. Values are expressed as percent distribution of [^{3}H]18:1 n-9 and represent mean ± S.D., n = 5. *Significant from control, $p < 0.05$. **Significant from 13.2 kDa SCP-2 expressing cells, $p < 0.05$.

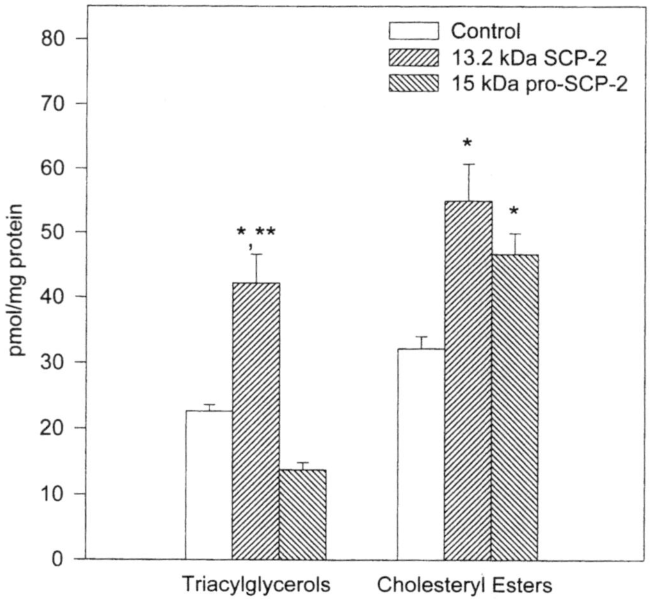

Fig. 4. proSCP-2 and SCP-2 expression causes a differential targeting of [^{3}H]18:1 n-9 into cholesteryl ester and triacylglycerol pools. Values are expressed as pmol of [^{3}H]18:1 n-9 esterified and represent mean ± S.D., n = 5. *Significant from control, $p < 0.05$. **Significant from 13.2 kDa SCP-2 expressing cells, $p < 0.05$. Adapted from Starodub *et al.* [14].

distribution of radiotracer into the esterified neutral lipid (Fig. 3). proSCP-2 expression also demonstrated a slight increase in targeting into the total phospholipid pool and a reduction in the unesterified fatty acid pool (Fig. 3). Resolving the esterified neutral lipid fraction into cholesteryl ester and triacylglycerol fractions, it is apparent that SCP-2 expression increased the formation of triacylglycerols containing radiotracer, while expression of proSCP-2 was without an affect (Fig. 4). However, both of these proteins increased the mass of the cholesteryl esters containing the radiotracer (Fig. 4). These observations are consistent with the data presented in Fig. 2 demonstrating that both of these proteins increased overall cholesteryl ester mass, but only SCP-2 expression enhanced triacylglycerol mass. Thus, expression of SCP-2 and proSCP-2 targets [^{3}H]18:1 n-9 into esterified neutral lipids, principally cholesteryl esters. There is a differential targeting of [^{3}H]18:1 n-9 into triacylglycerols by SCP-2 and into phospholipid pools by proSCP-2, suggesting that expression of the protein containing the presequence may have relevance to this function.

Discussion

One prevailing thought is that fatty acid binding proteins (FABP) facilitate fatty acid uptake, trafficking, and ultimate deposition into specific lipid pools within tissues expressing FABP. There is substantial evidence that different FABP have a different impact on fatty acid uptake [42–44] and target the fatty acids for esterification into distinct lipid pools [42–44]. Certainly, ample evidence indicates that FABP stimulate an increase in phosphatidic acid biosynthesis [45–47], suggesting a direct impact on phospholipid metabolism. My question is what about all of the tissues that don't express FABP? Do these cells possess a protein that aids in the uptake, trafficking, and targeting of fatty acids? I propose that the ubiquitously distributed SCP-2 functions as a universal protein that facilitates the uptake, trafficking, and targeting of fatty acids in a wide variety of tissue that lack an endogenous FABP.

SCP-2 increases the uptake of fluorescent fatty acids [18] and increases the apparent cytoplasmic diffusion coefficient of NBD-stearate [18]. These results indicate that SCP-2 increases not only fatty acid uptake, but also facilitates the movement of fatty acid within the cell. This is similar to what has been demonstrated for intestinal- and liver FABP, where the magnitude of increase in the apparent cytoplasmic diffusion coefficient is similar to that observed for SCP-2 [48], despite the fact that there was over 10 times more FABP expressed in the cells relative to SCP-2 levels. Data presented here, using [^{3}H]oleic acid clearly demonstrate that fatty acid uptake is increased in cells expressing physiologically relevant amounts of SCP-2. Uptake of [^{3}H]oleic acid was similar to that observed in L-FABP expressing cells [42]. Hence, evidence to date suggest that SCP-2 functions to facilitate fatty acid uptake when expressed in L-cell fibroblasts, supporting the supposition that SCP-2 functions as a fatty acid binding protein.

An increase in fatty acid uptake is of interest, but are there any lipid pools to which SCP-2 targets fatty acid for esterification? Data presented herein demonstrate a decreased proportion of fatty acid found in the unesterified fatty acid pool in SCP-2 expressing cells, with an increase in the proportion targeted and esterified into neutral lipids. An anomaly is the increased targeting of fatty acid into the phospholipid fraction in cells expressing the proSCP-2, which contains the 20 amino acid presequence that has several proposed functions [11, 14, 49]. Needless to say, the majority of the fatty acid is targeted for esterification into the cholesteryl esters and for SCP-2 expressing cells, into the triacylglycerol fraction. Hence, there appears to be a differential targeting of fatty acid into the phospholipid fraction for cells expressing proSCP-2 and into the triacylglycerol fraction for cells expressing SCP-2, with expression of either protein increasing targeting into the cholesteryl ester fraction.

In part, this targeting is consistent with the observed increase in the levels of cholesteryl esters in cells expressing either form of SCP-2. Further, only SCP-2 expressing cells have an increase in the levels of triacylglycerols, in agreement with the targeting data discussed above. Expression of either protein decreases the amount of total phospholipid mass, with decreases found in nearly all major phospholipid classes except sphingomyelin [19]. This is discrepant relative to the slight increase in [^{3}H]oleic acid targeting to the phospholipid pool in proSCP-2 expressing cells. However, an increase in targeting may represent an increase in acyl chain turnover that is independent of an increase in biosynthetic capacity. Further, the effect of SCP-2 on phospholipid levels is in direct contrast to what is observed with L-FABP and to a lessor extent I-FABP [50]. However, similar to the L- and I-FABP expression, expression of either SCP-2 increases the overall unsaturation of the phospholipids predominantly by increasing the proportion of polyunsaturated fatty acids, such as docosahexaenoic and arachidonic acid, found in the phospholipids [19, 50]. Thus, SCP-2 expression enhances an increase in lipid biosynthesis and modulates acyl chain composition, consistent, in part, with what is observed for some FABP.

In summary, evidence suggest that SCP-2 may function as a fatty acid binding protein. Because of its ubiquitous distribution this is an unique idea, providing all cells with a protein that is capable of facilitating cellular fatty acid uptake and trafficking. Recent structural studies substantiate this point, showing that SCP-2 has a hydrophobic binding pocket capable of binding fatty acids [31, 32]. Further, my proposal is consistent with studies demonstrating that SCP-2 binds fatty acids and fatty acylCoAs with a high affinity [23–26]. Future studies are required to provide more evidence to support this idea, however it is important to note that at this point there is ample evidence supporting the concept that SCP-2 serves as the universal fatty acid binding protein.

References

1. Murphy EJ, Schroeder F: Sterol carrier protein-2 mediated cholesterol esterification in transfected L-cell fibroblasts. Biochim Biophys Acta 1345: 283–292, 1997
2. Moncecchi D, Murphy EJ, Prows DR, Schroeder F: Sterol carrier protein-2 expression in mouse L-cell fibroblasts alters cholesterol uptake. Biochem Biophys A 1302: 110–116, 1996
3. Schroeder F, Frolov A, Schoer J, Gallegos A, Atshaves BP, Stolwich NJ, Scott AI, Kier AB: Intracellular cholesterol binding proteins, cholesterol transport, and membrane domains. In: D. Freeman, T.Y. Chang (eds). Intracellular Cholesterol Trafficking. Kluwer Academic Publishers, Boston, MA, 1998, pp 231–234
4. Ohba T, Rennert H, Pfeifer SM, He Z, Yamamoto R, Holt JA, Billheimer JT, Strauss JF III: The structure of the human sterol carrier protein x/sterol carrier protein 2 gene (SCP2). Genomics 24: 370–374, 1994
5. Seedorf U, Assmann G: Cloning, expression, and nucleotide sequence of rat liver sterol carrier protein 2 cDNAs. J Biol Chem 266: 630–636, 1991
6. Moncecchi D, Pastuszyn A, Scallen TJ: cDNA sequence and bacterial expression of mouse liver sterol carrier protein-2. J Biol Chem 266: 9885–9892, 1991

7. Yamamoto R, Kallen CB, Babalola GO, Rennert H, Billheimer JT, Strauss JF III: Cloning and expression of a cDNA encoding human sterol carrier protein 2. Proc Natl Acad Sci USA 88: 463–467, 1991
8. Ossendorp BC, van Heusden GPH, Wirtz KWA: The amino acid sequence of rat liver non-specific lipid transfer protein (sterol carrier protein 2) is present in a high molecular weight protein: Evidence from cDNA analysis. Biochem Biophys Res Commun 168: 631–636, 1990
9. Matsuura JE, George HJ, Ramachandran N, Alvarez JG, Strauss JF III, Billheimer JT: Expression of the mature and the pro-form of human sterol carrier protein 2 in *Escherichia coli* alters bacterial lipids. Biochemistry 32: 567–572, 1993
10. Seedorf U, Scheek S, Engel T, Steif C, Hinz H-J, Assmann G: Structure-activity studies on human sterol carrier protein 2. J Biol Chem 269: 2613–2618, 1994
11. Schroeder F, Frolov A, Starodub O, Atshaves BB, Russell W, Petrescu A, Huang H, Gallegos AM, McIntosh A, Tahotna D, Russell DH, Billheimer JT, Baum CL, Kier AB: Pro-sterol carrier protein-2. Role of the n-terminal presequence in structure, function, and peroxisomal targeting. J Biol Chem 275: 25547–25555, 2000
12. van Heusden GPH, Bos K, Raetz CRH, Wirtz KWA: Chinese hamster ovary cells deficient in peroxisomes lack the nonspecific lipid transfer protein (sterol carrier protein 2). J Biol Chem 265: 4105–4110, 1990
13. Keller GA, Scallen TJ, Clarke D, Maher PA, Krisans SK, Singer SJ: Subcellular localization of sterol carrier protein-2 in rat hepatocytes: Its primary localization to peroxisomes. J Cell Biol 108: 1353–1361, 1989
14. Starodub O, Jolly CA, Atshaves BP, Roths JB, Murphy EJ, Kier AB, Schroeder F: Sterol carrier protein-2 localization in endoplasmic reticulum and role in phospholipid formation. Am J Physiol Cell Physiol 279: C1259–C1269, 2000
15. Conneely OM, Headon DR, Olson CD, Ungar F, Dempsey ME: Intramitochondrial movement of adrenal sterol carrier protein with cholesterol in response to corticotropin. Proc Natl Acad Sci USA 81: 2970–2974, 1984
16. Vahouny GV, Dennis P, Chanderbhan R, Fiskum G, Noland BJ, Scallen TJ: Sterol carrier protein 2 (SCP2)-mediated transfer of cholesterol to mitochondrial inner membranes. Biochem Biophys Res Commun 122: 509–515, 1984
17. Van der Krift TP, Leunissen J, Teerlink T, van Heusden GPH, Verkleij AJ, Wirtz KWA: Ultrastructural localization of a peroxisomal protein in rat liver using the specific antibody against the non-specific lipid transfer protein (sterol carrier protein 2). Biochim Biophys Acta 812: 387–392, 1985
18. Murphy EJ: Sterol carrier protein-2 expression increases NBD-stearate uptake and cytoplasmic diffusion in L cells. Am J Physiol 275: G237–G243, 1998
19. Murphy EJ, Stiles T, Schroeder F: Sterol carrier protein-2 expression alters phospholipid content and fatty acyl composition in L-cell fibroblasts. J Lipid Res 41: 788–796, 2000
20. Colles SM, Woodford JK, Moncecchi D, Myers-Payne SC, McLean LR, Billheimer JT, Schroeder F: Cholesterol interaction with recombinant human sterol carrier protein-2. Lipids 30: 795–803, 1995
21. Gadella TWJ Jr, Bastiaens PIH, Visser AJWG, Wirtz KWA: Shape and lipid-binding site of the nonspecific lipid-transfer protein (sterol carrier protein 2): A steady-state and time-resolved fluorescence study. Biochemistry 30: 5555–5564, 1991
22. Stolwich NJ, Frolov A, Petrescu A, Scott AI, Billheimer JT, Schroeder F: Holo-sterol carrier protein-2: ^{13}C-NMR investigation of cholesterol and fatty acid binding sites. J Biol Chem 274: 35425–35433, 1999
23. Schroeder F, Myers-Payne SC, Billheimer JT, Wood WG: Probing the ligand binding sites of fatty acid and sterol carrier proteins: effects of ethanol. Biochemistry 34: 11919–11927, 1995
24. Stolwich NJ, Frolov A, Atshaves B, Murphy EJ, Jolly CA, Billheimer JT, Scott AI, Schroeder F: The sterol carrier protein-2 fatty acid binding site: An NMR, circular dichroic and fluorescence spectroscopic determination. Biochemistry 36: 1719–1729, 1997
25. Gallegos AM, Atshaves BP, Storey SM, Starodub O, Petrescu AD, Huang H, McIntosh AL, Martin GG, Chao H, Kier AB, Schroeder F: Gene structure, intracellular localization, and functional roles of sterol carrier protein-2. Prog Lipid Res 40: 498–563, 2001
26. Frolov A, Cho TH, Billheimer JT, Schroeder F: Sterol carrier protein-2, a new fatty acyl coenzyme A-binding protein. J Biol Chem 271: 31878–31884, 1996
27. Dansen TB, Westerman J, Wouters FS, Wanders JR, van Hoek A, Gadella TW Jr, Wirtz KW: High-affinity binding of very-long-chain fatty acyl-CoA esters to the peroxisomal non-specific lipid-transfer protein (sterol carrier protein-2). Biochem J 339: 193–199, 1999
28. van Amerongen A, Demel RA, Westerman J, Wirtz KWA: Transfer of cholesterol and oxysterol derivatives by the nonspecific lipid transfer protein (sterol carrier protein 2): A study on its mode of action. Biochim Biophys Acta 1004: 36–43, 1989
29. Butko P, Hapala I, Scallen TJ, Schroeder F: Acidic phospholipids strikingly potentiate sterol carrier protein 2 mediated intermembrane sterol transfer. Biochemistry 29: 4070–4077, 1990
30. Frolov A, Woodford JK, Murphy EJ, Billheimer JT, Schroeder F: Spontaneous and protein-mediated sterol transfer between intracellular membranes. J Biol Chem 271: 16075–16083, 1996
31. García FL, Szyperski T, Dyer JH, Choinowski T, Seedorf U, Hauser H, Wüthrich K: NMR structure of the sterol carrier protein-2: Implications for the biological role. J Mol Biol 295: 595–603, 2000
32. Choinowski T, Hauser H, Piontek K: Structure of sterol carrier protein 2 at 1.8 Å resolution reveals a hydrophobic tunnel suitable for lipid binding. Biochemistry 39: 1897–1902, 2000
33. Atshaves BP, Petrescu AD, Starodub O, Roths JB, Kier AB, Schroeder F: Expression and intracellular processing of the 58 kDa sterol carrier protein-2/3-oxoacyl-CoA thiolase in transfected mouse L-cell fibroblasts. J Lipid Res 40: 610–622, 1999
34. van Amerongen A, Teerlink T, van Heusden GPH, Wirtz KWA: The non-specific lipid transfer protein (sterol carrier protein 2) from rat and bovine liver. Chem Phys Lipids 38: 195–204, 1985
35. Higuchi K: An improved chemically define culture medium for strain L mouse cells based on growth responses to graded levels of nutrients including iron and zinc. J Cell Physiol 75: 65–72, 1970
36. Hara A, Radin NS: Lipid extraction of tissues with a low-toxicity solvent. Anal Biochem 90: 420–426, 1978
37. Murphy EJ, Rosenberger TA, Horrocks LA: Effects of maturation on the phospholipid and phospholipid fatty acid compositions in primary rat cortical astrocyte cell cultures. Neurochem Res 22: 1205–1213, 1997
38. Demediuk P, Anderson DK, Horrocks LA, Means ED: Mechanical damage to murine neuronal-enriched cultures during harvesting: Effects on free fatty acids, diglycerides, Na^+,K^+-ATPase, and lipid peroxidation. *In Vitro* Cell Dev Biol 21: 569–574, 1985
39. Bradford M: A rapid and sensitive method for the quantitation of microgram quantities of protein utilizing the principle of protein-dye binding. Anal Biochem 72: 248–254, 1976
40. Jones M, Keenan RW, Horowitz P: Use of 6-p-toluidino-2-naphthalenesulfonic acid to quantitate lipids after thin-layer chromatography. J Chromatogr 237: 522–524, 1982
41. Rosenberger, TA, Oki, J, Purdon, AD, Rapoport, SI, Murphy, EJ: Rapid synthesis and turnover of brain microsomal ether phospholipids in the adult rat. J Lipid Res 43: 59–68, 2002
42. Murphy EJ, Prows DR, Jefferson JR, Schroeder F: Liver fatty acid binding protein expression in transfected fibroblasts stimulates fatty

acid uptake and metabolism. Biochim Biophys Acta 1301: 191–196, 1996
43. Prows DR, Murphy EJ, Schroeder F: Intestinal and liver fatty acid binding proteins differentially affect fatty acid uptake and esterification in L-cells. Lipids 30: 907–910, 1995
44. Prows DR, Murphy EJ, Moncecchi D, Schroeder F: Intestinal fatty acid-binding protein expression stimulates fibroblast fatty acid esterification. Chem Phys Lipids 84: 47–56, 1996
45. Jolly CA, Hubbell T, Behnke WD, Schroeder F: Fatty acid binding protein: Stimulation of microsomal phosphatidic acid formation. Arch Biochem Biophys 341: 112–121, 1997
46. Jolly CA, Murphy EJ, Schroeder F: Differential influence of rat liver fatty acid binding protein isoforms on phospholipid fatty acid composition: Phosphatidic acid biosynthesis and phospholipid fatty acid remodeling. Biochim Biophys Acta 1390: 258–268, 1998
47. Hubbell T, Behnke WD, Woodford JK, Schroeder F: Recombinant liver fatty acid binding protein interacts with fatty acyl-coenzyme A. Biochemistry 33: 3327–3335, 1994
48. Murphy EJ: L-FABP and L-FABP expression increases NBD-stearate uptake and cytoplasmic diffusion in L cells. Am J Physiol 275: G244–G249, 1998
49. Gallegos AM, Schoer JK, Starodub O, Kier AB, Billheimer JT, Schroeder F: A potential role for sterol carrier protein-2 in cholesterol transfer to mitochondria. Chem Phys Lipids 105: 9–29, 2000
50. Murphy EJ, Prows D, Stiles T, Schroeder F: Phospholipid and phospholipid fatty acid composition of L-cell fibroblast: Effect of intestinal and liver fatty acid binding protein. Lipids 35: 729–738, 2000

Molecular and Cellular Biochemistry **239**: 95–100, 2002.

Fatty acid binding protein expression in different human adipose tissue depots in relation to rates of lipolysis and insulin concentration in obese individuals

R.M. Fisher,[1] A. Thörne,[2] A. Hamsten[1] and P. Arner[3]

[1]*Atherosclerosis Research Unit, King Gustaf V Research Institute, Karolinska Hospital, Karolinska Institute; Departments of* [2]*Surgery and* [3]*Medicine, Huddinge University Hospital, Karolinska Institute, Stockholm, Sweden*

Abstract

Two fatty acid binding proteins (FABPs) are expressed in adipose tissue, adipocyte lipid binding protein (ALBP) and keratinocyte lipid binding protein (KLBP). This study investigated FABP expression in visceral and subcutaneous human adipose tissue depots and associations with lipolytic differences between the depots and circulating insulin concentrations. ALBP and KLBP (protein and RNA) were quantified in subcutaneous and omental adipose tissue from obese individuals and expressed relative to actin. ALBP RNA and protein expression was significantly higher in subcutaneous compared to omental adipose tissue (both $p < 0.05$), whereas KLBP RNA and protein expression was no different between the two sites. There were significant inverse correlations between serum insulin concentrations and the ALBP/KLBP RNA ratio in both subcutaneous and omental adipose tissue (both $p < 0.02$). Basal rates of glycerol and fatty acid release measured in adipocytes isolated from subcutaneous and omental adipose tissue were significantly higher in the former ($p \leq 0.02$). Therefore the relative ALBP/KLBP content of human adipose tissue is different in different adipose tissue depots and at the RNA level is related to the circulating insulin concentration, at least in obese subjects. The higher rates of basal lipolysis in adipocytes isolated from subcutaneous compared to omental adipose tissue might be related to the increased ALBP content of the former. Therefore adipose tissue FABPs are interesting candidates for investigation to further our understanding of the insulin resistance syndrome and regulation of lipolysis. (Mol Cell Biochem **239**: 95–100, 2002)

Key words: adipocyte lipid binding protein, keratinocyte lipid binding protein, adipocyte, lipolysis, fatty acid

Introduction

Increased plasma non-esterified fatty acid (NEFA) concentrations are believed to play an important role in insulin resistance and the associated predisposition to coronary heart disease and non-insulin-dependent diabetes mellitus [1]. Long chain fatty acids (FAs) are required for a variety of cellular processes, but most tissues that utilise them have to take them up from the circulation. NEFAs are virtually insoluble in aqueous solutions and are highly toxic if not complexed with protein. In the circulation most NEFAs are bound to albumin and intracellularly to cytoplasmic fatty acid binding proteins (FABPs) [2]. FABPs can localise to the nucleus as well as the cytoplasm and can modulate ligand-dependent trans-activation by peroxisome proliferator-activated receptors (PPARs) [3].

White adipose tissue contains two different FABPs: adipocyte lipid binding protein (ALBP, also known as aP2) and keratinocyte lipid binding protein (KLBP). The ALBP:KLBP ratio in adipocytes isolated from normal mice is approximately 99:1 [4]. However, mice lacking ALBP have a strong induction of KLBP expression within their adipose tissue [4,

Address for offprints: R.M. Fisher, Atherosclerosis Research Unit, King Gustaf V Research Institute, Karolinska Hospital, SE-171 76 Stockholm, Sweden (E-mail: rachel.fisher@medks.ki.se)

5]. ALBP knock out mice have reduced adipocyte lipolysis [4, 5] and when fed a high fat diet, they fail to develop the insulin resistance normally associated with the subsequent obesity [6]. This suggests that ALBP might be of importance for the non-hormonal regulation of adipose tissue lipolysis and insulin resistance.

Adipose tissue from various depots within the body has different metabolic properties. Catecholamine-induced lipolysis is increased in visceral compared to subcutaneous adipocytes, but the reverse is true for basal (non-hormonally stimulated) rates of lipolysis [7, 8]. Differences in lipolytic rates between adipose tissue depots are believed to be clinically important, possibly linking abdominal obesity to metabolic complications such as glucose intolerance, hyperinsulinaemia and dyslipidaemia, because only visceral fat has a direct connection to the liver via the portal vein. High portal NEFA concentrations might disturb liver function and thereby induce hyperglycaemia, hyperinsulinaemia and dyslipidaemia [1, 7, 8]. Mechanisms linked to the actions of lipolysis regulating hormones explaining the metabolic differences between adipose tissue depots have been proposed [7, 8]. However, very little is known about the regulation of spontaneous (basal) lipolytic activity and it is possible that non-hormonal mechanisms, such as those involving FABPs, could be involved.

We recently reported on differences in ALBP and KLBP expression in subcutaneous and omental (visceral) adipose tissue in lean and obese individuals [9]. In this study we investigated more thoroughly the obese subjects. First we studied whether differences in expression between the sites occurred in parallel with lipolytic differences between these depots. Second we looked for associations of adipose tissue ALBP and KLBP expression with insulin resistance, here indirectly evaluated by measuring fasting plasma insulin concentrations. One hypothesis was that if adipose tissue FABPs are important for insulin sensitivity and/or lipolysis, their tissue expression should be related to circulating insulin concentrations and the lipolytic activities of fat cells.

Materials and methods

Subjects

The study subjects were the same as those investigated previously [9]. Clinical details and surgical procedures have been described in detail elsewhere [9]. In brief, 20 obese individuals (7 male, 13 female) were studied, age 39.0 ± 2.4 years, body mass index 44.0 ± 1.3 kg/m^2. They were Caucasian, born in Sweden and were undergoing weight reduction surgery with adjustable gastric banding or elective laparoscopic cholecystectomy, but were otherwise apparently healthy, had no previous medical history of diabetes and were not taking any regular medication. Operations started at 8 a.m. following an overnight fast. The study was approved by the ethics committee of the Karolinska Institute, Stockholm, Sweden. All subjects gave their informed consent.

Adipose tissue biopsies

Biopsies from the subcutaneous and omental fat tissue regions were collected under general anaesthesia as described [10]. Tissue was used immediately for determination of fat cell size, isolation of fat cells and measurement of rates of glycerol and FA release. Tissue for subsequent RNA and protein analysis was frozen immediately in liquid nitrogen in tissue pieces (~ 300 mg) and stored at –70°C.

Isolation of fat cells and determination of fat cell size

Isolated fat cells from both the subcutaneous and omental adipose tissue regions were prepared by collagenase treatment as described [10, 11]. Using direct microscopy, diameters of 100 fat cells were determined and the mean fat cell volume calculated (assuming that the measured cells were spheres). Fat cell weight was calculated using the density of human fat cell triacylglycerol (assuming that lipids constitute > 95% of a fat cell).

Incubation of fat cells and determination of glycerol and FA release

Incubations were performed exactly as described [11]. In brief, diluted fat cell suspensions were incubated in albumin-containing buffers with or without noradrenaline at increasing concentrations (10^{-12} to 10^{-4} mol/l). Experiments measuring glycerol and FA release into the medium were run in parallel. Glycerol and FA concentrations were determined by luminescence techniques. Either basal lipolytic rates were subtracted from the rate obtained with the maximum effective concentration of noradrenaline, or the maximum rate was expressed as a percentage of the basal value. Lipolysis was expressed either as a function of the lipid weight of the incubated adipocytes or per number of cells incubated.

ALBP, KLBP and actin mRNA analysis

Total RNA was extracted from 300 mg pieces of adipose tissue. ALBP, KLBP and β-actin mRNA were quantified by Northern blot analysis as described [9]. Differences in loading were adjusted to β-actin expression.

ALBP, KLBP and actin protein analysis

Total protein was extracted by homogenising 300 mg pieces of adipose tissue and ALBP, KLBP and actin proteins were quantified by Western blot analysis as described [9]. To account for differences in loading, ALBP and KLBP were expressed relative to actin.

Statistical analysis

In some cases, for technical reasons, it was not possible to perform all analyses on all pairs of omental and subcutaneous adipose tissue samples. The number of individuals included in all analyses is stated. Results are presented as mean ± S.E.M. Mean values were compared using a paired *t*-test. Correlations were performed using the Z test. ANOVA and ANCOVA were used to determine the influence of sex on the measured variables. Statistical significance was assigned to a value of $p < 0.05$.

Results

Summary of previously published results with this cohort

The subjects were hyperinsulinaemic (21.3 ± 1.9 μU/ml), but had normal mean blood glucose concentrations (5.7 ± 0.2 mmol/l) indicating that they were insulin resistant. Mean fasting plasma triacylglycerol and cholesterol concentrations were 2.2 ± 0.3 and 5.5 ± 0.2 mmol/l respectively. The detailed clinical data are presented in Table 1 in the previous study [9].

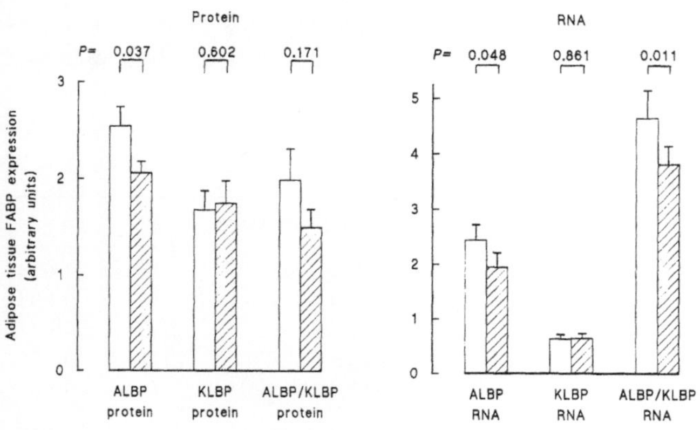

Fig. 1. ALBP and KLBP expression in subcutaneous (open bars) and omental (hatched bars) adipose tissue from obese individuals. FABP protein and RNA are expressed relative to actin protein and β-actin RNA respectively. Data are expressed in arbitrary units as mean ± S.E.M. Results are shown for 20 individuals, except for KLBP RNA, which was quantifiable in only 13 subjects. p-Values for comparison of subcutaneous and omental adipose tissue are shown. Data are modified from those previously published with this cohort [9].

FABP protein and RNA quantified in subcutaneous and omental adipose tissue were expressed relative to actin protein and β-actin RNA respectively. The details of the comparison are presented in Table 2 in the previous study [9]. A different mode of presentation of these data is shown here in Fig. 1. Expression of ALBP RNA and protein was on average 40 and 30% greater respectively in subcutaneous compared to omental adipose tissue (both $p < 0.05$). However, neither KLBP RNA nor protein expression differed between the two sites. Hence the ALBP/KLBP ratio for RNA and protein was on average 50% ($p = 0.01$) and 20% ($p = 0.17$) greater respectively in subcutaneous compared to omental adipose tissue. Actin protein and β-actin RNA expression showed no significant regional variation (data not shown).

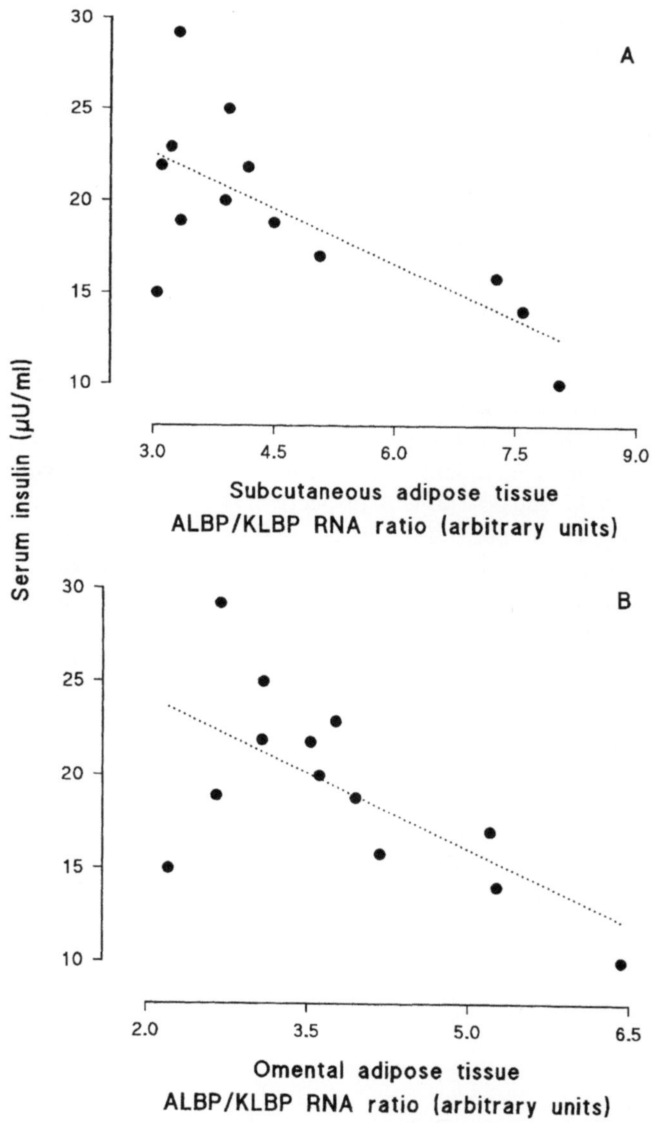

Fig. 2. Relationship between serum insulin concentrations and the ALBP/KLBP RNA ratio in subcutaneous (A) and omental (B) adipose tissue from 13 obese individuals. For subcutaneous adipose tissue, $r = -0.711$, $p = 0.005$; for omental adipose tissue, $r = -0.646$, $p = 0.015$.

New results

Significant inverse correlations were found between serum insulin concentration and the ALBP/KLBP RNA ratio in both subcutaneous (Fig. 2A) and omental (Fig. 2B) adipose tissue (both $p < 0.02$). In contrast, there was no correlation between insulin concentration and ALBP or KLBP RNA expression alone (data not shown). The ALBP/KLBP RNA ratio in the two adipose tissue depots was significantly correlated ($r = 0.854$, $p < 0.0001$). There was no relationship between the ALBP/KLBP RNA ratio in either subcutaneous or omental adipose tissue and plasma glucose concentration ($r = 0.089$ and $r = -0.124$ respectively).

When adipose tissue FABP protein expression was investigated, no correlations were identified between expression of adipose tissue ALBP or KLBP and serum insulin concentration. In neither subcutaneous nor omental adipose tissue was the ALBP/KLBP protein ratio tissue correlated with circulating insulin concentration ($r = -0.229$ and $r = -0.003$ respectively) or plasma glucose concentration ($r = 0.200$ and $r = 0.181$ respectively).

Both men and women were included in the study. However, ALBP and KLBP expression in subcutaneous and omental adipose tissue was not significantly different between men and women, and the correlation of the ALBP/KLBP RNA ratio with serum insulin concentrations was not influenced by sex (in the 13 individuals in whom KLBP RNA was measured, the male-to-female ratio was 4:9).

The basal rates of both glycerol and FA release (measured in independent experiments in isolated adipocytes from 17 and 15 subjects respectively) were significantly greater in adipocytes isolated from subcutaneous compared to omental adipose tissue (Table 1). The average mean cell volume of isolated subcutaneous adipocytes was 39% greater than that for omental adipocytes (937 ± 49 vs. 672 ± 43 pl, $p < 0.0001$). However, the regional differences in basal lipolysis rates were found whether glycerol and FA release were related to cell number or lipid weight. Maximum lipolysis rates (corrected for basal rates) following stimulation with noradrenaline tended to be higher ($p = 0.07$–0.09) in omental compared to subcutaneous adipocytes when related to the lipid weight of the incubated adipocytes, but were no different when expressed per number of cells incubated (Table 1). Noradrenaline stimulated rates of glycerol or FA release were significantly higher ($p < 0.02$) in omental compared to subcutaneous adipocytes when expressed as a percentage of basal rates (Table 1).

Discussion

Interest in FABPs expressed in adipose tissue has increased in recent years following the observations of the clinical phenotypes and disturbances of lipolysis in ALBP knock out mice [4–6, 12, 13]. Furthermore, ALBP and hormone sensitive lipase (HSL) form a complex, suggesting that HSL-derived FAs might be bound by ALBP to facilitate their efflux from the adipocyte [14].

The previously published data from this cohort [9] identified a greater ALBP/KLBP ratio in subcutaneous compared to omental adipose tissue from obese subjects, which appeared to be explained by differences in the ALBP content of the two depots. Although isolated subcutaneous adipocytes were significantly larger than omental adipocytes, the difference in ALBP expression between the two depots is unlikely simply to reflect cell size. ALBP RNA and protein were expressed relative to actin, which is expected to account for differences in cell size [15, 16]. Furthermore, actin expression was not subject to site variation. It is possible that some actin could have originated from cells other than adipocytes within the adipose tissue biopsies, but this is expected to be only a minor source of error. Both men and

Table 1. Basal and noradrenaline stimulated rates of lipolysis in isolated subcutaneous and omental adipocytes from obese individuals

			Subcutaneous adipocytes	Omental adipocytes	N	p
Glycerol release	Basal	μmol/g lipid/2 h	0.85 ± 0.16	0.50 ± 0.06	17	0.024
		μmol/10^7 cells/2 h	6.99 ± 1.18	2.96 ± 0.35	17	0.003
	Noradrenaline stimulated	μmol/g lipid/2 h	1.84 ± 0.19	2.42 ± 0.31	17	0.071
		μmol/10^7 cells/2 h	15.35 ± 1.47	14.53 ± 2.11	17	0.730
	NA as percentage of basal	%	383 ± 42	637 ± 77	17	0.005
Fatty acid release	Basal	μmol/g lipid/2 h	2.41 ± 0.37	1.52 ± 0.29	15	0.018
		μmol/10^7 cells/2 h	19.99 ± 2.88	9.09 ± 1.93	15	0 003
	Noradrenaline stimulated	μmol/g lipid/2 h	4.25 ± 0.76	5.74 ± 1.00	15	0.091
		μmol/10^7 cells/2 h	37.55 ± 7.04	34.58 ± 5.78	15	0.676
	NA as percentage of basal	%	350 ± 49	858 ± 209	15	0.016

In independent experiments either glycerol or fatty acid release from the isolated adipocytes was measured at 37°C over a 2-h period. The maximal rates of lipolysis obtained following stimulation with noradrenaline were corrected for basal (unstimulated) lipolysis rates or expressed as a percentage of the basal values. NA, noradrenaline. Data presented as mean ± S.E.M.

women were included in this study, but no influence of sex on the differences in ALBP expression was identified.

ALBP is an extremely abundant adipocyte protein, constituting 1–5% of total soluble protein [17], so even a small percentage-wise increase in ALBP protein in subcutaneous compared to omental adipose tissue represents a large absolute change in ALBP mass between the two sites. Furthermore, there are potentially significant consequences of altering the relative expression of ALBP and KLBP since they have been suggested to differ in their efficiencies as cofactors for lipolysis [5]. Differing FA-binding affinities [17, 18] and surface charge patterns [17, 19] might affect their interactions with intracellular targets such as phospholipid membranes, proteins such as HSL, or PPARs, thus modulating gene expression [20]. Indeed, differences in the effects of ALBP and KLBP on PPAR-mediated trans-activation have been reported [3].

As previously demonstrated [7, 8], basal rates of lipolysis (measured by both glycerol and FA release) were significantly greater in adipocytes isolated from subcutaneous compared to omental adipose tissue from the obese subjects, whether expressed relative to the number or lipid weight of the incubated cells. No attempts were made to correlate directly the FABP content of the adipocytes and the lipolysis rates since ALBP and KLBP measurements were made in extracts from intact adipose tissue pieces, not in the isolated adipocytes in which the lipolysis rates were measured. However, increased ALBP expression and an increased ALBP/KLBP RNA ratio in subcutaneous compared to omental adipose tissue were measured in the same subjects. Therefore it is a possibility that the decreased basal lipolysis rates in omental adipocytes were related to the reduced ALBP/KLBP ratio of this depot compared to subcutaneous tissue. However, experiments in which both lipolysis rates and FABP levels are measured in isolated adipocytes are required in order to confirm such a relationship. Furthermore, it would be of interest to look for associations of adipose tissue FABP expression level with plasma NEFA concentration.

Catecholamine stimulated rates of lipolysis are reported to be higher in omental compared to subcutaneous adipocytes [7, 8]. In the present study, maximum rates of noradrenaline simulated glycerol and FA release were higher (although not statistically significantly) in omental compared to subcutaneous adipocytes when expressed relative to the lipid weight of the incubated cells, but there was no difference when rates were expressed relative to the number of cells incubated. However, the percentage increase relative to basal rates was significantly greater in omental compared to subcutaneous adipocytes. Interpretation of stimulated rates of lipolysis is more complicated than basal rates. FA release from stimulated adipocytes is suggested to be regulated, at least in part, by FA transport across the cell membrane [21]. Therefore a possible regulatory role of intracellular FABPs in sequestering liberated FAs and shuttling them from sites of lipolysis to the membrane for transport out of the cell may be diminished in stimulated adipocytes. Additionally, the higher expression of the three different β-adrenoreceptor subtypes and a lower expression of the α2-adrenoreceptors in omental compared to subcutaneous adipose tissue [7], all of which are stimulated by noradrenaline [22], make it harder to consider the role(s) of ALBP and KLBP in the differences observed in catecholamine stimulated lipolysis between the two adipose tissue depots. However, it remains possible that differences in basal lipolysis rates are related to the adipocyte FABP content, but that other factors become more important under stimulated conditions when endogenous insulin and catecholamine production occurs, for example in connection with meals and during physical activity and mental stress.

Perhaps the most intriguing and novel observation in this study is that we identified a relationship between FABP gene expression in human adipose tissue from obese subjects and circulating insulin concentrations. The ALBP/KLBP RNA ratio in both subcutaneous and omental adipose tissue was inversely associated with serum insulin concentrations. However, no such relationship was found for the ratio of ALBP/KLBP proteins. This might suggest that, at least in obesity, transcription rates of adipose tissue FABPs are regulated by insulin, but that the subsequent translation steps to produce FABP proteins are subject to differential regulation by unknown mechanisms. The absence of statistically significant correlations between expression of ALBP or KLBP (protein or RNA) alone and serum insulin concentrations suggests that the relative abundance of ALBP and KLBP is co-ordinately regulated by insulin.

In summary, this study suggests that adipose tissue ALBP and KLBP may be involved in the pathogenesis of hyperinsulinaemia/insulin resistance. Alterations in FABP content might be related to metabolic differences, such as basal (non-hormonally stimulated) rates of lipolysis in different adipose tissue depots since adipose tissue FABP expression is subject to regional variation, with the ALBP/KLBP ratio being higher in subcutaneous compared to omental adipose tissue in obese subjects.

Acknowledgements

This work was supported by grants from the Swedish National Network and Graduate School for Cardiovascular Research, the Swedish Medical Research Council (projects 8691 and 1034), the Marianne and Marcus Wallenberg Foundation, the Swedish Institute, the Söderberg Foundation, the Novo Nordisk Foundation, the Swedish Diabetes Association and the Swedish Heart and Lung Foundation.

References

1. Frayn KN, Williams CM, Arner P: Are increased plasma non-esterified fatty acid concentrations a risk marker for coronary heart disease and other chronic diseases? Clin Sci 90: 243–253, 1996
2. Ribarik Coe N, Bernlohr DA: Physiological properties and functions of intracellular fatty acid-binding proteins. Biochim Biophys Acta 1391: 287–306, 1998
3. Helledie T, Antonius M, Sørensen RV, Hertzel AV, Bernlohr DA, Kølvraa S, Kristiansen K, Mandrup S: Lipid-binding proteins modulate ligand-dependent trans-activation by peroxisome proliferator-activated receptors and localize to the nucleus as well as the cytoplasm. J Lipid Res 41: 1740–1751, 2000
4. Ribarik Coe N, Simpson MA, Bernlohr DA: Targeted disruption of the adipocyte lipid-binding protein (aP2 protein) gene impairs fat cell lipolysis and increases cellular fatty acid levels. J Lipid Res 40: 967–972, 1999
5. Scheja L, Makowski L, Uysal T, Wiesbrock SM, Shimshek DR, Meyers DS, Morgan M, Parker RA, Hotamisligil GS: Altered insulin secretion associated with reduced lipolytic efficiency in *aP2*$^{-/-}$ mice. Diabetes 48: 1987–1994, 1999
6. Hotamisligil GS, Johnson RS, Distel RJ, Ellis R, Papaioannou VE, Spiegelman BM: Uncoupling of obesity from insulin resistance through a targeted mutation in aP2, the adipocyte fatty acid binding protein. Science 274: 1377–1379, 1996
7. Arner P: Differences in lipolysis between human subcutaneous and omental adipose tissues. Ann Med 27: 435–438, 1995
8. Kissebah AH, Krakower GR: Regional adiposity and morbidity. Physiol Rev 74: 761–811, 1994
9. Fisher RM, Eriksson P, Hoffstedt J, Hotamisligil G, Thörne A, Rydén M, Hamsten A, Arner P: Fatty acid binding protein expression in different adipose tissue depots from lean and obese individuals. Diabetologia 44:1268–1273, 2001
10. van Harmelen V, Reynisdottir S, Eriksson P, Thörne A, Hoffstedt J, Lönnqvist F, Arner P: Leptin secretion from subcutaneous and visceral adipose tissue in women. Diabetes 47: 913–917, 1998
11. Lönnqvist F, Thörne A, Nilsell K, Hoffstedt J, Arner P: A pathogenic role of visceral fat beta 3-adrenoceptors in obesity. J Clin Invest 95: 1109–1116, 1995
12. Bernlohr DA, Ribarik Coe N, Simpson MA, Vogel Hertzel A: Regulation of gene expression in adipose cells by polyinsaturated fatty acids. Adv Exp Med Biol 422: 145–156, 1997
13. Uysal KT, Scheja L, Wiesbrock SM, Bonner-Weir S, Hotamisligil GS: Improved glucose and lipid metabolism in genetically obese mice lacking aP2. Endocrinology 141: 3388–3396, 2000
14. Shen W-J, Sridhar K, Bernlohr DA, Kraemer FB: Interaction of rat hormone-sensitive lipase with adipocyte lipid-binding protein. Proc Natl Acad Sci USA 96: 5528–5532, 1999
15. Schmidt EE, Schibler U: Cell size regulation, a mechanism that controls cellular RNA accumulation: Consequences on regulation of the ubiquitous transcription factors Oct1 and NF-Y, and the liver-enriched transcription factor DPB. J Cell Biol 128: 467–483, 1995
16. Eriksson P, Reynisdottir S, Lönnqvist F, Stemme V, Hamsten A, Arner P: Adipose tissue secretion of plasminogen activator inhibitor-1 in non-obese and obese individuals. Diabetologia 41: 65–71, 1998
17. Simpson MA, LiCata VJ, Ribarik Coe N, Bernlohr DA: Biochemical and biophysical analysis of the intracellular lipid binding proteins of adipocytes. Mol Cell Biochem 192: 33–40, 1999
18. Shaughnessy S, Smith ER, Kodukula S, Storch J, Fried SK: Adipocyte metabolism in adipocyte fatty acid binding protein knockout (aP2$^{-/-}$) mice after short-term high-fat feeding. Functional compensation by the keratinocyte fatty acid binding protein. Diabetes 49: 904–911, 2000
19. LiCata VJ, Bernlohr DA: Surface properties of adipocyte lipid-binding protein: Response to lipid binding, and comparison with homologous proteins. Proteins 33: 577–589, 1998
20. Schoonjans K, Staels B, Auwerx J: The peroxisome proliferator activated receptors (PPARS) and their effects on lipid metabolism and adipocyte differentiation. Biochim Biophys Acta 1302: 93–109, 1996
21. Melki SA, Abumrad NA: Glycerolipid synthesis in isolated adipocytes: Substrate dependence and influence of norepinephrine. J Lipid Res 33: 669–678, 1992
22. Arner P: Catecholamine-induced lipolysis in obesity. Int J Obes 23(suppl 1): 10–13, 1999

Molecular and Cellular Biochemistry **239**: 101–112, 2002.

Protein acylation in the cardiac muscle like cell line, H9c2

Danny M. Hasselbaink, Theo H.M. Roemen and Ger J. van der Vusse
Department of Physiology, Cardiovascular Research Institute Maastricht, Maastricht University, Maastricht, The Netherlands

Abstract

Besides serving as oxidisable substrates, fatty acids (FA) are involved in co- and post-translational modification of proteins (protein acylation). Despite the high rate of fatty acid utilisation in the heart, information on protein acylation in cardiac muscle is scarce. To explore this subject in more detail, we used the H9c2 cell line as an experimental model. After incubation with ^{3}H-palmitate or ^{3}H-myristate, cells were lysed and proteins precipitated, followed by extensive delipidation. The delipidated proteins were subjected to SDS-PAGE and transferred to nitro-cellulose prior to autoradiography. In addition, TLC was used to separate the various lipid classes. The first aspect we addressed was the extent of protein acylation as a function of time, relative to fatty acid incorporation into various lipid classes. Cells were incubated for 30 min, 1 h and 2 h with 100 μCi palmitate (PA, 2.3 nmol) or 125 μCi myristate (MA, 2.5 nmol). Palmitoylation increased from 0.48 ± 0.25 to 1.25 ± 0.56 μCi/mg protein between 30 min to 2 h, while myristoylation increased from 0.25 ± 0.12 to 0.77 ± 0.36 μCi/mg protein. Furthermore, delipidated proteins subjected to autoradiography showed that a set of distinct proteins was labelled with ^{3}H-palmitate. Incorporation into phospholipids (PL) increased from 40–60% of the total amount of radio-labelled PA or MA supplied between 30 min and 2 h. Only the FA pool differed between MA and PA, with a higher FA content present after incubations with MA. Second, we investigated palmitoylation and incorporation into cellular lipids as a function of the amount of PA applied. Palmitoylation showed saturation at high PA concentrations. The percentage incorporation of ^{3}H-PA in the various lipids depended on the amount of PA added: a decline in the PL pool with a concomitant increase in the size of the diacylglycerol pool at high PA concentrations. Third, inhibition of palmitoylation by cerulenin and tunicamycin was investigated. While both were able to inhibit palmitoylation, cerulenin also inhibited the incorporation of PA into various lipid classes, indicating differences in inhibitory action. (Mol Cell Biochem **239**: 101–112, 2002)

Key words: long chain fatty acids, diacylglycerols, triacylglycerols, phospholipids

Introduction

Under normal physiological circumstances, blood-borne fatty acids are the major energy source for cardiac muscle, their oxidation providing about 70% of the energy demand of the heart [1]. In addition, they serve as building blocks of cellular membranes after esterification into phospholipids or they are incorporated in the intracellular triacylglycerol pool. Recently evidence is accumulating that fatty acids also partake in signal transduction pathways [2–5]. In this respect, it is noteworthy that fatty acids are involved in co- and post-translational modifications of proteins.

Covalent attachment of fatty acids to proteins is now a widely recognised form of protein modification. This modification is thought to be important for protein–protein interaction, conformation, protein localisation, and activity [6–10]. The two most common forms of protein acylation are modifications with myristate (C14:0) and palmitate (C16:0), both saturated fatty acids. Despite the high rate of fatty acid utilisation in the heart, information on the physiological and pathophysiological significance of protein acylation in myocardial tissue is scarce. In 1999 Van der Vusse and Roemen showed for the first time that acylated proteins are present in normoxic cardiac tissue and were

Address for offprints: G.J. van der Vusse, Department of Physiology, Cardiovascular Research Institute Maastricht, Maastricht University, P.O. Box 616, 6200 MD Maastricht, The Netherlands (E-mail: vandervusse@fys.unimaas.nl)

able to quantify the amount of acylated proteins in this tissue [11].

Protein acylation most likely depends on a number of factors which could include the availability of exogenous fatty acids, the energy demand of the cell, and incorporation and release of fatty acids into or from the endogenous phospholipid pool or triacylglycerol pool. This indicates that under circumstances in which fatty acid metabolism is altered, like hypertrophy, diabetes and ischemia/reperfusion, myocardial protein acylation may be changed.

In the present study, we explored various aspects of protein acylation in cardiac muscle cells. To this end, we used the clonal rat muscle cell line H9c2 as a cell model of cardiac myocytes. This cell line is derived from embryonic rat heart tissue and has properties of skeletal muscle [12] as well as cardiac muscle [13, 14]. As H9c2 cells possess properties of cardiac muscle, they are commonly used as a model system to study metabolic properties of myocardial tissue [15–18].

To further investigate protein acylation in cardiac tissue three separate goals were addressed in the present study. The first aspect of this study was to explore protein palmitoylation and myristoylation in H9c2 cells as function of time in relation to fatty acid incorporation into various lipid classes, such as triacylglycerol, phospholipids, and unesterified fatty acids. Second, the extent of protein palmitoylation and incorporation in cellular lipids as a function of the amount of exogenous palmitate applied was investigated. The final aspect was to explore whether protein acylation in the experimental cell line could be inhibited by cerulenin or tunicamycin, two known inhibitors of protein acylation [19, 20].

Materials and methods

Materials

[9,10-^{3}H] Palmitic acid (43 Ci/mmol) and [9,10-^{3}H] myristic acid (49 Ci/mmol) were obtained from NEN Life Science Products (Boston, MA, USA). Protease inhibitor cocktail for mammalian cell extracts (P8340), palmitate, cerulenin and tunicamycin were purchased from Sigma-Aldrich (St. Louis, MO, USA). Hyperfilm™-^{3}H was acquired from Amersham Pharmacia Biotech (Buckinghamshire, UK). The scintillation fluid Opti-fluor® was obtained from Packard Instruments (Meriden, CT, USA).

Cell culture

H9c2 cells were obtained from the American Type Culture Collection (ATCC CRL1446; Rockville, MD, USA) and were used from passages 22–25. These cells were cultured in DMEM (Gibco BRL, Life Technologies, Gaithersburg, MD, USA) supplemented with 10% foetal calf serum (FCS, Life Technologies). Cells were plated in a 25 cm^2 flask (Falcon, Becton Dickinson, Plymouth, UK) under 5% CO_2 in water saturated air. The experiments were performed 2 days after the cells were confluent, but still in the myoblast form.

Incubations of H9c2 cells

^{3}H-labelled fatty acid incorporation as a function of time
Prior to the start of the experiments, the H9c2 cells were washed 3 times with PBS. Subsequently, cells were incubated for 0, 0.5, 1 or 2 h with 100 μCi ^{3}H-labelled palmitate (43 Ci/mmol) or 125 μCi ^{3}H-labelled myristate (49 Ci/mmol), each dissolved in 10 μl ethanol and transferred to 1 ml DMEM per flask. At the end of the incubation, cells were washed 3 times with ice-cold PBS. Cells were lysed using 1 ml harvesting buffer (50 mM Tris-HCl pH7.6, 150 mM NaCl, 5 mM EDTA, 1% Triton X-100 (w/v), 10 μl protease inhibitor cocktail) for 30 min on ice. Subsequently, cell remnants were scraped from the bottom of the flask using a rubber-policeman, after which the lysate was transferred to a glass tube.

Addition of an increasing amounts of non-radiolabelled palmitate
H9c2 cells were incubated for 0, 1, or 2 h with 100 μCi (2.3 nmol) ^{3}H-labelled palmitate in the presence of 0, 25, 50, 75 or 150 nmol non-radiolabelled palmitate per incubation. The mixture of labelled and non-labelled palmitate was evaporated to complete dryness under nitrogen and subsequently dissolved in 10 μl ethanol and transferred to 1 ml DMEM per flask. The same procedure as described above was performed at the end of the incubation period.

Inhibitors (cerulenin and tunicamycin)
The stock solution of cerulenin in ethanol contained 1 mM inhibitor. An appropriate volume of the stock solution was transferred to a glass tube and dried under nitrogen. Subsequently, cerulenin was dissolved in 10 μl ethanol, leading to final concentrations of 0.1 mM and 0.3 mM cerulenin in 1 ml DMEM per incubation. H9c2 cells were washed 3 times with PBS, after which the cells were pre-incubated with medium supplemented with cerulenin for 90 min. At the end of the pre-incubation period, the medium was removed and cells were washed 3 times with PBS. Medium supplemented with 100 μCi ^{3}H-labelled palmitate per ml was added to the cells for up to 30 and 60 min. The same procedure as described above was performed at the end of the incubation period.

A stock solution of 1 mg/ml tunicamycin in 10 mM NaOH was prepared. H9c2 cells were incubated with 100 μCi ^{3}H-labelled palmitate in combination with 0, 10 or 20 μg/ml tunicamycin per 1 ml DMEM for up to 30 or 60 min. Addition of the tunicamycin solution to the medium did not led to an alteration of the pH (data not shown). The same procedure as described above was performed at the end of the incubation period.

Protein precipitation

Proteins were precipitated according to Wessel and Flügge [21]. In short, cell lysates were subjected to the following protocol. To the lysate, 4 ml methanol was added and the samples were vortexed and centrifuged (1 min, 4500 × g, 20°C). Then 2 ml of chloroform was added after which the samples were vortexed and centrifuged again (1 min, 4500 × g). For phase separation, 3 ml of milliQ water was added, and the samples were vortexed vigorously and centrifuged for 5 min at 4500 × g. The upper phase was carefully removed and transferred to a glass tube. A further 3 ml methanol was added to the lower chloroform phase and interphase. The sample was vortexed and centrifuged once more (15 min, 4500 × g, 20°C) to precipitate the protein. The supernatant was removed and combined with the earlier obtained upper phase. The precipitated proteins were subsequently subjected to extensive delipidation to remove all non-covalently bound fatty acids.

Removal of non-covalently bound fatty acids

The method for extraction of non-covalently bound fatty acids was adapted from Muszbek and Laposata [22]. The precipitated proteins were extracted with 3 ml chloroform/methanol (2:1). The sample was vigorously vortexed and incubated for 30 min at room temperature. To pellet the protein precipitates, 3 ml of methanol was added to the sample and centrifuged (15 min, 4500 × g, 20°C). The protein precipitate was extracted twice more with chloroform/methanol (2:1) and thereafter twice with chloroform/methanol (1:2). After the last extraction, the precipitated protein fraction contained no radioactivity removable by lipid extracting solvents. The supernatant of each extraction step was collected and combined with the previously obtained supernatants in a glass tube. The protein precipitates as well as the combined lipid extracts were dried completely under nitrogen.

Separation of lipid classes by thin layer chromatography (TLC)

The dried lipid extract was dissolved in 2 ml chloroform/methanol (2:1). A TLC plate coated with a 0.25 mm thick layer of silica gel 60 (Merck), 10 µl of the extract was applied. The lipid spots were pre-developed in chloroform/methanol/milliQ water/acetic acid (10:10:1:1, by volume) until the liquid front reached a level of 1 cm above the site of application of the lipid extract [23, 24]. After the plate was dried under nitrogen, the lipid spots were developed in petroleum ether /diethylether/acetic acid (120:25:1.5, by volume). After development the plate was dried under nitrogen and measured using Tracemaster 20 with detector LB2821 (Berthold, Benelux Analytical Instruments, Vilvoorde, Belgium). Lipid spot were scraped from the plate and counted with a scintillation counter.

Total lipid content of H9c2 cells

The technique for quantitative analysis of fatty acyl moieties present in the various lipid classes are based on analytical techniques published earlier by Roemen *et al.* [23]. In short, H9c2 cells were incubated with 77.3 nmol non-labelled palmitate for 0, 0.5, 1, and 2 h. Subsequently, cells were scraped from the plate with 2 ml suprasolve methanol using a rubber-policeman and transferred to a glass tube. Thereafter 4 ml of chloroform was added. At this point, a mixture of heptadecanoic acid (8 nmol) and triheptadecanoyl glycerol (125 nmol) was added to correct for losses during extraction and assay procedures. The chloroform/methanol (2:1, by volume) extraction step was repeated twice. Subsequently, the extracts were combined, dried under a stream of nitrogen and the lipid residue was taken up in 150 µl chloroform/methanol (2:1, by volume). Thereafter the concentrated lipid extract was subjected to the same TLC procedure as described above. The lipid moieties were made visible with rhodamine B-6G and fluorescein in methanol. The lipid spots corresponding to unesterified FA, triacylglycerols, and phospholipids were scraped from the plate and transmethylated according to Morrison and Smith [25].

Subsequently, the methylesters were substracted with pentane and after evaporation of pentane, dissolved in 2,2,4-trimethylpentane containing appropriate amounts of the methyl ester of pentadecanoic acid (recovery standard). Quantification of the fatty acyl moieties in the various lipid classes were achieved with the use of a standard solution of methyl esters of fatty acids, composed of (by their chemical notation) 14:0, 16:0, 16:1, 18:0, 18:1, 18:2, 18:3n-3, 18:3n-6, 20:0, 20:4, 22:0, 22:4, 22:6, 24:0 and 24:1. The methyl esters were analysed by gas chromatography using a fused-silica capillary column coated with 0.2 µm SP-2330 (30 × 0.25 mm ID; Supelco, Bellefonte, PA, USA). The starting temperature of the column was 150°C for 2 min and the column temperature was increased to 190°C at 4°C/min. Thereafter, temperature was kept at 190°C for 10 min and subsequently increased to 210°C at a rate of 10°C/min. The helium flow rate was 20 cm/sec. Detection was accomplished by flame ionisation.

Protein separation and autoradiography

Delipidated protein was solubilised in 90 μl lysis buffer (40 mM Tris, pH7.4, 2% SDS, 1 mM EDTA). A single aliquot of 10 μl was used to determine the protein content and the incorporated ^{3}H-labelled fatty acids in the sample. To the remaining fraction, 20 μl sample buffer was added (0.5 mM Tris-HCl, 20% glycerol (v/v), 4% SDS (w/v), 0.02% bromophenol blue, milliQ, 2.5% β-mercapto-ethanol(v/v)). Proteins were separated by SDS-PAGE in 4–15% gradient gels according to Laemmli [26].

Subsequently, proteins were blotted onto nitro-cellulose (pore size 0.45 μm) using the Electrophoresis System from Bio-Rad Laboratories (Hercules, CA, USA). The blots were dried and exposed to Hyperfilm™-^{3}H for 2–3 weeks.

Quantification

For quantification analysis, autoradiography films were scanned with the Fluor-S imaging system (Bio-Rad Laboratories) and images were analysed with the program Quantity One (Bio-Rad Laboratories) to obtain densitometric values. For each band detected with the Quantity One software, the optical density × mm was measured and from this the relative quantity of each detected band was assessed.

Miscellaneous techniques

Total protein content of H9c2 cells was quantified using the Micro BCA protein assay kit (Pierce, Rockford, IL, USA) with BSA as standard. Radioactivity was counted in a liquid scintillation counter.

Statistical analysis

Results are expressed as the mean ± S.D. for at least four data points, unless otherwise indicated. Analysis of experiments with two groups was performed with Two-sample Student's *t*-tests. Differences between various groups at one time point were assessed using a one-way ANOVA. For all analyses, the level of significance was set at 0.05.

Results

Protein acylation and ^{3}H-fatty acid incorporation in cellular lipids as a function of time

The amount of non-extractable radioactivity in the protein fraction after incubation with either ^{3}H-palmitate or ^{3}H-myristate was determined for several time points over a period of up to 2 h. Figure 1 shows that the amount of activity bound to the protein fraction after incubation with ^{3}H-palmitate increased in a non-linear fashion to 1.25 ± 0.56 μCi per mg protein. The time-related pattern of the covalent attachment of ^{3}H-myristate to proteins showed a more linear relationship. The amount of covalently bound ^{3}H-myristate increased from 0.36 to 0.77 μCi/mg protein after 1 and 2 h respectively. Over 99% of the labelled fatty acids present in the lysate were found in the combined lipid pools, while less than 0.3% became covalently attached to proteins.

Figures 2 and 3 show that both palmitate and myristate were preferentially incorporated in the phospholipid (PL) pool, increasing from 40% of the total activity found in this lipid fraction after 30 min to 60% after 2 h. Furthermore, there was a slight difference between the distribution over diacyl-

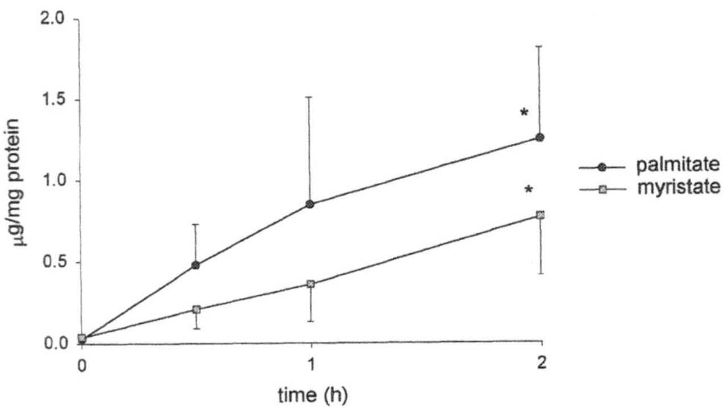

Fig. 1. Protein acylation as function of incubation time. H9c2 cells were incubated for up to 2 h with 100 μCi/ml ^{3}H-palmitate or 125 μCi/ml ^{3}H-myristate. Data are mean ± S.D. (n = 6). *p < 0.05; 0.5 vs. 2 h.

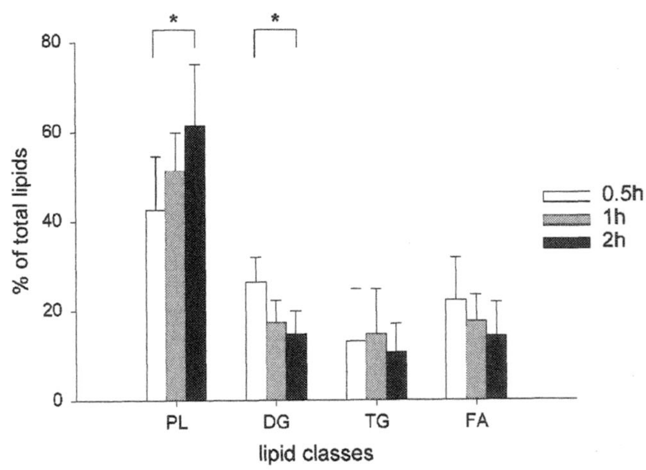

Fig. 2. ^{3}H-palmitate incorporation in various lipid classes in H9c2 cells. H9c2 cells were incubated with 100 μCi/ml ^{3}H-palmitate for up to 2 h. PL – phospholipids; DG – diacylglycerols; TG – triacylglycerols; FA – unesterified fatty acids. Data are mean ± S.D. (n = 4). *p < 0.05; 0.5 vs. 2 h.

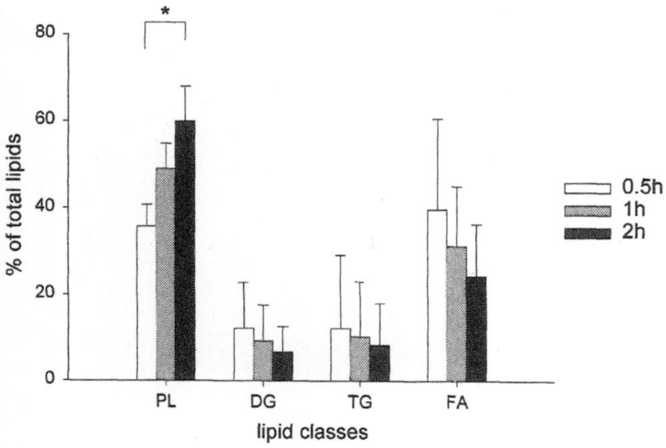

Fig. 3. ^{3}H-myristate incorporation in various lipid classes in H9c2 cells. H9c2 cells were incubated with 125 μCi/ml ^{3}H-myristate for up to 2 h. PL – phospholipids; DG – diacylglycerols; TG – triacylglycerols; FA – unesterified fatty acids. Data are mean ± S.D. (n = 4). *$p < 0.05$; 0.5 vs. 2 h.

glycerol (DG) and fatty acid (FA) pools when PA and MA are compared. Although not significant, incubations with ^{3}H-myristate led to a higher level of this fatty acid in the cellular unesterified fatty acid pool when compared with ^{3}H-palmitate, while after 0.5 h significantly less ^{3}H-myristate was incorporated in the DG pool.

Autoradiography of ^{3}H-palmitate labelled proteins

Figure 4 illustrates that routine incubation of H9c2 cells with ^{3}H-labelled palmitate resulted in a pattern in which distinct

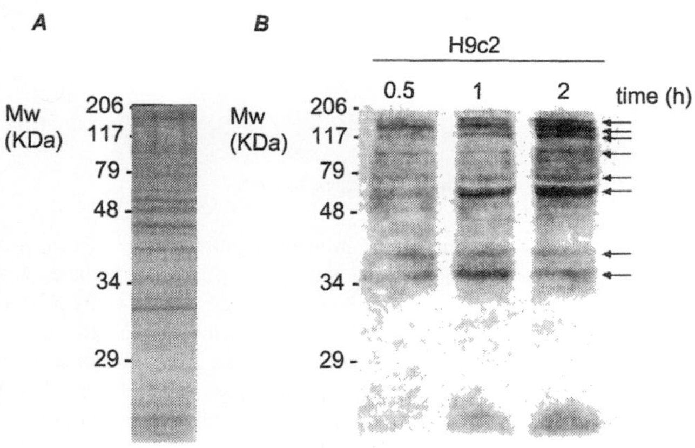

Fig. 4. A representative Coomassie blue staining of H9c2 proteins and a autoradiogram of H9c2 cells incubated with ^{3}H-palmitate for 0.5, 1 and 2 h. H9c2 cells were labelled with 100 μCi/ml ^{3}H-palmitate for up to 2 h. Panel A shows a Coomassie Blue staining of a 12% gel and Panel B shows an autoradiogram obtained after separation of h9c2 proteins using a 12% gel, blotted onto nitro-cellulose, and subjected to 3 weeks exposure to a film. Arrows indicate bands quantified as shown in Table 1.

Table 1. Quantification of covalently bound 3H-palmitate to a distinct set of proteins in H9c2 cells

Distinct proteins	Incubation time (h)		
	0.5	1	2
Mw (kDa)	Relative quantity (%)		
194	7.1	4.0	4.5
168	3.4	3.3	3.4
123	2.6	2.1	2.9
93	6.9	4.3	4.4
78	5.6	6.4	7.2
69	4.9	4.0	4.2
42	5.0	7.9	3.6
36	4.8	6.5	5.3

The autoradiogram was scanned with the Fluor-S imaging system (Biorad) and analysed with the software program Quantity-One (Biorad). The values shown represent the relative quantity (%) of the OD × mm of each selected band, after background substraction. The sum of OD × mm of all detectable bands is set at 100%. Only the most prominent bands with a relative quantity > 1.5% are shown.

proteins could be identified. Comparison of a Coomassie blue stained gel (panel A) with the autoradiogram (panel B) shows that only a selected number of proteins present in the H9c2 cell became modified by covalent attachment of ^{3}H-palmitate. Furthermore, it can be seen from Fig. 4, panel B that the density of several distinct proteins varied in time. The variation in relative quantity of the most prominent bands is shown in Table 1. It can be seen that, for instance, the relative quantity of the 78 kDa band showed a gradual increase in time, while the quantity of the 168 and 123 kDa bands remained stable.

The effect of the concentration of exogenous palmitate on palmitoylation

To study whether protein acylation could be saturated, we incubated H9c2 cells with 100 μCi ^{3}H-palmitate (2.3 nmol) in the presence of increasing amounts of non-labelled palmitate, varying from 0 nmol palmitate to 150 nmol palmitate. Table 2 shows the metabolic fate of the ^{3}H-fatty acids supplied with increasing concentrations of non-radiolabelled palmitate. The amount of radioactive label present in the lysate diminished with increasing concentrations of non-radiolabelled palmitate. At the same time, the amount of radioactive label present in the medium at the end of the incubation period was enlarged.

Taking into account the differences in specific activity, the amount of nmol palmitate per incubation can be calculated. Figure 5 shows that palmitoylation of cellular proteins was not only time dependent (as already shown in Fig. 1) but more importantly, that the amount of fatty acids that became covalently attached to proteins increased with the supplied amount of palmitate. At increasing concentrations

Table 2. Quantitative distribution of ^{3}H-palmitate with increasing concentrations of non-labelled palmitate

Amount of palmitate (nmol)	2.3	27.3	52.3	77.3	152.3
Fraction	Amount of recovered ^{3}H-palmitate (μCi)				
Medium	11.5 ± 0.9	16.2 ± 2.1	41.5 ± 9.3	43.5 ± 4.6	58.9 ± 6.8
Lysate	91.1 ± 11.6	74.6 ± 9.6	72.6 ± 12.8	67.5 ± 8.3	54.4 ± 9.9
Lipid	87.3 ± 2.3	77.5 ± 3.5	66.9 ± 1.9	60.7 ± 1.1	52.8 ± 2.4
Protein	0.23 ± 0.04	0.11 ± 0.01	0.09 ± 0.01	0.07 ± 0.01	0.05 ± 0.01

H9c2 cells were incubated with 100 μCi ^{3}H-palmitate in the presence of increasing amounts of non-labelled palmitate. Alliquots were taken from the medium at the end of the incubation period. Furthermore, alliquots were taken from the lysate, lipid fraction, and protein fraction. The lipid fraction includes the phospholipids, diacylglycerols, triacylglycerols, and unesterified fatty acids (PL, DG, TG, and FA). The values shown represent the mean ± S.D. of ^{3}H-label recovered in each fraction in μCi (n = 4).

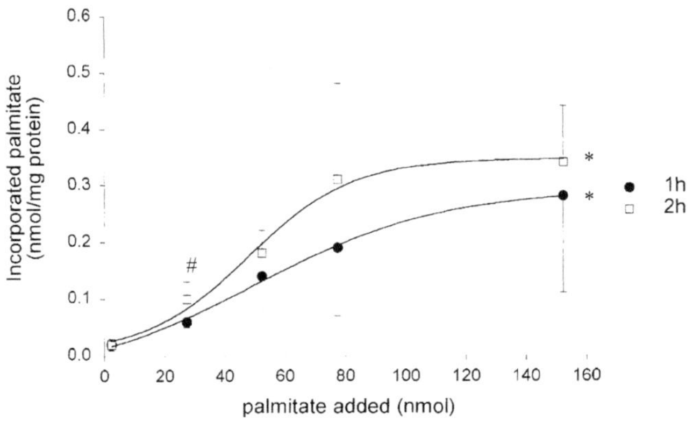

Fig. 5. Protein palmitoylation as function of exogenous palmitate applied. H9c2 cells were incubated with 100 μCi ^{3}H-palmitate in the presence of increasing amounts of non-labelled palmitate, varying from 0–150 nmol. The amount (nmol/incubation) of covalently attached palmitate per mg protein was calculated by taking into account the specific activity for each condition. Data are mean ± S.D. (n = 4). *p < 0.05; 2.3 and 27.3 nmol vs. 152.3 nmol. #p < 0.05; 1 vs. 2 h.

of exogenously supplied palmitate, the amount of palmitate incorporated in the protein fraction levelled off at approximately 0.35 nmol/mg protein.

Figure 6 shows that the amount of palmitate present in the incubation medium also influenced the amount of palmitate incorporated into the several lipid fractions. Panel A shows that increasing amounts of palmitate led to an increased incorporation of palmitate into the PL, DG and TG pool. Panel B shows that the amount of fatty acid, which remained unesterified in the intracellular lipid pool, increased linear with the amount of exogenously supplied palmitate.

In parallel experiments, H9c2 cells were incubated solely with 77.3 nmol non-labelled palmitate after which the content of PL, TG, and FA was measured by capillary gas chromatography (Fig. 7, panels A, C and E). It is shown that the amount of fatty acyl moieties present in the PL pool remained in the order of 220 nmol per incubation. The amount of unlabelled palmitate present in the PL pool remained relatively

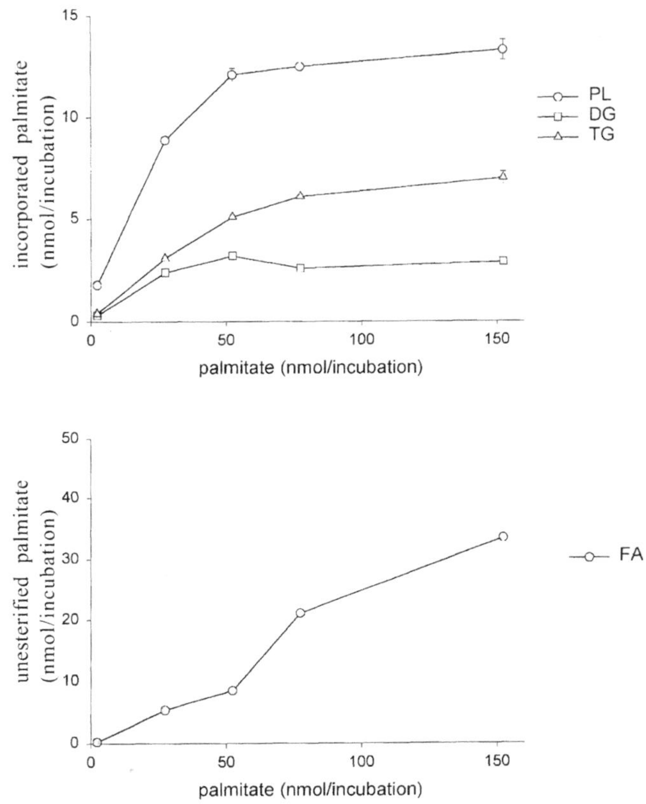

Fig. 6. Effect of amount of non-labelled palmitate on the rate of palmitate incorporation in lipid classes. H9c2 cells were incubated with 100 μCi ^{3}H-palmitate in the presence of increasing amounts of non-labelled palmitate, varying from 0–150 nmol non-radiolabelled palmitate for 2 h. Panel A depicts the amount of esterified palmitate in the phospholipids (PL), diacylglycerols (DG) and triacylglycerols (TG) pools; Panel B shows the amount of palmitate in the unesterified fatty acid (FA) pool. Data are mean ± S.D. (n = 4).

stable, varying from 47.5–53.9 nmol between 0 and 2 h, respectively (panel A). Furthermore, the total fatty acid content present in the TG pool was raised from 7.4 to 14.0 nmol per incubation after 0 and 2 h, respectively. The amount of

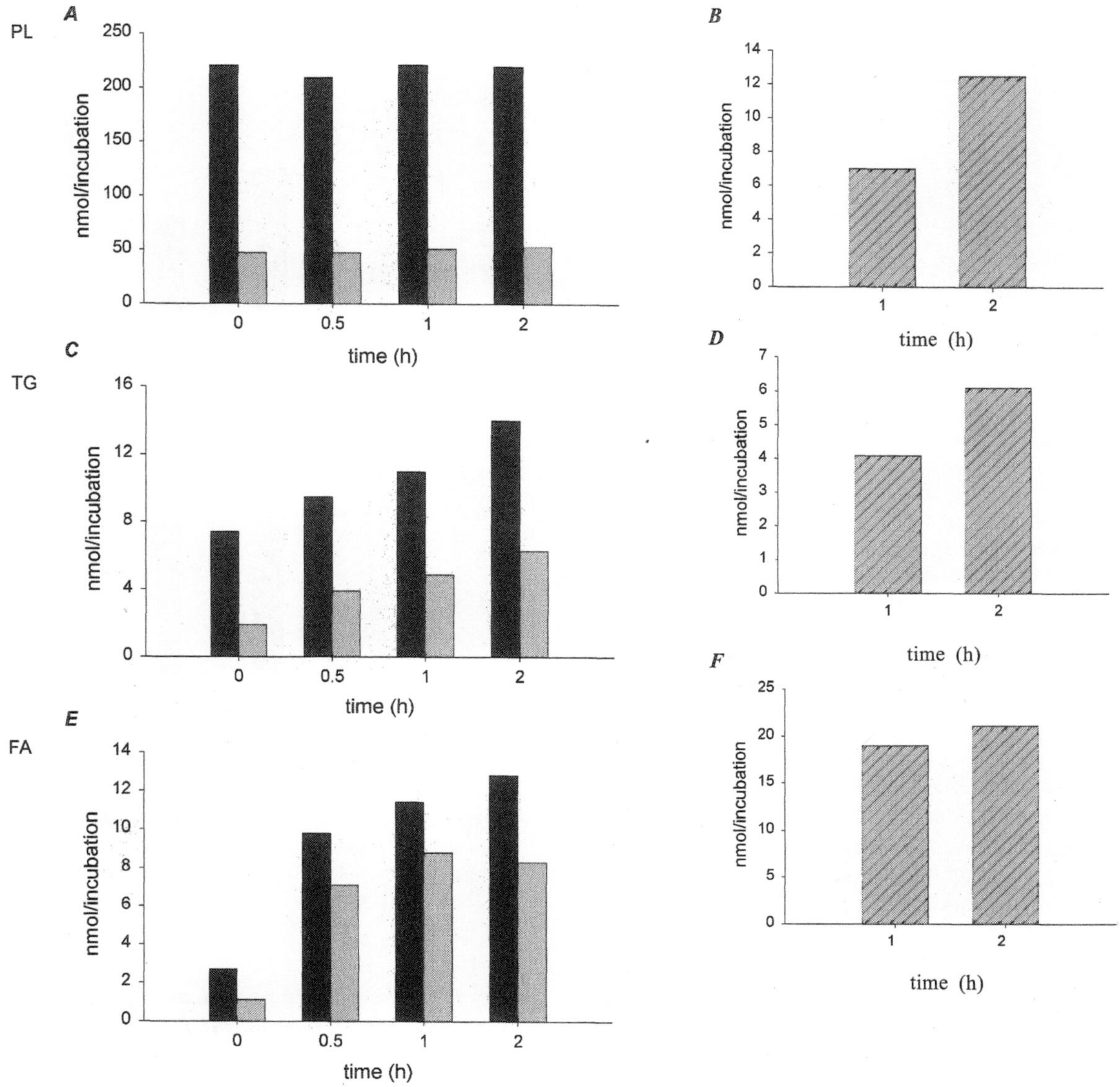

Fig. 7. Total fatty acid and palmitate content vs. ^{3}H-fatty acid incorporation in several lipid classes. Panels A, C and E show the total amount of fatty acyl moieties (black bars) and that of palmitate (grey bars) present in a particular lipid class per incubation. H9c2 cells were incubated in the presence of 77.3 nmol non-labelled palmitate for 0, 0.5, 1 and 2 h (no radioactivity added). For the phospholipids (PL), triacylglycerols (TG) and unesterified fatty acids (FA) the fatty acyl composition was determined by capillary gas chromatography. Panels B, D and F represent the amount of palmitate that became incorporated in a particular lipid class per incubation during 1 and 2 h. The amount of palmitate incorporated was determined after incubation of H9c2 cells with 100 μCi ^{3}H-palmitate (2.3 nmol) in the presence of 75 nmol non-labelled palmitate, taking into account the final specific activity of the palmitate present in the incubation mixture. Data represent the mean of two independent experiments.

unlabelled palmitate increased from 1.9 to 6.3 nmol between 0 and 2 h (panel C). The fatty acyl moieties in the FA pool increased from 2.7 to 12.8 nmol per incubation after 0 and 2 h, respectively. The amount palmitate in the FA pool first increased from 1.1 to 8.9 nmol palmitate after 0 and 1 h, respectively, and stabilised thereafter (panel E).

These results were compared with the amounts of palmitate incorporated per incubation of H9c2 cells in the presence of 77.3 nmol (75 nmol non-labelled + 2.3 nmol ^{3}H-labelled) palmitate after taking into account specific activity (Fig. 7, panels B, D and F). The amount of palmitate incorporated into the PL pool increased from 7 to 12.5 nmol after 1 and 2 h, respectively (panel B). Panel D shows that the amount of palmitate incorporated into the TG pool increased from 4.1 to 6.1 nmol after 1 and 2 h, respectively. The amount of palmitate remaining in the lipid pool as unesterified palmitate remained stable, varying from 19.0 to 21.1 nmol after 1 and 2 h, respectively (panel F).

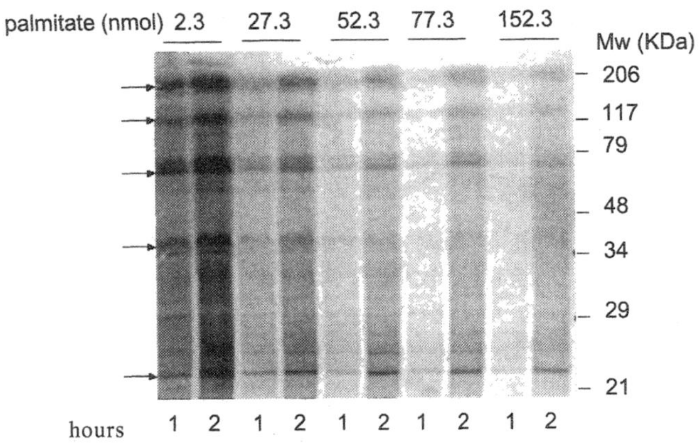

Fig. 8. Autoradiogram of H9c2 cells incubated with ^{3}H-palmitate in the presence of increasing amounts of non-labelled palmitate, varying from 0–150 nmol. Shown is an autoradiogram obtained after separation of H9c2 proteins using a 4–15% gradient gel, blotted onto nitrocellulose, and subjected to 3 weeks exposure to a film. Arrows indicate distinct bands, the quantification of which are shown in Table 2.

Autoradiography of ^{3}H-palmitate labelled proteins after incubation with increasing amounts of exogenous palmitate

Figure 8 illustrates that the amount of radioactivity covalently attached to proteins decreased with increasing amounts of non-labelled palmitate. Although this is most prominently seen between 2.3 and 27.3 nmol palmitate, higher amounts of non-labelled palmitate further diminished the number of bands that remained visible.

Table 3 shows the changes in density of the most prominent bands. The intensity of most protein bands became substantially weaker with increasing palmitate concentrations. The 22 kDa protein band was the most striking exception. The band of this protein showed only a minor decline from 0.60 to 0.45 arbitrary units (a.u.) when the exogenous palmitate concentration was increased from 2.3 to 152.3 nmol.

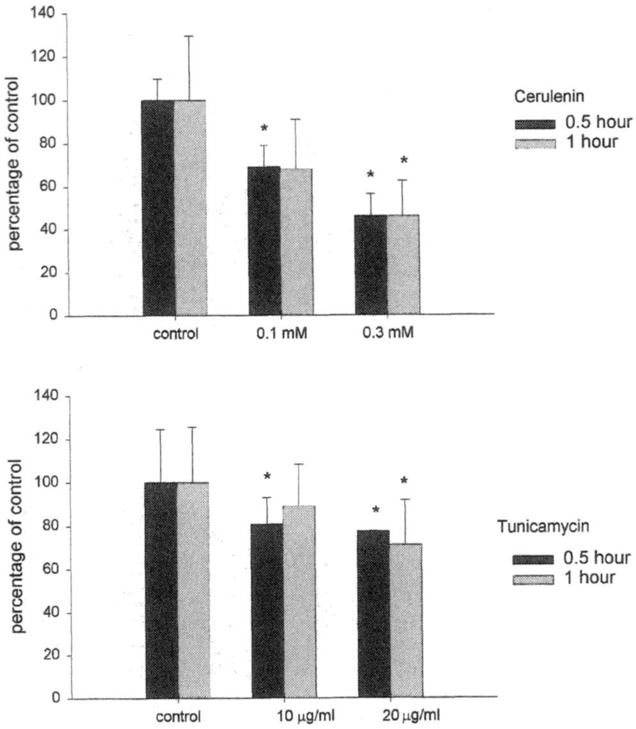

Fig. 9. Inhibition of protein palmitoylation by cerulenin and tunicamycin. (Panel A) H9c2 cells were pre-incubated for 90 min with either 0.1 or 0.3 mM cerulenin followed by incubation with 100 µCi ^{3}H-palmitate for 30 and 60 min. (Panel B) H9c2 cells were incubated with 100 µCi ^{3}H-palmitate in the presence of either 10 or 20 µg/ml tunicamycin for 30 and 60 min. The values shown represent the mean ± S.D. and are expressed as percentage of the respective control incubations. (n = 4). *$p < 0.05$. vs. control.

The effect of cerulenin or tunicamycin on inhibition of protein palmitoylation and ^{3}H-fatty acid incorporation in cellular lipids

Figure 9 shows that both cerulenin and tunicamycin were capable of inhibiting protein acylation in the H9c2 cell line.

Table 3. Quantification of covalently bound ^{3}H-palmitate to a distinct set of proteins and the influence of increasing amounts of non-labelled palmitate

	Distinct proteins										
Amount of palmitate (nmol)		2.3		27.3		52.3		77.3		152.3	
Time Quantity (a.u.)	Mw (kDa)	0.5 h	1 h	0.5 h	1 h	0.5 h	1 h	0.5 h	1 h	0.5 h	1 h
	194	0.58	1.82	0.41	1.51	0.47	0.78	0.20	0.48	0.14	0.22
	123	0.68	0.87	0.41	0.70	0.17	0.47	0.16	0.43	0.10	0.25
	69	0.47	0.66	0.38	0.54	0.25	0.48	0.13	0.24	0.13	0.30
	36	0.64	0.65	0.32	0.43	0.13	0.17	0.07	0.03	0.10	0.18
	22	0.30	0.60	0.28	0.46	0.16	0.52	0.18	0.52	0.10	0.45

H9c2 cells were incubated with 100 µCi ^{3}H-palmitate in the presence of increasing amounts, varying from 0–150 nmol, of non-labelled palmitate. The protein fraction was subjected to SDS-PAGE and blotted on nitro-cellulose. The autoradiogram was scanned with the Fluor-S imaging system (Biorad) and analysed with the software program Quantity-One (Biorad). The values shown here represent the quantity, in arbitrary units, of the OD × mm of the distinct protein bands, after background substraction. Note that only a selection of the most prominent bands are shown.

Increasing the concentration of cerulenin from 0.1 to 0.3 mM led to an inhibition of 32 and 54%, respectively, after 1 h, while increasing the tunicamycin concentration from 10 to 20 μg/ml inhibited protein acylation by 16 and 29%, respectively, after 1 h.

The effect of cerulenin and tunicamycin on ^{3}H-fatty acid on incorporation in cellular lipids was determined by measuring the radioactive label present in each lipid class. Incubations with increasing concentrations tunicamycin did not have any effect on the degree of incorporation (Fig. 10). This is in contrast with incubations in the presence of cerulenin. With increasing cerulenin concentrations, a decreasing percentage of the applied ^{3}H-palmitate was incorporated into the PL pool, while the percentage of this fatty acid remaining in the FA pool increased. There was no difference observed in the percentage of incorporation in the DG and TG pool with increasing cerulenin concentrations.

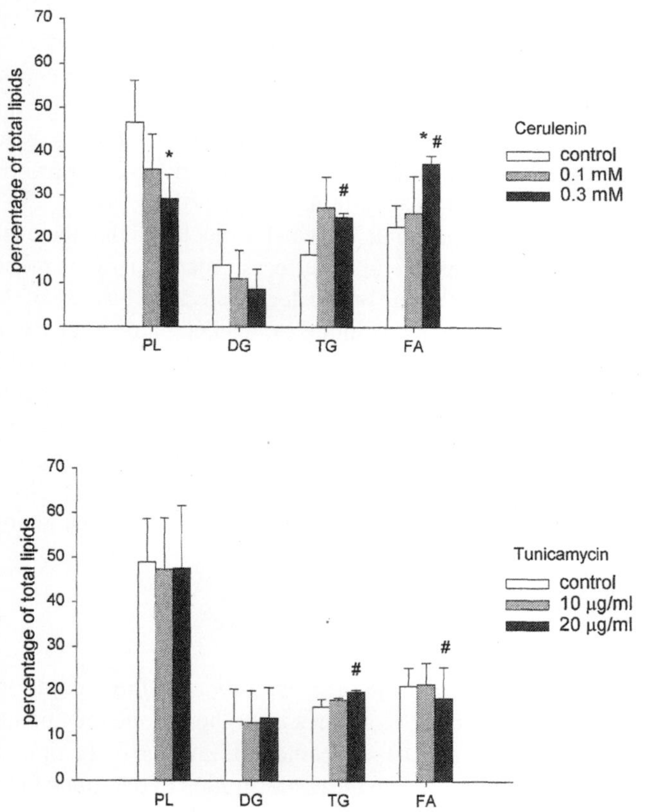

Fig. 10. Effect of cerulenin and tunicamycin on incorporation of ^{3}H-palmitate in endogenous lipid pools. (Panel A) H9c2 cells were pre-incubated for 90 min with either 0.1 or 0.3 mM cerulenin followed by incubation with 100 μCi ^{3}H-palmitate for 60 min. (Panel B) H9c2 cells were incubated with 100 μCi ^{3}H-palmitate in the presence of either 10 or 20 μg/ml tunicamycin for 60 min. PL – phospholipids; DG – diacylglycerols; TG – triacylglycerols; FA – unesterified fatty acids. *$p < 0.05$ vs. control; #$p < 0.05$, 0.3 mM cerulenin vs. 20 μg/ml tunicamycin.

Discussion

The choice of the H9c2 cell line

The main aim of this study was to investigate some basic properties of protein acylation in cardiac muscle. The H9c2 cell line was chosen as experimental model, since this cell line is derived from embryonic rat heart tissue and thus exhibits several characteristics of cardiomyocytes, such as the expression of the L-type calcium channel [13, 27], and the tissue specific splicing protein SmN [14]. Besides properties of the cardiac muscle, this cell line also possesses properties that are present in skeletal muscle [12]. An additional advantage of this cell line is that the amount of cells that can be obtained is virtually unlimited, in contrast with freshly isolated cardiomyocytes.

Experimental conditions

For this study, we chose to incubate the cells with 2.3 nmol ^{3}H-labelled palmitate, corresponding with 100 μCi. This amount was sufficient to enable us to reliably measure incorporation of radioactivity into the protein fraction.

Although myristate is not as abundantly present in cells as is palmitate, we chose to incubate the cells with an amount of myristate comparable with that of palmitate, namely 2.5 nmol myristate, corresponding with 125 μCi.

From a pilot study it was known that 2 h of incubation was an adequate time point to measure protein acylation, while there was still a measurable amount of radioactively labelled unesterified fatty acids left (unpublished result). Therefore, we routinely incubated H9c2 cells for 2 h or less.

In the experiment with increasing amounts of exogenous palmitate, we chose for total palmitate amounts ranging from 0–150 nmol, to explore whether acylation of proteins shows saturation characteristics.

Protein acylation and ^{3}H-fatty acid incorporation in cellular lipids as function of time

Protein acylation

After the tritiated fatty acids entered the cell only a small percentage of the fatty acids, namely 0.3% or less, became covalently attached to proteins. The bulk of ^{3}H-fatty acids was found to be incorporated in the phospholipids (PL), diacylglycerols (DG), and triacylglycerols (TG), or stayed unmetabolised as unesterified fatty acids (FA).

Several lines of evidence proved the covalent binding of ^{3}H-labelled fatty acids to proteins. First, the tritiated fatty acid was recovered with the protein fraction after extensive delipidation. Furthermore, the tritiated fatty acids could not be separated from the proteins by SDS-PAGE.

Protein synthesis influences palmitoylation as well as myristoylation, although the impact of protein synthesis of the latter is of more importance due to the co-translation aspects of this process. The finding that comparable amounts of myristate and palmitate became covalently attached to proteins after incubation with these tritiated fatty acids, strongly suggests a relatively high rate of protein synthesis in these cells, even though the cells are two days after reaching confluency.

Autoradiography

The differences observed in the degree of covalent attachment of tritiated fatty acids to distinct proteins bands is remarkable. The relative quantity fluctuates not only in time, but also between the different proteins. When reviewing the development of the relative quantity in time, it is most likely that different proteins have different steady state situations. Besides differences in the turnover rate of the covalent attached fatty acid, preferential incorporation into certain proteins and/or the availability of multiple acylation sites could also explain the observed results.

Incorporation into cellular lipids

After fatty acids are taken up, they are converted to fatty acyl-CoAs. These on their turn can be used for protein acylation or for incorporation of the fatty acyl moiety into diacylglycerols, triacylglycerols or phospholipids via the formation of phosphatidate (diacylglycerol 3-phosphate) [28]. After incubation of H9c2 cells with ^{3}H-labelled fatty acids the majority of the radioactive label will therefore be recovered in the PL, DG, or TG pool. When no conversion of the ^{3}H-fatty acid to fatty acyl-CoA takes place, and thus the fatty acid remains unesterified, it can be recovered in the FA pool.

Incubation of H9c2 cells with ^{3}H-palmitate as well as with ^{3}H-myristate led to incorporation in the PL-pool from 40–60% between 30 min to 2 h. It is rather surprising that the incorporation of myristate in the PL pool is on the same order of magnitude as palmitate. This is most striking, especially when taking into account the relative fatty acyl composition of the PL pool in the rat heart [11], which shows that myristate is a relatively uncommon fatty acid to be found in the PL pool. The unexpected high percentage of myristate in the PL pool could be due to the surplus of myristate in our experimental set-up as compared to normal *in vivo* circumstances in the rat heart. This would indicate that the enzymes involved in phospholipid synthesis make no distinction between palmitate and myristate, and thus that the relative fatty acid composition of the PL pool strongly depends on the species of fatty acids that are available.

The effect of the concentration of exogenous palmitate

Palmitoylation and autoradiography

Increasing exogenous palmitate concentrations does not only lead to an increased amount of palmitate covalently attached to proteins, but more importantly, it shows that protein palmitoylation is also a saturatable process. At 0.35 nmol palmitate per mg protein, the protein palmitoylation seems to be saturated. This could mean that the acylation of the majority of proteins that can be modified by covalent attachment of fatty acids have reached a steady state situation. It is of interest to note that proteins may have multiple acylation sites [29–35]. Each acylation site within a protein could have a differential palmitate turnover rate. This is supported by an observation from Little *et al.* [33] who showed that three cysteines present in the N-CAM protein are acylated to different extents. The difference in turnover rate of the different acylation sites could be due to the amino acids surrounding the acylation site or protein structure, thus favouring one acylation site over an other by inducing differences in accessibility to enzymes or mechanisms involved in protein acylation and/or deacylation.

From Fig. 8 and Table 3 it can be concluded that there is a remarkable difference in the effect of increasing the amount of exogenous palmitate on the degree covalent binding of ^{3}H-palmitate to proteins. Dilution of labelled palmitate by non-labelled fatty acids resulted, as expected, in an appreciable decline of intensity of the protein bands under investigation. The 22 kDa band, however, showed to be relatively insensitive to the increasing amount of non-labelled palmitate. This indicates that this particular protein has a relatively high capacity for covalent binding of fatty acids such as palmitate.

Incorporation into cellular lipids

Figure 6 shows that the incorporation of palmitate into the various lipid classes saturates at higher exogenous palmitate levels. In contrast, the content of unesterified palmitate linearly increased when the concentration of palmitate in the incubation mixture was enhanced. This indicates that the capacity of the H9c2 cells to incorporate exogenous fatty acids into their esterified lipid pools is exceeded when the supply of extracellular fatty acids is increased. The increase in FA and DG content at high extracellular palmitate levels in the H9c2 cells could have a major impact on intracellular signal transduction pathways leading for example to apoptosis [36–38], activation of protein kinase C [39, 40] as well as numerous other signal transduction cascades [2, 41–45].

The increase of the amount of palmitate in the TG pool during 2 h of incubation with a concomitant increase in the total amount of fatty acids in this particular pool (Fig. 7, panel C) can be explained by the increase in incorporation of exogenous palmitate (Fig. 7, panel D). In contrast, the fact that the palmitate content in the phospholipid pool did not measurable changed during the 2 h of incubation (Fig. 7, panel A) while the amount of exogenous palmitate incorporated increased substantially during the incubation period (Fig. 7 panel B), suggests that in this particular esterified lipid pool the palmitoyl moieties are subjected to a turnover process without a net increase in the amount of cellular phospholipids.

Inhibition of protein acylation by cerulenin or tunicamycin

Palmitoylation

In order to study whether protein palmitoylation could be inhibited in the H9c2 cell line we used cerulenin and tunicamycin, previously described as inhibitors of protein palmitoylation [19, 20, 46] in, for instance adipocytes and PC12 cells. Although cerulenin as well as tunicamycin are capable of inhibiting protein acylation, neither of them is capable to inhibit protein acylation for 100%. Furthermore, the percentage of inhibition of both cerulenin as tunicamycin is not linear with increasing concentrations. These two facts could indicate that there are multiple enzymes or mechanisms responsible for protein acylation. However, it is difficult to compare the effectiveness of these inhibitors. This is due to the fact that the tunicamycin used in our experiments is a mixture of four homologues of which the composition may vary from lot to lot.

Incorporation into cellular lipids

The ^{3}H-palmitate incorporation in the PL, DG, and TG pool differed between cerulenin and tunicamycin. There was no difference in pattern and percentage of incorporation between control incubations and incubations with increasing tunicamycin concentrations. This was in contrast with incubations with increasing concentration of cerulenin which led to a decrease of incorporation in the PL and DG pool, and increase of incorporation in the TG pool, while more fatty acids remained unesterified. The differences in the pattern of fatty acid incorporation in the several lipid pools could indicate that the underlying mechanism concerning the inhibition of protein acylation differs between cerulenin and tunicamycin. There are two possible ways in which cerulenin may prevent protein acylation. The first option is that cerulenin prevents the conversion of palmitate to fatty acyl-CoA. The second option is that cerulenin inhibits the synthetic pathway beyond phosphatidate (diacylglycerol 3-phosphate). This is a common intermediate in the synthesis of phospholipids and triacylglycerols, since the incorporation in label of the former was decreased but not into triacylglycerols. Both would lead to an altered degree of ^{3}H-palmitate esterification into the several lipid pools and to an increase in the unesterified FA pool. On the other hand, neither of the tunicamycin concentrations applied, altered the degree of ^{3}H-palmitate incorporation into the several lipid pools. This indicates that the mechanism of action differs between tunicamycin and cerulenin. This could mean that tunicamycin is capable to inhibit protein acylation at a point beyond fatty acyl-CoA formation or that inhibition is achieved via an other, as yet unknown mechanism.

In summary, the present study showed that the H9c2 model system is suited for studying basic properties of protein acylation in cardiac-like muscle cells. We showed that the amount of fatty acid covalently attached to proteins increased with time. The majority of the applied fatty acids were incorporated into the various lipid classes with phospholipids as the most prominent pool. Furthermore, protein acylation can be saturated by increasing the amount of fatty acids applied to the cells. This also leads to redistribution of the fatty acid over the various lipid pools. Incubations with two known protein acylation inhibitors, cerulenin and tunicamycin, in the H9c2 cell line also led to a significant decrease in covalent attachment of fatty acids to proteins. Inhibition of protein acylation by these inhibitors most likely occurs via separate mechanisms, as indicated by differential effects of these inhibitors on the incorporation of the fatty acids in the various lipid classes.

References

1. Van der Vusse GJ, Glatz JF, Stam HC, Reneman RS: Fatty acid homeostasis in the normoxic and ischemic heart. Physiol Rev 72: 881–940, 1992
2. Graber R, Sumida C, Nunez EA: Fatty acids and cell signal transduction. J Lipid Mediat Cell Signal 9: 91–116, 1994
3. Milligan G, Parenti M, Magee AI: The dynamic role of palmitoylation in signal transduction. Trends Biochem Sci 20: 181–187, 1995
4. Van Bilsen M, de Vries JE, Van der Vusse GJ: Long-term effects of fatty acids on cell viability and gene expression of neonatal cardiac myocytes. Prostagl Leukot Essent Fatty Acids 57: 39–45, 1997
5. Ordway RW, Singer JJ, Walsh JV: Direct regulation of ion channels by fatty acids. Trends Neurosci 14: 96–100, 1991
6. Dunphy JT, Linder ME: Signalling functions of protein palmitoylation. Biochim Biophys Acta 1436: 245–261, 1998
7. McCabe JB, Berthiaume LG: Functional roles for fatty acylated amino-terminal domains in subcellular localization. Mol Biol Cell 10: 3771–3786, 1999
8. Resh MD: Fatty acylation of proteins: New insights into membrane targeting of myristoylated and palmitoylated proteins. Biochim Biophys Acta 1451: 1–16, 1999

9. Stoffel RH, Inglese J, Macrae AD, Lefkowitz RJ, Premont RT: Palmitoylation increases the kinase activity of the G protein-coupled receptor kinase, GRK6. Biochemistry 37: 16053–16059, 1998
10. Yang S, Zhang L, Huang Y: Membrane association and conformational change of palmitoylated G(o)alpha. FEBS Lett 498: 76–81, 2001
11. Van der Vusse GJ, Roemen TH: Protein acylation in normoxic and ischemic/reperfused cardiac tissue. Prostagl Leukot Essent Fatty Acids 60: 299–305, 1999
12. Kimes BW, Brandt BL: Properties of a clonal muscle cell line from rat heart. Exp Cell Res 98: 367–381, 1976
13. Mejia-Alvarez R, Tomaselli GF, Marban E: Simultaneous expression of cardiac and skeletal muscle isoforms of the L-type Ca^{2+} channel in a rat heart muscle cell line. J Physiol (Lond) 478: 315–329, 1994
14. Gerrelli D, Huntriss JD, Latchman DS: Antagonistic effects of retinoic acid and thyroid hormone on the expression of the tissue-specific splicing protein SmN in a clonal cell line derived from rat heart. J Mol Cell Cardiol 26: 713–719, 1994
15. Brostrom MA, Reilly BA, Wilson FJ, Brostrom CO: Vasopressin-induced hypertrophy in H9c2 heart-derived myocytes. Int J Biochem Cell Biol 32: 993–1006, 2000
16. Moon CH, Jung YS, Kim MH, Park RM, Lee SH, Baik EJ: Protein Kinase C Inhibitors Attenuate Protective Effect of High Glucose against Hypoxic Injury in H9c2 Cardiac Cells. Jpn J Physiol 50: 645–649, 2000
17. van der Lee KA, Willemsen PH, van der Vusse GJ, van Bilsen M: Effects of fatty acids on uncoupling protein-2 expression in the rat heart. FASEB J 14: 495–502, 2000
18. Dransfeld O, Uphues I, Sasson S, Schurmann A, Joost HG, Eckel J: Regulation of subcellular distribution of GLUT4 in cardiomyocytes: Rab4A reduces basal glucose transport and augments insulin responsiveness. Exp Clin Endocrinol Diabetes 108: 26–36, 2000
19. Patterson SI, Skene JH: Inhibition of dynamic protein palmitoylation in intact cells with tunicamycin. Methods Enzymol 250: 284–300, 1995
20. Schlesinger MJ, Malfer C: Cerulenin blocks fatty acid acylation of glycoproteins and inhibits vesicular stomatitis and Sindbis virus particle formation. J Biol Chem 257: 9887–9890, 1982
21. Wessel D, Flugge UI: A method for the quantitative recovery of protein in dilute solution in the presence of detergents and lipids. Anal Biochem 138: 141–143, 1984
22. Muszbek L, Laposata M: Covalent modification of platelet proteins by palmitate. Blood 74: 1339–1347, 1989
23. Roemen TH, Keizer H, van der Vusse GJ: Quantitative assessment of gamma-linolenic acid in human blood and plasma with capillary gas chromatography. J Chromatogr 528: 447–452, 1990
24. Van der Vusse GJ, Roemen TH, Prinzen FW, Coumans WA, Reneman RS: Uptake and tissue content of fatty acids in dog myocardium under normoxic and ischemic conditions. Circ Res 50: 538–546, 1982
25. Morrison WR, Smith LM: Preparation of fatty acid methyl esters and dimethylacetals from lipids with boron fluoridemethanol. J Lipid Res 27: 600–608, 1964
26. Laemmli UK: Cleavage of structural proteins during the assembly of the head of bacteriophage T4. Nature 227: 680–685, 1970
27. Hescheler J, Meyer R, Plant S, Krautwurst D, Rosenthal W, Schultz G: Morphological, biochemical, and electrophysiological characterization of a clonal cell (H9c2) line from rat heart. Circ Res 69: 1476–1486, 1991
28. Jackowski S, Wang J, Baburina I: Activity of the phosphatidylcholine biosynthetic pathway modulates the distribution of fatty acids into glycerolipids in proliferating cells. Biochim Biophys Acta 1483: 301–315, 2000
29. Blanpain C, Wittamer V, vanderwinden JM, Boom A, Renneboog B, Lee B, Le Poul E, El Asmar L, Govaerts C, Vassart G, Doms RW, Parmentier M: Palmitoylation of CCR5 is critical for receptor trafficking and efficient activation of intracellular signaling pathways. J Biol Chem 276: 23795–23804, 2001
30. Chien AJ, Carr KM, Shirokov RE, Rios E, Hosey MM: Identification of palmitoylation sites within the L-type calcium channel beta2a subunit and effects on channel function. J Biol Chem 271: 26465–26468, 1996
31. Dietzen DJ, Hastings WR, Lublin DM: Caveolin is palmitoylated on multiple cysteine residues. Palmitoylation is not necessary for localization of caveolin to caveolae. J Biol Chem 270: 6838–6842, 1995
32. Jin H, Xie Z, George SR, O'Dowd BF: Palmitoylation occurs at cysteine 347 and cysteine 351 of the dopamine D(1) receptor. Eur J Pharmacol 386: 305–312, 1999
33. Little EB, Edelman GM, Cunningham BA: Palmitoylation of the cytoplasmic domain of the neural cell adhesion molecule N-CAM serves as an anchor to cellular membranes. Cell Adhes Commun 6: 415–430, 1998
34. Tao N, Wagner SJ, Lublin DM: CD36 is palmitoylated on both N- and C-terminal cytoplasmic tails. J Biol Chem 271: 22315–22320, 1996
35. Veit M, Sollner TH, Rothman JE: Multiple palmitoylation of synaptotagmin and the t-SNARE SNAP-25. FEBS Lett 385: 119–123, 1996
36. De Vries JE, Vork MM, Roemen TH, de Jong YF, Cleutjens JP, van der Vusse GJ, Van Bilsen M: Saturated but not mono-unsaturated fatty acids induce apoptotic cell death in neonatal rat ventricular myocytes. J Lipid Res 38: 1384–1394, 1997
37. Listenberger LL, Ory DS, Schaffer JE: Palmitate-induced apoptosis can occur through a ceramide-independent pathway. J Biol Chem 276: 14890–14895, 2001
38. Sparagna GC, Hickson-Bick DL: Cardiac fatty acid metabolism and the induction of apoptosis. Am J Med Sci 318: 15–21, 1999
39. Huang KP The mechanism of protein kinase C activation. Trends Neurosci 12: 425–432, 1989
40. Bell RM, Hannun YA, Loomis CR: Mechanism of regulation of protein kinase C by lipid second messengers. Symp Fundam Cancer Res 39: 145–156, 1986
41. Bishop WR, Bell RM: Functions of diacylglycerol in glycerolipid metabolism, signal transduction and cellular transformation. Oncogene Res 2: 205–218, 1988
42. Aragones J, Jones DR, Martin S, San Juan MA, Alfranca A, Vidal F, Vara A, Merida I, Landazuri MO: Evidence for the involvement of diacylglycerol kinase in the activation of hypoxia-inducible transcription factor 1 by low oxygen tension. J Biol Chem 276: 10548–10555, 2001
43. Ohanian J, Ohanian V: Lipid second messenger regulation: The role of diacylglycerol kinases and their relevance to hypertension. J Hum Hypertens 15: 93–98, 2001
44. Rodriguez de Turco EB, Tang W, Topham MK, Sakane F, Marcheselli VL, Chen C, Taketomi A, Prescott SM, Bazan NG: Diacylglycerol kinase epsilon regulates seizure susceptibility and long-term potentiation through arachidonoyl-inositol lipid signaling. Proc Natl Acad Sci U S A 98: 4740–4745, 2001
45. Topham MK, Prescott SM: Diacylglycerol kinase zeta regulates Ras activation by a novel mechanism. J Cell Biol 152: 1135–1143, 2001
46. Jochen AL, Hays J, Mick G: Inhibitory effects of cerulenin on protein palmitoylation and insulin internalization in rat adipocytes. Biochim Biophys Acta 1259: 65–72, 1995

Molecular and Cellular Biochemistry **239**: 113–119, 2002.

Alternative lipid mobilization: The insect shuttle system

Dick J. van der Horst, Dennis van Hoof, Wil J.A. van Marrewijk and Kees W. Rodenburg

Department of Biochemical Physiology and Institute of Biomembranes, Utrecht University, Utrecht, The Netherlands

Abstract

Lipid mobilization in long-distance flying insects has revealed a novel concept for lipid transport in the circulatory system during exercise. Similar to energy generation for sustained locomotion in mammals, the work accomplished by non-stop flight activity is powered by oxidation of free fatty acids (FFA) derived from endogenous reserves of triacylglycerol. The transport form of the lipid, however, is diacylglycerol (DAG), which is delivered to the flight muscles associated with lipoproteins. In the insect system, the multifunctional lipoprotein, high-density lipophorin (HDLp) is loaded with DAG while additionally, multiple copies of the exchangeable apolipoprotein, apoLp-III, associate with the expanding particle. As a result, lipid-enriched low-density lipophorin (LDLp) is formed. At the flight muscles, LDLp-carried DAG is hydrolyzed and FFA are imported into the muscle cells for energy generation. The depletion of DAG from LDLp results in the recovery of both HDLp and apoLp-III, which are reutilized for another cycle of DAG transport. A receptor for HDLp, identified as a novel member of the vertebrate low-density lipoprotein (LDL) receptor family, does not seem to be involved in the lipophorin shuttle mechanism operative during flight activity. In addition, endocytosis of HDLp mediated by the insect receptor does not seem to follow the classical mammalian LDL pathway.

Many structural elements of the lipid mobilization system in insects are similar to those in mammals. Domain structures of apoLp-I and apoLp-II, the non-exchangeable apolipoprotein components of HDLp, are related to apoB100. ApoLp-III is a bundle of five amphipathic α-helices that binds to a lipid surface very similar to the four-helix bundle of the N-terminal domain of human apoE. Despite these similarities, the functioning of the insect lipoprotein in energy transport during flight activity is intriguingly different, since the TAG-rich mammalian lipoproteins play no role as a carrier of mobilized lipids during exercise and besides, these lipoproteins are not functioning as a reusable shuttle for lipid transport. On the other hand, the deviant behavior of similar molecules in a different biological system may provide a useful alternative model for studying the molecular basis of processes related to human disorders and disease. (Mol Cell Biochem **239**: 113–119, 2002)

Key words: lipid transport, lipoprotein, lipophorin, apolipophorin III, apoE, low-density lipoprotein receptor, insect lipophorin receptor, diacylglycerol, exercise, insect flight

Introduction

Generally, animal organisms (including man) accumulate lipid reserves in body depots. These reserves are usually composed of triacylglycerol (TAG), the fatty acids of which are derived from lipids that are either absorbed from the food or resulting from *de novo* biosynthesis, and transported in the circulation associated with lipoproteins. Under a variety of circumstances, such as during prolonged muscular exercise, lipid is mobilized from the storage tissue and subsequently transported in the circulation to provide the metabolic fuel for the energy-generating pathways in the working muscles. Hydrolysis of TAG from mammalian adipose tissue results in the release of free fatty acids (FFA), which are transported in the circulatory system bound to albumin, an abundant serum protein harboring several fatty acid binding sites [25].

Address for offprints: D.J. Van der Horst, Department of Biochemical Physiology and Institute of Biomembranes, Utrecht University, Padualaan 8, 3584 CH Utrecht, The Netherlands (E-mail: d.j.vanderhorst@bio.uu.nl)

Flight activity of insects has been shown to provide a fascinating, yet relatively simple model system for studying the regulation of processes involved in energy metabolism during exercise. Active insect flight muscles are among the most energy-demanding tissues known, which is reflected in the extremely high metabolic rates sustained during long-distance flight. Despite many similarities with the mammalian system, including the storage of lipid reserves as TAG and their utilization for the energy-generating processes in the muscles as FFA, particularly the exercise-induced processes of lipid mobilization and transport of lipids appear to be different, as FFA-albumin complexes are not involved. Instead, diacylglycerol (DAG) is released, and transported in the insect blood (hemolymph) associated with the same lipoprotein system that is also used for the transport of lipids during resting conditions, for instance following absorption of lipids from the intestine (reviewed in [20, 21, 26]).

In this short review, recent advances on the molecular and cellular aspects of flight-related lipid mobilization and transport are discussed from the perspective that many structural elements of the lipid binding proteins of mammals and insects are (very) similar, but that their functioning is apparently different. Additionally, since insects constitute the largest – and very successful – animal group on earth, understanding of their solutions for lipid transport has a clear intrinsic value.

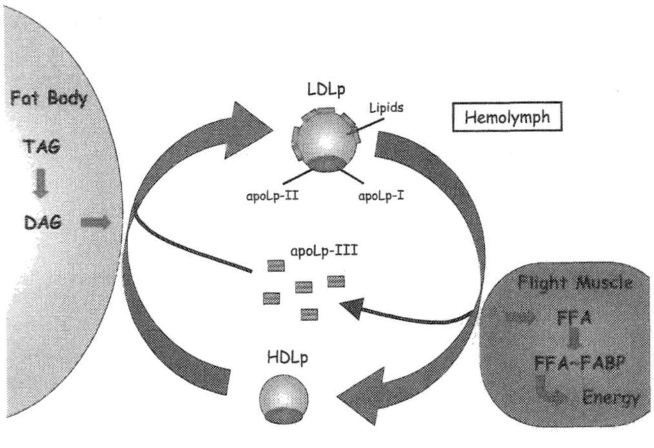

Fig. 1. Schematic overview of the insect lipoprotein shuttle system during flight activity. DAG mobilized from TAG reserves are loaded onto HDLp; by concomitant association of apoLp-III to the expanding particle, LDLp is formed. The FFA resulting from DAG hydrolysis at the flight muscles results are imported into the cell and transported by FABP to the mitochondria to power energy generation. Both HDLp and apoLp-III are recovered and used for another cycle of DAG uptake. The non-exchangeable apolipoproteins of HDLp and LDLp (apoLp-I and apoLp-II) and lipid are schematically indicated. The exchangeable apoLp-III is indicated in its lipid-free form (double-boxed rectangles) and its lipid-associated form in LDLp (two connected single-boxed rectangles) to mark the lipid-induced conformational change of the protein.

Lipid transport during sustained flight activity

Lipid mobilization in long-distance flying insects such as the migratory locust has revealed a novel concept for lipid transport involving the multifunctional lipoprotein in the insect circulatory system (hemolymph), high-density lipophorin (HDLp) (Fig. 1). Flight activity induces lipid mobilization in the adipose (fat body) cells and ultimately results in the release of DAG from TAG stores. The lipid is loaded onto preexisting HDLp particles circulating in the hemolymph. Concomitant with this loading process, multiple copies of an exchangeable apolipoprotein, apolipophorin III (apoLp-III) reversibly associate with the expanding particle surface and finally, the particle is converted into lipid-enriched low-density lipophorin (LDLp), which has a high capacity to transport DAG molecules to the flight muscles. Hydrolysis of LDLp-carried DAG at the flight muscles results in dissociation of apoLp-III from the particle until, eventually, HDLp is recovered. Both HDLp and apoLp-III are reutilized for another cycle of lipid uptake and transport, thus acting as an efficient lipid shuttle mechanism (see Fig. 1). This lipid loading and unloading of lipophorin particles during sustained flight activity does not require their internalization by either lipid donor or recipient cells and consequently, the increased capacity for lipid transport is achieved without additional lipophorin biosynthesis (reviewed in [20, 21, 24, 26–28]).

The FFA produced are oxidized for energy generation; the mechanism by which the extracellularly liberated FFA are translocated across the plasma membrane may involve (trans)-membrane protein(s) similar to those in mammals (see [9] for a recent review) but is as yet unknown. The intracellular transport of FFA is mediated by a fatty acid binding protein (FABP) (reviewed in [12]); the high amino acid sequence similarity (82%) between the FABP of locust flight muscle and that of human skeletal muscles is reflected in a strong structural similarity (Fig. 2).

Structure of insect lipoproteins

Insect hemolymph generally contains abundant amounts of a single lipoprotein particle, HDLp. A characteristic feature of HDLp is its ability to function as a reusable shuttle for a variety of lipids by the selective loading and unloading of lipid components at target tissues (see [21, 27] for recent reviews). Typically, HDLp is a spherical particle of 450–600 kDa with a density similar to that of mammalian HDL (~ 1.12 g/ml), which usually contains DAG as its prevalent lipid component in addition to phospholipids, sterols and hydrocarbons. The lipoprotein generally contains one copy of each of

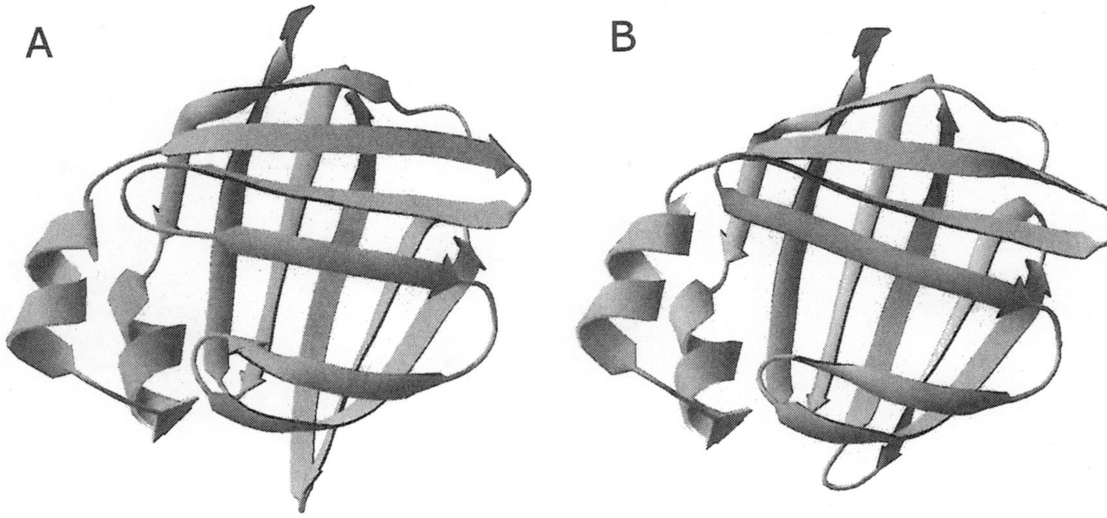

Fig. 2. Three dimensional structures of human and locust FABP. (A) The x-ray crystal structure of human FABP [34]. (B) A model of locust (*Locusta migratoria*) FABP, build by homology using the amino acid sequence of the locust FABP [14] and the model-building facility at the expasy internet site (http:// www.expasy.ch/swissmod/swiss-model.html). The figure is a Swiss-PDB viewer ribbon display, based on the coordinates of the human FABP structure (PDB entry: 1HMS) and the locust model.

the two non-exchangeable apolipoproteins, apolipophorin I (apoLp-I) and apolipophorin II (apoLp-II) (~ 240 and ~ 80 kDa, respectively), which are derived from a common precursor protein through post-translational cleavage [30]. Molecular characterization of the apolipophorin precursor has recently been disclosed for a few insect species (reviewed in [27]), showing that the protein is arranged with apoLp-II at the N-terminal end and apoLp-I at the C-terminal end (hence also termed apoLp-II/I). Similar to the other apolipophorin precursors known to date, in locust apoLp-II/I (3359 amino acid residues), the N-terminus of apoLp-I is preceded by the amino acid sequence RQKR [2], suggesting that in the generation of apoLp-II and apoLp-I by proteolytic processing of the precursor protein, dibasic processing endoproteases of the subtilisin family are likely involved.

The N-terminal ~ 900 amino acid residues of apoLp-II/I show significant homology with the same part of human apolipoprotein B100 (apoB100), its non-exchangeable counterpart in human (very) low-density lipoproteins ((V)LDL), invertebrate and vertebrate vitellogenins, and the large subunit of mammalian microsomal triglyceride transfer protein (MTP) [1, 16]. These homologies indicate that the apolipophorin precursors are members of a large lipid transfer protein (LLTP) superfamily that emerged from an ancestral molecule designed to ensure a pivotal event in the intra- and extracellular transfer of lipids and liposoluble substances [1]. In addition, the amino acid residues 2800–3050 show homology to domain D of the von Willebrand factor [1, 2]; the function of this domain is unknown. Further sequence analysis of the known insect apoLp-II/I sequences comparable to that described [1] (K.W. Rodenburg, unpublished data) revealed the presence of a large β-strand containing domain between residues ~ 900 and ~ 2200, which potentially relates to the large lipid-binding domain in apoB100 termed β-strand motif, two of which are found in the apoB100 protein [22]. Between residues ~ 2250 and ~ 2750 in apoLp-II/I, an α-helical domain shows sequence homology to the amphipathic α-helical cluster α2 in apoB100 [22].

The domain structures of apoLp-II/I and apoB100 are summarized in Fig. 3. Interestingly, whereas apoB and MTP interact in human lipoprotein assembly (reviewed in [10, 32]) and the N-terminal region of apoB is necessary for enabling MTP responsiveness [11, 32], the cleavage of the insect apoLp-II/I precursor protein occurs within the MTP-homologous domain. Most vitellogenins contain a similar dibasic processing site in their N-terminal domain. The function of the processing step is unknown, but may be indispensable for the intracellular assembly of the resulting apolipoproteins to macromolecular particles. Recently, we have constructed a protein expression system using insect *Sf*9 cells, secreting truncation variants of processed locust apoLp-II/I (K.W. Rodenburg, M.A.M. Kasperaitis and D.J. Van der Horst, unpublished data). This system will allow us to study the role of the two apolipophorins individually in the mechanism of macromolecular assembly of lipophorin particles, which constitutes one of the major issues in insect lipoprotein metabolism. Interestingly, the elucidation of the *Drosophila* genome sequence indicates the presence of an MTP gene in insects. This strongly suggests that MTP is involved in the assembly of insect lipoproteins similar that of mammals (see [10] for

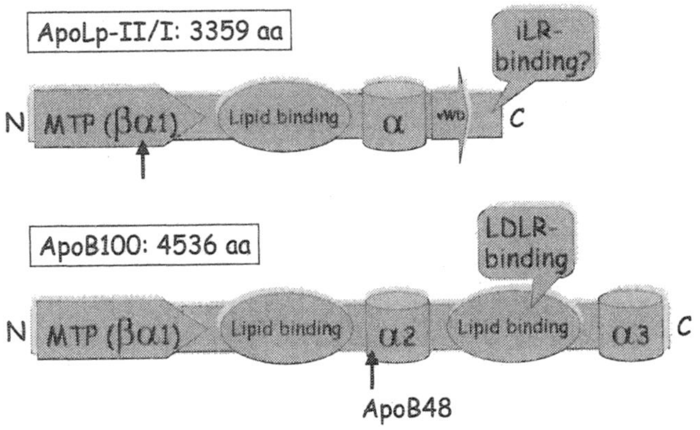

Fig. 3. Linear domain structure organization of locust apoLp-II/I and human apoB100. The domains are depicted in the N- to C-terminal direction, indicated by N and C, respectively. The MTP-homologous domain or βα1 domain [22], indicated by MTP (βα1), is shown by a block arrow, the lipid-binding domains or β-strand motifs are oval-shaped, the amphipathic α-helical domains, indicated by α, α2, or α3, are barrel-shaped, and the von Willebrand factor domain D (vWD) is displayed by the wide-sized block arrow in the C-terminus of apoLp-II/I. The vertical arrow in bold in the MTP (βα1) domain of apoLp-II/I marks the dibasic processing site. A similar arrow in apoB100 marks the C-terminal position of apoB48, the non-exchangeable apolipoprotein of chylomicrons. The putative LDL receptor (LDLR) and iLR-binding sites in apoB and apoLp-II/I are indicated by the callouts above the respective bars. Note that the iLR-binding site in apoLp-II/I is unknown, as implied by the question mark in the callout. The domains are not drawn to scale. aa, amino acid.

a review). Human MTP has a higher affinity for TAG than for DAG [32]. It will be imperative to assess whether the reverse is true for insect MTP.

Apolipophorin III structure and binding to a lipoprotein surface

An intriguing phenomenon pertinent to the shuttle mechanism for transporting lipids during exercise is the association of the exchangeable apoLp-III with the expanding surface of the lipophorin particle during lipid loading. Insect apoLp-III (~ 18–20 kDa) is one of the best examples of the reversible existence of exchangeable apolipoproteins in lipid-free and lipoprotein-associated states and serves as a model for studies of apolipoproteins with lipoprotein surfaces (recently reviewed in [18, 21, 27]). ApoLp-III is rich in amphipathic α-helices, characterized by well-defined polar and non-polar faces, which are proposed to interact with the aqueous milieu (blood or hemolymph) and hydrophobic lipids, respectively, at the lipoprotein surface (reviewed in [18]). As is apparent from comparison of their x-ray structures, there is considerable similarity between the structural organization of apoLp-III [4] and the 22 kDa N-terminal domain of human apolipoprotein E (apoE) [33] (Fig. 4). ApoLp-III represents the only full-length apolipoprotein of which the atomic three-dimensional structure has been determined [4]. The three-dimensional structures, which were determined in the lipid-free state, reveal that both proteins consist of elongated amphipathic α-helices, organized as an antiparallel α-helix bundle. ApoLp-III is a five-helix bundle while the N-terminal domain of human apoE is a four-helix bundle. These helices are organized such that their hydrophobic faces are oriented toward the center of the bundle while their hydrophilic faces are exposed to the aqueous environment. This molecular architecture explains the water solubility of these proteins in the absence of lipid and additionally allows for postulating that conformational changes of the protein may accompany its association with lipid. Toward this end it has been proposed that both apoLp-III and the N-terminal domain of human apoE undergo a pronounced lipid-triggered opening of the helix bundle, allowing the hydrophobic interior of the helices to interact with a lipid surface [3, 4, 18, 23] (see Fig. 4). Over the last two years, based on the interaction with lipoproteins or phospholipid vesicles, different models for the lipid-associated conformation of these apolipoproteins have been proposed [8, 19, 23, 29]. The different results obtained, however, may depend on the obvious differences in the hydrophobic surfaces exposed in these lipid-binding assays, imposing the apolipoproteins to adopt different conformations upon interaction.

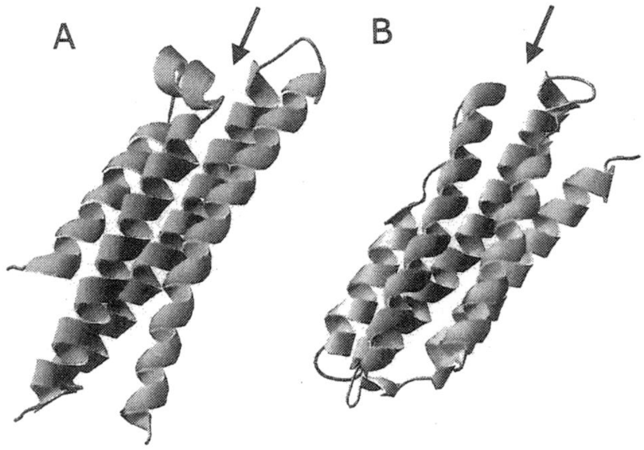

Fig. 4. Amphipathic α-helix bundle structure of apoE and apoLp-III. The 4 and 5 helix bundle organization of the x-ray crystal structures of the N-terminal 22 kDa fragment of apoE [33] (A) and full length apoLp-III [4] (B), respectively, are presented in a ribbon display. The arrows indicate the putative initial lipid-binding sites in the proteins.

Receptor-mediated endocytosis of lipophorin

In the concept of lipid transport during exercise in insects, a major difference between the functioning of lipoproteins of mammals and insects is the selective mechanism by which the insect lipoproteins transfer their hydrophobic cargo. Circulating HDLp particles may serve as a DAG donor or acceptor, dependent on the physiological situation, and function in a reusable lipid shuttle without additional synthesis or increased degradation of the apolipoprotein matrix, as discussed above (see also Fig. 1). In apparent contrast to this concept, in fat body tissue of larval and young adult locusts an endocytic receptor involved in receptor-mediated uptake of HDLp was demonstrated [6]. An endocytic receptor has been cloned and sequenced, and identified as a novel member of the LDL receptor family [7]. Clones of an LDL receptor-deficient Chinese hamster ovary (CHO) cell line stably transfected with this insect lipophorin receptor (iLR) revealed a punctate fluorescent staining upon incubation with fluorescently-labeled HDLp, which was absent from mock-transfected cells. These data suggest that iLR may function *in vivo* as an endocytic receptor [7].

This insect receptor represents the first identification of an invertebrate LDL receptor family member with an extracellular domain composed of a single ligand-binding domain, harboring multiple imperfect cysteine-rich repeats, and an epidermal growth factor (EGF)-precursor homology domain [7]. This type of lipoprotein receptor has been found in many vertebrates, including mammals. Three-dimensional structures of elements representing these two iLR domains (Figs 5A and 5B) bear a striking resemblance to those of mammalian LDL receptors. Despite their pronounced structural similarity, however, the functioning of both receptors in lipid transport in the two animal groups seems to be intriguingly different (see below). Recently, in mosquito oocytes, the presence of a receptor homologous to the locust fat body lipophorin receptor was demonstrated [5].

In the binding of LDL to its receptor, the most C-terminal 1000 amino acids in apoB are involved [3]. Remarkably, whereas iLR has 57% amino acid sequence similarity to the human LDL receptor and the domain structure of apoLp-II/I resembles that of apoB100, apoLp-II/I does not show homology to this C-terminal part of apoB100 (see Fig. 3). Therefore, it will be a challenge to localize the receptor-binding domain of apoLp-II/I.

Expression of iLR mRNA in fat body cells is down-regulated during adult development [7], which is consistent with the down-regulation of receptor-mediated endocytosis of lipophorins in fat body tissue [6] and suggests that this receptor is not involved in the lipophorin shuttle mechanism operative in the flying insect. Nevertheless, an endocytic uptake of HDLp seems to conflict with the selective process of lipid transport between HDLp and fat body cells without degradation of the lipophorin matrix. However, the pathway followed by the internalized HDLp may be different from the classical receptor-mediated lysosomal pathway typical of LDL receptor-internalized ligands. Therefore, an important question is whether this novel iLR, in contrast to the other members of the LDL receptor family, is able to recycle its ligand after intracellular trafficking. Recent observations on the uptake and trafficking of protein-labeled HDLp in CHO cells, mediated by iLR, indicate that, in contrast to the classical mammalian LDL pathway, HDLp accumulates in one large juxtanuclear organelle per cell. Co-localization of HDLp with transferrin identified these organelles as recycling compartments (Fig. 6; D. Van Hoof, K. W. Rodenburg and D. J. Van der Horst, in preparation), suggesting that HDLp is returned to the cell membrane in a manner that is similar to the transferrin recycling pathway [17] and thus escapes from the lysosomal pathway. Such a deviation from the classical LDL pathway would imply that, upon endocytosis, mammalian and insect lipoproteins have distinct fates. Indeed, the recycling of HDLp highlights the functioning of the lipoprotein as an alternative shuttle.

Concluding remarks and perspectives

While a number of interesting similarities exist in the processes in the metabolic processes operative in mammals and insects, the latter have evolved evolutionarily to different mechanisms for substrate mobilization and make use of other carrier molecules for substrate transport, offering adaptations to their specific requirements. The data presented above highlight both the parallels and the differences in structure and functioning of lipid binding proteins pertinent to exercise. Several topics related to this field, for instance TAG mobilization in insects vs. vertebrates, and the regulation of hormone-sensitive lipase in vertebrate adipocytes the much less understood process of lipolysis in insect fat body cells, are covered in a recent review [27].

Several structural elements of the lipid mobilization system in insects are similar to those in mammals. For instance, domain structures of the non-exchangeable apolipoprotein components of HDLp are related to apoB100, and the exchangeable apoLp-III is a bundle of five amphipathic α-helices that binds to a lipid surface very similar to the four-helix bundle of the N-terminal part of human apoE. Additionally, an insect HDLp receptor was identified as a novel member of the mammalian LDL receptor family. Despite these similarities, the functioning of the insect lipoprotein in energy transport during flight activity is remarkably different. Mammalian lipoproteins are not used to carry mobilized lipids during exercise, nor do the apoB-containing lipoproteins (including chylomicrons) act as a reusable lipid shuttle since their remnant particles are finally subjected to receptor-mediated endocytosis by members of the LDL receptor family, resulting in complete degradation of the particle. Intriguingly, apoE plays a critical role in the metabolism of these lipoproteins as it mediates clearance of their remnant particles from

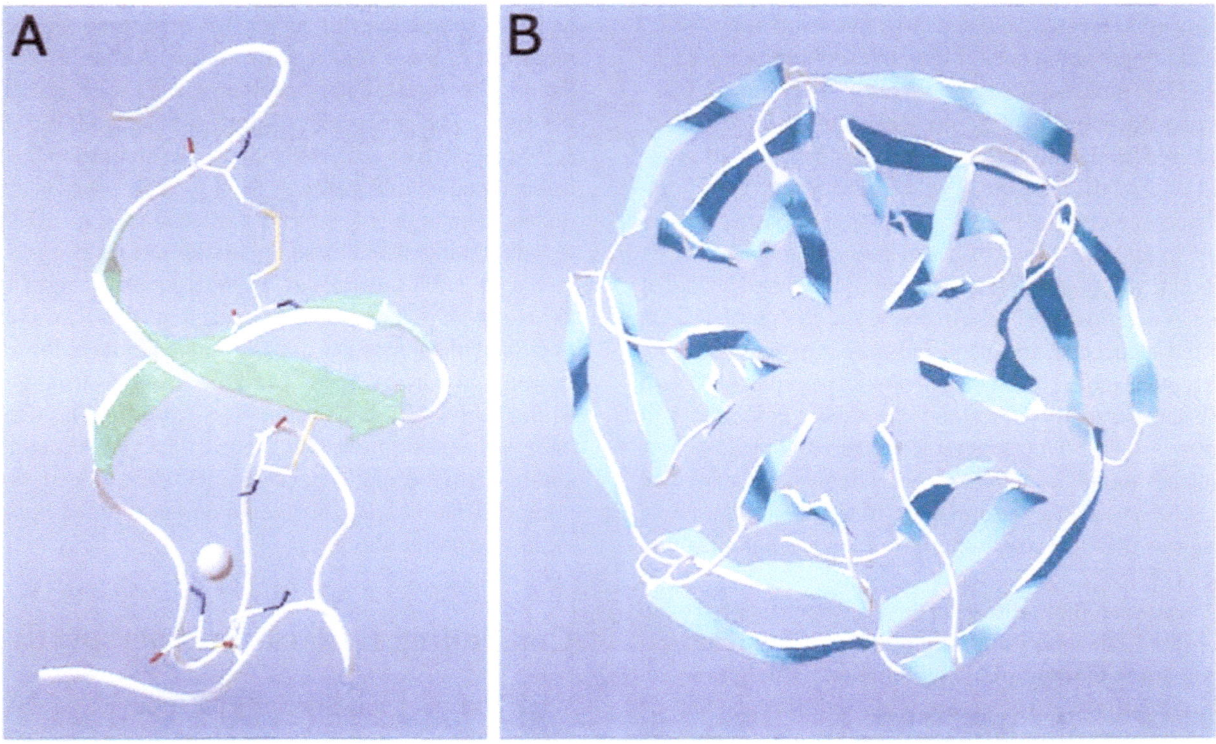

Fig. 5. Protein models of extracellular domains of iLR based on elucidated atomic structures of the human LDL receptor. (A) The first cysteine-rich repeat of the ligand-binding domain, harboring two strands, forming a β-sheet (in green), six cysteine residues (including the backbone atoms, O in red and N in blue) forming disulfide bonds (in yellow) indicated in wire frame representations, and a possible Ca^{2+}-ion (grey sphere) incorporated at the C-terminal region of the repeat. (B) The six-bladed β-propeller of the EGF-precursor homology domain. It is noteworthy that, based on the sequence homology of iLR with the protein database (PDB), models can be build for 87% of the complete insect receptor and 96% of its extracellular part. See the legend of Fig. 2 for details on model building and display representation.

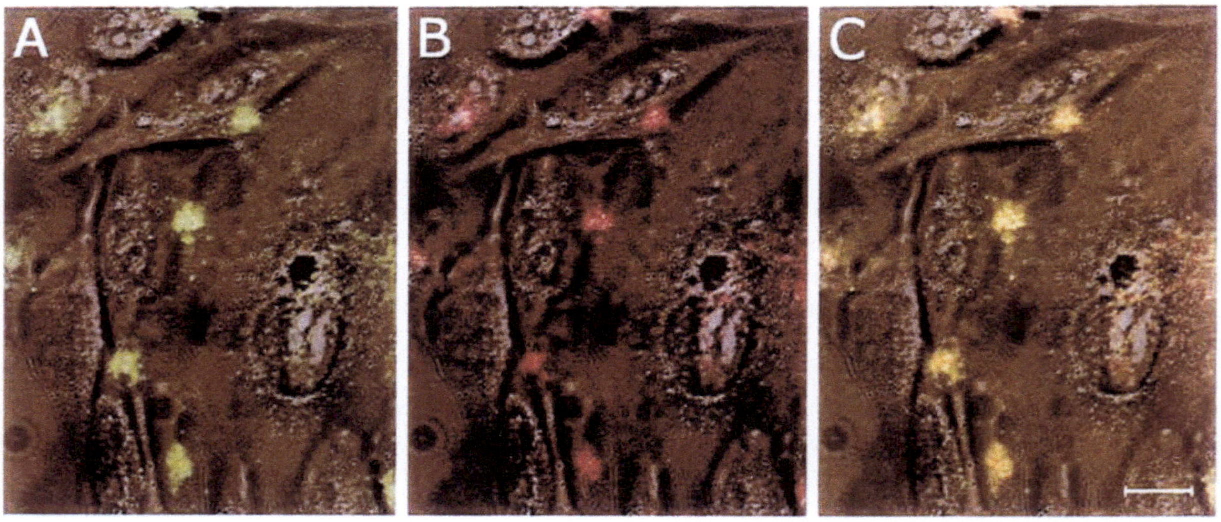

Fig. 6. Confocal laser microscopic digital images of CHO cells incubated with fluorescently-labeled HDLp (A) and transferrin (B) after a chase period of 20 min. Co-localization of both ligands is visualized in yellow when images A and B are merged (C). Scale bar: 10 μm.

the circulation by acting as a ligand for members of the LDL receptor family (reviewed in [15]). Recycling of apoE occurs after dissociation of the lipoproteins in intracellular endosomal compartments (reviewed in [13]). In contrast, apoLp-III dissociates from LDLp after extracellular lipid depletion at the flight muscles, resulting in the recovery of HDLp. Nonetheless, the different functioning of similar molecules in the insect system may render the latter system a useful alternative model for studying the molecular basis of processes related to human disorders and disease.

Acknowledgements

We thank Dr. Willem J. Hage (Netherlands Institute for Developmental Biology, Utrecht, The Netherlands) for discussions on and technical help with the confocal laser scanning microscope.

References

1. Babin PJ, Bogerd J, Kooiman FP, Van Marrewijk WJA, Van der Horst DJ: Apolipophorin II/I, apolipoprotein B, vitellogenin, and microsomal triglyceride transfer protein genes are derived from a common ancestor. J Mol Evol 49: 150–160, 1999
2. Bogerd J, Babin PJ, Kooiman FP, André M, Ballagny C, Van Marrewijk WJA, Van der Horst DJ: Molecular characterization and gene expression in the eye of the apolipophorin II/I precursor from *Locusta migratoria*. J Comp Neurol 427: 546–558, 2000
3. Borén J, Lee I, Zhu W, Arnold K, Taylor S, Innerarity TL: Identification of the low density lipoprotein receptor-binding site in apolipoprotein B100 and the modulation of its binding activity by the carboxyl terminus in familial defective apo-B100. J Clin Invest 101: 1084–1093, 1998
4. Breiter DB, Kanost MR, Benning MM, Wesenberg G, Law JH, Wells MA, Rayment I, Holden HM: Molecular structure of an apolipoprotein determined at 2.5-Å resolution. Biochemistry 30: 603–608, 1991
5. Cheon H-M, Seo S-J, Sun J, Sappington TW, Raikhel AS: Molecular characterization of the VLDL receptor homolog mediating uptake of lipophorin in oocytes of the mosquito *Aedes aegypti*. Insect Biochem Mol Biol 31: 753–760, 2001
6. Dantuma NP, Pijnenburg MAP, Diederen JHB, Van der Horst DJ: Developmental down-regulation of receptor-mediated endocytosis of an insect lipoprotein. J Lipid Res 38: 254–265, 1997
7. Dantuma NP, Potters M, De Winther MPJ, Tensen CP, Kooiman FP, Bogerd J, Van der Horst DJ: An insect homolog of the vertebrate low density lipoprotein receptor mediates endocytosis of lipophorins. J Lipid Res 40: 973–978, 1999
8. Fisher CA, Narayanaswami V, Ryan RO: The lipid-associated conformation of the low density lipoprotein receptor binding domain of human apolipoprotein E. J Biol Chem 275: 33601–33606, 2000
9. Glatz JFC, Van der Vusse GJ: Cellular fatty acid binding proteins: Their function and physiological significance. Prog Lipid Res 35: 243–282, 1996
10. Gordon DA, Jamil H: Progress towards understanding the role of microsomal triglyceride transfer protein in apolipoprotein-B lipoprotein assembly. Biochim Biophys Acta 1486: 72–83, 2000
11. Gretch DG, Sturley SL, Wang L, Lipton BA, Dunning A, Grunwald KAA, Wetterau JR, Yao Z, Talmundi P, Attie AD: The amino terminus of apolipoprotein B is necessary but not sufficient for microsomal triglyceride transfer protein responsiveness. J Biol Chem 271: 8682–8691, 1996
12. Haunerland NH: Transport and utilization of lipids in insect flight muscle. Comp Biochem Physiol 117B: 475–482, 1997
13. Heeren J, Beisiegel U: Intracellular metabolism of triglyceride-rich lipoproteins. Curr Opin Lipidol 12: 255–260, 2001
14. Maatman RGHJ, Degano M, Van Moerkerk HTB, Van Marrewijk WJA, Van der Horst DJ, Sacchettini JC, Veerkamp JH: Primary structure and binding characteristics of locust and human muscle fatty-acid-binding proteins. Eur J Biochem 221: 801–810, 1994
15. Mahley RW, Huang Y: Apolipoprotein E: From atherosclerosis to Alzheimer's disease and beyond. Curr Opin Lipidol 10: 207–217, 1999
16. Mann CJ, Anderson TA, Read J, Chester SA, Harrison GB, Köchl S, Ritchie PJ, Bradbury P, Hussain FS, Almey J, Vanloo B, Rosseneu M, Infante R, Hancock JM, Levitt DG, Banaszak LJ, Scott J, Shoulders C: The structure of vitellogenin provides a molecular model for the assembly and secretion of atherogenic lipoproteins. J Mol Biol 285: 391–408, 1999
17. Mukherjee S, Ghosh RN, Maxfield FR: Endocytosis. Physiol Rev 77: 759–803, 1997
18. Narayanaswami V, Ryan RO: Molecular basis of exchangeable apolipoprotein function. Biochim Biophys Acta 1483: 15–36, 2000
19. Narayanaswami V, Wang J, Schieve D, Kay CM, Ryan RO: A molecular trigger of lipid-binding induced opening of a helix bundle exchangeable apolipoprotein. Proc Natl Acad Sci USA 96: 4366–4371, 1999
20. Ryan RO: Dynamics in insect lipophorin metabolism. J Lipid Res 31: 1725–1739, 1990
21. Ryan RO, Van der Horst DJ: Lipid transport biochemistry and its role in energy production. Annu Rev Entomol 45: 233–260, 2000
22. Segrest JP, Jones MK, De Loof H, Dashti N: Structure of apolipoprotein B-100 in low density lipoproteins. J Lipid Res 42: 1346–1367, 2001
23. Soulages JL, Arrese EL: Dynamics and hydration of the α-helices of apolipophorin III. J Biol Chem 275: 17501–17509, 2000
24. Soulages JL, Wells MA: Lipophorin: the structure of an insect lipoprotein and its role in lipid transport in insects. Adv Prot Chem 45: 371–415, 1994
25. Spector AA: Fatty acid binding to plasma albumin. J Lipid Res 16: 165–179, 1995
26. Van der Horst DJ: Lipid transport function of lipoproteins in flying insects. Biochim Biophys Acta 1047: 195–211, 1990
27. Van der Horst DJ, Van Marrewijk WJA, Diederen JHB: Adipokinetic hormones of insect: Release, signal transduction, and responses. Int Rev Cytol 211: 179–240, 2001
28. Van der Horst DJ, Weers PMM, Van Marrewijk WJA: Lipoproteins and lipid transport. In: D.W. Stanley-Samuelson, D.R. Nelson (eds). Insect Lipids: Chemistry, Biochemistry and Biology. University of Nebraska Press, Lincoln, NE, 1993, pp 1–24
29. Weers PMM, Van der Horst DJ, Ryan RO: Interaction of locust apolipophorin III with lipoproteins and phospholipid vesicles: Effect of glycosylation. J Lipid Res 41: 416–423, 2000
30. Weers PMM, Van Marrewijk WJA, Beenakkers AMT, Van der Horst DJ: Biosynthesis of locust lipophorin: Apolipophorin I and II originate from a common precursor. J Biol Chem 268: 4300–4303, 1993
31. Weisgraber KH: Apolipoprotein E: Structure-function relationships. Adv Prot Chem 45: 249–302, 1994
32. Wetterau JR, Lin MCM, Jamil H: Microsomal triglyceride transfer protein. Biochim Biophys Acta 1345: 136–150, 1997
33. Wilson C, Wardell MR, Weisgraber KH, Mahley RW, Agard DA: Three dimensional structure of the LDL-receptor binding domain of human apolipoprotein E. Science 252: 1817–1822, 1991
34. Zanotti G, Scapin G, Spadon P, Veerkamp JH, Sacchettini JC: Three-dimensional structure of recombinant human muscle fatty acid-binding protein. J Biol Chem 267: 18541–18550, 1992

Molecular and Cellular Biochemistry **239**: 121–130, 2002.

Giant membrane vesicles as a model to study cellular substrate uptake dissected from metabolism

D.P.Y. Koonen,[1] W.A. Coumans,[1] Y. Arumugam,[2] A. Bonen,[2] J.F.C. Glatz[1] and J.J.F.P. Luiken[1]

[1]*Department of Physiology, Cardiovascular Research Institute Maastricht (CARIM), Maastricht University, Maastricht, the Netherlands;* [2]*Department of Kinesiology, University of Waterloo, Waterloo, Ontario, Canada*

Abstract

In order to use giant vesicles for substrate uptake studies in metabolically important tissues, we characterized giant vesicles isolated from heart, liver, skeletal muscle and adipose tissue. We investigated which cell types and which plasma membrane regions are involved in giant vesicle formation and we examined the presence of transporters for metabolic substrates. Analysis of giant vesicles with markers specific for distinct cell types and distinct domains of the plasma membrane reveals that the plasma membrane of parenchymal cells, but not endothelial cells, are the source of the vesicle membranes. In addition, plasma membrane regions enriched in caveolae and involved in docking of recycling vesicles from the endosomal compartment are retained in giant vesicles, indicating that KCl-induced alterations in recycling processes are involved in giant vesicle formation. Giant vesicles contain vesicular lumen consisting of the soluble constituents of the cytoplasm including, fatty-acid binding proteins. Furthermore, giant vesicles isolated from heart, liver, skeletal muscle and adipose tissue are similar in size (10–15 μm) and shape and do not contain subcellular organelles, providing the advantage that substrate fluxes in the different organs can be studied independently of the surface/volume ratio but most importantly in the absence of intracellular metabolism. (Mol Cell Biochem **239**: 121–130, 2002)

Key words: giant membrane vesicles, fatty acid uptake, glucose uptake, fatty acid transporters, plasma membrane regions

Introduction

Studies on the cellular uptake of metabolic substrates, such as long-chain fatty acids (FA) [1–3] glucose [4, 5] and lactate [6] are usually carried out with isolated cells. A major difficulty of these studies, however, is that the uptake process cannot be separated from subsequent metabolism. Therefore, plasma membrane vesicles have been proposed as an alternative model to study substrate uptake. In contrast to isolated cells, membrane vesicles consisting of plasma membranes do not contain subcellular organelles, thus providing the advantage that ligand uptake can be studied in the absence of metabolism. Routinely, plasma membranes have been prepared from a crude membrane fraction obtained by extensive homogenization, followed by differential centrifugation and purification on density gradients. Such membrane vesicles are relatively small (< 1 μm) [7–10] and are oriented only 50–60% right-side out. Moreover, small vesicles have a variable degree of contamination with intracellular membranes [8, 10–13].

In an attempt to overcome these disadvantages of the use of small vesicles for uptake studies, a preparative method was developed to obtain giant membrane vesicles [14]. Their discovery was the result of a coincidental observation in patch-clamp studies with frog skeletal muscle [14]. It was found that large membrane structures cleave off from cutaneous pectoris muscle when treated with collagenase in the presence of high (120 mM) KCl [14]. The molecular mecha-

Address for offprints: D.P.Y. Koonen, Department of Physiology, CARIM, Maastricht University, P.O. Box 616, NL-6200 MD Maastricht, The Netherlands (E-mail: d.koonen@fys.unimaas.nl)

nism leading to excision of large membrane spheres from the surface of muscle cells is not yet understood. In this respect, it has been observed that the proteolytic treatment by itself is not sufficient for formation of vesicles from the sarcolemma, but that the combination of a collagenase with KCl, as iso-osmotic medium, is crucial to initiate this budding process [14]. It has been speculated that a high extracellular KCl concentration causes a disequilibrium between cellular endocytotic and exocytotic pathways resulting in an increased delivery of membranes to the sarcolemma [14]. Collagenase activity is probably solely required to degrade the collagen matrix for creating space for the budding event. It is likely that the budding-off of large membrane structures from the sarcolemma might be a compensatory mechanism for the sarcolemma to prevent package problems due to excessive supply of membranes from the recycling compartment.

Initially, giant vesicles were obtained from rat and human skeletal muscle in sufficient yield to allow the measurement of transport rates of ions and substrates [9, 14, 15]. Applying the established procedure for isolation of giant vesicles from rat hindlimb muscle but using a different type of collagenase, we succeeded in harvesting giant vesicular structures also from heart, epididymal fat pads and, liver [3, 16]. Giant vesicles have two major advantages when compared to small vesicles, as they are fully right-side-out and, on average, have a 10-fold greater diameter. The latter leads to a 10^3-fold greater volume over small vesicles and a surface-to-volume ratio similar to that of mammalian cells, which allows a more accurate study of initial uptake rates of substrates because of a volume-dependent delay in their efflux [14]. Furthermore, studies analyzing possible contamination of the vesicle preparation with intracellular membranes concluded that giant vesicle membranes were largely, if not completely, of sarcolemmal origin [9, 14, 17]. However, it is not known whether the membranes of the giant vesicles are randomly build up of plasma membrane, or, alternatively, that specialized regions of the plasma membrane contribute to the formation of giant vesicles. Furthermore, considering the heterogeneity of cell types present in the different mammalian organs, it might be possible that both parenchymal (myocytes, hepatocytes, adipocytes) and endothelial cells deliver plasma membranes for the formation of giant vesicles. Although microscopic studies indicate that the process of giant vesicle formation in heart and skeletal muscle occurs at the level of myocytes [14], we cannot exclude that endothelial cells are involved in giant vesicle formation.

In order to use giant vesicles for substrate uptake studies in metabolically important tissues, a more complete characterization is appropriate. Therefore, we investigated (i) which cell types and (ii) which plasma membrane regions are involved in giant vesicle formation from rat heart, skeletal muscle, liver and epididymal fat pads. These study aims were addressed by measuring marker proteins in the various preparations. In addition, we investigated (iii) the presence of transporters for metabolic substrates in giant vesicles from these tissues, namely two membrane FA transporters, FA translocase (FAT/CD36) and plasmalemmal FA binding protein ($FABP_{pm}$), the cytoplasmic FA transporter $FABP_c$, and the membrane glucose transporter GLUT4.

Materials and methods

Materials

Bovine serum albumin (BSA) (fraction V) and collagenases type 2a and type VII were purchased from Sigma, St. Louis, MO, USA. Collagenase type II was from Worthington (Freehold, NJ, USA). Non-fat dry milk (Marvel) was obtained from Premier Brands, Moreton, UK. Western blot reagents were from Bio-Rad Laboratories (Hercules, CA, USA) and the enhanced chemiluminescence (ECL) kit was from Amersham Pharmacia Biotech (Buckingham, UK). Gp60 was detected with a polyclonal antibody, kindly provided by Dr. C. Tiruppathi, University of Illinois, Chicago, IL, USA. Caveolae were detected with a rabbit polyclonal (cav-1) and a monoclonal (cav-3) antibody purchased from Transduction Laboratories (Lexington, KY, USA). Dihydropyridine receptor (DHPR) was detected with a mouse monoclonal (IgG2a) antibody (Affinity Bioreagents) against the α2-subunit of rat skeletal muscle DHPR, a subunit also common to the cardiac channel isoform [18]. Anti-Pan Cadherin, a rabbit monoclonal antibody (Sigma, St. Louis, MO, USA) was used for detection of N-Cadherin, whereas the insulin receptor was detected with a mouse monoclonal (IgG1) antibody (Transduction Laboratories, Lexington, KY, USA) directed against the β-chain of the insulin receptor (IRβ). A mouse monoclonal (IgG2a) antibody (Biogenesis, Poole, UK) was used to recognise the rat transferrin receptor protein. FAT/CD36 was detected with a monoclonal antibody (MO25) directed against human CD36, kindly provided by Dr. N. Tandon, Otsuka Pharmaceuticals, Bethesda, MD, USA. A rabbit polyclonal against rat hepatic membrane fatty acid binding protein [2] was used to detect FABPpm (gift from Dr. D. Sorentino, Mount Sinai Medical Center, New York City, USA). A polyclonal immuno-A purified GLUT4 antibody was obtained from East Acres Biologicals (Southbridge, MA, USA). Rabbit anti-mouse immunoglobulin horseradish peroxidase and porc anti-rabbit immunoglobulin horseradish peroxidase were obtained from DAKO (Glostrup, Denmark).

Isolation of giant vesicles

Giant vesicles were prepared from skeletal muscle, heart, liver and adipose fat pads of adult Sprague–Dawley rats. In general, tissues were cut into thin layers (~ 1–3 mm thick) and incubated in vesicle preparation medium (140 mM KCl/10 mM MOPS (pH 7.4)) supplemented with aprotinin (10 mg/ml) and collagenase type II (150 U/ml) for 1 h at 34°C in a shaking waterbath. Following incubation, tissues were washed with KCl/MOPS and 10 mM EDTA. The supernatant was collected and Percoll® (final concentrations of 16% (v/v)) and aprotinin (10 mg/ml) were added. The resulting suspension was placed at the bottom of a density gradient consisting of a 3-ml middle layer of 4% Nycodenz (w/v) and a 1-ml KCl/MOPS upper layer. This sample was centrifuged at 60 × g for 45 min at room temperature. Subsequently, the vesicles were harvested from the interface of the upper and middle layer, diluted in KCl/MOPS, recentrifuged at 12,000 × g for 5 min, and finally resuspended in KCl/MOPS. Minor adjustments were made to this method to prepare giant vesicles from heart and epididymal fat pads. Collagenase type II (0.3%, w/v) was used for heart tissue, while collagenase type 2a (0.05%, w/v) was used for adipose tissues. Furthermore, 100 mM $CaCl_2$ was added to the incubation medium of heart and adipose tissues, and heart tissues were incubated for 90 min. Protein concentration was determined and vesicles were immediately used for glucose and FA transport experiments. Vesicles were photographed and sized from the photomicrograph. Thereafter, vesicle samples were stored at –80°C for immunoanalyses.

Fatty acid and glucose uptake by giant vesicles

Palmitate uptake was measured by addition of unlabeled and radiolabeled 0.3 μCi [9,10-^{3}H]palmitate and 0.06 μCi [^{14}C]mannitol in a 0.1% BSA KCL/MOPS solution to 40 μl of vesicles (~ 80 μg protein). The reaction was carried out at room temperature for precisely 15 sec. Palmitate uptake was terminated by addition of 1.4 ml of ice-cold KCL/MOPS-2.5 mM HgCl and 0.1% BSA. The sample was quickly recentrifuged at maximal speed in a microcentrifuge for 2 min. The supernatant was discarded, and radioactivity was measured in the tip of the tube. Non-specific uptake was measured by adding the stop solution prior to the addition of the radiolabeled palmitate solution.

Glucose uptake studies were carried out by addition of 40 μl 0.1% BSA in KCl/MOPS containing 0.3 μCi [^{3}H]D-glucose (200 μM) and 0.06 μCi [^{14}C] mannitol. Incubations were carried out at room temperature for precisely 2 min and radioactivity was determined as described above.

Sample preparation for Western blotting

Total tissue homogenates were prepared from heart, liver, skeletal muscle and adipose tissues. Briefly, tissues (~ 60 mg) were homogenized in 1 ml of TES-buffer (in mM: 10 Tris-HCL, 2 diNaEDTA, 250 sucrose, pH 7.4) supplemented with leupeptine (mg/100ml) and PMSF-DMSO for 5 interrupted 5-sec bursts with an Ultra Turrax homogenizer (Janke and Kunkel, Ika-Labortechnik) set at 24,000 rpm. Thereafter, cells were disintegrated in 4 cycles (5-sec on, 15-sec off) with an amplitude of 10–12 using an Ultrasonic Disintigrater (Soniprep 150, Sanyo Gallenkamp). Protein concentration was determined in triplicate by the Bicinchonic acid assay (Sigma, St. Louis, MO, USA) with the use of BSA as standard and samples were stored at –80°C.

Western blotting

Total tissue homogenates (10 μg) and vesicle (5 μg) samples were separated on 10 and 12% SDS-polyacrylamide gels (150 v for 1 h). GLUT4 was detected by loading 10 μg of homogenate and vesicle protein. Proteins were then transferred to Trans-Blot® pure nitrocellulose membranes (Bio-Rad Laboratories, Hercules, CA, USA) (100 v for 90 min). The membranes were blocked with TBS-T (20 mM Tris-base, 137 mM NaCl, 0.1 M HCl (pH 7.5), 0.1% Tween-20) containing either 7.5% BSA (FAT/CD36), or 5% non-fat dry milk (FABPpm, cav-1, cav-3, GP60, IRβ, DHPR, transferrin receptor) for 1 h at room temperature. Membranes were incubated overnight at 4°C using the relevant primary antibodies diluted in (i) TBS-T (FAT/CD36), (ii) TBS-T containing 5% non-fat dry milk (FABPpm, cav-1, GLUT-4, N-cadherin, IRβ, DHPR), (iii) 1% non-fat dry milk (cav-3, GP60), and (iv) TBS-T with 1% BSA (transferrin receptor). The following concentrations of primary antibodies were used: FAT/CD36 1:20.000, FABPpm 1:3000, cav-1 1:5000, cav-3 1:6000, GP60 1:500, GLUT-4 1:7000, N-cadherin 1:3000, IRβ 1:3000, DHPR 1:6000, and transferrin 1:1000. Following primary antibody incubation membranes were washed 5 × 5 min with TBS-T and incubated for 1 h at room temperature with either rabbit-anti mouse immunoglobulin G-horseradish peroxidase-conjugated secondary antibody (FAT/CD36, IRβ, DHPR: 1:20.000; cav-3: 1:24.000; transferrin receptor: 1:2000) or porc-anti rabbit G-horseradish peroxidase-conjugated secondary antibody (FABPpm, GP60, N-cadherin: 1:3000; cav-1: 1:5000; GLUT-4: 1:7000). Thereafter, membranes were washed 5 × 5 min with TBS-T and 1 × 5 min with TBS (20 mM Tris-base, 137 mM NaCl, 0.1 M HCl (pH 7.5)). Immune complexes were detected with enhanced chemiluminescence (ECL) by exposing the membranes to film (Hyperfilm-ECL) at room temperature according to the instructions of the manufacturer. Protein band densities were obtained by scanning the films with a MultiImager (FluorS™, Bio-Rad) connected to a personal computer followed by densitometry using the computer program Quantity One (Bio-Rad Laboratories, Hercules, CA, USA).

Assay of amount of H-FABP and L-FABP

The content of heart-type cytoplasmic FABP (H-FABP) in homogenates and giant vesicles from heart and skeletal muscles were determined by a sandwich-type enzyme-linked immunosorbent assay (ELISA) as previously described [19]. The content of liver-type cytoplasmic FABP (L-FABP) in homogenates and liver giant vesicles was also determined by a sandwich ELISA [19].

Statistics

Data are represented as mean ± S.E.M. Statistical significance was calculated with paired *t*-tests ($p \leq 0.05$).

Results and discussion

Size determination of giant membrane vesicles

The average sizes of giant vesicles from the various tissues were measured to be between 10 and 15 μm (Table 1), which is in agreement with earlier studies [16, 20]. The striking similarity in vesicles sizes among the different organs (only liver vesicles were slightly, but significantly smaller than adipose and heart vesicles, $p < 0.05$) are in marked contrast with the difference in average sizes of the cells from which the vesicles originate. For example, the average cardiac myocyte is a rod shaped cell, measuring 100 μm in length and 30 μm in width, while the average hepatocyte measures 20 μm in diameter. Moreover, skeletal muscle cells are not single-cell entities but form multinucleated muscle fibers of considerable length upon cell fusion processes. Another example is the similarity in diameter of adipose vesicles obtained from lean and from obese Zucker rats [16], despite the fact that the cell diameter of obese adipocytes is considerably greater [21]. Taken together, these findings suggest that the molecular mechanism of vesicle formation is not dependent on cell form or size, but probably involves protein complexes in the plasma membrane at distinct distances of each other. These protein complexes could then be involved in the vesicle budding event in a manner resembling other subcellular vesicle formation processes such as endocytosis and exocytosis.

Table 1. Sizes of giant vesicles prepared from different tissues

Tissue	Vesicle diameter (mm)
Heart	12.6 ± 0.5
Skeletal muscle	14.3 ± 0.7
Liver	11.4 ± 0.5*
Adipose tissue	14.8 ± 0.5

Comparison in heart, skeletal muscle and liver are based on 125 vesicles measured in each tissue from 5 rats. Adipose tissue vesicles were measured in 3 rats (75 vesicles). Data are represented as mean ± S.E.M. *Significantly different from hindlimb muscle and adipose tissue ($p < 0.05$).

Characterization of giant vesicles

The degree of enrichment of cell-specific and plasma membrane region-specific markers within the plasma membranes of giant vesicles was assessed by a comparison of their relative abundance in these vesicles and in the original tissue. For a proper comparison one must know the relative contribution of plasma membrane proteins to the total protein content in both giant vesicles and tissue crude homogenates. A fractionation study of giant vesicles from skeletal muscle into membranes and lumen indicated that 10% of the total vesicular protein amount was plasma membrane-associated [14, 22]. Fractionation of heart giant vesicles yielded a similar ratio [20]. In most mammalian cells 5–10% of the total cellular protein resides in the plasma membrane. Thus, the contribution of plasma membrane protein to total protein is similar in giant vesicles and in the cells from which they are derived. By inference, when a cell-specific or membrane region-specific marker protein shows a similar abundance in the giant vesicle total protein pool and in the protein pool of a total homogenate of the corresponding organ, that cell type and/or particular membrane region is involved in the formation of giant vesicles.

Cell-specific and membrane domain-specific marker proteins

To assess the presence of membranes from different cell types in giant vesicle membranes, we have used the endothelium-specific markers Gp60 and caveolin type 1 (cav-1), and the myocyte-specific marker caveolin type 3 (cav-3). The presence of distinct plasma membrane domains was evaluated using cav-1 and cav-3 as markers for caveolae, the dihydropyridine receptor as marker for T-tubuli (in muscle), and neural cadherin was used as marker for membrane regions involved in cell-cell interactions. Furthermore, the insulin receptor was used as marker of insulin-sensitive tissues, whereas the transferrin receptor was used to mark the endosomal compartment.

Gp60, a glycoprotein of 60 kD, is a major constituent of the endothelial glycocalyx and functions as a receptor for albumin. Schnitzer *et al.* reported that its expression is restricted to tissues with microvascular beds lined with continuous endothelium [23]. Accordingly, we found it to be present in homogenates from heart, skeletal muscle and adi-

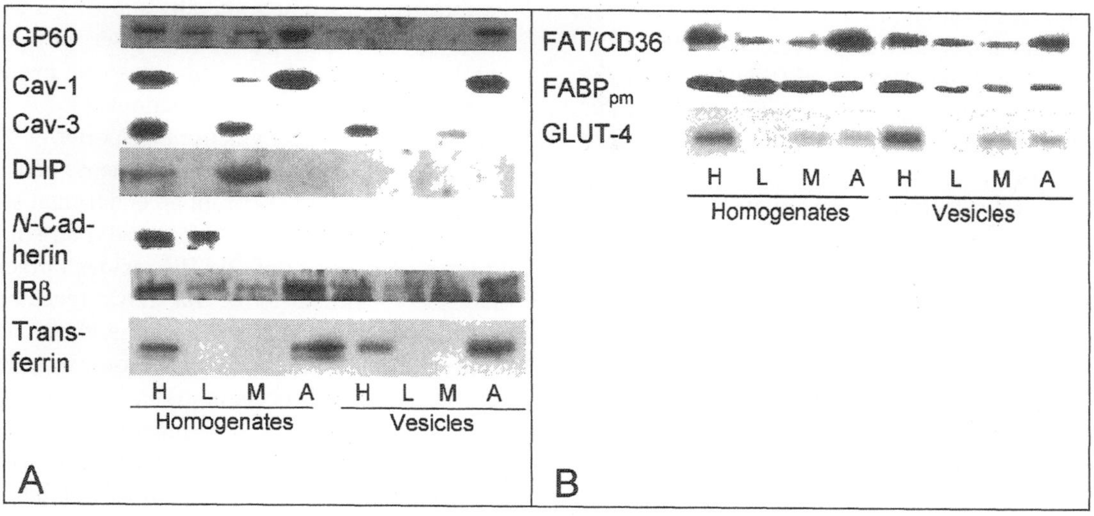

Fig. 1. Representative Western blots of albumin-binding protein (gp60), caveolin type 1 and 3 (cav-1, cav-3), dihydropyridine receptor (DHP), neural-cadherin, insulin receptor beta (IRβ) and transferrin receptor (A) and of the substrate transporters FAT/CD36, FABPpm and GLUT4 (B) in homogenates and giant vesicles. H – heart; L – liver; M – skeletal muscle; A – adipose tissue.

pose tissue, in abundance decreasing in this order (Fig. 1A). However, we also detected Gp60 in liver, which tissue has a sinusoidal endothelium. Gp60 was not detected in giant vesicles from heart, muscle and liver (Fig. 1A), indicating that in these tissues giant vesicle membranes are not derived from endothelial cells. The presence of Gp60 in giant vesicles from adipose tissue (Fig. 1A) may indicate that in this case endothelial cells participate in giant vesicle formation or that adipocytes themselves express Gp60.

Cav-1 and cav-3 are members of the protein family of caveolins, which play major roles in the architecture of caveolae. Caveolae are membrane domains with specialized functions, including sorting processes, transmembrane signalling, and molecular transport across membranes [24, 25]. The expression of cav-1 is restricted to endothelium, adipocytes, smooth muscle cells, and fibroblasts [25–27], while cav-3 specifically occurs in (all types of) muscle cells [28]. In line with this, both cav-1 (endothelium) and cav-3 (myocytes) were observed in heart and skeletal muscle homogenates, whereas only cav-3 was present in giant vesicles (in similar amounts as in the homogenates) (Fig. 1A). These findings indicate that heart and muscle giant vesicles are solely derived from myocytes and include caveolae. Cav-1 and cav-3 were not detectable in liver tissue, which is in agreement with previous results [25]. In epididymal fat pads, similar amounts of cav-1 were found in homogenates and giant vesicles, indicating that in adipose tissue endothelial cells and/or caveolar plasma membrane regions are involved in giant vesicle formation.

The dihydropyridine receptor (DHPR), or L-type Ca^{2+} channel, is localized in transverse invaginations of the muscle cell surface, the T-tubuli. Since the proportion of junctional T-tubules is higher in fast-contracting than in slow fibres [24], we observed the DHPR in greater amounts in skeletal muscle than in heart, and absent in liver and adipose tissue which lack T-tubuli (Fig. 1A). Importantly, DHPR was not detectable in giant vesicles (Fig. 1A) confirming previous results of Ploug *et al.* (1993) and indicating that T-tubuli are not involved in giant vesicle formation from muscle tissue.

Neural cadherin (N-cadherin) belongs to the cadherins, a family of cell adhesion proteins, and was used for the detection of plasma membrane regions involved in cell–cell interaction. In cardiac myocytes N-cadherin is localized in the intercalated disks [29, 30], the membrane regions involved in mediating the electric pulse from cell to cell, but skeletal muscle fibers, which are a longitudinal continuum of multiple fused cells, lack this membrane domain [31]. N-cadherin was found in homogenates from heart and liver, not in skeletal muscle and adipose tissue (Fig. 1A). Its presence in liver was reported earlier [32], and may relate to a role for N-cadherin in selective cell–cell adhesion [33]. N-cadherin was neither detected in heart nor in liver giant vesicles, suggesting that this specialized portion of the plasma membrane is not involved in the formation of giant vesicles.

The insulin receptor is an important hallmark of the plasma membrane of insulin-sensitive tissues such as heart, skeletal muscle, liver and adipose tissue [34]. The insulin receptor β-subunit was found in each of the tissue homogenates, and most abundantly in heart and adipose tissue (Fig. 1A). This observation confirms other studies on tissue characteristics of insulin binding [35]. The β-subunit was also present in giant vesicles derived from heart, liver, muscle, and epididymal fat pads in similar amounts as in the total tissue homogenates. This observation provides additional evidence for the plasmalemmal origin of giant vesicles.

The transferrin receptor is a commonly used marker of the endosomal compartment, because the vast majority of this protein is stored intracellularly. In contrast to the glucose transporter GLUT4 (see below), insulin recruits only a small portion of the transferrin receptor to the sarcolemma [36]. The transferrin receptor was detected in heart and muscle, but not in liver and adipose tissue (Fig. 1A). Others also reported that the transferrin receptor is absent from liver and occurs in only minute quantities in adipose tissue [36, 37]. The observation that in heart and muscle this receptor is present in similar amounts in total tissue homogenates and giant vesicles indicates that the endosomal compartment contributes to the delivery of membranes for giant vesicles.

Transporters for long-chain fatty acids and for glucose

FAT/CD36 and FABPpm are functionally important membrane FA transporters in heart, skeletal muscle and adipose tissue, and FABPpm also in liver (FAT/CD36 is virtually absent in liver [38–41], but inducible in this organ [42]. In heart the total amount of FAT/CD36 is about equally distributed between myocytes and endothelial cells [43]. At least in myocytes, we have demonstrated that FAT/CD36 is present both at the sarcolemma and in an intracellular (endosomal) compartment from which it can be translocated to the sarcolemma in response to cellular contractions [44, 45] or insulin [46]. FABPpm is highly homologous, if not identical, to mitochondrial aspartate aminotransferase (mAAT) [47], and, apart from this, may also be present in an intracellular (endosomal, non-mitochondrial) pool [16]. FAT/CD36 and FABPpm were present in total homogenates and in giant vesicles of each of the four tissues studied, thereby showing a similar tissue distribution pattern (Fig. 1B). This observation also indicates that the plasma membrane and endosomal membranes each contribute to giant vesicle formation.

GLUT4 is the major glucose transporter in insulin-sensitive tissues [48]. Insulin stimulates glucose transport through translocation of GLUT4 from an endosomal storage depot to the plasma membrane. Under basal (non-stimulated conditions) the majority of GLUT4 resides intracellularly, so that GLUT4, similar to the transferrin receptor, can be regarded as an endosomal protein rather than a plasma membrane marker [36]. Total tissue homogenates and giant vesicles displayed the same tissue distribution pattern of GLUT4 in that it is most abundant in heart, followed by adipose tissue and muscle, and is absent in liver (Fig. 1B). In each tissue, GLUT4 was equally abundant in giant vesicles and total tissue homogenates, indicating that GLUT4 containing endosomal compartments are involved in vesicle formation.

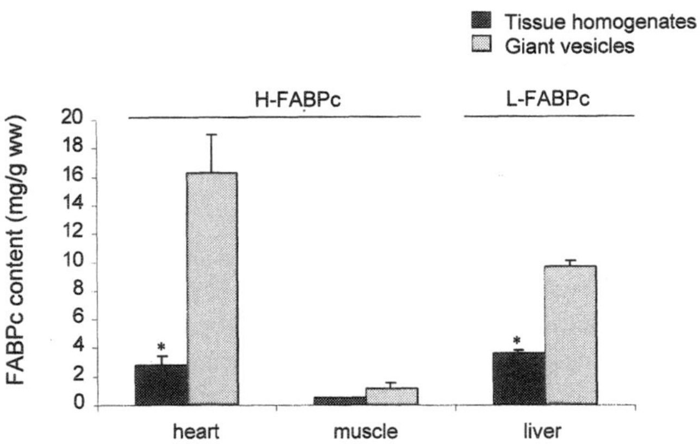

Fig. 2. Cytoplasmic FABP content in homogenates and giant vesicles from heart and skeletal muscle tissues (H-FABP) and liver (L-FABP). H-FABP and L-FABP were determined by a sandwich-type enzyme-linked immunosorbent assay (ELISA) and are based upon 5–6 sample sizes. *p < 0.05.

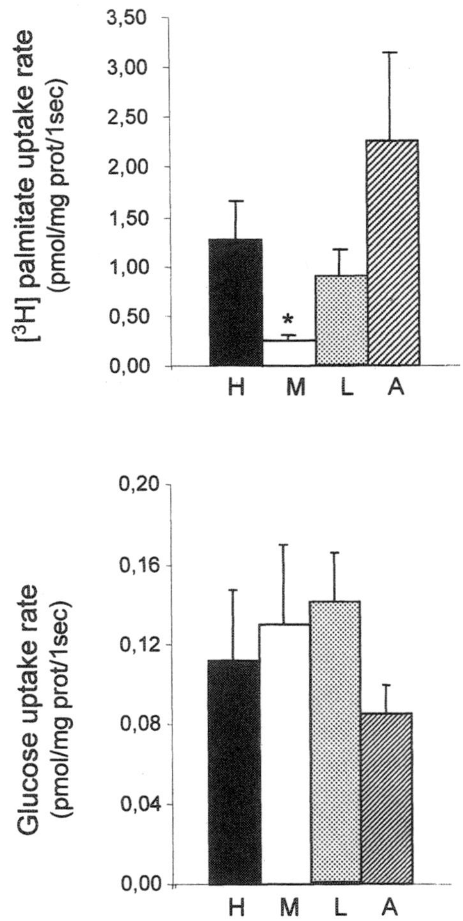

Fig. 3. Palmitate (top) and glucose (bottem) uptake rate into giant sarcolemmal vesicles obtained from heart (H), skeletal muscle (M), liver (L), and adipose tissue (A). Data are the means of of 3–4 experiments carried out with the vesicle preparations as described in the Materials and methods section. *p < 0.05.

$FABP_c$, a 15 kD cytoplasmic protein acting as intracellular acceptor and carrier of FA, occurs in several distinct types each showing a characteristic tissue distribution [41, 49]. Heart-type $FABP_c$ (H-$FABP_c$) is present in heart and to a lesser extent in muscle, and is absent in liver and adipose tissue (Fig. 2). Liver-type $FABP_c$ (L-FABPc) is specifically expressed in liver (Fig. 2). For both H-$FABP_c$ and L-$FABP_c$ we observed a 3–4-fold enrichment in giant vesicles compared to total tissue homogenates. Previously, we established that the vesicular lumen consists of the soluble portion of the cytoplasm [19]. The enrichment of $FABP_c$ isoforms in giant vesicles can then be explained by exclusion of other portions of the cytoplasm, because myosin [9], components of cytoskeletal and myofibrillar network [14], and mitochondria [14, 20] are absent in giant vesicles. Giant vesicles from adipose tissue were previously found to contain adipocyte-type $FABP_c$ (usually referred to as adipocyte lipid binding protein, ALBP) [16]. The presence of FABPc inside giant vesicles permits the proper study of FA uptake as $FABP_c$ then can act as a (natural) sink for these hydrophobic substrates.

Application of giant membrane vesicles for substrate uptake studies

The great similarity in size (10–15 μm) and shape (spherical) of giant vesicles from different tissues indicates that it is possible for the first time to compare metabolic substrate fluxes among these organs independently of the surface/volume ratio. This could provide additional insight into the organ-differences in functioning of membrane transporters involved in cellular uptake of metabolic substrates.

The rate of FA uptake by giant vesicles prepared from heart, muscle, adipose tissue and liver varied considerably (Fig. 3). For heart, muscle and adipose tissue the content of FAT/CD36 per organ correlated well with the FA uptake rate (r for linear fit = 0.85; Fig. 4A). In contrast, no significant correlation among the tissues was found between FA uptake and the contents of FABPpm (r for linear fit = 0.27; Fig. 4B). These findings confirm the view that FABPpm is necessary, but not rate limiting in FA uptake, since we [20] and others [1, 50] previously showed that FA uptake in heart could be largely inhibited by a polyclonal anti-FABPpm antiserum, indicating the involvement of this transporter in their bulk uptake. It appears, therefore, that FAT/CD36 and FABPpm cooperate in a joint fashion to translocate long-chain fatty acids across the sarcolemma. Whether this means that FAT/CD36 and FABPpm are two protein-components of a single LCFA transport system remains to be established.

In contrast to FA uptake, glucose uptake among the various tissues is remarkably similar (only adipose fat pads display a slightly lower uptake rate than the other tissues; Fig. 3). However, we did not find a significant correlation between glucose uptake rate and heart, muscle, and adipose tissue content of GLUT4 (r for linear fit = 0.0; data not shown). Since heart, skeletal muscle, liver and adipose tissues are known to express different glucose transporter isoforms [51], this lack of correlation might be explained by the simultaneous expression of other glucose transporter isoforms. For instance, GLUT1 is recognized to be the most ubiquitously expressed glucose transporter, and is often expressed in conjunction with a more tissue-restricted glucose transporter isoform, that is GLUT2 in liver, and GLUT4 in heart, muscle and adipose tissue. We therefore speculate that GLUT1 is

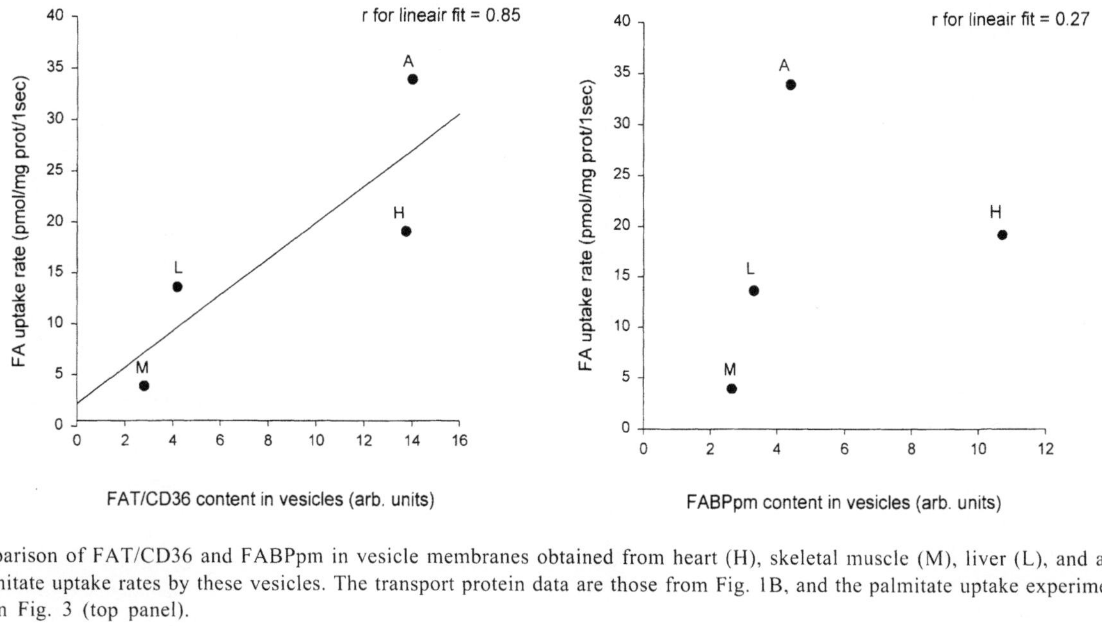

Fig. 4. Comparison of FAT/CD36 and FABPpm in vesicle membranes obtained from heart (H), skeletal muscle (M), liver (L), and adipose tissue (A) with palmitate uptake rates by these vesicles. The transport protein data are those from Fig. 1B, and the palmitate uptake experiments are those represented in Fig. 3 (top panel).

responsible for the bulk of glucose uptake in giant vesicles. This notion is strengthened by observations in heart and skeletal muscle, where GLUT1 protein levels were found to be ≥ 3-fold enriched in giant vesicles compared to total tissue homogenates (data not shown).

Conclusion

This is the first study to compare the characteristics of giant membrane vesicles obtained from rat heart, skeletal muscle, liver and adipose tissue. This comparison not only provides knowledge on the properties and composition of the vesicles, which information is needed for proper interpretation of experimental studies using this model system, but also gives further insight into the mechanism of giant vesicle formation.

Using plasma membrane markers from different cell types, we could establish that at least in heart, muscle and liver parenchymal cells, and not endothelial cells, are involved in vesicle formation. For adipose tissue, conclusive evidence could not be obtained, but given the findings on the other tissues, adipocyte giant vesicles most likely consist only of adipocyte plasma membranes. The presence of tissue-specific types of $FABP_c$ in the lumen of the giant vesicles also points towards parenchymal cells as the predominant source of giant vesicles.

Our data show that selected plasma membrane regions are involved in vesicle formation. In heart and muscle, the T-tubuli are not involved in giant vesicle formation, whereas in heart and liver, membrane regions involved in specific cell–cell interactions are not present in giant vesicles. Therefore, we speculate that membrane regions shared by the various parenchymal cell types, such as those involved in binding of ligands, transport of substrates, and regions containing caveolae, are the source of giant vesicle membranes, while regions with cell-specific functions are not. For example, it can be speculated that the hepatocyte-specific canalicular membranes, involved in mediating hepatic bile flow, also would not participate in giant vesicle formation.

Furthermore, we obtained additional evidence that endosomal membranes contribute to the formation of giant vesicles, as suggested by Ploug *et al.* (1993). Previously, the presence of an endosomal protein involved in the regulation of vesicular trafficking, vesicle-associated membrane protein-2 (VAMP-2), was demonstrated in giant vesicles [52]. In this study, both GLUT4 and the transferrin receptor were found to be present in giant vesicles. Moreover, caveolin, which was abundant in giant vesicles, is supposed to be involved in endosomal-vesicular trafficking [24]. As established in adipocytes, a small portion (≈ 20%) of the cellular amount of caveolin is present in an insulin-sensitive recycling compartment. Since caveolae are enriched in GLUT4, a colocalization of surface caveolin with GLUT4 was suggested [24, 53, 54]. There is however no general consensus on this view [24, 53, 54]. Taken together, the notion arises that the giant vesicle membranes are a continuum of combined endosomal regions with plasma membrane regions and are involved in docking of recycling endosomal vesicles at the plasma membrane. The large size of the vesicles raises the question whether the endosomal compartment can recruit sufficient membranes and suggests that extensive plasma membrane regions must be included in giant vesicles. In agreement with this suggestion, heart and muscle giant vesicles have been found in earlier studies to be enriched in well-established plasma membrane markers p-nitrophenylphosphatase and 5′nucleotidase [17, 20, 22]. Moreover, giant vesicles from all selected tissues were found to contain the insulin receptor, GLUT1 and caveolins, all of which are predominantly plasmalemmal proteins.

The putative involvement of plasma membrane fusion and excision events in the KCl-induced giant vesicle formation process, allows the use of giant vesicles as a model to study vesicular trafficking processes. In addition, based on the absence of metabolism, giant vesicles from heart, liver, skeletal muscle and adipose tissue are very suitable to study substrate uptake.

Acknowledgements

Antibodies for FABPpm and FAT/CD36 were kindly provided by Dr. D. Sorrentino, Mount Sinai Medical Center, New York City, USA, and by Dr. N. Tandon, Otsuka Pharmaceuticals, Bethesda, MD, USA. Gp60 antibody was a kind gift from Dr. C. Tirrupathi, University of Illinois, Chicago, IL, USA. This study was supported by NWO-MW: 903-39-194, the Canadian Institutes of Heart Research (CIHR), and the Heart and Stroke Foundation of Ontario.

References

1. Stremmel W: Fatty acid uptake by isolated heart myocytes represents a carrier-mediated transport process. J Clin Invest 81: 844–852, 1988
2. Sorrentino D, Stump D, Potter J, Robinson B, White R, Kiang C-L, Berk PD: Oleate uptake by cardiac myocytes is carrier-mediated and involves a 40-kDa plasma membrane, fatty acid binding protein similar to that in liver, adipose tissue, and gut. J Clin Invest 82: 928–835, 1988
3. Luiken JJFP, van Nieuwenhoven FA, America G, van der Vusse GJ, Glatz JFC: Uptake and metabolism of palmitate by isolated cardiac myocytes from adult rats: Involvement of sarcolemmal proteins. J Lipid Res 38: 745–758, 1997
4. Fischer Y, Thomas J, Sevilla L, Muñoz P, Becker C, Holman G, Kozka I, Palacin M, Testar X, Kammermeier H, Zorzano A: Insulin-induced recruitment of glucose transporter 4 (GLUT4) and

GLUT1 in isolated rat cardiac myocytes. J Biol Chem 272: 7085–7092, 1997
5. Russell RR, Bergeron R, Shulman GI, Young LH: Translocation of myocardial GLUT4 and increased glucose uptake through activation of AMPK by AICAR. Am J Physiol 277: H643–H649, 1999
6. Juel C: Intracellular pH recovery and lactate efflux in mouse soleus muscles stimulated *in vitro*: The involvement of sodium/proton exchange and a lactate carrier. Acta Physiol Scand 132: 363–371, 1988
7. Grimdtlich GK, Barnard RJ, Kaplan SA, Sternlicht E: Insulin binding and glucose transport in rat skeletal muscle sarcolemmal vesicles. Am J Physiol 249: E398–E408, 1985
8. King PA, Hirshman MF, Horton ED, Horton ES: Glucose transport in skeletal muscle membrane vesicles from control and exercised rats. Am J Physiol 257: C1128–C1134, 1989
9. Juel C: Muscle lactate transport studies in sarcolemmal giant vesicles. Biochim Biophys Acta 1065: 15–20, 1991
10. Ahmed A, Taylor PM, Rennie MJ: Characteristics of glutamine transport in sarcolemmal vesicles from rat skeletal muscle. Am J Physiol 259: E284–E291, 1990
11. Van Amelsvoort JMM, Sips HJ, Van Dam K: Sodium-dependent alanine transport in plasma-membrane vesicles from rat liver. Biochem J 174: 1083–1086, 1978
12. Bersohn MM: Sodium pump inhibition in sarcolemma from ischemic rats. J Mol Cell Cardiol 27: 1483–1489, 1995
13. Berardi S, Stieger B, Hagenbuch B, Carafoli E, Krähenbühl S: Characterization of L-carnitine transport into rat skeletal muscle plasma membrane vesicles. Eur J Biochem 267: 1985–1994, 2000
14. Ploug T, Wojtazewski J, Kristiansen S, Hespel P, Galbo H, Richter EA: Glucose transport and transporters in muscle giant vesicles: Differential effects of insulin and contractions. Am J Physiol 264: E270–E278, 1993
15. Standen NB, Stanfield PR, Ward TA: Single-channel recording from membrane vesicles formed from skeletal muscle fibres of the frog (*Rana temporaria*). J Physiol 346: 81P, 1984
16. Luiken JJFP, Arumugam Y, Dyck DJ, Bell RC, Pelsers MML, Turcotte LP, Tandon NN, Glatz JFC, Bonen A: Increased rates of fatty acid uptake and plasmalemmal fatty acid transporters in obese Zucker rats. J Biol Chem 276: 40567–40573, 2001
17. McCullagh KJA, Juel C, O'Brien M, Bonen A: Chronic muscle stimulation increases lactate transport in rat skeletal muscle. Mol Cell Biochem 156: 51–57, 1996
18. Tokumaru H, Anzai K, Abe T, Kirino Y: Purification of the cardiac 1,4-dihydropyridine receptor using immunoaffinity chromatography with a monoclonal antibody against the alpha 2 delta subunit of the skeletal muscle dihydropyridine receptor. Eur J Pharmacol 227: 363–370, 1992
19. Van Nieuwenhoven FA, Verstijnen CP, van Eys GJ, van Breda E, De Jong, YF, van der Vusse GJ, Glatz JFCL: Fatty acid transfer across the myocardial capillary wall: No evidence of a substantial role for cytoplasmic fatty acid-binding protein. J Mol Cell Cardiol 26: 1635–1647, 1994
20. Luiken JJFP, Turcotte LP, Bonen A: Protein-mediated palmitate uptake and expression of fatty acid transport proteins in heart giant vesicles. J Lipid Res 40: 1007–1016, 1999
21. Berk PD, Zhou S-L, Kiang C-L, Stump D, Bradbury M, Isola L: Uptake of long chain free fatty acids is selectively up-regulated in adipocytes of Zucker rats with genetic obesity and non-insulin-dependent diabetes mellitus. J Biol Chem 272: 8830–8835, 1997
22. Bonen A, Luiken JJFP, Liu S, Dyck DJ, Kiens B, Kristiansen S, van der Vusse GJ, Glatz JFC: Palmitate transport and fatty acid transporters in red and white muscles. Am J Physiol 275: E471–E478, 1998
23. Schnitzer JE: Gp60 is an albumin-binding glycoprotein expressed by continuous endothelium involved in albumin transcytosis. Am J Physiol 262: H246–H254, 1992
24. Muñoz P, Mora S, Sevilla L, Kaliman P, Tomàs E, Gumà A, Testar X, Palacín, Zorzano A: Expression and insulin-regulated distribution of caveolin in skeletal muscle. J Biol Chem 271: 8133–8139, 1996
25. Kogo H, Fujimoto T: Caveolin-1 isoforms are encoded by distinct mRNAs. Identification of mouse caveolin-1 mRNA variants caused by alternative transcription initiation and splicing. FEBS Lett 465: 119–123, 2000
26. Oka N, Asai K, Kudej RK, Edwards JG, Toya Y, Schwencke C, Vatner DE, Vatner SF, Ishikawa Y: Downregulation of caveolin by chronic β-adrenergic receptor stimulation in mice. Am J Physiol 273: C1957–C1962, 1997
27. Song KS, Li S, Okamoto T, Quilliam LA, Sargiacomo M, Lisant MP: Co-purification and direct interaction of ras with caveolin, an integral membrane protein of caveolae microdomains, detergent-free purification of caveolae membranes. J Biol Chem 271: 9690–9697, 1996
28. Song KS, Scherer PE, Tang ZL, Okamoto T, Li S, Chafel M, Chu C, Kohtz DS, Lisanti MP: Expression of caveolin-3 in skeletal, cardiac, and smooth muscle cells. J Biol Chem 271: 15160–15165, 1996
29. Knudsen KA, Soler AP, Johnson KR, Wheelock MJ: Interaction of α-catenin with the cadherin/catenin cell/cell adhesion complex via α-catenin. J Cell Biol 130: 67–77, 1995
30. Rimm DL, Koslov ER, Kebriaei P, Ciani CD, Morrow JS: Alpha1(E)-catenin is an actin-binding and -bundling protein mediating the attachement of F-actin to the membrane adhesion complex. Proc Natl Acad Sci USA 92: 8813–8817, 1995
31. Imanaka-Yoshida K, Knudsen KA, Linask KK: N-cadherin is required for the differentiation and initial myofibrillogenesis of chick cardiomyocytes. Cell Motil Cytoskel 39: 52–62, 1998
32. Linnemann D, Gaardsvoll H, Dalseg AM, Zhernosekov D, Lundgren T, Edvardsen K, Bock E: Characterization of N-cadherin messenger RNA and polypeptide expression in rat. Int J Dev Neurosci 12: 441–450, 1994
33. Miyatani S, Shimamura K, Hatta M, Nagafuchi A, Nose A, Matsunaga M, Hatta K, Takeichi M: Neural cadherin: Role in selective cell–cell adhesion. Science 245: 631–635, 1989
34. Gustavsson J, Parpal S, Karlsson M, Ramsing C, Thorn H, Borg M, Lindroth M, Holmgren-Peterson K, Magnusson K-E, Strålfors P: Localisation of the insulin receptor in caveolae of adipocyte plasma membrane. FASEB J 13: 1961–1971, 1999
35. Torlinska T, Mackowiak P, Nogowski L, Nowak KW, Madry E, Perz M: Characteristics of insulin receptor binding to various rat tissues. J Physiol Pharmacol 49: 261–270, 1998
36. Kandror KV, Pilch PF: Compartmentalization of protein traffic in insulin-sensitive cells. Am J Physiol 271: E1–E14, 1996
37. Jefferies WA, Brandon MR, Williams AF, Hunt SV: Analysis of lymphopoietic stem cells with a monoclonal antibody to the rat transferrin receptor. Immunology 54, 333–341, 1985
38. Abumrad N, Coburn C, Ibrahimi A: Membrane proteins implicated in long-chain fatty acid uptake by mammalian cells: CD36, FATP and FABPm. Biochim Biophys Acta 1441: 4–13, 1999
39. Luiken JJFP, Schaap FG, van Nieuwenhoven FA, van der Vusse GJ, Bonen A, Glatz JFC: Cellular fatty acid transport in heart and skeletal muscle as facilitated by proteins. Lipids 34: S169–S175
40. Febbraio M, Hajjar DP, Silverstein RL: CD36: A class B scavenger receptor involved in angiogenesis, artherosclerosis, inflam-

mation, and lipid metabolism. J Clin Invest 108: 785–791, 2001
41. Glatz JFC, Storch J: Unravelling the significance of cellular fatty acid-binding proteins. Lipidology 12: 267–274, 2001
42. Motojima K, Passilly P, Peters JM, Gonzalez FJ, Latruffe N: Expression of putative fatty acid transporter genes are regulated by peroxisome proliferator-activated receptor alpha and gamma activators in a tissue- and inducer-specific manner. J Biol Chem 273: 16710–16714, 1998
43. Brinkmann JFF: unpublished observations
44. Bonen A, Luiken JJFP, Arumugam Y, Glatz JFC, Tandon NN: Acute regulation of fatty acid uptake involves the cellular redistribution of fatty acid translocase. J Biol Chem 276: 14501–14508, 2000
45. Luiken JJFP, Willems J, van der Vusse GJ, Glatz JFC: Electrostimulation enhances FAT/CD36-mediated long-chain fatty acid uptake by isolated rat cardiac myocytes. Am J Physiol 281: E704–E712, 2001
46. Luiken JJFP, Dyck DJ, Tandon NN, Arumugam Y, Glatz JFC, Bonen A: Insulin translocates the fatty acid transporter FAT/CD36 to the plasma membrane. Am J Physiol (in press)
47. Stump DD, Zhou SL, Berk PD: Comparison of plasma membrane FABP and mitochondrial isoform of aspartate aminotransferase from rat liver. Am J Physiol 265: G894–902, 1993
48. Zorzano A, Sevilla L, Camps M, Becker C, Meyer J, Kammermeier H, Munoz P, Guma A, Testar X, Palacin M, Blasi J, Fischer Y: Regulation of glucose transport, and glucose transporters expression and trafficking in the heart: Studies in cardiac myocytes. Am J Cardiol 80: 65A–76A, 1997
49. Storch J, Thumser AE: The fatty acid transport function of fatty acid-binding proteins. Biochim Biophys Acta 1486: 28–44, 2000
50. Stremmel W, Berk PD: Hepatocellular influx of [^{14}C]oleate reflects membrane transport rather than intracellular metabolism or binding. Proc Natl Acad Sci USA 83: 3086–3090, 1986
51. Mueckler M: Facilitative glucose transporters. Eur J Biochem. 219: 713–725, 1994
52. Kristiansen S, Hargreaves M, Richter EA: Exercise-induced increase in glucose transport, GLUT-4, and VAMP-2 in plasma membrane from human muscle. Am J Physiol 270: E197–201, 1996
53. Scherer PE, Lisanti MP, Baldini G, Sargiacomo M, Mastick CC, Lodish HF: Induction of caveolin during adipogenesis and association of GLUT4 with caveolin-rich vesicles. J Cell Biol 127: 1233–1243, 1994
54. Kandror KV, Stephens JM, Pilch PF: Expression and compartmentalization of caveolin in adipose cells: Coordinate regulation with and structural segregation from GLUT4. J Cell Biol 129: 999–1006, 1995

Molecular and Cellular Biochemistry **239:** 131–138, 2002.

Peroxisome proliferator-activated receptors: Lipid binding proteins controling gene expression

Marc van Bilsen, Ger J. van der Vusse, Andries J. Gilde, Martijn Lindhout and Karin A.J.M. van der Lee

Department of Physiology, Cardiovascular Research Institute Maastricht (CARIM), Maastricht University, Maastricht, The Netherlands

Abstract

The peroxisome proliferator-activated receptors (PPARs) are members of the nuclear hormone receptor superfamily. Since their discovery in the beginning of the nineties the three isoforms (PPARα, β/δ and γ, encoded by different genes) have been implicated in the regulation of almost every single aspect of lipid metabolism and, consequently, in diseases that involve disturbances in lipid metabolism (obesity, diabetes, atherosclerosis, cardiac failure). Although their prominent role in these processes has hardly been disputed, the way in which the activity of these transcription factors is regulated under physiological and pathological conditions awaits further clarification. An unresolved issue has been the nature of the natural ligand of these receptors. Biochemical studies have shown that the PPAR isoforms are rather promiscuous with respect to ligand binding, with a large variety of naturally occurring lipid-like substances acting as low-affinity ligands. More recently this concept has been confirmed by crystallographic studies on the ligand-binding pocket. In addition to ligand availability, the trans-activating capacity likely depends on phosphorylation status of the PPARs and on the recruitment of auxiliary proteins (co-activators and co-repressors). Accordingly, the biological activity of these key-regulators of metabolism is controlled at multiple levels, which enables each tissue to fine tune its metabolic machinery to the demands of the body in a specific fashion. (Mol Cell Biochem **239**: 131–138, 2002)

Key words: fatty acids, ligands, PPAR, heart, gene expression

Introduction

Only a decade ago Issemann and Green [1] reported the discovery of the a ligand-activated transcription factor, that interacted with various drugs that were known for their ability to cause peroxisomal proliferation in liver. Accordingly the transcription factor was named peroxisome proliferator-activated receptor (PPAR). Since the discovery of this PPAR, later referred to as PPARα, two other PPAR subtypes have been cloned from different species, i.e. PPARβ/δ (also called NUC-1 or FAAR) and PPARγ. Each of the three subtypes was found to be encoded by a different gene.

The PPARs are members of the family of nuclear hormone receptors. All nuclear hormone receptors share a close structural homology, reflected by the presence of several structural and functional domains (Fig. 1) [2]. The amino-terminal A/B domain is least conserved and includes a ligand-independent transactivation function (AF-1), containing putative phosphorylation sites. The DNA-binding C domain is highly conserved and includes two zinc finger motifs, containing two conserved boxes, the P- and D-box. The P-box is specifically involved in the interaction with specific DNA sequences, whereas the D-box participates in receptor dimerisation. The variable hinge region (D domain) allows conformational changes. The ligand-binding domain (E domain) encompasses a ligand-dependent transactivation function (AF-2). In addition to ligand binding, the E domain is required for nuclear localisation, receptor dimerisation and the interaction with

Address for offprints: M. van Bilsen, Department of Physiology, Cardiovascular Research Institute Maastricht, Maastricht University, P.O. Box 616, 6200 MD Maastricht, The Netherlands (E-mail: marc.vanbilsen@fys.unimaas.nl)

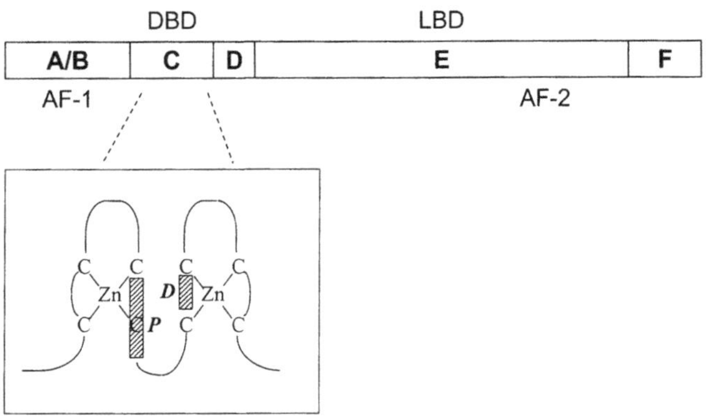

Fig. 1. Schematic presentation of modular composition of the PPARs. The A/B and E-region encompass the ligand-independent (AF-1) and ligand-dependent transactivation function (AF-2), respectively. DBD – DNA-binding domain; LBD – ligand binding domain; the two Zinc-fingers in the C-region, containing the P and D-boxes, are shown separately. For more details see text.

auxiliary proteins. At this moment, no specific function has been assigned to the F domain, that forms the carboxy-terminal end of the nuclear receptors.

The PPARs form heterodimers with the retinoid-X-receptor (RXR) [3]. The PPAR/RXR heterodimer is able to bind to a DNA consensus sequence, the peroxisome proliferator response element (PPRE). A PPRE is formed by a direct repeat of 6 nucleotides, separated by one spacer nucleotide (DR-1) and has the consensus sequence AGGTCA n AGGTCA. Functional PPREs have been found in a large number of genes encoding proteins involved in lipid metabolism. This includes proteins involved in inter-organ lipid transport (apolipoproteins A-I, A-II, C-III), in lipid processing and uptake (lipoprotein lipase (LPL), fatty acid translocase (FAT/CD36), fatty acid transport protein (FATP), and metabolism (e.g. acyl-CoA Oxidase (AOX), acyl-CoA synthetase (ACS), carnitine palmitoyl-transferase-I (CPT-I) and long-chain acyl-CoA dehydrogenase (LCAD)). The list of PPAR-responsive genes is still expanding and also includes genes, for which the link with fatty acid metabolism is less well established (e.g. uncoupling proteins 2 and 3) [4]. Interestingly, one of the pyruvate dehydrogenase kinase isoforms (PDK4) is also induced by PPARs [5]. By controling the degree of phosphorylation of the pyruvate dehydrogenase (PDH) complex, PDK4 plays a regulatory role in the control of glucose oxidation. This implicates that the PPARs are also involved in aspects of cellular glucose homeostasis. Just recently it was shown that the expression of a number of genes involved in hepatic amino acid metabolism is also regulated by PPARα [6]. In addition to the regulation of energy metabolism, PPARs have been described to be involved in processes such as adipogenesis, inflammation, atherosclerosis, and carcinogenesis (see [7–9] for recent reviews).

Localization of PPARs

The expression of the three PPAR subtypes is tissue restricted. PPARα is primarily expressed in tissues with a high oxidative capacity, like liver, kidney, and heart. The expression of PPARβ/δ appears to be ubiquitous, whereas that of PPARγ is confined to a limited number of tissues, primarily adipose tissue, and vascular cells. The existence of splice variants of PPARγ and, more recently, of PPARα has been demonstrated. PPARγ1 and PPARγ2 differ at their N-terminus [10]. The PPARα splice variant lacks exon 6 and gives rise to a premature stopcodon as a result of which a truncated protein, merely consisting of the A/B, C and part of the D domain is being formed [11]. To date there is some controversy whether the truncated splice variant is formed in all species [11, 12]. In support of the findings of Gervois *et al.* [11], we could show its presence in adipose and cardiac tissue of humans, but not of rats (Fig. 2).

The expression of both PPARα and PPARβ/δ in the cardiac muscle has been firmly established. Whether PPARγ is present in the heart, however, is still a matter of debate. Some research groups reported the presence of PPARγ at both the mRNA and protein level in cardiac tissue [13, 14]. Others were unable to detect PPARγ in the cardiac muscle using either *in situ* hybridisation or immunohistochemistry [15]. By means of northern blotting we observed low levels of PPARγ mRNA in adult rat heart tissue and in isolated cardiac fibroblasts, but not in isolated adult and neonatal cardiomyocytes. This would all be consistent with the idea that, if present at all, the levels of PPARγ in the heart are likely to be low and most likely confined to non-parenchymal cells of the heart. However, this notion was recently challenged by Takano *et al.* [16] who reported the presence of PPARγ in isolated neonatal cardiomyocytes at the mRNA and protein level.

Unlike other ligand-activated transcription factors, such as the oestrogen receptor, which are translocated to the nucleus upon the binding of their cognate ligands, it is generally believed that the PPARs are constitutively localized in the nucleus, independent of ligand binding. Recently, the PPARα splice variant was also found in he cytoplasm in relatively large amounts [11]. Although this may be a unique feature of this splice variant, it nonetheless challenges the concept that PPARs are exclusively localised in the nuclear compartment.

PPAR ligands: In vitro *and* in vivo *observations and considerations*

PPARα is the only subtype interacting with classical peroxisome proliferators, such as fenofibrate and clofibrate. The antidiabetic thiazolidinediones (TZDs) are specific synthetic

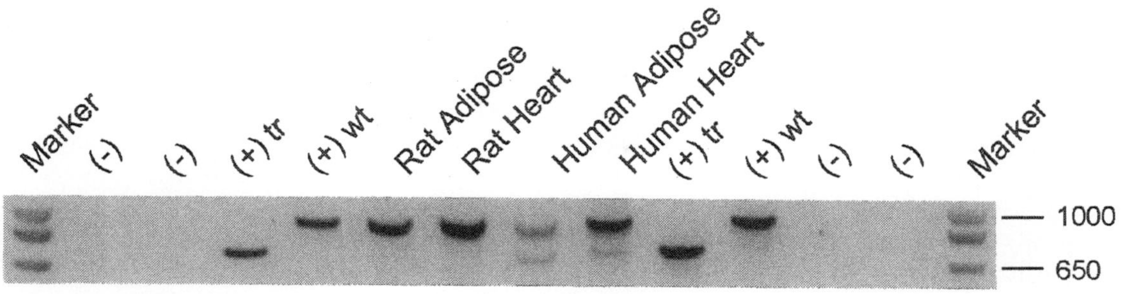

Fig. 2. Expression of the truncated PPARα splice variant, lacking exon 6, in human adipose and cardiac tissue, but not in rat tissue. The presence of $PPAR\alpha_{wt}$ (922 bp product) and $PPAR\alpha_{tr}$ (719 bp) was analysed by RT-PCR analysis using different primers sets for each species. The forward primers correspond to nucleotide 193–217 of the human gene and differs at 4 positions between rat and human. The reverse primers corresponding nucleotide 995–1115 of the human gene were identical. (+)tr and (+)wt refer to positive controls (plasmid DNA) for $PPAR\alpha_{wt}$ and $PPAR\alpha_{tr}$, respectively. (–) refers to negative controls.

activators for PPARγ. Only recently a specific synthetic ligand for PPARβ/δ was developed, the phenoxyacetic acid derivative L-165041 [17]. The isoform specific ligands are extensively being used to unravel the biological roles of each of the PPAR isoforms. Notwithstanding the isoform specificity of the synthetic ligands, it should be noted that PPAR-independent effects of the ligands have also been reported. This is particularly true for the TZD's, for which evidence has been provided that they modulate the inflammatory response, in part via NF-κB mediated mechanisms [18]. This additional feature of the TZD's should be kept in mind in evaluating the biological effects of these PPARγ ligands. For instance, the effect of TZDs on skeletal muscle metabolism was found to be fast and was not linked to changes in gene expression [19]. Also in light of the discussion of the presence or absence of PPARγ in the cardiac muscle cell, it could well be that the reported effects of TZD's on cardiac function and phenotype are primarily accomplished via PPARγ-independent mechanisms. Indeed, the anti-hypertrophic effect of PPARγ ligands in neonatal cardiomyocytes was associated with inhibition of NF-κB activation [20]. In addition, we were unable to detect any effect of TZD's, neither of Ciglitazone nor of Rosiglitazone, on the mRNA level or promoter activity of PPRE containing genes in neonatal cardiomyocytes (A. Gilde and M. van Bilsen, unpublished observations). Moreover, in a recent study [21] isolated adult myocytes were challenged with specific PPARα ligands, resulting in changes in LPL activity, but not in LPL protein content. The latter findings are indicative for post-translational rather than transcriptional effects of the PPARα ligands.

In addition to these synthetic compounds, a large variety of natural compounds has been shown to interact with the PPARs. Resolution of the 3-D structures of the three PPAR isotypes by crystallographic techniques is consistent with the idea that the ligand-binding pocket shows little specificity towards lipophilic compounds. Long-chain fatty acids are able to interact with all three PPAR isotypes. Next to fatty acids, many fatty acid derivatives, such as fatty acid methyl esters and branched-chain fatty acids like phytanic acid, and arachidonic acid metabolites, have been put forward as natural ligands of PPARα. With the exception of fatty acids PPARβ/δ ligands have not identified so far. The prostaglandin J2 derivative15-deoxy-Δ12,14-PGJ2 is considered one of the natural ligands of the PPARγ receptor [22].

In particular the capacity of individual fatty acid species, differing in chain length and/or degree of saturation, to act as ligands for the three PPAR isotypes has received widespread attention [23–27]. Various types of *in vitro* techniques, including classical competition binding assays [28] as well as gel mobility shift assays, making use of the ability of ligands to enhance PPAR-DNA interactions [26], point to fatty acid-binding properties of PPARs. Although there clearly are some differences in outcome between the various studies, the general picture that emerges is that both saturated, mono- and polyunsaturated fatty acids interact with PPARα, with only modest differences in affinity between the different fatty acid species. The same holds for PPARβ/δ, albeit that the affinity of the various fatty acid species for PPARβ/δ appears to be lower than that for PPARα [28]. PPARγ , on the other hand, displays a clear preference for unsaturated fatty acids [26–28]. The conclusions based on *in vitro* data are largely confirmed by transactivation studies in intact cells [29]. It is noteworthy that all of the fatty acids and fatty acid derivatives mentioned earlier are working as activators of the PPARs. Interestingly, recent studies [30, 31] indicate that long-chain acyl-CoA esters also interact with PPARα, but they were shown to act as inhibitors instead of activators. This opens the intriguing possibility that the ratio between fatty acids and fatty acyl-CoA esters affects the trans-activating activity of PPARα within the cell.

In the majority of studies the dissociation constants measured are in the micromolar range [25, 28, 32, 33]. Obviously, the dissociation constants of fatty acids for PPARs are of relevance to the question whether fatty acids are indeed able to function as natural ligands for PPARs. In this respect it is noteworthy that the actual intracellular fatty acid concentration is still subject of discussion. The total concentration of unesterified fatty acids in the cytoplasmic compartment of cardiomyocytes has been estimated to have an upper limit of 10 μM [34]. Even less is known about the actual concentration of fatty acids within the nuclear compartment. As the majority of the poorly water-soluble fatty acids is intercalated in phospholipid bilayers or bound to cytoplasmic fatty acid-binding proteins, the actual free fatty acid concentration is likely to be much lower. Accordingly, binding constants of fatty acids to PPAR should rather be in the nanomolar than micromolar range in order to be able to function as true natural ligands within the cellular context. Conversely, evidence is accumulating that transfer of fatty acids to PPAR in the nucleus may be facilitated by the intracellular fatty acid transporter FABP. Indeed H-FABP is present, albeit in low amounts, in nuclei of cardiomyocytes [35] and for liver-type FABP it has been shown that it physically interacts with PPARα [29]. Although the involvement of FABPs in this process provides an attractive hypothesis, at the same time, it raises questions as to how the fatty acids which bind with high affinity to FABPs are donated to PPARs, which have a much lower affinity for fatty acids. Moreover, it is even subject of speculation if fatty acids that function as ligands within the nucleus are derived from exogenous sources, or generated locally as a result of activation of phospholipases via signalling cascades. Altogether, these considerations indicate that there is still a gap between *in vitro* assays that demonstrate that fatty acids are able to bind to PPARs and the contention that fatty acids themselves function as ligands *in vivo*.

Fatty acids and cardiac gene expression

Effects of exogenous long-chain fatty acids on gene expression were first recognised in studies on adipocytes and hepatocytes [36, 37]. More recently it was shown that fatty acids also modulate gene expression in cardiac myocytes [38, 39]. When exposed to physiological concentrations of palmitic and oleic acid, the capacity of the neonatal cardiomyocytes to oxidise fatty acids increased significantly, a response which was associated with up to 4-fold increases in the mRNA levels of proteins involved in both fatty acid uptake and intracellular transport [39]. Using transfection assays with CPT-1 promoter/reporter constructs it was subsequently shown that this response required a functional PPRE site [40, 41]. In search for the PPAR species responsible, the cells were treated with PPAR isotype specific synthetic ligands. These studies revealed that PPARα, and to a lesser extent PPARβ/δ are involved in the fatty acid mediated upregulation of genes in the cardiac muscle cell. PPARγ specific ligands were found to be ineffective in this respect, arguing against a functional role of PPARγ in the regulation of cardiac lipid metabolism. Based on the ability of fatty acids to alter cardiac gene expression in a PPAR-dependent manner in isolated cell system one may infer that also under *in vivo* conditions fatty acids modulate the expression of proteins involved in fatty acid transport and metabolism.

Indeed, the relationship between plasma fatty acid levels and cardiac fatty acid utilisation has long been recognized. An elevation of plasma fatty acid levels, as occurs during perinatal development, fasting and diabetes, coincides with an increased cardiac fatty acid utilisation. Theoretically, the increased uptake and metabolism of fatty acids could be simply due to the rise in plasma concentration and, hence, an enhanced supply of this substrate to the organs. However, there are clear indications that in addition to mass effects, the cardiac muscle also adapts by adjusting the tissue content of proteins required for fatty acid uptake and the activity of enzymes involved in fatty acid oxidation via transcriptional mechanisms [38, 42]. It is tempting to speculate that the increased levels of blood-borne fatty acids are directly responsible for the alterations in cardiac gene expression. Studies with PPARα deficient mice provide strong evidence for the involvement of PPARα in cardiac gene expression. First, basic mRNA levels of mCPT-1, LCAD, MCAD and acyl-CoA oxidase were found to be lower in hearts of PPARα$^{-/-}$ mice than in wild-type animals [43]. Furthermore, in contrast to wild-type mice, fasting of PPARα$^{-/-}$ mice did not increase the expression of mCPT-1, MCAD, or acyl-CoA oxidase in the heart [44]. The latter findings provide compelling evidence for the regulatory role of PPARα in the expression of genes encoding proteins involved in cardiac fatty acid metabolism. On the other hand, despite the abundance of PPARα in both liver and heart, it has been consistently observed that the administration of PPARα ligands to adult rodents has marked effects on gene expression in liver, but hardly any in heart. Treatment of rats with various fibrates did not affect cardiac mRNA levels of proteins involved in lipid metabolism, including LPL, FATP and ACS [45–48]

Presently, no satisfying explanation can be offered for these apparently conflicting results. Although it is feasible that aspects of bio-availability of the ligands are involved, it is more likely that the answer will be found in the complex manner by which PPARα activity is being regulated in different tissues and species (Table 1).

Trans-activating capacity of PPAR

A first clue to differences in sensitivity between mammalian species comes from observations that the treatment of rats and mice with the hypolipidaemic fibrates causes severe peroxi-

Table 1. Factors modulating trans-activating capacity of PPARs

1.	PPAR expression (tissue specificity, splice variants, polymorphisms)
2.	Ligand specificity (natural agonists and antagonists)
3.	Ligand availability (role of FABPs, ACBPs)
4.	RXR availability (competition with other nuclear hormone receptors)
5.	Co-activators & co-repressors
6.	Phosphorylation of PPARs
7.	Competition for PPRE-binding sequences (e.g. PPAR vs. COUP-TF)
8.	PPRE flanking sequences

some proliferation in liver, ultimately leading to tumorigenesis, whereas no such side-effects were observed in hamsters and humans. The molecular basis for this difference in response remains largely enigmatic. In this context it is noteworthy that the truncated slice variant of PPARα is found in human, but not in rat tissue. Since the truncated PPARα has been shown to act in a dominant negative fashion [11], it is tempting to speculate that the ratio of splice variants determined the response of PPRE-containing genes and that changes in expression of the two splice variants allows for one level of transcriptional control in responsive species. Differences in behaviour of PPARα in different tissues are apparent from the fact that the same stimulus (fasting) markedly increased PPARα mRNA levels in liver, but actually lowered mRNA levels in rat cardiac tissue [42].

As many nuclear hormone receptors interact with RXR competition for this heterodimerisation partner may occur. Indeed, PPAR was shown to decrease the activity of thyroid hormone receptor by sequestration of RXR [49, 50]. Vice versa, PPAR activity was found to be diminished when cells were cotransfected with the thyroid hormone receptor or with COUP-TF (chicken ovalbumin upstream promoter transcription factor) [51, 52]. In addition to sequestration of RXR by other nuclear receptors, evidence was provided that under hypoxic conditions RXRα is rapidly degraded and that this mechanism may also be responsible for reducing the rate of transcription of PPRE containing genes as mCPTI under these conditions [53]. Furthermore, some of the nuclear receptors were shown to interact directly with DR1-sites. For COUP-TF is was shown that its binding to PPRE sequences represses transcriptional activity [52, 54]. The observation that during normal postnatal cardiac growth the level of COUP-TF decreases and that of PPARα increases, has been put forward as the molecular mechanism explaining the postnatal shift in cardiac substrate preference from glucose to fatty acids as primary energy source.

The PPAR/RXR dimer is known to interact with various other proteins, referred to as coactivators and corepressors depending on their ability to stimulate or inhibit receptor activity, respectively (for review see [55] and references therein). These cofactors are thought to modulate transcriptional activity by interacting both with nuclear receptors and basal transcription factors, thereby affecting the rate of assembly of the basal transcription initiation complex. Furthermore, some cofactors can alter the accessibility of genes through acetylation/deacetylation of histones. Co-activators known to interact with PPARs include the steroid receptor co-activator SRC-1, the PPAR binding protein (PBP), the PPARγ coactivators PGC-1 and PGC-2, and the integrator protein p300 [56–59]. Co-repressors for PPARs include the receptor-interacting protein-140 (RIP-140), the silencing mediator of retinoid and thyroid receptors (SMRT), and the nuclear receptor co-repressor (NCoR) [60–62]. Many of these cofactors have been shown to be expressed in cardiac muscle and are able to interact with various nuclear hormone receptors. However, for some of these proteins there are indications that their action may be PPAR subtype dependent. For instance, NCoR did not affect PPARγ activity, but repressed PPARα activity [62, 63]. Another level of complexity is added by the fact that the recruitment of (some of the) cofactors is ligand dependent. The inhibitory effect of acyl-CoA esters on PPARα is likely to be based on their ability to enhance recruitment of the co-repressor NCoR at the expense of the co-activator SRC1 [30].

Finally, in addition to regulatory mechanisms that are executed via protein-proteins interactions, there is solid evidence that both PPARα and PPARγ are phospho-proteins. To our knowledge this has not been established for PPARβ/δ. Phosphorylation affects the trans-activating capacity of PPARs and again the process may be influenced by ligand binding [64, 65]. Both PPARγ and PPARα have been shown to be phosphorylated in a MAP kinase dependent manner [66]. It remains to be seen whether phosphorylation of PPARα, either through ligand dependent or independent mechanisms, affects cardiac gene expression.

In conclusion, over the past few years the original concept of a simple regulation of PPAR activity via ligand interaction has been replaced by an exceedingly complex model by which the activity of each PPAR subtype is controlled by multiple factors. The presence and availability of activating and inhibitory ligands, competition for the mutual dimerisation partner RXR and for PPRE-sites, recruitment of co-activators and repressors, as well as the degree of phosphorylation, will all determine the ultimate rate of transcription of PPRE-containing genes. The multi-level control of this process probably enables fine tuning and adjustment of lipid metabolism to variations in energy demand.

Acknowledgements

Supported by grants 97.092 and D98.015 from the Netherlands Heart Foundation.

References

1. Issemann I, Green S: Activation of a member of the steroid hormone receptor superfamily by peroxisome proliferators. Nature 347: 645–650, 1990
2. Tsai MJ, O'Malley BW: Molecular mechanisms of action of steroid/thyroid receptor superfamily members. Ann Rev Biochem 63: 451–486, 1994
3. Kliewer SA, Umesono K, Noonan DJ, Heyman RA, Evans RM: Convergence of 9-*cis* retinoic acid and peroxisome proliferator signalling pathways through heterodimer formation of their receptors. Nature 358: 771–774, 1992
4. Van der Lee KAJM, Willemsen PHM, van der Vusse GJ, van Bilsen M: Effects of fatty acids on uncoupling protein-2 expression in the rat heart. FASEB J 14: 495–502, 2000
5. Wu P, Peters JM, Harris RA: Adaptive increase in pyruvate dehydrogenase kinase 4 during starvation is mediated by peroxisome proliferator-activated receptor α. Biochem Biophys Res Commun 287: 391–396, 2001
6. Kersten S, Mandard S, Escher P, Gonzalez FJ, Tafuri S, Desvergne B, Wahli W: The peroxisome proliferator-activated receptor α regulates amino acid metabolism. FASEB J 15: 1971–1978, 2001
7. Fruchart J-C, Duriez P, Staels B: Peroxisome proliferator-activated receptor-α activators regulate genes governing lipoprotein metabolism, vascular inflammation and atherosclerosis. Curr Opin Lipidol 10: 245–257, 1999
8. Grimaldi PA: Roles of lipid-activated receptors in the adipogenic action of fatty acids. Lipids 34: S205–S208, 1999
9. Vanden Heuvel JP: Peroxisome proliferator-activated receptors: A critical link among fatty acids, gene expression and carcinogenesis. J Nutr 129: 575S–580S, 1999
10. Fajas L, Auboeuf D, Raspe E, Schoonjans K, Lefebvre AM, Saladin R, Najib J, Laville M, Fruchart JC, Deeb S, Vidal Puig A, Flier J, Briggs MR, Staels B, Vidal H, Auwerx J: The organization, promoter analysis, and expression of the human PPARγ gene. J Biol Chem 272: 18779–18789, 1997
11. Gervois P, Torra IP, Chinetti G, Grötzinger T, Dubois G, Fruchart J-C, Fruchart-Najib J, Leitersdorf E, Staels B: A truncated human peroxisome proliferator-activated receptor α splice variant with dominant negative activity. Mol Endocrinol 13: 1535–1549, 1999
12. Hanselman JC, Vartanian MA, Koester BP, Gray SA, Essenburg AD, Rea TJ, Bisgaier CL, Pape ME: Expression of the mRNA encoding truncated PPARα does not correlate with hepatic insensitivity to peroxisome proliferators. Mol Cell Biochem 217: 91–97, 2001
13. Kliewer SA, Forman BM, Blumberg B, Ong ES, Borgmeyer U, Mangelsdorf DJ, Umesono K, Evans RM: Differential expression and activation of a family of murine peroxisome proliferator-activated receptors. Proc Natl Acad Sci USA 91: 7355–7359, 1994
14. Aperlo C, Pognonec P, Saladin R, Auwerx J, Boulukos KE: cDNA cloning and characterization of the transcriptional activities of the hamster peroxisome proliferator-activated receptor haPPARγ. Gene 162: 297–302, 1995
15. Braissant O, Foufelle F, Scotto C, Dauça M, Wahli W: Differential expression of peroxisome proliferator-activated receptors (PPARs): tissue distribution of PPAR-α, -β, and -γ in the adult rat. Endocrinology 137: 354–366, 1996
16. Takano H, Nagai T, Asakawa M, Toyozaki T, Oka T, Komuro I, Saito T, Masuda Y: Peroxisome proliferator-activated receptor activators inhibit lipopolysaccharide-induced tumor necrosis factor-α expression in neonatal rat cardiac myocytes. Circ Res 87: 596–602, 2000
17. Berger J, Leibowitz MD, Doebber TW, Elbrecht A, Zhang B, Zhou G, Biswas C, Cullinan CA, Hayes NS, Li Y, Tanen M, Ventre J, Wu MS, Berger GD, Mosley R, Marquis R, Santini C, Sahoo SP, Tolman RL, Smith RG, Moller DE: Novel peroxisome proliferator-activated receptor (PPAR) γ and PPARδ ligands produce distinct biological effects. J Biol Chem 274: 6718–6725, 1999
18. Chawla A, Barak Y, Nagy L, Liao D, Tontonoz P, Evans RM: PPAR-γ dependent and independent effects on macrophage-gene expression in lipid metabolism and inflammation. Nat Med 7: 48–52, 2001
19. Brunmair B, Gras F, Neschen S, Roden M, Wagner L, Waldhausl W, Furnsinn C: Direct thiazolidinedione action on isolated rat skeletal muscle fuel handling is independent of peroxisome proliferator-activated receptor-γ-mediated changes in gene expression. Diabetes 50: 2309–2315, 2001
20. Yamamoto K, Ohki R, Lee RT, Ikeda U, Shimada K: Peroxisome proliferator-activated receptor γ activators inhibit cardiac hypertrophy in cardiac myocytes. Circulation 104: 1670–1675, 2001
21. Carroll R, Severson DL: Peroxisome proliferator-activated receptor-α ligands inhibit cardiac lipoprotein lipase activity. Am J Physiol Heart Circ Physiol 281: H888–894, 2001
22. Forman BM, Tontonoz P, Chen J, Brun RP, Spiegelman BM, Evans RM: 15-Deoxy-$\Delta^{12,14}$-prostaglandin J_2 is a ligand for the adipocyte determination factor PPARγ. Cell 83: 803–812, 1995
23. Gottlicher M, Widmark E, Li Q, Gustafsson JA: Fatty acids activate a chimera of the clofibric acid-activated receptor and the glucocorticoid receptor. Proc Natl Acad Sci USA 89: 4653–4657, 1992
24. Issemann I, Prince RA, Tugwood JD, Green S: The peroxisome proliferator-activated receptor: retinoid X receptor heterodimer is activated by fatty acids and fibrate hypolipidaemic drugs. J Mol Endocrinol 11: 37–47, 1993
25. Kliewer SA, Sundseth SS, Jones SA, Brown PJ, Wisely GB, Koble CS, Devchand P, Wahli W, Willson TM, Lenhard JM, Lehmann JM: Fatty acids and eicosanoids regulate gene expression through direct interactions with peroxisome proliferator-activated receptors α and γ. Proc Natl Acad Sci USA 94: 4318–4323, 1997
26. Forman BM, Chen J, Evans RM: Hypolipidemic drugs, polyunsaturated fatty acids, and eicosanoids are ligands for peroxisome proliferator-activated receptors α and δ. Proc Natl Acad Sci USA 94: 4312–4317, 1997
27. Johnson TE, Holloway MK, Vogel R, Rutledge SJ, Perkins JJ, Rodan GA, Schmidt A: Structural requirements and cell-type specificity for ligand activation of peroxisome proliferator-activated receptors. J Steroid Biochem Mol Biol 63: 1–8, 1997
28. Xu HE, Lambert MH, Montana VG, Parks DJ, Blanchard SG, Brown PJ, Sternbach DD, Lehmann JM, Wisely GB, Willson TM, Kliewer SA, Milburn MV: Molecular recognition of fatty acids by peroxisome proliferator-activated receptors. Mol Cell 3: 397–403, 1999
29. Wolfrum C, Borrmann CM, Borchers T, Spener F: Fatty acids and hypolipidemic drugs regulate peroxisome proliferator-activated receptors α- and γ-mediated gene expression via liver fatty acid binding protein: A signaling path to the nucleus. Proc Natl Acad Sci USA 98: 2323–2328, 2001
30. Elholm M, Dam I, Jorgensen C, Krogsdam AM, Holst D, Kratchmarova I, Gottlicher M, Gustafsson JA, Berge R, Flatmark T, Knudsen J, Mandrup S, Kristiansen K: Acyl-CoA esters antagonize the effects of ligands on peroxisome proliferator-activated receptor α conformation, DNA binding, and interaction with Co-factors. J Biol Chem 276: 21410–21416, 2001
31. Murakami K, Ide T, Nakazawa T, Okazaki T, Mochizuki T, Kadowaki T: Fatty-acyl-CoA thioesters inhibit recruitment of steroid receptor co-activator 1 to α and γ isoforms of peroxisome-proliferator-activated receptors by competing with agonists. Biochem J 353: 231–238, 2001
32. Krey G, Braissant O, L-Horset F, Kalkhoven E, Perroud M, Parker MG, Wahli W: Fatty acids, eicosanoids, and hypolipidemic agents identified as ligands of peroxisome proliferator-activated receptors by

coactivator-dependent receptor ligand assay. Mol Endocrinol 11: 779–791, 1997
33. Murakami K, Ide T, Suzuki M, Mochizuki T, Kadowaki T: Evidence for direct binding of fatty acids and eicosanoids to human peroxisome proliferators-activated receptor α. Biochem Biophys Res Commun 260: 609–613, 1999
34. Van der Vusse GJ, Glatz JFC, Stam HCG, Reneman RS: Fatty acid homeostasis in the normoxic and ischemic heart. Physiol Rev 72: 881–940, 1992
35. Borchers T, Unterberg C, Rudel H, Robenek H, Spener F: Subcellular distribution of cardiac fatty acid-binding protein in bovine heart muscle and quantitation with an enzyme-linked immunosorbent assay. Biochim Biophys Acta 1002: 54–61, 1989
36. Grimaldi PA, Knobel SM, Whitesell RR, Abumrad NA: Induction of aP2 gene expression by nonmetabolized long-chain fatty acids. Proc Natl Acad Sci USA 89: 10930–10934, 1992
37. Jump DB, Clarke SD, Thelen A, Liimatta M: Coordinate regulation of glycolytic and lipogenic gene expression by polyunsaturated fatty acids. J Lipid Res 35: 1076–1084, 1994
38. van Bilsen M, van der Vusse GJ, Reneman RS: Transcriptional regulation of metabolic processes: Implications for cardiac metabolism. Pflügers Arch 437: 2–14, 1998
39. Van der Lee KAJM, Vork MM, de Vries JE, Willemsen PHM, Glatz JFC, Reneman RS, van der Vusse GJ, van Bilsen M: Long-chain fatty acid-induced changes in gene expression in neonatal cardiac myocytes. J Lipid Res 41: 41–47, 2000
40. Mascaro C, Acosta E, Ortiz JA, Marrero PF, Hegardt FG, Haro D: Control of human muscle-type carnitine palmitoyltransferase I gene transcription by peroxisome proliferator-activated receptor. J Biol Chem 273: 8560–8563, 1998
41. Brandt JM, Djouadi F, Kelly DP: Fatty acids activate transcription of the muscle carnitine palmitoyltransferase I gene in cardiac myocytes via the peroxisome proliferator-activated receptor α. J Biol Chem 273: 23786–23792, 1998
42. Van der Lee KAJM, Willemsen PHM, Samec S, Seydoux J, Dulloo AG, Pelsers MMAL, Glatz JFC, van der Vusse GJ, van Bilsen M: Fasting-induced changes in the expression of genes controlling substrate metabolism in the rat heart. J Lipid Res 42: 1752–1758, 2001
43. Watanabe K, Fujii H, Takahashi T, Kodama M, Aizawa Y, Ohta Y, Ono T, Hasegawa G, Naito M, Nakajima T, Kamijo Y, Gonzalez FJ, Aoyama T: Constitutive regulation of cardiac fatty acid metabolism through peroxisome proliferator-activated receptor α associated with age-dependent cardiac toxicity. J Biol Chem 275: 22293–22299, 2000
44. Leone TC, Weinheimer CJ, Kelly DP: A critical role for the peroxisome proliferator-activated receptor α (PPARα) in the cellular fasting response: The PPARα-null mouse as a model of fatty acid oxidation disorders. Proc Natl Acad Sci USA 96: 7473–7478, 1999
45. Schoonjans K, Staels B, Grimaldi P, Auwerx J: Acyl-CoA synthetase mRNA expression is controlled by fibric-acid derivatives, feeding and liver proliferation. Eur J Biochem 216: 615–622, 1993
46. Schoonjans K, Staels B, Auwerx J: Role of the peroxisome proliferator-activated receptor (PPAR) in mediating the effects of fibrates and fatty acids on gene expression. J Lipid Res 37: 907–925, 1996
47. Schoonjans K, Peinado-Onsurbe J, Lefebvre A-M, Heyman RA, Briggs M, Deeb S, Staels B, Auwerx J: PPARα and PPARγ activators direct a distinct tissue-specific transcriptional response via a PPRE in the lipoprotein lipase gene. EMBO J 15: 5336–5348, 1996
48. Cook WS, Yeldandi AV, Rao MS, Hashimoto T, Reddy JK: Less extrahepatic induction of fatty acid beta-oxidation enzymes by PPARα. Biochem Biophys Res Commun 278: 250–257, 2000
49. Juge-Aubry CE, Gorla-Bajszczak A, Pernin A, Lemberger T, Wahli W, Burger AG, Meier CA: Peroxisome proliferator-activated receptor mediates cross-talk with thyroid hormone receptor by competition for retinoid X receptor. Possible role of a leucine zipper-like heptad repeat. J Biol Chem 270: 18117–18122, 1995
50. Meier-Heusler SC, Zhu X, Juge-Aubry C, Pernin A, Burger AG, Cheng S-y, Meier CA: Modulation of thyroid hormone action by mutant thyroid hormone receptors, c-erbAα2 and peroxisome proliferator-activated receptor: evidence for different mechanisms of inhibition. Mol Cell Endocrin 107: 55–66, 1995
51. Winrow CJ, Kassam A, Miyata KS, Marcus SL, Hunter J, Capone JP, Rachubinski RA: Interplay of the peroxisome proliferator-activated receptor and the thyroid hormone receptor-signaling pathways in regulating peroxisome proliferator-responsive genes. Ann NY Acad Sci 804: 214–230, 1996
52. Marcus SL, Capone JP, Rachubinski RA: Identification of COUP-TFII as a peroxisome proliferator response element binding factor using genetic selection in yeast: COUP-TFII activates transcription in yeast but antagonizes PPAR signaling in mammalian cells. Mol Cell Endocrinol 120: 31–39, 1996
53. Huss JM, Levy FH, Kelly DP: Hypoxia inhibits the peroxisome proliferator-activated receptor α/retinoid X receptor gene regulatory pathway in cardiac myocytes: A mechanism for O_2-dependent modulation of mitochondrial fatty acid oxidation. J Biol Chem 276: 27605–27612, 2001
54. Miyata KS, Zhang B, Marcus SL, Capone JP, Rachubinski RA: Chicken ovalbumin upstream promoter transcription factor (COUP-TF) binds to a peroxisome proliferator-responsive element and antagonizes peroxisome proliferator-mediated sigaling. J Biol Chem 268: 19169–19172, 1993
55. McKenna NJ, Lanz RB, O'Malley BW: Nuclear receptor coregulators: Cellular and molecular biology. Endocr Rev 20: 321–344, 1999
56. Zhu Y, Qi C, Calandra C, Rao MS, Reddy JK: Cloning and identification of mouse steroid receptor coactivator-1 (mSRC-1), as a coactivator of peroxisome proliferator-activated receptor γ. Gene Expres 6: 185–195, 1996
57. Zhu Y, Qi C, Jain S, Rao MS, Reddy JK: Isolation and characterization of PBP, a protein that interacts with peroxisome proliferator-activated receptor. J Biol Chem 272: 25500–25506, 1997
58. Puigserver P, Wu Z, Park CW, Graves R, Wright M, Spiegelman BM: A cold-inducible coactivator of nuclear receptors linked to adaptive thermogenesis. Cell 92: 829–839, 1998
59. Gelman L, Zhou G, Fajas L, Raspé E, Fruchart J-C, Auwerx J: p300 interacts with the N- and C-terminal part of PPARγ2 in a ligand-independent and -dependent manner, respectively. J Biol Chem 274: 7681–7688, 1999
60. Miyata KS, McCaw SE, Meertens LM, Patel HV, Rachubinski RA, Capone JP: Receptor-interacting protein 140 interacts with and inhibits its transactivation by, peroxisome proliferator-activated receptor α and liver-X-receptor α. Mol Cell Endocrinol 146: 69–76, 1998
61. Robinson CE, Wu X, Nawaz Z, Oñate SA, Gimble JM: A corepressor and chicken ovalbumin upstream promoter transcriptional factor proteins modulate peroxisome proliferator-activated receptor-γ_2/retinoid X receptor α-activated transcription from the murine lipoprotein lipase promoter. Endocrinology 140: 1586–1593, 1999
62. Dowell P, Ishmael JE, Avram D, Peterson VJ, Nevrivy DJ, Leid M: Identification of nuclear receptor corepressor as a peroxisome proliferator-activated receptor α interacting protein. J Biol Chem 274: 15901–15907, 1999
63. Lavinsky RM, Jepsen K, Heinzel T, Torchia J, Mullen T-M, Schiff R, Del-Rio AL, Ricote M, Ngo S, Gemsch J, Hisenbeck SG, Osborne CK, Glass CK, Rosenfeld MG, Rose DW: Diverse signaling pathways modulate nuclear receptor recruitment of N-CoR and SMRT complexes. Proc Natl Acad Sci USA 95: 2920–2925, 1998
64. Camp HS, Tafuri SR: Regulation of peroxisome proliferator-activated receptor γ activity by mitogen-activated protein kinase. J Biol Chem 272: 10811–10816, 1997

65. Camp HS, Tafuri SR, Leff T: *c*-Jun N-terminal kinase phosphorylates peroxisome proliferator-activated receptor-γ1 and negatively regulates its transcriptional activity. Endocrinology 140: 392–397, 1999
66. Juge-Aubry CE, Hammar E, Siegrist-Kaiser C, Pernin A, Takeshita A, Chin WW, Burger AG, Meier CA: Regulation of the transcriptional activity of the peroxisome proliferator-activated receptor α by phosphorylation of a ligand-independent *trans*-activating domain. J Biol Chem 274: 10505–10510, 1999

Molecular and Cellular Biochemistry **239:** 139–147, 2002.

New insights into the fatty acid-binding protein (FABP) family in the small intestine

Philippe Besnard, Isabelle Niot, Hélène Poirier, Lionel Clément and André Bernard

Laboratoire de Physiologie de la Nutrition, Ecole Nationale Supérieure de Biologie Appliquée à la Nutrition et à l'Alimentation (ENSBANA), FRE 2328 CNRS-CESG/Université de Bourgogne, Dijon, France

Abstract

The fatty acid-binding protein (FABP) superfamily is constituted by 14–15 kDa soluble proteins which bind with a high affinity either long-chain fatty acids (LCFAs), bile acids (BAs) or retinoids. In the small intestine, three different FABP isoforms exhibiting a high affinity for LCFAs and/or BAs are expressed: the intestinal and the liver-type (I-FABP and L-FABP) and the ileal bile acid-binding protein (I-BABP). Despite of extensive investigations, their respective physiological function(s) are not clearly established. In contrast to the I-FABP, L-FABP and I-BABP share several common structural features (shape, size and volume of the hydrophobic pocket). Moreover, L-FABP and I-BABP genes are also specifically regulated by their respective preferential ligands through a very similar molecular mechanism. Although, they exhibit differences in their binding specificities and location along the small intestine supporting a specialization, it is likely that L-FABP and I-BABP genes exert the same type of basic function(s) in the enterocyte, in contrast to I-FABP. (Mol Cell Biochem **239**: 139–147, 2002)

Key words: soluble lipid-binding proteins, fatty acid-binding proteins, small intestine

Abbreviations: LCFA – long-chain fatty acids; LCA – long-chain acyl-CoA; BA – bile acids; CS – cholesterol; PL – phospholipids; TG – triglycerides; LBP – lipid-binding proteins; FABP – fatty acid-binding proteins; I-BABP – ileal bile acid-binding protein; PPAR – peroxisome proliferator-activated receptor; PPRE – peroxisome proliferator-responsive element; RXR – retinoid-X-receptor; FXR – farnesoid-X-receptor; BARE – bile acid-responsive element; PYY – peptide YY; EGF – epidermal growth factor; AOFA – anthroyloxy-fatty acid

In the Western diet, lipids represent more than 40% of the daily caloric intake, while the nutritional advice is 10% lower. This high fat supply associated with a qualitative imbalance (excess of saturated fatty acids and cholesterol, high ω6/ω3 ratio for poly-unsaturated fatty acids) greatly contributes not only to the increase of obesity prevalence in the population, but also to the appearance of diseases of plethora such as artheroclerosis, non insulin-dependent diabetes, breast and colon cancers, whom the human and social costs are dramatic. Therefore, a considerable energy is devoted since several years to better understand how lipids are involved in the etiology of these pathologies. Nevertheless, although the gut is responsible for fat absorption, little is known about the mechanisms involved in the uptake and metabolic fate of lipids or their metabolic derivatives, such as bile acids (BA), in the intestinal cell. The functional importance of different soluble lipid-binding proteins (LBP), which all belong to the fatty acid-binding proteins (FABP) superfamily, is especially unclear in the small intestine.

The dietary fat is the major source of lipids found in the intestinal lumen. It is mainly constituted by triglycerides (TG), and in a lesser extent by free or esterified cholesterol (CS) and phospholipids (PL). Lipid hydrolysis by the pancreatic juice releases great quantities of long-chain fatty acid (LCFA) whom the intestinal uptake is complex. It is a dual phenomenon involving both a passive diffusion and a proteins-mediated transport [1–9]. The small intestine being an organ daily subjected to large variations in the lipid supply,

Address for offprints: P. Besnard, Laboratoire de Physiologie de la Nutrition, Ecole Nationale Supérieure de Biologie Appliquée à la Nutrition et à l'Alimentation, (ENSBANA), FRE 2328 CNRS-CESG/Université de Bourgogne, 1, esplanade Erasme, F-21000 Dijon, France (E-mail: pbesnard@u-bourgogne.fr)

the passive diffusion might constitute a high capacity, low affinity transport system very efficient during lipid loads (postprandial period), while membrane lipid-binding proteins might constitute a low capacity, high affinity system that might predominate at low substrate concentration (interprandial periods and fasting) [9]. Once into the cell, LCFA are reversibly bound to two different FABPs: the liver type (L-FABP), which is also found in the liver and kidneys, and the intestinal type (I-FABP) which is strictly confined to the small intestine. The acylation of LCFA by the acyl-CoA synthetases (ACS) is the first step of the cellular FA metabolism. During the postprandial period, LCFA are mainly re-esterified in TG and PL before to be associated with apolipoproteins to form TG-rich lipoproteins (mainly chylomicrons) which are secreted into lymph (Fig. 1A) [9].

Primary BAs are synthesized from CS in the liver where they are conjugated with taurine or glycine prior to be secreted into bile. In the intestinal lumen, they facilitate the hydrolysis then the absorption of dietary lipids and fat soluble vitamins. Conjugated-BAs are efficiently reabsorbed in the ileum by an active transport system constituted by a couple of BA carriers, the ileal bile acid transporter (IBAT), which is

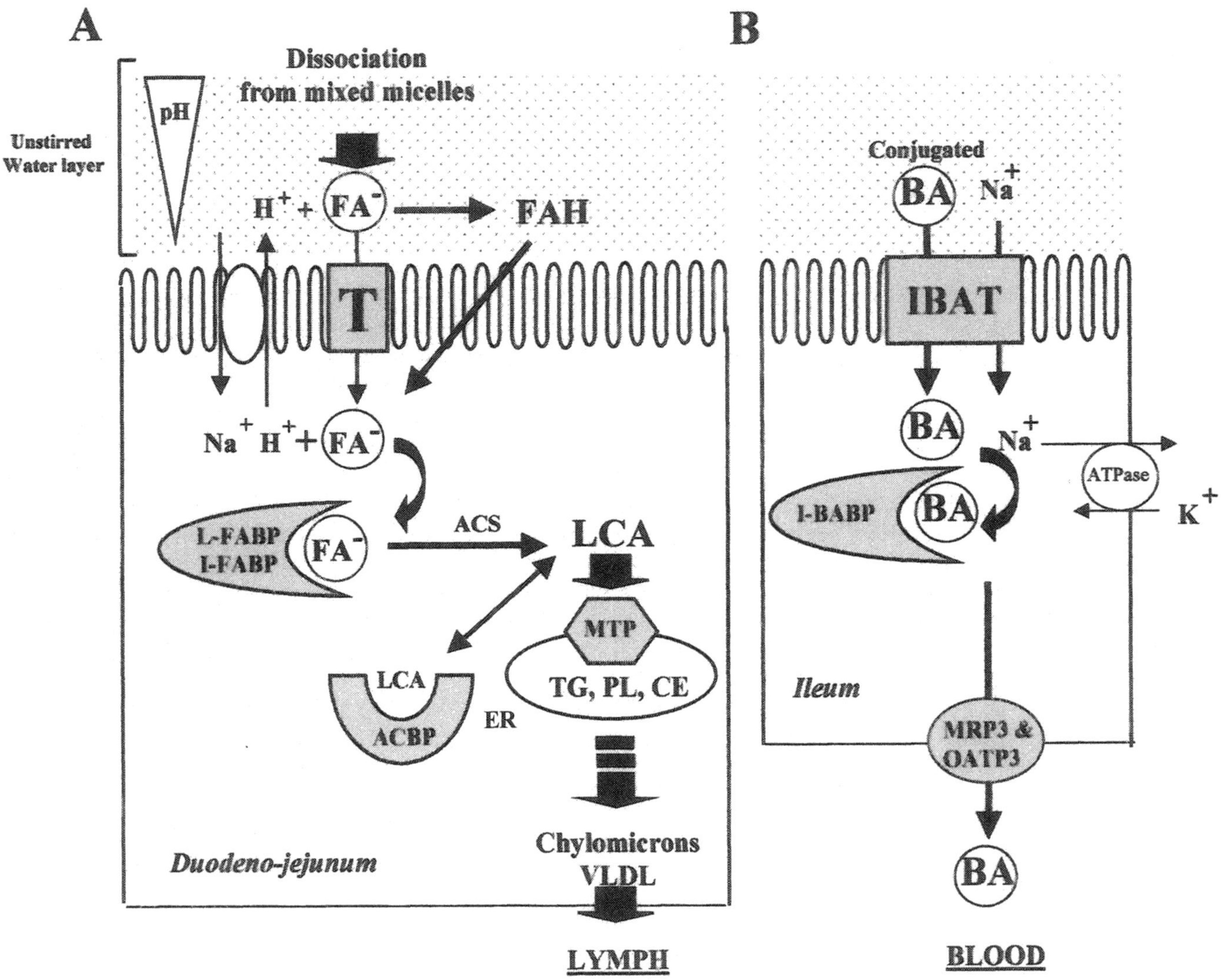

Fig. 1. Main steps of intestinal absorption of long chain fatty acids (A) and bile acids (B). The plasma membrane and soluble lipid-binding proteins involved in the absorption of both long-chain fatty acids and conjugated bile acids are shown in grey. (A) FA^- – ionised long-chain fatty acids; FAH – protonated long-chain fatty acids ; T – protein-mediated transport; L-FABP and I-FABP – liver and intestinal fatty acid-binding proteins; ACS – Acyl-CoA synthetases; ACBP – acyl-CoA-binding protein; LCA – long-chain acyl-CoA; MTP – microsomal triglyceride transfer protein; TG – triglycerides; PL – phospholipids; CE – cholesterol esters, ER – endoplasmic reticulum; VLDL – very-low-density lipoprotein. (B) BA – bile acids; IBAT – ileal bile acid transporter; I-BABP – ileal bile acid-binding protein; MRP3 – multidrug-resistance protein 3; OATP3 – organic anion exchange transport protein 3.

a 38 kDa protein located in the brush border membrane of ileocytes [10] and the ileal bile acid-binding protein (I-BABP), which belongs to the FABP family [11]. BAs are secreted without any transformation into blood by a basolateral transport system constituted of two unrelated membrane proteins : the multidrug resistance protein 3 (MRP-3) [12] and the organic anion exchange transport protein 3 (OATP-3) [13, 14] (Fig. 1B). More than 90% BAs are reabsorbed along the small intestine in healthy human and return to the liver, through the portal blood, before to be secreted again into bile [15]. This enterohepatic circulation of BAs is essential for the maintenance of CS balance. Indeed, BAs not reclaimed by the intestinal absorption constitute the main physiological way to eliminate a CS excess.

Therefore, three different FABP types exhibiting a high affinity for LCFA and/or BAs are found in the small intestine. However, despite of an extensive investigation, little is known about their respective physiological action(s).

Specific locations and binding properties supporting a specialization

The respective localization along the small intestine and the binding properties of I-FABP, L-FABP and I-BABP are at the origin of a cellular specialization. It is well known that the small intestine is a heterogeneous organ characterized by a variable genotypic expression along the cephalocaudal axis producing a regional specialization. According to this characteristic, the ileum, which plays a critical role in the

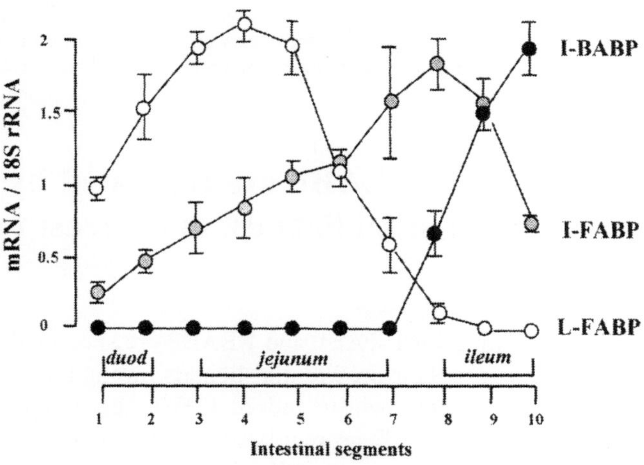

Fig. 2. Distribution of FABPs along the cephalo-caudal axis of the small intestine in the mouse. The intestine was divided in 10 equal segments from the pylorus to the ileocecal valvula, then the mucosa was scraped off. Steady state levels of I-FABP (shaded circle), L-FABP (open circle) and I-BABP (black circle) mRNA were determined by Northern blotting. The results are expressed after normalization to 18S rRNA.

enterohepatic circulation of BA, is the exclusive intestinal segment where the I-BABP gene is expressed (Fig. 2). Even if I-FABP and L-FABP are co-expressed in the enterocyte, a slightly different pattern of expression is found along the small intestine in the mouse. Indeed, I-FABP mRNA levels increase from the duodenum to reach the highest levels in the proximal ileum, while the highest L-FABP expression occur in the proximal jejunum then progressively fall to an undetectable level in the distal ileum (Fig. 2) [16].

Similarly, the comparison of the binding properties (stoichiometry, specificity and affinity) reveals strong differences between these 3 FABPs. Indeed, I-FABP only binds LCFA with a ratio of one FA for one protein. By contrast, L-FABP can bind 2 LCFA as well as a large number of bulky hydrophobic molecules including BAs, heme, various xenobiotics and carcinogens, but generally with a ratio of 1 for 1. The binding affinities of these 2 FABPs have accurately been determined by Richieri and collaborators for LCFA [17]. If L-FABP and I-FABP have similar dissociation constant (kD) values for saturated LCFA, by contrast, I-FABP exhibits a lower avidity for unsaturated FA than L-FABP. The binding properties of I-BABP are not yet well known. Its affinity for BAs (mono-, di- or tri-hydroxylated, conjugated or unconjugated) has not been determined. LCFA can also be bound by I-BABP but with a low affinity [18, 19]. The binding capacity of I-BABP is not definitively determined since ratios of 1 mol protein for 1 or 2 mol BAs have been reported [20, 21].

These specificities clearly lead to a specialization, I-BABP being involved in the ileal fate of BAs, while I-FABP and L-FABP take in charge LCFA in the gut. Nevertheless, the comparison of their respective molecular structure and gene regulation demonstrate that I-BABP and L-FABP share several common features, in contrast to I-FABP.

L-FABP and I-BABP are highly structurally-related molecules

For one FABP given, there is an extensive inter-species conservation of the peptide sequence. For instance, more than 80% homology exists between human and rat L-FABP. By contrast, two different FABP are poorly related, the highest sequence identity being found for L-FABP and I-BABP. Despite of different peptidic sequence, FABPs exhibit a similar tertiary structure that consists of two α helices (αI, αII) and ten anti parallel β strands (βA-βJ) organised in two almost orthogonal β sheets forming an hydrophobic pocket (Fig. 3). A 'portal region' constituted by the α helices connected to the βC/βD and βE/βF strands allows the entry and the exit of one LCFA [22]. The fact that the peptide sequences of L-FABP and I-BABP consist of 127–128 amino-acid residues instead of 131–133 for the other FABPs produces a shortening of β

A

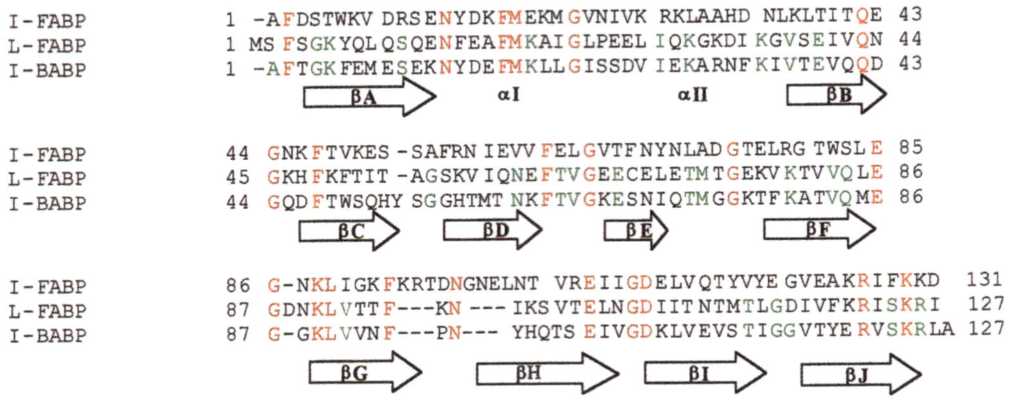

B

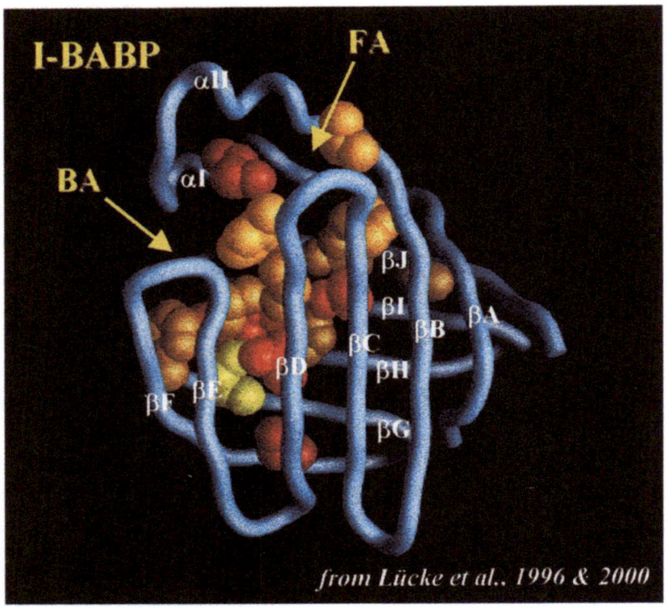

C

Comparison of structural properties

	I-FABP	L-FABP	I-BABP
Portal region(s)	1	2	2
Flexibility	+	+++	+++
Volume of binding pocket	+	+++	+++

Fig. 3. Main structural features of FABPs expressed in the small intestine. (A) Amino-acid sequence alignment of human I-FABP, L-FABP and I-BABP. Amino-acid sequence identity between the 3 FABPs (in red) and between L-FABP and I-BABP (in green), as well as the localization of α helices and β strands are shown. (B) Example of a typical tertiary structure of a FABPs (reproduced with the permission of Dr C. Lücke, J.W. Goethe-Universität, Frankfurt, Germany). (C) Comparison of structural properties of FABPs.

strands G and H at the origin of a relatively large opening in the protein structure (Fig. 3) [23]. This gap constitutes a second portal for a ligand entry and exit. Moreover, the structure of L-FABP and I-BABP is characterized by a reduced stability of hydrogen bonding network in the β sheets leading to an unusual flexibility of the protein backbone [20, 24]. Finally, their ligand binding pockets are the largest of all family members, (around 450 Å^3 for the L-FABP and I-BABP vs. 230 Å^3 for the I-FABP) [20, 25, 26]. These specific structural features shared by both L-FABP and I-BABP allow the binding of the bulky and rigid ligands such as BAs, and explain why the L-FABP can bound two LCFA in contrast to the I- FABP. The fact that L-FABP and I-BABP are closely related from an evolutionary point of view [27], likely account for this structural likeness.

L-FABP and I-BABP are regulated by their preferential ligands, in contrast to I-FABP

Genes encoding for L-FABP and I-BABP are specifically up-regulated by their preferential ligands through a very similar molecular mechanism. Indeed, L-FABP gene expression is transcriptionally up-regulated by LCFA in both the liver and the small intestine [28]. This regulation is mediated by a family of nuclear receptors called peroxisome proliferator-activated receptors (PPARs). After activation by LCFA, PPAR binds, as an heterodimer with the retinoid-X-receptor (RXR), to a specific responsive element (PPRE) generally located in the promoter of target genes [29]. A functional PPRE sequence has been identified in the proximal part of the rat L-FABP promoter [28]. Similarly, a BA-mediated induction of I-BABP gene expression has been recently depicted. This regulation takes place after BA activation of the farnesoid-X-receptor (FXR) and its interaction, as a heterodimer with RXR, to a BA-responsive element (BARE) found in the promoter of human, rabbit and mouse I-BABP promoter [30].

Different regulatory systems seem involved in the control the I-FABP expression. Indeed, a direct FA-mediated modulation of I-FABP gene expression has never been highlighted as well as *in vivo* than *in vitro* [29, 31, 32]. By contrast, in enterocyte-like hBRIE 380 cells, the peptide YY (PYY) induces I-FABP mRNA level [33]. This gut regulatory peptide is secreted mainly by ileal endocrine cells particularly when fat reach the distal part of the gut secondary to a dietary overload [34]. Through an inhibitory action on gastric emptying and intestinal motility, PYY also contributes to the 'ileal brake', which ensure efficient utilisation of ingested food. The fact that PYY can act as paracrine agent [35] might explain why I-FABP induction, in rat subjected to a high fat diet, occurs only in ileum [36]. However, the physiological relevance of this finding must be confirmed in intact animals. More recently, a down-regulation of I-FABP expression by the epidermal growth factor (EGF) has been reported in highly differentiated enterocyte-like Caco2 cells. This growth factor, found in the intestinal lumen, contribute to the re-epithelialization after intestinal mucosa injury. It is noteworthy that EGF does not affect the L-FABP expression [37].

Therefore, L-FABP and I-BABP appear to be highly related molecules, in contrast to I-FABP. Their structural specificities as well as their gene regulation by their preferential ligand, strongly suggest that L-FABP and I-BABP, in spite of a binding specialization, exert the same type of functions in the enterocyte.

What function(s) for FABPs in the small intestine?

The roles of soluble FABPs seem more complex than those which are generally assigned, i.e. ligand desorption from plasma membrane and facilitation of their intracellular diffusion. Because I-FABP and L-FABP are co-expressed in the enterocyte the question a redundant or complementary transport system is raised. Their respective specificities speak for a specialization, at least in the standard dietary fat supply : I-FABP being involved in the TG synthesis and secretion while L-FABP might be preferentially implicated in PL syn-

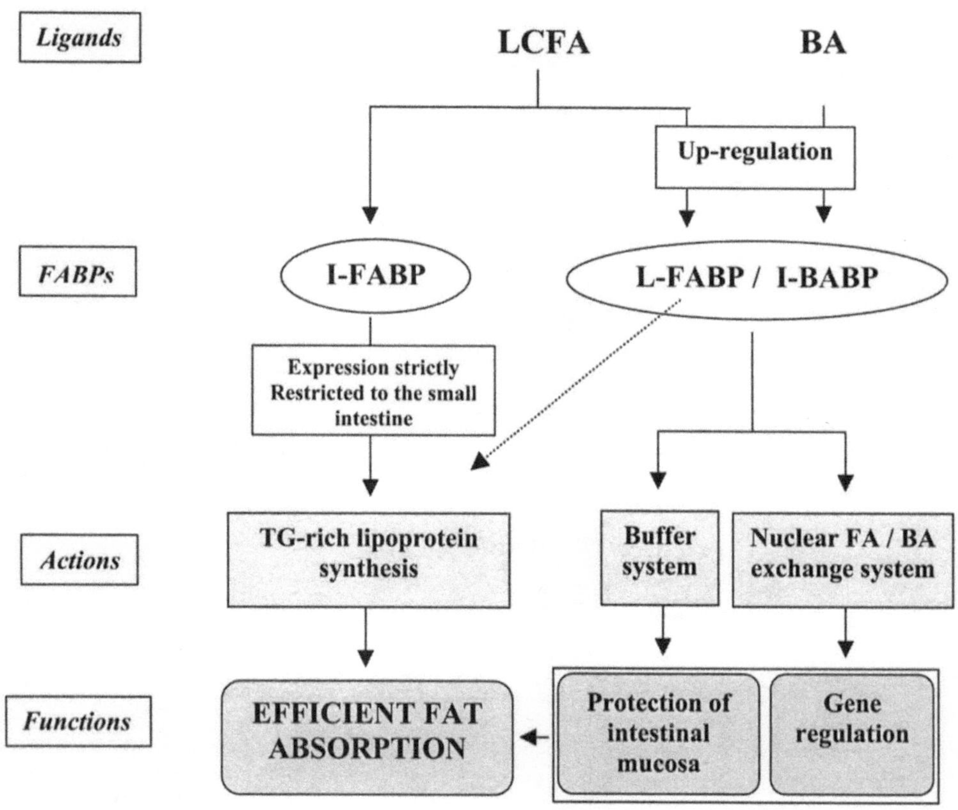

Fig. 4. Putative functions of I-FABP, L-FABP and I-BABP in the small intestine.

thesis, membrane protection and gene regulation (Fig. 4).

The highlighting of a human polymorphism in the gene encoding for I-FABP (FABP2) has greatly contributed to better understand its physiological importance in TG-rich lipoprotein synthesis. One base substitution in the codon 54 of FABP2 leads to the change of an Ala by a Thr. Initially found in the Pima indians, this substitution is associated with a high TG plasma level, an insulin resistance [38], and an increase in the body mass index [39]. The fact that the muted Thr[54]-I-FABP has a 2-fold greater affinity for LCFA than the wild Ala[54]-protein may explain these metabolic disturbances [38]. Indeed, a greater avidity of the mutant I-FABP for LCFA could lead to an increase in both cellular FA uptake and TG-rich lipoproteins synthesis [38]. According to this assumption, a dramatic rise in LCFA transport and TG secretion is found in Thr[54]-I-FABP-transfected Caco-2 cells as compared to cells transfected the wild isoform [40]. Similar data have been recently reported in the human jejunal organ culture model in which the Thr[54]-encoding allele appear to be associated with a dramatic rise in TG synthesis and chylomicrons output [41]. It noteworthy that these metabolic disturbances occur without modification in I-FABP and L-FABP levels and in absence of neuro-endocrine influences. Taken together these data strongly suggest that I-FABP is involved in the targeting of dietary LCFA towards the endoplasmic reticulum, where they participate to the synthesis of TG-rich lipoproteins. The fact that an EGF-mediated inhibition of I-FABP is associated with a decrease in [^{14}C] palmitate uptake, TG synthesis and secretion in Caco-2 cells is in good agreement with this assumption [37].

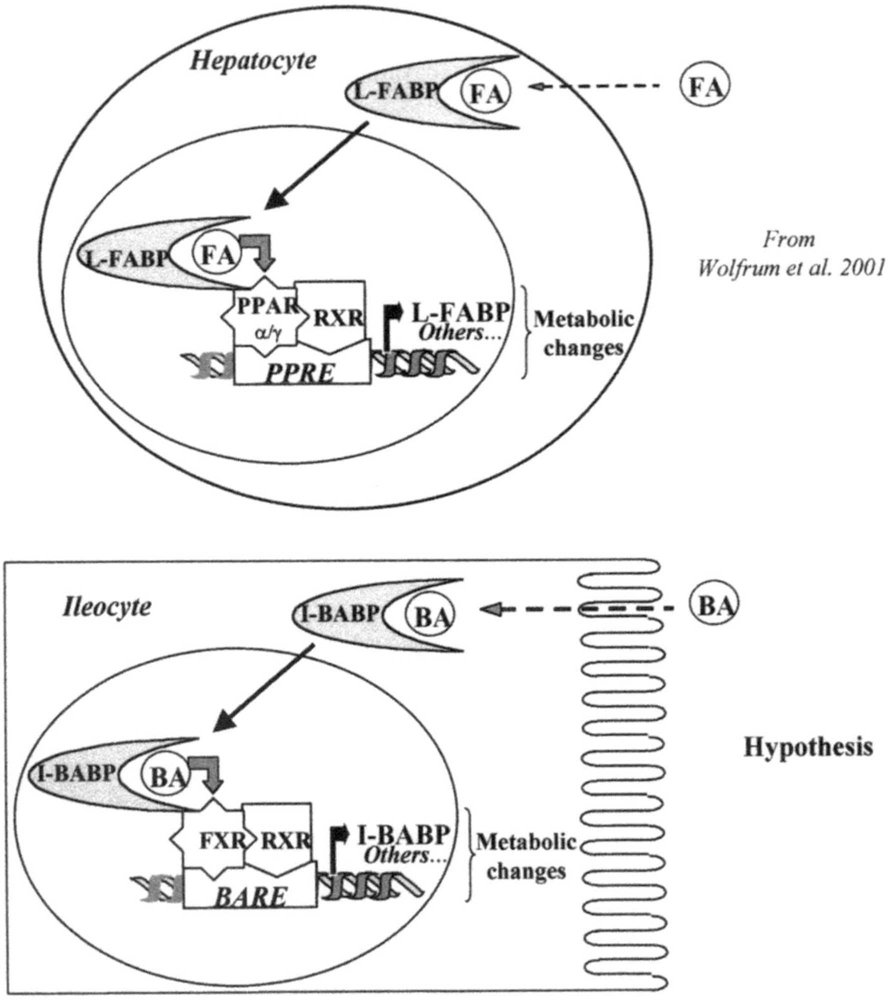

Fig. 5. Gene regulation by FABPs. (A) A protein-protein interaction between L-FABP and PPARα or PPARγ leads to the regulation of PPAR target genes in the liver. FA – long-chain fatty acids; PPAR – peroxisome proliferator-activated receptor; RXR – retinoid-X-receptor; PPRE – peroxisome proliferator-responsive element. (B) Hypothesis of work: Because I-BABP and L-FABP share common structural features and are similarly regulated by their respective preferential ligands, it could be hypothesized that I-BABP could contribute to BA-mediated gene regulation through the direct activation of FXR. BA – bile acids; FXR – farnesoid-X-receptor; BARE – bile acid-responsive element.

From different transfection studies carried out either in cell lines within I-FABP or L-FABP are lacking as the fibroblastic L cells [42] and pluripotent mouse embryonic stem ES cells [43], or within only one FABP is constitutively expressed as hBRIE 380i (+I-FABP/–L-FABP) [44] and HepG2 (+L-FABP/ –I-FABP) [45], it can be concluded that both common and specific functions are exerted by these 2 soluble LBPs in the intestinal cell. Indeed, the LCFA uptake and diffusion in cell are increased by both I-FABP and L-FABP. By contrast, convergent data indicate that I-FABP targets LCFA preferentially towards the TG pathway, whereas L-FABP preferentially increase the PL synthesis.

A part of these functional differences can be explained by structural specificities. Using a resonance energy transfer assay, Hsu and Storch have demonstrated that the transfer of fluorescent anthroyloxy-FA (AOFA) from I-FABP or L-FABP to acceptor membrane vesicles occurred by different molecular mechanisms [46]. For I-FABP, the transfer of AOFA requires a direct collisional interaction with the PL bilayer, while the FA exchanges between L-FABP and membranes are carried out by an aqueous diffusion. Collisional exchanges occur by ionic interactions between few amino-acid residues of the helicoidal domain of I-FABP with anionic PL of membranes. Indeed, in helix-less mutant I-FABP obtained by site-directed mutagenesis, the collisional exchanges are totally suppressed. In these conditions, the transfer of the AOFA occurs by aqueous diffusion, as for L-FABP [47]. It is thought that the collision with a target membrane yields a conformational change in the flexible region of I-FABP backbone constituted by the distal half of α-II helix and βC-βD turns producing the 'hinged opening' of the portal domain and the release of FA [48]. This dynamic domain seems to play a critical role in the binding characteristics of I-FABP. Therefore, a single modification of its amino-acid sequence can have significant physiological consequences. A good illustration of this assumption is that a punctual mutation in the portal domain such as the Ala^{54}-Thr mutation deeply affects the binding affinity of I-FABP leading to a dramatic alteration in the lipid metabolism. The fact that LCFA delivery by the apical side of the enterocyte are preferentially bound to I-FABP [49] ensures their dynamic transfer towards a cellular compartment devoted to TG synthesis.

If compelling evidences demonstrate the I-FABP involvement in the TG-rich lipoprotein synthesis, an implication of L-FABP in this pathway is not excluded as it is suggested by the lack of dramatic phenotype in I-FABP null mice [50]. Such a redundant function might be physiologically useful especially during a high fat supply (Fig. 4). Nevertheless, LCFA are mainly bound to L-FABP (predicted ration of binding of FA to L-FABP/I-FABP = 3.3) in the intestinal cell [49]. Therefore, it is likely that L-FABP can act as a buffer protein protecting the cell against the harmful effect of an excess of free FAs. The fact that L-FABP gene expression is modulated by the lipid content of the diet supports this assumption. This positive regulatory loop might be essential to the maintain of a functional integrity of the intestinal mucosa. A such buffering action, likely also shared by I-BABP, might greatly contribute to the maintain of an efficient FA absorption and enterohepatic circulation of BAs (Fig. 4).

Finally, L-FABP protein seems to exert an active role in the gene regulation as a partner of PPAR [51]. Indeed, it has recently provided evidences showing that L-FABP through a direct protein-protein interaction with PPARα or PPARγ induces the FA transfer to these nuclear receptors leading to the regulation of specific PPAR target genes in hepatocytes (Fig. 5A). In the small intestine, this L-FABP-mediated PPAR activation might lead to the induction of several genes encoding for lipid-binding proteins putatively implicated in fat absorption and known to be PPAR target genes in other tissues (FAT/CD36, ACS, FATP). Because L-FABP and I-BABP are structurally related molecules and are similarly regulated by their preferential ligands, it is tempting to speculate that I-BABP might also play a role in BA signalling to affect FXR activation (Fig. 5B). Further experiments are required to explore this possibility.

In conclusion, although FABPs found in the gut have a comparable tertiary structure, the specificity of their amino-acid sequence confers different physiological action. I-FABP, whom the expression is strictly restricted to the small intestine, seems preferentially involved in the supply of dietary lipids to the organism. L-FABP and I-BABP, through the control of the intracellular unesterified FA and BA levels, would rather have a buffering action. They might not only protect the cell against the detergent effects of free FA and BA, but also play an indirect, but crucial, role in gene regulation. Through these functions, soluble LBPs expressed in the gut could actively participate to the intestinal adaptation to changes in lipid contents of the diet, and BA pool size.

Addendum

During the submission of this manuscript, the implication of other members of FABP family in the activation of PPARs has been reported (Tan et al., Molec Cell Biol 22: 5114–5127, 2002) supporting the implication of FABPs in the regulation of gene experiments.

Acknowledgements

This work is supported by funds from Danone institute, Conseil Régional de Bourgogne, Glaxo-Smith-Kline.

References

1. Mangroo D, Trigatti BL, Gerber GE: Membrane permeation and intracellular trafficking of long-chain fatty acids: Insights from *Escherichia coli* and 3T3-L1 adipocytes. Biochem Cell Biol 73: 223–234, 1995
2. Berk PD: How do long-chain free fatty acids cross cell membrane? Proc Soc Exp Biol Med 212: 1–4, 1996
3. Fitscher BA, Elsing C, Riedel HD, Gorski J, Stremmel W: Protein-mediated facilitated uptake processes for fatty acids, bilirubin, and other amphipathic compounds. Proc Soc Exp Biol Med 212: 15–23, 1996
4. Zakim D: Fatty acids enter cells by simple diffusion. Proc Soc Exp Biol Med 212: 5–14, 1996
5. Glatz JFC, Luiken JJFP, van Nieuwenhoven FA, van der Vusse GJ: Molecular mechanism of cellular uptake and intracellular translocation of fatty acids. Prost Leuk Essent Fatty Acids 57: 3–9, 1997
6. Abumrad N, Harmon C, Ibrahimi A: Membrane transport of long-chain fatty acids: Evidence for a facilated process. J Lipid Res 39: 2309–2318, 1998
7. Hamilton JA: Fatty acid transport: Difficult or easy? J Lipid Res 39: 467–481, 1998
8. McArthur MJ, Atshaves BP, Frolov A, Foxworth WD, Kier AB, Schroeder F: Cellular uptake and intracellular trafficking of long-chain fatty acids. J Lipid Res 40: 1371–1383, 1999
9. Besnard P, Niot I: Role of lipid-binding proteins in intestinal absorption of long-chain fatty acid. In: A.B. Christophe, S. De Vriese (eds). Fat Digestion and Absorption. AOCS Press, Champaign, IL, USA, 2000, pp 96–118
10. Kramer W, Girbig F, Gutjahr U, Kowalewski S, Jouvenal K, Müller G, Tripier D, Wess G: Intestinal bile acid absorption. Na^{+}-dependent bile acid transport activity in rabbit small intestine correlates with the coexpression of an integral 93-kDa and a peripheral 14-kDa bile acid-binding membrane protein along the duodenum-ileum axis. J Biol Chem 268: 18035–18046, 1993
11. Veerkamp JH, Maatman RGHJ: Cytoplasmic fatty acid-binding proteins: Their structures and genes. Prog Lipids Res 34: 17–52, 1995
12. Inokuchi A, Hinoshita E, Iwamoto Y, Kohno K, Kuwano M, Uchiumi T: Enhanced expression of the human multidrug resistance protein 3 by bile salt in human enterocytes: A transcriptional control of a plausible bile acid transporter. J Biol Chem 4: 4, 2001
13. Weinberg SL, Burckhardt G, Wilson FA: Taurocholate transport by rat intestinal basolateral membrane vesicles. J Clin Invest 78: 44–50, 1996
14. Abe T, Kakyo M, Sakagami H, Tokui T, Nishio T, Tanemoto M, Nomura H, Hebert SC, Matsuno S, Kondo H, Yawo H: Molecular characterization and tissue distribution of a new organic anion transporter subtype (oatp3) that transports thyroid hormones and taurocholate and comparison with oatp2. J Biol Chem 273: 22395–22401, 1998
15. Russell DW, Setchell KD: Bile acid biosynthesis. Biochemistry 31: 4737–4749, 1992
16. Poirier H, Niot I, Degrace P, Monnot MC, Bernard A, Besnard P: Fatty acid regulation of fatty acid-binding proteins expression in the small intestine. Am J Physiol 273: G289–G295, 1997
17. Richieri GV, Ogata RT, Kleinfeld AM: Kinetics of fatty acid interactions with fatty acid-binding proteins from adipocyte, heart and intestine. J Biol Chem 271: 11291–11300, 1996
18. Sacchettini JC, Scapin G, Gopaul D, Gordon JI: Refinement of the structure of *Escherichia coli*-derived rat intestinal fatty acid binding protein with bound oleate to 1.75 A resolution. J Biol Chem 267: 23534–23545, 1992
19. Fujita M, Fujii H, Kanda T, Sato E, Hatakeyama K, Ono T: Molecular cloning, expression, and characterization of a human intestinal 15-kDa protein. Eur J Biochem 233: 406–413, 1995
20. Lücke C, Zhang F, Rüterjans H, Hamilton JA, Sacchettini JC: Flexibility is a likely determinant of binding specificity in the case of ileal lipid binding protein. Structure 4: 785–800, 1996
21. Kramer W, Corsiero D, Friedrich M, Girbig F, Stengelin S, Weyland C: Intestinal absorption of bile acids: Paradoxical behaviour of the 14 kDa ileal lipid-binding protein in differential photoaffinity labelling. Biochem J 333: 335–341, 1998
22. Sacchettini JC, Gordon JI: Rat intestinal fatty acid-binding protein. A model system for analyzing the forces that can bind fatty acids to proteins. J Biol Chem 268: 18399–18402, 1993
23. Thompson J, Ory J, Reese-Wagoner A, Banaszak L: The liver fatty acid-binding protein: Comparison of cavity properties of intracellular lipid-binding proteins. Mol Cell Biochem 192: 9–16, 1999
24. Lücke C, Fushman D, Ludwig C, Hamilton JA, Sacchettini JC, Rüterjans HA: Comparative study of the backbone dynamics of two closely related lipid binding proteins: Bovine heart fatty acid binding protein and porcine ileal lipid binding protein. Mol Cell Biochem 192: 109–121, 1999
25. Lücke C, Zhang F, Hamilton JA, Sacchettini JC, Ruterjans H: Solution structure of ileal lipid binding protein in complex with glycocholate. Eur J Biochem 267: 2929–2938, 2000
26. Thompson J, Winter N, Terwey D, Bratt J, Banaszak L: The crystal structure of the liver fatty acid-binding protein. J Biol Chem 272: 7140–7150, 1997
27. Santomé JA, Di Pietro SM, Cavagnari BM, Cordoba OL, Dell' Angelica EC: Fatty acid-binding proteins. Chronological description and discussion of hypotheses involving their molecular evolution. Trends Comp Biochem Physiol 4: 23–38, 1998
28. Poirier H, Niot I, Monnot MC, Braissant O, Meunier-Durmort C, Costet P, Pineau T, Wahli W, Willson TM, Besnard P: Differential involvement of peroxisome-proliferator-activated receptors alpha and delta in fibrate and fatty-acid-mediated inductions of the gene encoding liver fatty-acid-binding protein in the liver and the small intestine. Biochem J 355: 481–488, 2001
29. Poirier H, Braissant O, Niot I, Wahli W, Besnard P: 9-*cis*-retinoic acid enhances fatty acid-induced expression of the liver fatty acid-binding protein gene. FEBS Lett 412: 480–484, 1997
30. Grober J, Zaghini I, Fujii H, Jones SA, Kliewer SA, Willson TM, Ono T, Besnard P: Identification of a bile acid-responsive element in the human ileal bile acid-binding protein gene. Involvement of the farnesoid X receptor/9-*cis*-retinoic acid receptor heterodimer J Biol Chem 274: 29749–29754, 1999
31. Hallden G, Holehouse EL, Dong X, Aponte GW: Expression of intestinal fatty acid-binding protein in intestinal epithelial cell lines, hBRIE 380 cells. Am J Physiol 267: G730–G743, 1994
32. Le Beyec J, Delers F, Jourdant F, Schreider C, Chambaz J, Cardot P, Pinçon-Raymond MA: Complete epithelial organization of Caco-2 cells induces I-FABP and potentializes apolipoprotein gene expression. Exp Cell Res 236: 311–320, 1997
33. Hallden G, Aponte GW: Evidence for a role of the gut hormone PYY in the regulation of intestinal fatty acid-binding protein transcripts in differentiated subpopulations of intestinal epithelial cell hybrids. J Biol Chem 272: 12591–12600, 1997
34. Aponte GW, Park K, Hess R, Garcia R, Taylor IL: Meal-induced peptide tyrosine tyrosine inhibition of pancreatic secretion in the rat. Faseb J 3: 1949–1955, 1989
35. Laburthe M, Chenut B, Rouyer-Fessard C, Tatemoto K, Couvineau

A, Servin A, Amiranoff B: Interaction of peptide YY with rat intestinal epithelial plasma membranes: Binding of the radioiodinated peptide. Endocrinology 118: 1910–1917, 1986
36. Bass NM: The cellular fatty acid-binding proteins: Aspects of structure, regulation, and function. Int Rev Cytol 111: 143–184, 1988
37. Darimont C, Gradoux N, De Pover A: Epidermal growth factor regulates fatty acid uptake and metabolism in Caco-2 cells. Am J Physiol 276: G606–G612, 1999
38. Baier LJ, Sacchettini JC, Knowler WC, Eads J, Paolisso G, Tataranni PA, Mochizuki H, Bennett PH, Bogardus C, Prochazka M: An amino acid substitution in the human intestinal fatty acid-binding protein is associated with increased fatty acid binding, increased fat oxidation, and insulin resistance. J Clin Invest 95: 1281–1287, 1995
39. Hegele RA, Harris SB, Hanley AJG, Sadikian S, Connelly PW, Zinman B: Genetic variation of intestinal fatty acid-binding protein associated with variation in body mass in aboriginal canadians. J Clin Endocrinol Metab 81: 4334–4337, 1996
40. Baier LJ, Bogardus C, Sacchettini JC: A polymorphism in the human intestinal fatty acid-binding protein alters fatty acid transport across Caco-2 cells. J Biol Chem 271: 10892–10896, 1996
41. Levy E, Menard D, Delvin E, Stan S, Mitchell G, Lambert M, Ziv E, Feoli-Fonseca JC, Seidman E: The polymorphism at codon 54 of the FABP2 gene increases fat absorption in human intestinal explants. J Biol Chem 276: 39679–39684, 2001
42. Prows DR, Schroeder F: Metallothionein-II_A promoter induction alters rat intestinal fatty acid-binding protein expression, fatty acid uptake, and lipid metabolism in transfected L-cells. Arch Biochem Biophys 340: 135–143, 1997
43. Atshaves BP, Foxworth WB, Frolov A, Roths JB, Kier AB, Oetama BK, Piedrahita JA, Schroeder F: Cellular differentiation and I-FABP protein expression modulate fatty acid uptake and diffusion. Am J Physiol 274: C633–C644, 1998
44. Holehouse E, Liu M-L, Aponte GW: Oleic acid distribution in small intestinal epithelial cells expressing intestinal-fatty acid binding protein. Biochim Biophys Acta 1390: 52–64, 1998
45. Wolfrum C, Buhlmann C, Rolf B, Börchers T, Spener F: Variation of liver-type fatty acid-binding protein content in the human hepatoma cell line HepG2 by peroxisome proliferators and antisense RNA affects the rate of fatty acid uptake. Biochim Biophys Acta 1437: 194–201, 1999
46. Hsu K-T, Storch J: Fatty acid transfer from liver and intestinal fatty acid-binding proteins to membranes occurs by different mechanisms. J Biol Chem 271: 13317–13323, 1996
47. Corsico B, Cistola DP, Frieden C, Storch J: The helical domain of intestinal fatty acid-binding protein is critical for collisional transfer of fatty acid to phospholipid membranes. Proc Natl Acad Sci USA 95: 12174–12178, 1998
48. Hodsdon ME, Cistola DP: Discrete backbone disorder in the nuclear magnetic resonance structure of apo intestinal fatty acid-binding protein: Implications for the mechanism of ligand entry. Biochemistry 36: 1450–1460, 1997
49. Alpers DH, Bass NM, Engle MJ, DeSchryver-Kecskemeti K: Intestinal fatty acid binding protein may favor differential apical fatty acid binding in the intestine. Biochim Biophys Acta 1483: 352–362, 2000
50. Vassileva G, Huwyler L, Poirier K, Agellon LB, Toth MJ: The intestinal fatty acid binding protein is not essential for dietary fat absorption in mice. Faseb J 14: 2040–2046, 2000
51. Wolfrum C, Borrmann CM, Borchers T, Spener F: Fatty acids and hypolipidemic drugs regulate peroxisome proliferator-activated receptors alpha- and gamma-mediated gene expression via liver fatty acid binding protein: A signaling path to the nucleus. Proc Natl Acad Sci USA 98: 2323–2328, 2001

Molecular and Cellular Biochemistry **239**: 149–155, 2002.

Regulation of the ileal bile acid-binding protein gene: An approach to determine its physiological function(s)

Jean-François Landrier, Jacques Grober, Isabelle Zaghini and Philippe Besnard

Laboratoire de Physiologie de la Nutrition, Ecole Nationale Supérieure de Biologie Appliquée à la Nutrition et à l'Alimentation (ENSBANA), FRE 2328 CNRS-CESG/Université de Bourgogne, Dijon, France

Abstract

Ileal bile acid-binding protein (I-BABP) is a soluble bile acids (BA) carrier protein which belongs to the fatty acid-binding protein (FABP) family. In the gut, its expression is strictly restricted to the ileum, where it is thought to be involved in the active BA reabsorption. Therefore, I-BABP gene expression levels might be rate limiting for the BA enterohepatic circulation, and hence, might be crucial for cholesterol (CS) homeostasis. Indeed, BA not reclaimed by intestinal absorption constitute the main way to eliminate a CS excess. However, such a function is not yet established. Because generally rate limiting genes are tightly controlled, we have undertaken the study of the I-BABP gene regulation. It was found that both BA and CS, probably via oxysterols, are able to up-regulate the transcription rate of I-BABP gene. The fact that intracellular sterol sensors (FXR, LXR and SREBP1c) are involved in the control of I-BABP gene expression strongly suggest a crucial role for I-BABP in the ileum. (Mol Cell Biochem **239**: 149–155, 2002)

Key words: ileal bile acid-binding protein, farnesoid-X-receptor, sterol regulatory element-binding proteins, liver-X-receptor, bile acids, cholesterol, small intestine

Abbreviations: BA – bile acids; CS – cholesterol; CDCA – chenodeoxycholic acid; DCA – deoxycholic acid; IBAT – ileal bile acid transporter; I-BABP – ileal bile acid-binding protein; $OATP_3$ – Sodium-independent organic anion exchange system; MRP_3 – multidrug resistance protein 3; L-FABP – liver-type fatty acid-binding protein; CYP7A1 – cholesterol 7α-hydroxylase; FXR – farnesoid-X-receptor; LXR – liver-X-receptor; SREBPs – sterol regulatory element-binding proteins; BARE – bile acid responsive element; SRE – sterol-responsive element; ABCA1 – ATP-binding cassette A1; SCD-1 – stearoyl-CoA desaturase 1; SHP-1 – small heterodimer partner 1; LRH-1 – liver receptor homolog 1; CAT - chloramphenicol acetyl transferase

Introduction

It is well established that the excessive cholesterol (CS) consumption in Western countries greatly contributes to the appearance and development of atherosclerosis and leads to vascular strokes. In the body, CS balance results from an equilibrium between supplies (diet and cellular *de novo* synthesis), and losses (cellular use and elimination in feces, essentially as bile acids). Bile acids (BA) are synthesized from CS in the liver. After conjugation with glycine or taurine, BA are secreted with bile in the intestinal lumen where they actively participate to the digestion and absorption of dietary fat and lipid-soluble vitamins [1]. In healthy man, more than 90% of BA are reabsorbed throughout the small intestine and then return by the portal vein to the liver, where they are secreted again into bile. This enterohepatic circulation is essential for maintenance of a BA balance, and hence, for CS homeostasis. Indeed, the fecal excretion of BA represents the

Address for offprints: P. Besnard, Laboratoire de Physiologie de la Nutrition, Ecole Nationale Supérieure de Biologie Appliquée à la Nutrition et à l'Alimentation (ENSBANA), FRE 2328 CNRS-CESG/Université de Bourgogne, 1 esplanade Erasme, F21000 Dijon, France (E-mail: pbesnard@u-bourgogne.fr)

main physiological way for CS elimination. Therefore, liver and small intestine appear to play an essential role in the maintenance of CS balance.

Although the regulation of the hepatic BA biosynthesis is well documented [1], by contrast, little is known about the molecular mechanisms controlling BA reabsorption by the small intestine. Intestinal BA uptake mainly takes place through an active transport located in the distal part of the small intestine. To date, four unrelated proteins exhibiting a high affinity for BA have been identified in the ileum. The first member of this putative transport system is a protein found in the brush border membrane of ileocytes, called ileal bile acid transporter (IBAT) [2], or apical sodium-dependent bile acid transporter (ASBT). The Na^+-dependent BA transfer assumed by the IBAT is energy dependent [3]. Once into the ileocyte, BA are bound to the ileal bile acid-binding protein (I-BABP), also termed ileal lipid binding protein (ILBP) or ileal-15 kDa protein (I-15P). This BA carrier is a soluble protein which belongs to the fatty acid binding protein (FABP) superfamily [4]. Finally, BA are secreted into the portal blood by a Na^+-independent organic anion exchange system (OATP3) located in the basolateral membrane of ileocyte [5, 6]. Recent data strongly suggest that the multidrug resistance protein 3 (MRP3) might also be involved in this cellular BA efflux [7]. In this BA transport system, only I-BABP is a soluble protein. Therefore, it is likely that I-BABP is essential for an efficient BA desorption from the apical plasma membrane, as well as for the BA intracellular trafficking and targeting on OATP3 and/or MRP3. If this assumption is correct, I-BABP expression level might be rate limiting for the enterohepatic BA circulation, and hence, for the maintenance of CS homeostasis.

I-BABP: A protein in search of a function

I-BABP is a relatively well characterized protein. In the digestive tract, I-BABP mRNA signal and immunoreactivity have been detected in the ileum of different species including pig [8], mice [9], rat [10], rabbit [11], and man [12]. In this tissue, its expression is relatively high since I-BABP represents more than 0.1% of total proteins [13]. In the liver, I-BABP has been found in the large cholangiocytes from the bile duct epithelium [14]. I-BABP mRNA signals have also been found in various tissues synthesizing steroid hormones such as ovaries [10, 15], adrenal glands [10, 16], and placenta [12]. Ontogenicity of I-BABP has been extensively studied in mice [9, 17] and rat [18, 19]. I-BABP mRNA are detectable in ileum at early step of embryonic development, then mRNA levels rise rapidly at the suckling/weaning transition and remain constant throughout weaning.

The tertiary structure of I-BABP has been elucidated [20–22]. Similar to the other members of the FABP superfamily, I-BABP conformation consists in 10 antiparallel β strands forming a water-filled 'clam-shell' structure, covered on one side by two short, nearly parallel α helices [20, 21]. Specific features, only shared with liver FABP (L-FABP), i.e. high molecular flexibility, large volume of the hydrophobic pocket, presence of a second portal region [20–22], explain why I-BABP is able to bind bulky and rigid molecules as BA [9, 12]. I-BABP can also bind long-chain fatty acids (LCFA) [9, 12], but only with a low affinity [23]. The binding capacity of I-BABP is not definitively determined since ratios of 1 mol protein for 1 or 2 mol BA have been reported [20, 24]. Paradoxically, I-BABP remains a protein in search of a function. Several characteristics (ileal location, binding affinity) suggest that I-BABP expression level might be rate limiting for enterohepatic BA circulation. To explore this assumption, we have undertaken a study of I-BABP gene regulation. The rate limiting genes indeed were found generally to be tightly regulated.

BA regulate I-BABP gene expression

If I-BABP is critical for enterohepatic BA circulation, its expression must be directly modulated by BA. Two opposite scenari supporting two distinct functions for I-BABP can be envisioned consecutively to a rise in the BA pool size. First, a down-regulation of I-BABP gene expression leading to an increase of the BAs lost in feces. Such a negative regulatory loop has been previously reported for the cholesterol 7α-hydroxylase (CYP7A1), which is the rate limiting enzyme of the hepatic BAs biosynthetic pathway [1]. A BA-mediated inhibition of I-BABP expression supposes an active implication of this protein in the maintenance of the BA pool size. The second possibility is a BA-mediated induction of I-BABP gene expression. This scenario would support the idea that I-BABP exerts a buffering action protecting the ileocyte against the detergent effect of free BAs .

An ileal by-pass realized in rats has showed that I-BABP gene expression is effectively bile sensitive [25]. Indeed, the low cytosolic I-BABP concentration and mRNA levels found in diverted ileum was totally reversed to normal when bile was directly added in the ileal lumen. This effect is independent of a neuroendocrine control since a bile addition in the culture medium leads to the induction of I-BABP expression in both organ culture of ileal explants and in enterocyte-like CaCo-2 cells [25]. Among the different biliary components, only BA are able to stimulate the expression of I-BABP in CaCo-2 cells [26] (Fig. 1A). The screening of various BA has revealed that chenodeoxycholic acid (CDCA) and deoxycholic acid (DCA) are the strongest I-BABP inducers. This

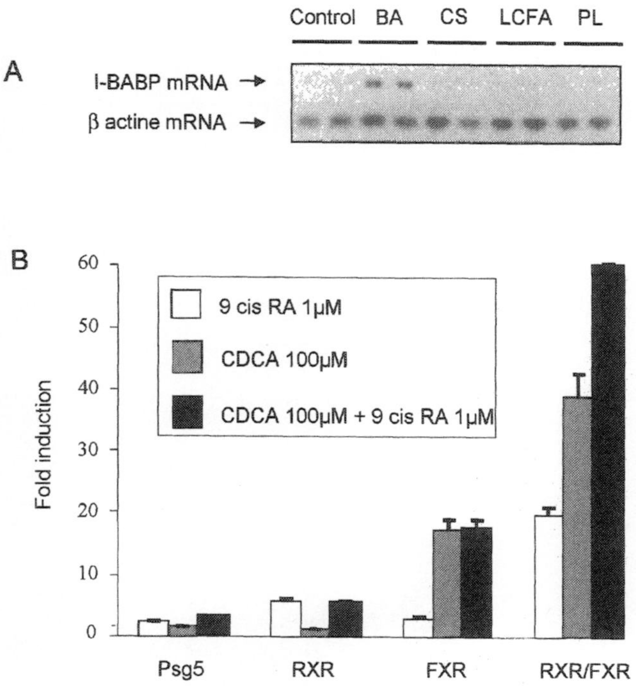

Fig. 1. I-BABP gene expression is regulated by BAs via FXR. (A) CaCo-2 cells were incubated for 24 h in the medium containing either a bile acid (BA, 1 mM glycocholic acid), cholesterol (CS, 100 µM), long chain fatty acid (LCFA, 100 µM palmitic acid), or phospholipids (PL, phosphatidylchline 100 µM). (B) The I-BABP-CAT construct was cotransfected in CaCo-2 cells with an empty vector (pSG5) or with expression vectors for human FXR and/or human RXR. Cells were treated either with the RXR ligand 9-cis-retinoic acid (9-cis-RA or the FXR ligand chenodeoxycholate (CDCA) or both ligands for 24 h.

BA up-regulation has a transcriptional origin and requires a *de novo* protein synthesis [26]. It is noteworthy that this finding has constituted the first demonstration that BA can be transcriptional inducers.

By homology with the gene regulation by LCFA, which requires the intervention of peroxisome proliferator-activated receptors (PPAR/NR1C) [27], it was tempting to speculate that BA up-regulate the I-BABP gene through a similar mechanism. The discovery in 1999 that the farnesoid-X-Receptor (FXR; NR1H4, also known as HRR1 and RIP14) is activated by biologically relevant concentrations of BA [28–30] has allowed to show that this orphan nuclear receptor was responsible for BA-mediated gene regulation. Northern blot analysis and *in situ* hybridization revealed that FXR expression is restricted to tissues involved in the metabolic fate of BA, i.e. small intestine, colon, kidney and liver [31]. It is well known that the small intestine is a heterogeneous organ characterized by a variable genotypic expression along the gastro-colic axis at the origin of a functional specialization. Interestingly, FXR expression appears to be strictly restricted to the ileum in the mouse, where it is co-expressed with I-BABP [32]. FXR binds DNA as permissive heterodimer with the retinoid X receptor (RXR) [31]. Co-transfection assays realized in CaCo-2 cells with the human I-BABP promoter fused to the chloramphenicol acetyl transferase (CAT) gene shows that induction by BA of the CAT reporter gene is strictly FXR dependent (Fig. 1B), the transactivation being maximal in the presence of the heterodimer FXR/RXR. Interestingly, the strongest FXR-mediated transactivation of the reporter gene expression was found with CDCA which is the main I-BABP gene inducer in CaCo2 cells [26]. This result also correlates quite well with the high binding affinity of FXR for CDCA [30]. The fact that the FXR DNA binding site exhibits a high homology with those of the ecdysone receptor has contributed to the rapid identification of a bile acid-responsive element (BARE) in the I-BABP gene. Indeed, as this insect receptor, the FXR/RXR heterodimer binds to a DNA sequence that consists of an inverted repeat (IR) of the half-site sequence AG(G/T)TCA separated by one nucleotide (IR-1). Sequence alignment of the proximal promoters of the human, rabbit and mouse I-BABP reveals a highly conserved, but imperfect IR-1 sequence. Electrophoretic mobility shift assay (EMSA) shows that this IR-1 sequence is a bile acid-responsive element (BARE). Transfection assays realized with a mutated or deleted BARE have proved its functionality [32]. The recent generation of a FXR null mice has allowed to establish that the basal as well BA-induced I-BABP gene expression is highly dependent of FXR [33].

I-BABP is a sterol target gene

BA biosynthesis is directly dependent of the CS pool size. This relationship leads us to hypothesize that I-BABP gene might also be controlled by changes in CS levels. It is now well established that CS is a gene modulator. This action requires the proteolytic activation of specific transcriptional factors, termed sterol-responsive element-binding proteins or SREBPs which, in turn, become able to interact with a sterol-responsive element (SRE) located in the promoter of sterol target genes [34]. The fact that computer analysis has revealed the existence of a SRE-like sequence in the proximal part of the human I-BABP promoter supported our hypothesis. To check this, mice were subjected for 14 days to a high CS diet. As shown in Fig. 2A, a supply with 2% CS leads to a slight, but significant rise in I-BABP mRNA levels. Interestingly, the level of mRNA encoding for SREBP1c isotype is also specifically induced, while SREBP1a is unchanged and SREBP2 decreased. Co-transfection assays in CaCo-2 cells demonstrate the involvement of SREBP1c in this regulation (Fig. 2B). The direct binding of SREBP1c upon the I-BABP^{SRE} has been finally established by EMSA [35].

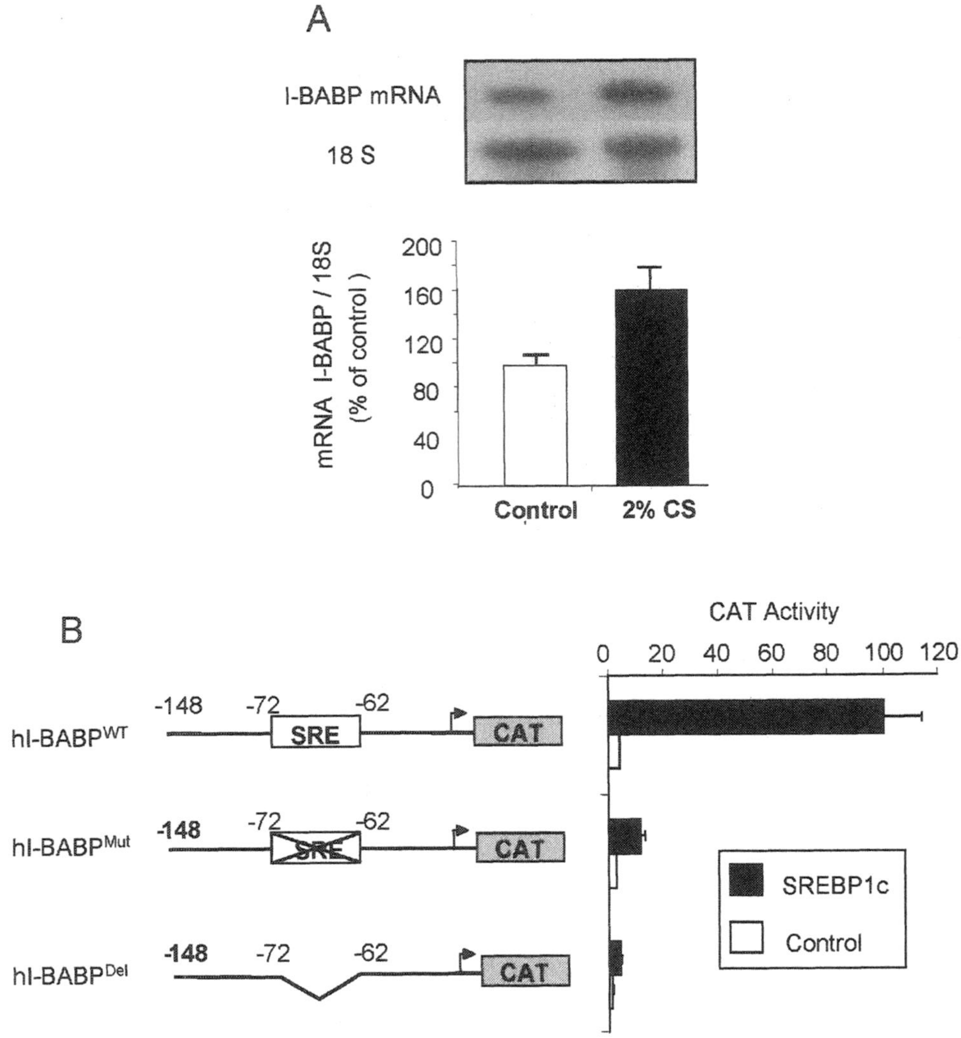

Fig. 2. I-BABP gene expression is regulated by CS via SREBP-1c. (A) Mice were fed *ad libitum* for 14 days either a standard laboratory chow containing either 0.2% (w/w) cholesterol (control) or 2% cholesterol (2% CS). Autoradiograms and bar graph representation of I-BABP data normalized to 18 S rRNA are shown. (B) The I-BABP-CAT construct was cotransfected in Caco-2 cells with an empty vector (control) or with expression vector for active human SREBP1c (SREBP1c). Cells were treated with cholesterol and 25-hydroxy-cholesterol to inhibit maturation of endogenous SREBPs for 24 h.

Recently, it has been established that the liver-X-receptor (LXR; NR1H2 and NR1H3), once activated by the CS derivatives oxysterols, induces the transcription and maturation of SREBP1c [36–38]. Therefore, we have postulated that CS-mediated up-regulation of the I-BABP gene found *in vivo* might be secondary to the LXR-dependent induction and activation of SREBP1c. Such a mechanism has been recently described for the stearoyl-CoA desaturase 1 (SCD-1) gene [36]. In agreement with our assumption, both I-BABP and SREBP1c expressions are significantly enhanced in mice force-fed with the specific LXR agonist GW3965. Similar data have been reproduced in organ culture of ileal explants, excluding an indirect effect of LXR on I-BABP gene expression, via a change in the BA pool size [35]. Indeed, it is well established that CYP7A1, which is the key enzyme of the BA biosynthetic pathway, is directly up-regulated by LXR.

Taken together, our data show a high responsiveness of I-BABP gene to cellular sterol changes. They demonstrate that the regulation of I-BABP expression by CS and metabolic derivatives (BA and oxysterols) takes place through the activation of various sterol sensors (FXR, LRX/SREBP1c). From a functional point of view, they exclude the hypothesis of an involvement of I-BABP in BA elimination in feces consecutively to an increase of the BA pool size. Indeed, such a function is inconsistent with a BA and CS up-regulation of I-BABP gene expression.

What putative function(s) for I-BABP?

The knowledge of I-BABP gene regulation is insufficient to determine the physiological role of this protein. By contrast, putative functions emerge when I-BABP is reintroduced in a more integrated regulatory pathway including the control of other hepatic and intestinal genes recently described as sterol-target genes (Fig. 3). As already shown, a CS overload activates the sterol sensors FXR, LXR, and SREBP1c. FXR, as heterodimer with RXR, exerts two main controls. Through a regulatory cascade involving the small heterodimer partner 1 (SHP-1; NR0B2) and the liver receptor homolog 1 (LRH-1; NR5A2), it represses the CYP7A1 in the liver [39, 40], while it induces I-BABP gene in the small intestine. LXR, in presence of both oxysterols and RXR, induces the expression and the maturation of SREBP1c which, in turn, increases the gene expression of both I-BABP and stearoyl-coA desaturase 1 (SCD-1) [36], a microsomal enzyme catalysing the transformation of stearoyl-CoA in oleoyl-CoA. LXR also up-regulates the expression of both CYP7A1 yielding a rise in BAs biosynthesis, and ATP-binding cassette transporter A1 (ABCA-1), a plasma membrane protein responsible for the cellular efflux of CS [41].

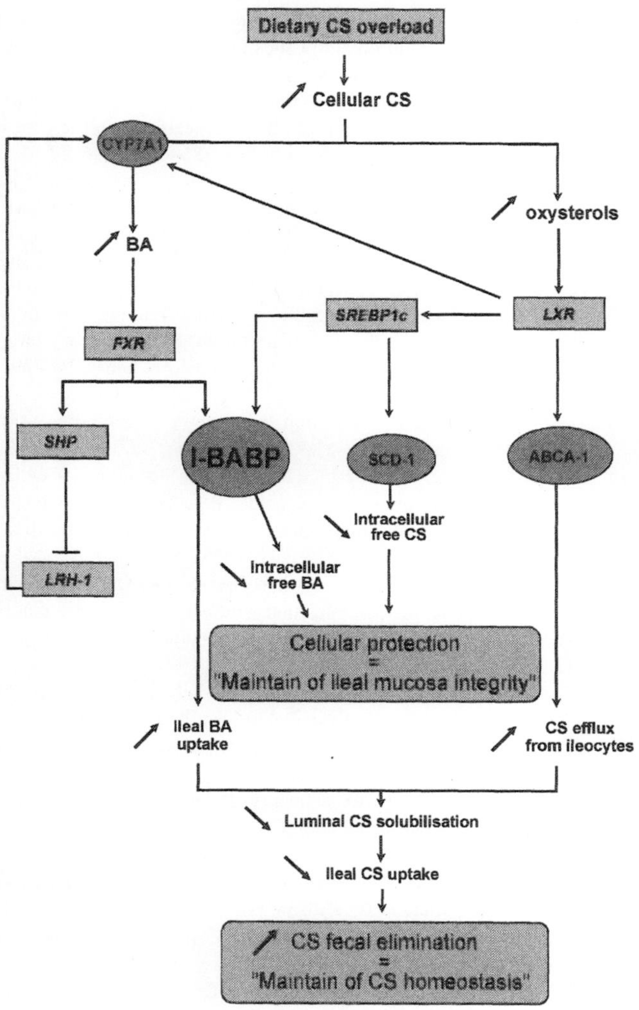

Fig. 3. Effect of a cholesterol (CS) overload on the regulation of some hepatic and intestinal sterol target genes. BAs – bile acids; CS – cholesterol; I-BABP – ileal bile acid-binding protein; CYP7A1 – cholesterol 7α-hydroxylase; FXR – farnesoid-X-receptor; LXR – liver-X-receptor; SREBP1c – sterol regulatory element-binding protein 1c; ABCA1 – ATP-binding cassette A1; SCD-1 – stearoyl-CoA desaturase 1; SHP-1 – small heterodimer partner 1; LRH-1 – liver receptor homolog 1.

This short overview shows that at least three distinct genes (I-BABP, SCD-1 and ABCA-1) found in the ileum are up-regulated secondary to a CS load. The coordinate regulation of the I-BABP gene with SCD-1 and ABCA-1 might present, at least, two immediate physiological interests : first, as suggested for SCD-1, whom the induction would promote the esterification of free CS to protect the cell against its harmful effects [36], an increase of BA-binding capacity secondary to the rise in I-BABP expression might protect the ileal mucosa against the cytotoxic effect of the large quantity of 'free' BA produced during CS feeding. The maintenance of the functional integrity of ileal mucosa is essential for an efficient enterohepatic circulation.

Second, the rise in ileal BA uptake might further enhance the fecal elimination of CS induced by the reverse CS transporter ABCA1 [42]. Indeed, it was recently shown that a high CS feeding induces the ABCA1 expression throughout the small intestine increasing the CS efflux from enterocytes [42]. Since CS absorption depends on BA levels in intestinal lumen, the CS-mediated rise in I-BABP levels ensuring a quantitative extraction of BAs from the lumen might lead to a failure in CS solubilization and hence, to favour fecal CS elimination. Moreover, the driving force for ileal BA uptake secondary to the up-regulation of I-BABP gene is reinforced by the fact that BA enhance also the binding affinity and capacity of I-BABP leading to a nearly quantitative extraction of BA from ileal lumen [22]. Such a system which associates gene regulation and change in binding properties of I-BABP would allow a efficient adaptation of ileal BAs transport to change in substrate levels.

In conclusion, the I-BABP gene expression appears to be tightly controlled by cellular sterol sensors. Although, its precise function in ileocyte is not yet definitively established, this finding strongly suggests that I-BABP gene expression is important for the maintain of CS homeostasis.

Addendum

During the submission of this manuscript, an excellent review on bile salt transporters has been published [P.J. Meierawl, B. Stieger Annu. Rev. Physiol 64: 635–661, 2002].

Acknowledgements

This work was supported by funds from GlaxoSmithKline, Arcol, Conseil Régional de Bourgogne. J.-F.L. is supported by a fellowship from the Ministère de l'Education Nationale, de la Recherche et de la Technologie.

References

1. Russell DW, Setchell KD: Bile acid biosynthesis. Biochemistry 31: 4737–4749, 1992
2. Kramer W, Girbig F, Gutjahr U, Kowalewski S, Jouvenal K, Muller G, Tripier D, Wess G: Intestinal bile acid absorption. Na(+)-dependent bile acid transport activity in rabbit small intestine correlates with the coexpression of an integral 93-kDa and a peripheral 14-kDa bile acid-binding membrane protein along the duodenum-ileum axis. J Biol Chem 268: 18035–18046, 1993
3. Saeki T, Matoba K, Furukawa H, Kirifuji K, Kanamoto R, Iwami K: Characterization, cDNA cloning, and functional expression of mouse ileal sodium-dependent bile acid transporter. J Biochem 125: 846–851, 1999
4. Veerkamp JH, Maatman RG: Cytoplasmic fatty acid-binding proteins: Their structure and genes. Prog Lipid Res 34: 17–52, 1995
5. Weinberg SL, Burckhardt G, Wilson FA: Taurocholate transport by rat intestinal basolateral membrane vesicles. Evidence for the presence of an anion exchange transport system. J Clin Invest 78: 44–50, 1986
6. Abe T, Kakyo M, Sakagami H, Tokui T, Nishio T, Tanemoto M, Nomura H, Hebert SC, Matsuno S, Kondo H, Yawo H: Molecular characterization and tissue distribution of a new organic anion transporter subtype (oatp3) that transports thyroid hormones and taurocholate and comparison with oatp2. J Biol Chem 273: 22395–22401, 1998
7. Inokuchi A, Hinoshita E, Iwamoto Y, Kohno K, Kuwano M, Uchiumi T: Enhanced expression of the human multidrug resistance protein 3 by bile salt in human enterocytes: A transcriptional control of a plausible bile acid transporter. J Biol Chem 2001 (in press)
8. Gantz I, Nothwehr SF, Lucey M, Sacchettini JC, DelValle J, Banaszak LJ, Naud M, Gordon JI, Yamada T: Gastrotropin: Not an enterooxyntin but a member of a family of cytoplasmic hydrophobic ligand binding proteins. J Biol Chem 264: 20248–20254, 1989
9. Sacchettini JC, Hauft SM, Van Camp SL, Cistola DP, Gordon JI: Developmental and structural studies of an intracellular lipid binding protein expressed in the ileal epithelium. J Biol Chem 265: 19199–19207, 1990
10. Amano O, Kanda T, Ono T, Iseki S: Immunocytochemical localization of rat intestinal 15 kDa protein, a member of cytoplasmic fatty acid-binding proteins. Anat Rec 234: 215–222, 1992
11. Stengelin S, Apel S, Becker W, Maier M, Rosenberger J, Bewersdorf U, Girbig F, Weyland C, Wess G, Kramer W: The rabbit ileal lipid-binding protein. Gene cloning and functional expression of the recombinant protein. Eur J Biochem 239: 887–896, 1996
12. Fujita M, Fujii H, Kanda T, Sato E, Hatakeyama K, Ono T: Molecular cloning, expression, and characterization of a human intestinal 15-kDa protein. Eur J Biochem 233: 406–413, 1995
13. Walz DA, Wider MD, Snow JW, Dass C, Desiderio DM: The complete amino acid sequence of porcine gastrotropin, an ileal protein which stimulates gastric acid and pepsinogen secretion. J Biol Chem 263: 14189–14195, 1988
14. Alpini G, Glaser SS, Rodgers R, Phinizy JL, Robertson WE, Lasater J, Caligiuri A, Tretjak Z, LeSage GD: Functional expression of the apical Na^+-dependent bile acid transporter in large but not small rat cholangiocytes. Gastroenterology 113: 1734–1740, 1997
15. Watanabe K, Hoshi N, Tsuura Y, Kanda T, Fujita M, Fujii H, Ono T, Suzuki T: Immunohistochemical distribution of intestinal 15 kDa protein in human tissues. Arch Histol Cytol 58: 303–306, 1995
16. Iseki S, Amano O, Kanda T, Fujii H, Ono T: Expression and localization of intestinal 15 kDa protein in the rat. Mol Cell Biochem 123: 113–120, 1993
17. Crossman MW, Hauft SM, Gordon JI: The mouse ileal lipid-binding protein gene: A model for studying axial patterning during gut morphogenesis. J Cell Biol 126: 1547–1564, 1994
18. Gong YZ, Kato T, Schwartz DA, Norris JS, Wilson FA: Ontogenic and glucocorticoid-accelerated expression of rat 14 kDa bile acid-binding protein. Anat Rec 245: 532–538, 1996
19. Hwang ST, Henning SJ: Hormonal regulation of expression of ileal bile acid binding protein in suckling rats. Am J Physiol Regul Integr Comp Physiol 278: R1555–R1563, 2000
20. Lucke C, Zhang F, Ruterjans H, Hamilton JA, Sacchettini JC: Flexibility is a likely determinant of binding specificity in the case of ileal lipid binding protein. Structure 4: 785–800, 1996
21. Lucke C, Zhang F, Hamilton JA, Sacchettini JC, Ruterjans H: Solution structure of ileal lipid binding protein in complex with glycocholate. Eur J Biochem 267: 2929–2938, 2000
22. Kramer W, Sauber K, Baringhaus KH, Kurz M, Stengelin S, Lange G, Corsiero D, Girbig F, Konig W, Weyland C: Identification of the bile acid-binding site of the ileal lipid-binding protein by photoaffinity labeling, matrix-assisted laser desorption ionization-mass spectrometry, and nmr structure. J Biol Chem 276: 7291–7301, 2001
23. Miller KR, Cistola DP: Titration calorimetry as a binding assay for lipid-binding proteins. Mol Cell Biochem 123: 29–37, 1993
24. Kramer W, Corsiero D, Friedrich M, Girbig F, Stengelin S, Weyland C: Intestinal absorption of bile acids: Paradoxical behaviour of the 14 kDa ileal lipid-binding protein in differential photoaffinity labelling. Biochem J 333: 335–341, 1998
25. Kanda T, Niot I, Foucaud L, Fujii H, Bernard A, Ono T, Besnard P: Effect of bile on the intestinal bile-acid binding protein (I-BABP) expression. *In vitro* and *in vivo* studies. FEBS Lett 384: 131–134, 1996
26. Kanda T, Foucand L, Nakamura Y, Niot I, Besnard P, Fujita M, Sakai Y, Hatakeyama K, Ono T, Fujii H: Regulation of expression of human intestinal bile acid-binding protein in Caco-2 cells. Biochem J 330: 261–265, 1998
27. Escher P, Wahli W: Peroxisome proliferator-activated receptors: Insight into multiple cellular functions. Mutat Res 448: 121–138, 2000
28. Parks DJ, Blanchard SG, Bledsoe RK, Chandra G, Consler TG, Kliewer SA, Stimmel JB, Willson TM, Zavacki AM, Moore DD, Lehmann JM: Bile acids: Natural ligands for an orphan nuclear receptor. Science 284: 1365–1368, 1999
29. Makishima M, Okamoto AY, Repa JJ, Tu H, Learned RM, Luk A, Hull MV, Lustig KD, Mangelsdorf DJ, Shan B: Identification of a nuclear receptor for bile acids. Science 284: 1362–1365, 1999
30. Wang H, Chen J, Hollister K, Sowers LC, Forman BM: Endogenous bile acids are ligands for the nuclear receptor FXR/BAR. Mol Cell 3: 543–553, 1999
31. Forman BM, Goode E, Chen J, Oro AE, Bradley DJ, Perlmann T, Noonan DJ, Burka LT, McMorris T, Lamph WW *et al.*: Identifica-

tion of a nuclear receptor that is activated by farnesol metabolites. Cell 81: 687–693, 1995
32. Grober J, Zaghini I, Fujii H, Jones SA, Kliewer SA, Willson TM, Ono T, Besnard P: Identification of a bile acid-responsive element in the human ileal bile acid-binding protein gene. Involvement of the farnesoid x receptor/9-cis-retinoic acid receptor heterodimer. J Biol Chem 274: 29749–29754, 1999
33. Sinal CJ, Tohkin M, Miyata M, Ward JM, Lambert G, Gonzalez FJ: Targeted disruption of the nuclear receptor FXR/BAR impairs bile acid and lipid homeostasis. Cell 102: 731–744, 2000
34. Shimano H: Sterol regulatory element-binding proteins (SREBPs): Transcriptional regulators of lipid synthetic genes. Prog Lipid Res 40: 439–452, 2001
35. Zaghini I, Landrier JF, Grober J, Krief S, Jones SA, Monnot MC, Lefrere I, Watson MA, Collins JL, Fujii H, Besnard P: Sterol regulatory element-binding protein-1c is responsible for cholesterol regulation of ileal bile-acid binding protein gene *in vivo*. 2001 (submitted)
36. Repa JJ, Liang G, Ou J, Bashmakov Y, Lobaccaro JM, Shimomura I, Shan B, Brown MS, Goldstein JL, Mangelsdorf DJ: Regulation of mouse sterol regulatory element-binding protein-1c gene (SREBP-1c) by oxysterol receptors, LXRalpha and LXRbeta. Genes Dev 14: 2819–2830, 2000
37. DeBose-Boyd RA, Ou J, Goldstein JL, Brown MS: Expression of sterol regulatory element-binding protein 1c (SREBP-1c) mRNA in rat hepatoma cells requires endogenous LXR ligands. Proc Natl Acad Sci USA 98: 1477–1482, 2001
38. Yoshikawa T, Shimano H, Amemiya-Kudo M, Yahagi N, Hasty AH, Matsuzaka T, Okazaki H, Tamura Y, Iizuka Y, Ohashi K, Osuga J, Harada K, Gotoda T, Kimura S, Ishibashi S, Yamada N: Identification of liver X receptor-retinoid X receptor as an activator of the sterol regulatory element-binding protein 1c gene promoter. Mol Cell Biol 21: 2991–3000, 2001
39. Lu TT, Makishima M, Repa JJ, Schoonjans K, Kerr TA, Auwerx J, Mangelsdorf DJ: Molecular basis for feedback regulation of bile acid synthesis by nuclear receptors. Mol Cell 6: 507–515, 2000
40. Goodwin B, Jones SA, Price RR, Watson MA, McKee DD, Moore LB, Galardi C, Wilson JG, Lewis MC, Roth ME, Maloney PR, Willson TM, Kliewer SA: A regulatory cascade of the nuclear receptors FXR, SHP-1, and LRH-1 represses bile acid biosynthesis. Mol Cell 6: 517–526, 2000
41. Schwartz K, Lawn RM, Wade DP: ABC1 gene expression and ApoA-I-mediated cholesterol efflux are regulated by LXR. Biochem Biophys Res Commun 274: 794–802, 2000
42. Repa JJ, Turley SD, Lobaccaro JA, Medina J, Li L, Lustig K, Shan B, Heyman RA, Dietschy JM, Mangelsdorf DJ: Regulation of absorption and ABC1-mediated efflux of cholesterol by RXR heterodimers. Science 289: 1524–1529, 2000

Molecular and Cellular Biochemistry **239**: 157–164, 2002.

Role of adipocyte lipid-binding protein (ALBP) and acyl-CoA binding protein (ACBP) in PPAR-mediated transactivation

Torben Helledie, Claus Jørgensen, Marianne Antonius, Anne-M. Krogsdam, Irina Kratchmarova, Karsten Kristiansen and Susanne Mandrup

Department of Biochemistry and Molecular Biology, University of Southern Denmark, Odense, Denmark

Abstract

Peroxisome proliferator-activated receptors (PPARs) are nuclear hormone receptors that are activated by a number of fatty acids and fatty acid derivatives. By contrast, we have recently shown that acyl-CoA esters display PPAR antagonistic properties *in vitro*. We have also shown that the adipocyte lipid binding protein (ALBP), the keratinocyte lipid binding protein (KLBP) and the acyl-CoA binding protein (ACBP) exhibit a prominent nuclear localization in differentiating 3T3-L1 adipocytes. Similarly, ectopic expression of these proteins in CV-1 cells resulted in a primarily nuclear localization. We therefore speculated that FABPs and ACBP might regulate the availability of PPAR agonists and antagonists by affecting not only their esterification in the cytoplasm but also their transport to and availability in the nucleus. We show here that coexpression of ALBP or ACBP exerts a negative effect on ligand-dependent PPAR transactivation, when tetradecylthioacetic (TTA) is used as ligand but not when the thiazolidinedione BRL49653 is used as ligand. The results presented here do not support the hypothesis that ALBP facilitates the transport of the fatty acid-type ligands to the nucleus, rather ALBP appears to sequester or increase the turn-over of the agonist. Similarly, our results are in keeping with a model in which ACBP increase the metabolism of these ligands. (Mol Cell Biochem **239**: 157–164, 2002)

Key words: peroxisome proliferator activated receptor, acyl-CoA binding protein, nuclear hormone receptor, adipocyte lipid binding protein

Introduction

The peroxisome proliferator-activated receptors (PPARs) are transcription factors belonging to the large superfamily of nuclear hormone receptors. The transcriptional activity of most members of this family is controlled by binding of small lipophilic molecules, which may derive either from exogenous sources or may be synthesized endogenously in the cell. Interestingly, a large number of lipid metabolites have recently been shown to specifically activate nuclear hormone receptors, which in turn regulate the metabolic pathways of which these metabolites are part [1]. Thus, these receptors are part of intricate feed-forward and feed-backward mechanisms regulating cellular metabolism. Members of the PPAR family are involved in the regulation of fatty acid metabolism – i.e. fatty acid uptake, transport, esterification, oxidation and modification. The different PPAR subtypes have distinct roles with PPARα being involved in induction of fatty acid oxidation, whereas PPARγ activity is associated with increased lipid storage. In addition, these receptors are involved the regulation of cell growth and differentiation [2–4].

The ligand binding pockets of the PPARs are large and flexible and appear to be relatively promiscuous possessing a general affinity for a broad range of fatty acids and fatty

Address for offprints: S. Mandrup, Department of Biochemistry and Molecular Biology, University of Southern Denmark, Odense, DK-5230 Odense M, Denmark (E-mail: s.mandrup@bmb.sdu.dk)

acid derivatives [4]. Some fatty acid metabolites, most notably products of the cycloxygenase and lipoxygenase pathways, have been shown to be ligands of the PPARs. Although these metabolites efficiently activate PPARs when administered to cell cultures in micromolar concentrations, the nature of and the processes involved in the synthesis of the natural PPAR ligands remain elusive. Activation of the PPARs is likely to be the result of cumulative effects of exogenously and endogenously synthesized ligands, the relative importance of which may vary between cell types. In addition to the naturally occurring fatty acid derivatives, a large number of synthetic compounds are known to be potent and subtype specific PPAR ligands. Thus, thiazolidinedione compounds used as insulin sensitizers in the treatment of type II diabetes are high affinity PPARγ ligands, while fibrates used as hypolipidemic drugs are selective PPARα ligands [4].

The intracellular fatty acid binding proteins (FABPs) comprise a family of 9 structurally and functionally similar proteins [5], which bind a broad range of fatty acids with affinities in the low to high nanomolar range depending on the FABP subtype [6]. In addition, some FABPs bind retinoic acids and fatty acid derivatives such as eicosanoids [7]. Structurally, FABPs are composed of two five-stranded β-sheets and two α-helices. The two β-sheets are almost perpendicular to each other and form the ligand-binding cavity, whereas the two α-helices have been found to control access to the ligand-binding cavity and to be involved in association between FABPs and membranes [8]. Members of the FABP family are thought to be involved in intracellular fatty acid trafficking but the exact function of these proteins remains largely unresolved.

Several attempts to unravel the function of FABPs have been made by overexpressing them in various cell lines. Although the results are variable and seem to be subtype and cell type specific, it appears that increasing the cellular FABP concentration increases fatty acid uptake and trafficking (reviewed by [8]). In addition, a number of studies have indicated that ectopic expression of FABPs affects metabolic pathways in different cell lines [7–9]. A major problem with many of these studies is that they rely on the isolation of individual clones. Recently, the phenotypes of mice with targeted disruption of the heart-type FABP, the ALBP and the intestinal-type FABP, respectively, were reported. In keeping with the notion that FABPs are involved in intracellular trafficking of fatty acids, heart muscle cells from mice with targeted disruption of the heart-type FABP (H-FABP) exhibit a pronounced decrease in fatty acid uptake and β-oxidation [10, 11]. Furthermore, adipocyte lipolysis in response to β-adrenergic stimulation was decreased in mice with targeted deletion of ALBP [12, 13]. The relatively limited effect of genetically targeting a single FABP is probably due to functional redundancy between the individual FABPs.

The ubiquitously expressed acyl-CoA binding protein (ACBP) belongs to a different group of intracellular lipid binding proteins encompassing also the testis specific endozepine like protein (ELP), a putative brain specific isoform and some larger presumably membrane-bound proteins with sequence similarity to ACBP [14]. In contrast to the rather promiscuous ligand binding capacity of the FABPs, the ACBP proteins bind specifically medium to long chain acyl-CoA esters with high affinity (Kd's in the 1 nM range) and do not bind fatty acids. The ligand-binding cavity is formed by 4 α-helices, and the CoA head functions as a lid over this cavity thereby protecting the acyl chain from interaction with the solvent. ACBP is known to be able to function as a pool former and transporter of acyl-CoA esters, and to relieve product inhibition of acyl-CoA synthetase *in vitro*, but like for the FABPs, the exact biochemical functions carried out by ACBP remain elusive [15]. ACBP can donate acyl-CoA esters to mitochondrial β-oxidation and glycerolipid synthesis *in vitro* [16], and recent experiments indicate that ACBP can interact directly with carnitoyl-palmitoyl-transferase I [17], but no direct evidence of an effect of ACBP on mitochondrial β-oxidation *in vivo* has been reported. Targeted disruption of the ACBP gene in yeast did not result in general defects in glycerolipid synthesis, but fatty acid chain elongation was significantly decreased [18].

FABPs as well as ACBP were originally thought to be exclusively cytoplasmic proteins, but we and others have demonstrated that a large fraction of these proteins reside in the nucleus of several cell types. Thus, subcellular fractionation showed that L-FABP and H-FABP are also present in the nucleus [19, 20], and recently we showed by using immunofluorescence microscopy that a significant fraction of ACBP, ALBP and keratinocyte lipid binding protein (KLBP) localize to the nucleus when ectopically expressed in CV-1 cells. Furthermore, using confocal microscopy we reported a pronounced nuclear localization of endogenous ACBP, ALBP and KLBP in differentiating 3T3-L1 preadipocytes and adipocytes [21]. This suggests that FABPs and ACBP control availability and trafficking of their ligands not only in the cytosol but also in the nucleus.

The thought that FABPs might regulate PPAR transactivation by regulating ligand availability is not new. Early studies demonstrated that liver FABP (L-FABP) potentiated the mitogenic effect of peroxisome proliferators [22]. L-FABP binds peroxisome proliferators, and so it was hypothesized that L-FABP increase the transport of these to the nucleus. Recently, this view was further supported by the observation that the ability of different exogenously added fatty acids or synthetic PPARα agonists to activate PPARα in HepG2 cells was positively correlated with the concentration of L-FABP in the cells [23]. L-FABP is with respect to structure and ligand binding quite different from the other FABPs. In addition, transfer of fatty acids between L-FABP and membranes is through aqueous diffusion, whereas transfer between membranes and other FABPs occurs through collisional transfer

[24]. Thus, the data from experiments with L-FABP do not necessarily apply to other FABPs.

We were interested in investigating how overexpression of ALBP and ACBP affected the ability of exogenous ligands to activate the different PPAR subtypes In addition to the increased trafficking suggested by the results from experiments with L-FABP, we envisioned that FABPs and ACBP might be involved in regulating the metabolic turn-over of exogenous ligands. The results summarized here suggest that ALBP and ACBP can indeed increase the metabolic turn-over and decrease the availability of exogenous fatty acid type ligands. Although we report that acyl-CoA esters have the potential to function as PPARα and -δ antagonists, our results do not support a model where ACBP decrease PPAR-dependent transactivation by increasing the acyl-CoA concentration in the nucleus.

Materials and methods

Differential protease sensitivity assay

In vitro transcription/translation, protease protection, gel electrophoresis and visualization were conducted as previously described [25]. Linoleic acid was from Sigma, and S-hexadecyl-CoA was synthesized as described in [26].

Palmitoyl-CoA agarose bead assay

Expression of mPPARα, rRXRα, agarose palmitoyl-CoA pull-down experiments, gel electrophoresis and Western blotting was conducted as described in [25]. Wy14653 and 9-cis retinoic acid were from Biomol and BRL49653 was kindly provided by Novo Nordisk A/S.

GST-pull-down

Expression and purification of GST-PPARα (ligand binding domain) and GST-N-CoR (aa. 2239–2453), *in vitro* transcription/translation, pull-down experiments, gel electrophoresis and visualization was conducted as described previously [25].

Transient transfections

For transient transfections, $PPRE_3$-TK-LUC [27] was used as reporter and pSV-β-galactosidase-control (Promega) was used for normalization. The vectors pcDNA1-ACBP and pcDNA3-ALBP contain the ACBP and ALBP cDNA respectively. In addition, the following plasmids were used: pCMX-mRXR [28], pSG5-PPARα [29], pSG5-PPARδ [30] and pSPORT-PPARγ2 [31]. CV-1 cells were grown to 50–70% confluency in 60 mm dishes and transfected with a total of 6 μg DNA per dish using the DC-Chol lipofection procedure [32]. The amounts of vectors used in each individual transfection are indicated in the legend. When necessary pcDNA3.1 was added to ensure equal DNA/promoter load. All transfections were performed as triplicates. The DNA was left on the cells for 5–7 h before the medium was changed to MEM containing 10% resin-charcoal stripped CS (v/v) supplemented with either activator (tetradecylthioacetic acid (TTA) or BRL49653) in DMSO or with DMSO alone. After 48 h, the CV-1 cells were washed in PBS and scraped off the plate in lysis buffer (100 mM potassium phosphate pH 7.8, 0.2% Triton X-100, 1 mM DTE) (Tropix). Lysates were stored at –80°C. Luciferase and β-galactosidase (Tropix) activities were determined by a Berthold MicroLumat LB96B luminometer using 96 well microtiter plates and 25 and 10 μl lysate for luciferase and β-galcatosidase assay, respectively. Results were tested by single classification ANOVA.

Results and discussion

Acyl-CoA esters act as PPARα antagonists in vitro

Fatty acids are rapidly activated to CoA esters upon entry into cells. The process is reverted by cellular thioesterases and so equilibrium exists between the FA and its CoA ester [33]. While FAs can serve as PPAR agonists, previous experiments have suggested that acyl-CoA esters are unable to activate PPARs [34, 35]. To investigate whether PPARα bind acyl-CoA esters, we incubated recombinant full length PPARα or RXRα with palmitoyl-CoA-coupled beads. As shown in Fig. 1a, PPARα but not RXRα was retained significantly better on palmitoyl-CoA-coupled beads than on protein A beads. This retention was diminished by addition of the PPARα ligand Wy14643, whereas the PPARγ ligand BRL49653 did not affect the retention. These results suggest that palmitoyl-CoA and Wy14643 compete for binding to the ligand-binding domain of PPARα.

Binding of ligand to nuclear hormone receptors is known to stabilize the structure of the ligand-binding domain. Thus, in differential protease sensitivity assays, binding of agonists to PPARs has been shown to decrease sensitivity towards chymotrypsin [36, 37]. The effect of antagonists on the protease sensitivity of PPARs has not been reported, but for a number of other nuclear receptors it has been shown that the conformational changes induced by antagonists result in profiles in a protease sensitivity assays which differ significantly from those observed with agonists [38, 39]. In accordance with linoleic acid being a PPARα agonist, we found that this fatty acid decreased the protease sensitivity of *in vitro*

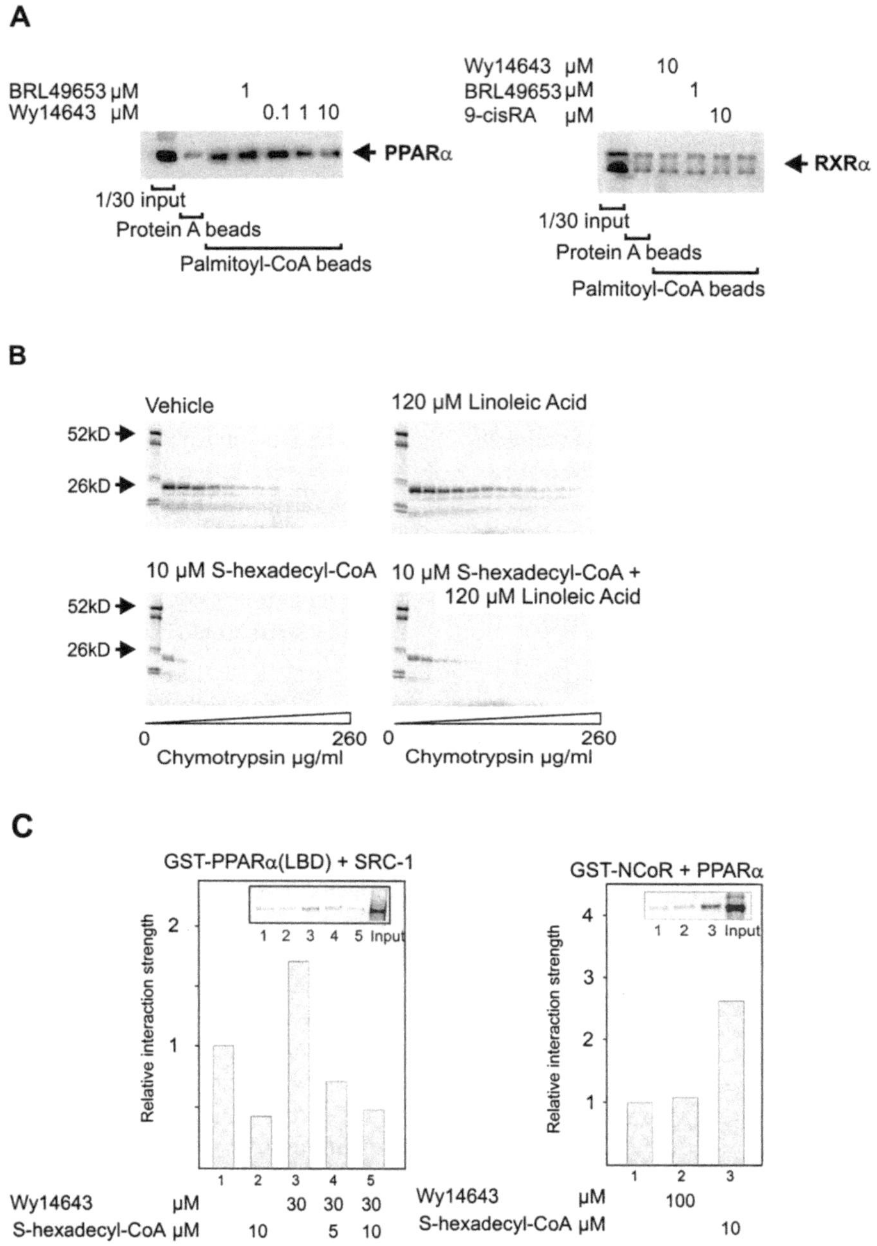

Fig. 1. Acyl-CoA esters act as PPARα antagonists *in vitro.* (A) Immunoblot analysis of proteins retained on palmitoyl-CoA- or protein A-coupled agarose beads incubated with whole cell extracts of yeast expressing either mPPARα and rRXRα, (left), or rRXRα alone, (right). Bound proteins were recovered by boiling in SDS sample buffer, and analyzed by immunoblotting. For competition, the PPARα-selective ligand, Wy14643, the PPARγ-selective ligand, BRL49653, or the RXR ligand, 9-cis retinoic acid, was added to the binding reactions as indicated. (B) *In vitro* transcribed/translated ^{35}S-labeled mPPARα was preincubated with either S-hexadecyl-CoA (10 μM) or vehicle for 20 min. Then each preincubation mixture was further incubated for 20 min with either linoleic acid (120 μM) or vehicle. The mixtures were subjected to limited digestion with chymotrypsin and the digests were analyzed by SDS-PAGE followed by phosphorimaging. (C) (Left) Bacterially expressed GST-PPARα (ligand binding domain) immobilized on glutathione-sepharose was incubated with *in vitro* transcribed/translated ^{35}S-labeled SRC-1 in the presence of vehicle, Wy14643 or S-hexadecyl-CoA. (Right) Bacterially expressed GST-mNCoR (aa 2239–2453) immobilized on glutathione-sepharose was incubated with *in vitro* transcribed/translated ^{35}S-labeled mPPARα in the presence of vehicle, Wy14643 or S-hexadecyl-CoA. Bound proteins were recovered by boiling in SDS sample buffer, separated by SDS-PAGE and quantified by phosphorimaging. This figure is reproduced from [25].

translated PPARα (Fig. 1b). By contrast, the non-hydrolyzable acyl-CoA ester increased the protease sensitivity (Fig. 1b) and did so in a dose-dependent manner [25]. To ascertain that acyl-CoA esters bound reversibly to PPARα and did not result in an irreversible denaturation of PPARα, we preincubated *in vitro* translated PPARα with the non-hydrolyzable acyl-CoA ester before adding linoleic acid. As seen in Fig. 1b linoleic acid is able to partially reverse the effect of the acyl-CoA ester on PPARα conformation. In additional experiments we showed that the effect of the acyl-CoA ester was strictly dependent on the C-terminal region of the ligand-binding domain of PPARα, i.e. a functional ligand-binding domain, whereas the N-terminal AB-domain was dispensable [25]. These results are in keeping with findings from the binding assay with the palmitoyl-CoA beads indicating that acyl-CoA esters compete with agonists for binding to the ligand-binding domain. Furthermore, the increase in protease sensitivity by binding of the acyl-CoA ester suggests that acyl-CoA esters might be antagonists of PPARα. Interestingly, we recently found that the protease sensitivity of PPARδ was also significantly increased by acyl-CoA esters, while the protease sensitivity of PPARγ was only marginally affected [40].

To further explore the possibility that acyl-CoA esters are PPAR antagonists, we investigated the effect of acyl-CoA esters on the interaction between PPARα and the steroid receptor co-activator 1 (SRC-1) or the nuclear receptor co-repressor (N-CoR) in GST pull-down experiments. Whereas the agonist Wy14643 resulted in increased interaction between PPARα and SRC-1, the non-hydrolyzable acyl-CoA ester decreased this interaction (Fig. 1c). Furthermore, the acyl-CoA ester also increased the interaction between PPARα and the co-repressor N-CoR (Fig. 1c). In similar experiments with PPARδ and PPARγ we recently showed that the ligand-enhanced interaction of PPARδ but not PPARγ with SRC-1 is also antagonized by acyl-CoA esters [40]. Thus, by all these *in vitro* criteria, acyl-CoA esters behave as PPARα and PPARδ antagonists. Although it remains to be shown that acyl-CoA esters also in an *in vivo* context can antagonize the effects of agonists, our findings indicate that the conversion of fatty acids to CoA esters not only represents an inactivation of a potential PPAR agonist, but also the generation of a PPARα and -δ antagonist.

Effect of ALBP and ACBP on ligand-induced PPAR transactivation

Given that ALBP binds potential PPAR agonists, that ACBP binds potential PPAR antagonists and that endogenous proteins of both localize to the nucleus as well as the cytoplasm, we decided to investigate how these lipid-binding proteins (LBPs) affected the ability of an exogenous fatty acid ligand to induce PPAR-dependent transactivation. A model illustrating the different ways by which LBPs might influence the ability of exogenous fatty acid-type ligands to activate PPARs is shown Fig. 2. From a very naïve point of view and disregarding the effect of these LBPs on the metabolism of the ligand, one could rationalize that if the LBPs were primarily transporting ligands to the nucleus and increasing the availability of ligands there, ALBP would be expected to

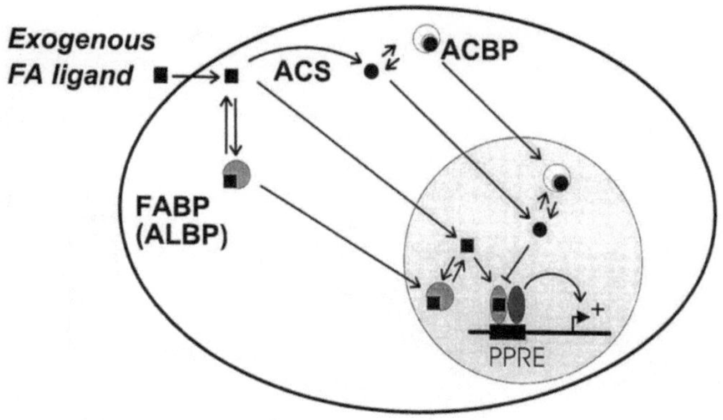

Fig. 2. Model indicating how lipid-binding proteins might affect the ability of exogenous ligands to activate PPARs. FABPs might increase trafficking of exogenous PPAR ligands to the nucleus thereby increasing the availability of the ligand. Alternatively, FABPs might sequester ligands thereby decreasing the availability. In addition, these lipid-binding proteins might affect metabolic turnover of ligands.

positively and ACBP to negatively modulate PPAR transactivation. On the other hand, if the LBPs were merely sequestering ligands and thereby decreasing their availability in the nucleus, one would expect ACBP to positively and ALBP to negatively affect PPAR transactivation. From this point of view, one would expect opposite effects of ACBP and ALBP on the ability of an exogenously added fatty acid to activate PPARs.

To test this hypothesis we transiently transfected CV-1 cells with PPAR reporter constructs and vectors encoding RXRα and the different subtypes of the PPARs. To activate the PPARs we added the thia-substituted fatty acid tetradecylthioacetic acid (TTA). Like normal fatty acids this fatty acid is rapidly activated to its CoA ester and incorporated into phospholipids, but due to the sulphur substitution TTA cannot be β-oxidized. TTA is a potent activator of PPARα but can also activate PPARδ and PPARγ at higher concentrations. At the 100 μM concentration used in our assays it activates all PPAR subtypes.

To our surprise co-transfection with ALBP and ACBP had similar effects on the ability of TTA to induce PPAR transactivation. Both LBPs inhibited TTA-induced PPAR transactivation (Fig. 3), whereas co-transfection with a leptin expression vector had no effect [21]. In contrast, neither ALBP nor ACBP co-transfection affected the ability of the thiazolidinedione BRL49653 to activate PPARγ. These results combined with results using the ligand-binding domains of the PPARs fused to the Gal4 DNA binding domain in the transactivation assays [21] clearly showed that ALBP did not increase the availability of the TTA ligand in the nucleus. Rather processes furthered by high levels of ALBP and ACBP appeared to compete with the PPARs for binding of TTA.

We cannot at this point determine whether ALBP sequesters or increases the metabolism of TTA. However, since ALBP

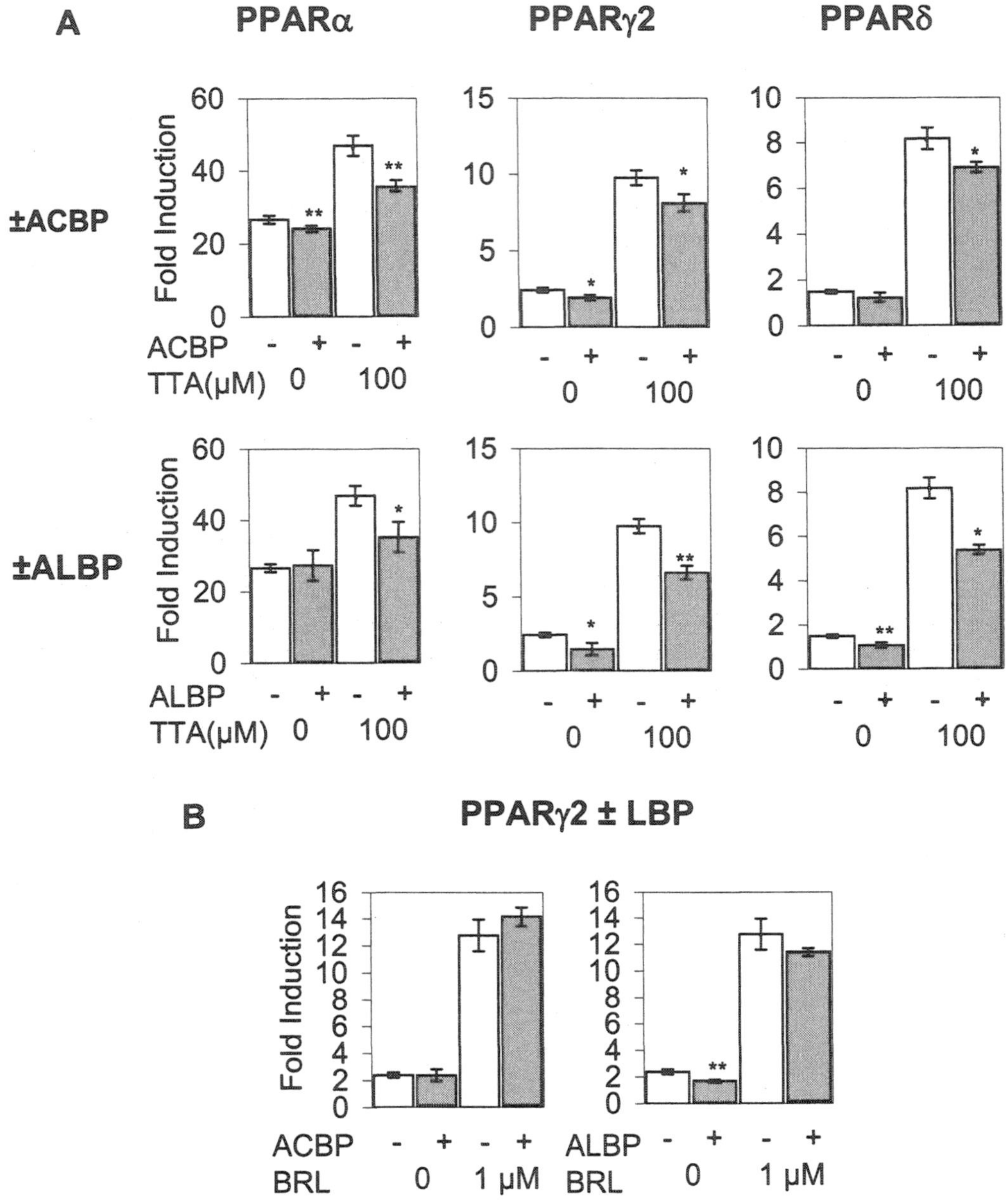

Fig. 3. Co-transfection of ALBP and ACBP decreases TTA-dependent transactivation by PPARs but not BRL49653-dependent transactivation by PPARγ. CV-1 cells were transiently transfected with the reporter plasmid $PPRE_3$-TK-LUC (1 μg) pSV-β-galactosidase-control (0.5 μg), pCMX-mRXRα (0.5 μg) and either pSG5-PPARα (0.5 μg), pSG5-PPARδ (0.5 μg) or pSPORT-PPARγ2 (0.5 μg) expressing the full-length PPAR subtypes. Cells were treated or not with 100 μM TTA (A) or 1 μM BRL49653 (B) and co-transfected or not with pcDNA1-ACBP (1 μg) or pcDNA3-ALBP (1 μg), as indicated. Equal DNA/promoter load was obtained by the addition of the empty pcDNA 3.1 vector. Luciferase activities were normalized to β-galactosidase activities. The normalized luciferase activities are shown as fold induction relative to that of transfection with the reporter plasmid alone. Each column and error bar represents the average and the standard deviation, respectively, of 3 independent transfections. Results were tested by single classification ANOVA. $*p < 0.05$; $**p < 0.01$. The results are representative of 3 or more experiments. This figure is reproduced from [21].

is equally efficient in decreasing TTA activation of PPARα and PPARγ for which TTA has very different EC_{50} values [21] it appears more likely that ALBP increases the metabolic turnover of TTA. Furthermore, since acyl-CoA esters antagonize only PPARα and PPARδ *in vitro*, whereas ACBP inhibits TTA-dependent transactivation of all PPAR subtypes, it is unlikely that inhibition of PPAR activity is achieved by an ACBP-mediated increase in the acyl-CoA availability in the nucleus. Thus, we favor the hypothesis that both ALBP and ACBP decrease TTA-induced PPAR transactivation by increasing the esterification of TTA. The finding that BRL49653-induced PPARγ activation is not affected by ALBP and ACBP is fully in keeping with this hypothesis since this compound is not readily metabolized. In addition, sequestration of BRL49653 by ALBP would be unlikely since ALBP binds thiazolidinediones with a Kd in the 10 μM range (D. Bernlohr, personal communication), whereas PPARγ binds these compounds with affinities orders of magnitude higher.

Interestingly, the expression of many LBPs is activated by members of the PPAR family. Thus, expression of H-FABP, L-FABP and intestinal FABP (I-FABP) is increased by PPARα and PPARδ activators [41–43], and ALBP is known to be transcriptionally activated by PPARγ in adipocytes [31]. Furthermore, we have recently shown the ACBP gene is transcriptionally activated by PPARγ in adipocytes [44]. Thus, induction of ALBP and ACBP in response to high levels of PPARγ ligands may represent a negative feedback on PPARγ transcriptional activity as illustrated in Fig. 4.

While the results summarized in this paper demonstrate that processes furthered by high levels of ALBP or ACBP can compete with PPARs for exogenously added fatty acid type ligands, the interplay between PPARs and LBPs is likely to be much more complex and highly dependent on the type of ligand, cell type and the concentration and type of the lipid-binding protein. Thus, as mentioned above, results from other investigators indicate that L-FABP potentiates the ability of PPARα agonists to activate PPARα [22, 23]. In addition, it is possible that FABPs and ACBP also regulate the synthesis of endogenous PPAR ligands by regulating access to different metabolic pathways. Our finding that ACBP antisense RNA inhibits differentiation of 3T3-L1 cells, and that this inhibition can be rescued by PPARγ ligands suggested that ACBP might be involved in the synthesis of endogenous PPARγ ligands [45].

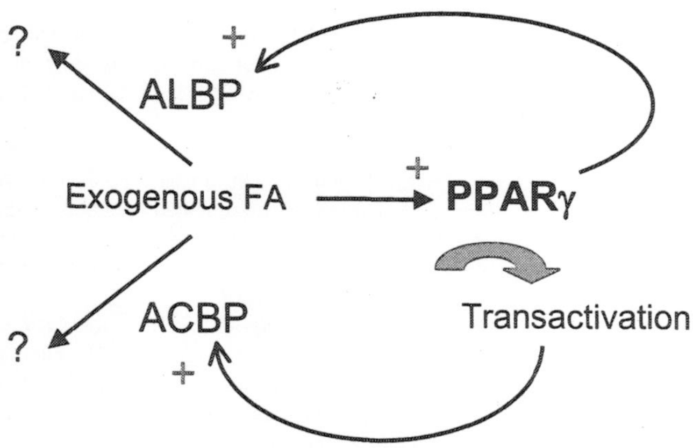

Fig. 4. Model showing how ACBP and ALBP proteins might be involved in a negative feed-back regulation of the activation of PPARγ by exogenous fatty acid-type ligands. PPARγ activates the expression of both ACBP and ALBP, which in turn stimulate processes that compete with PPARγ for binding of exogenous fatty acid-type ligands.

Acknowledgements

We thank Dr. D.A. Bernlohr for valuable discussions, Dr. R.K. Berge for supply of TTA, Dr. J. Knudsen for supply of S-hexadecyl-CoA, Novo Nordisk A/S for supply of BRL49653, Dr. U.B. Jensen for preparation of DC-Chol for transfections, Drs. P. Grimaldi, E.-Z. Zoubir, B.M. Spiegelman, and J.D. Tugwood, for the PPARδ, PPARγ, and PPARα expression plasmids, respectively, Dr. R. M. Evans for the $PPRE_3$-TK-LUC reporter and the RXRα expression plasmids, Dr. B.W. O'Malley for the SRC-1 plasmid and Dr. M. A. Lazar for the N-CoR plasmid. This work was supported by the Danish Biotechnology Program, the Danish Natural Science Research Council, the Novo-Nordisk Foundation and the Carlsberg Foundation.

References

1. Kliewer SA, Lehmann JM, Willson TM: Orphan nuclear receptors: Shifting endocrinology into reverse. Science 284: 757–760, 1999
2. Schoonjans K, Staels B, Auwerx J: The peroxisome proliferator activated receptors (PPARS) and their effects on lipid metabolism and adipocyte differentiation. Biochim Biophys Acta 1302: 93–109, 1996
3. Kersten S, Desvergne B, Wahli W: Roles of PPARs in health and disease. Nature 405: 421–424, 2000
4. Willson TM, Brown PJ, Sternbach DD, Henke BR: The PPARs: From orphan receptors to drug discovery. J Med Chem 43: 527–550, 2000
5. Hertzel AV, Bernlohr DA: The mammalian fatty acid-binding protein multigene family: Molecular and genetic insights into function. Trends Endocrinol Metab 11: 175–180, 2000
6. Richieri GV, Ogata RT, Zimmerman AW, Veerkamp JH, Kleinfeld AM: Fatty acid binding proteins from different tissues show distinct patterns of fatty acid interactions. Biochemistry 39: 7197–7204, 2000
7. Glatz JF, van der Vusse GJ: Cellular fatty acid-binding proteins: Their function and physiological significance. Prog Lipid Res 35: 243–282, 1996
8. Glatz JF, Storch J: Unravelling the significance of cellular fatty acid-binding proteins. Curr Opin Lipidol 12: 267–274, 2001
9. Prinsen CF, Veerkamp JH: Transfection of L6 myoblasts with adipocyte fatty acid-binding protein cDNA does not affect fatty acid uptake but disturbs lipid metabolism and fusion. Biochem J 329: 265–273, 1998
10. Binas B, Danneberg H, McWhir J, Mullins L, Clark AJ: Requirement for the heart-type fatty acid binding protein in cardiac fatty acid utilization. FASEB J 13: 805–812, 1999
11. Schaap FG, Binas B, Danneberg H, van D V, Glatz JF: Impaired long-chain fatty acid utilization by cardiac myocytes isolated from mice lacking the heart-type fatty acid binding protein gene. Circ Res 85: 329–337, 1999
12. Coe NR, Simpson MA, Bernlohr DA: Targeted disruption of the adipocyte lipid-binding protein (aP2 protein) gene impairs fat cell lipolysis and increases cellular fatty acid levels. J Lipid Res 40: 967–972, 1999
13. Scheja L, Makowski L, Uysal KT, Wiesbrock SM, Shimshek DR, Meyers DS, Morgan M, Parker RA, Hotamisligil GS: Altered insulin secretion associated with reduced lipolytic efficiency in $aP2^{-/-}$ mice. Diabetes 48: 1987–1994, 1999

14. Kragelund BB, Knudsen J, Poulsen FM: Acyl-coenzyme A binding protein (ACBP). Biochim Biophys Acta 1441: 150–161, 1999
15. Knudsen J, Neergaard TB, Gaigg B, Jensen MV, Hansen JK: Role of acyl-CoA binding protein in acyl-CoA metabolism and acyl-CoA-mediated cell signaling. J Nutr 130: 294S–298S, 2000
16. Rasmussen JT, Faergeman NJ, Kristiansen K, Knudsen J: Acyl-CoA-binding protein (ACBP) can mediate intermembrane acyl-CoA transport and donate acyl-CoA for beta-oxidation and glycerolipid synthesis. Biochem J 299: 165–170, 1994
17. Abo-Hashema KA, Cake MH, Lukas MA, Knudsen J: The interaction of acyl-CoA with acyl-CoA binding protein and carnitine palmitoyltransferase I. Int J Biochem Cell Biol 33: 807–815, 2001
18. Gaigg B, Neergaard TB, Schneiter R, Hansen JK, Faergeman NJ, Jensen NA, Andersen JR, Friis J, Sandhoff R, Schroder HD, Knudsen J: Depletion of acyl-coenzyme A-binding protein affects sphingolipid synthesis and causes vesicle accumulation and membrane defects in *Saccharomyces cerevisiae*. Mol Biol Cell 12: 1147–1160, 2001
19. Bordewick U, Heese M, Borchers T, Robenek H, Spener F: Compartmentation of hepatic fatty-acid-binding protein in liver cells and its effect on microsomal phosphatidic acid biosynthesis. Biol Chem Hoppe Seyler 370: 229–238, 1989
20. Borchers T, Unterberg C, Rudel H, Robenek H, Spener F: Subcellular distribution of cardiac fatty acid-binding protein in bovine heart muscle and quantitation with an enzyme-linked immunosorbent assay. Biochim Biophys Acta 1002: 54–61, 1989
21. Helledie T, Antonius M, Sorensen RV, Hertzel AV, Bernlohr DA, Kolvraa S, Kristiansen K, Mandrup S: Lipid-binding proteins modulate ligand-dependent trans-activation by peroxisome proliferator-activated receptors and localize to the nucleus as well as the cytoplasm. J Lipid Res 41: 1740–1751, 2000
22. Khan SH, Sorof S: Liver fatty acid-binding protein: Specific mediator of the mitogenesis induced by two classes of carcinogenic peroxisome proliferators. Proc Natl Acad Sci USA 91: 848–852, 1994
23. Wolfrum C, Borrmann CM, Borchers T, Spener F: Fatty acids and hypolipidemic drugs regulate peroxisome proliferator-activated receptors α- and γ-mediated gene expression via liver fatty acid binding protein: A signaling path to the nucleus. Proc Natl Acad Sci USA 98: 2323–2328, 2001
24. Storch J, Thumser AE: The fatty acid transport function of fatty acid-binding proteins. Biochim Biophys Acta 1486: 28–44, 2000
25. Elholm M, Dam I, Jørgensen C, Krogsdam AM, Holst D, Kratchmarova I, Gottlicher M, Gustafsson JA, Berge R, Flatmark T, Knudsen J, Mandrup S, Kristiansen K: Acyl-CoA esters antagonize the effects of ligands on peroxisome proliferator-activated receptor α conformation, DNA binding, and interaction with cofactors. J Biol Chem 276: 21410–21416, 2001
26. Rosendal J, Ertbjerg P, Knudsen J: Characterization of ligand binding to acyl-CoA-binding protein. Biochem J 290: 321–326, 1993
27. Kliewer SA, Umesono K, Noonan DJ, Heyman RA, Evans RM: Convergence of 9-cis retinoic acid and peroxisome proliferator signalling pathways through heterodimer formation of their receptors. Nature 358: 771–774, 1992
28. Blumberg B, Mangelsdorf DJ, Dyck JA, Bittner DA, Evans RM, De RE: Multiple retinoid-responsive receptors in a single cell: Families of retinoid 'X' receptors and retinoic acid receptors in the *Xenopus* egg. Proc Natl Acad Sci USA 89: 2321–2325, 1992
29. Tugwood JD, Issemann I, Anderson RG, Bundell KR, McPheat WL, Green S: The mouse peroxisome proliferator activated receptor recognizes a response element in the 5′ flanking sequence of the rat acyl CoA oxidase gene. EMBO J. 11: 433–439, 1992
30. Amri EZ, Bonino F, Ailhaud G, Abumrad NA, Grimaldi PA: Cloning of a protein that mediates transcriptional effects of fatty acids in preadipocytes. Homology to peroxisome proliferator-activated receptors. J Biol Chem 270: 2367–2371, 1995
31. Tontonoz P, Hu E, Graves RA, Budavari AI, Spiegelman BM: mPPAR gamma 2: Tissue-specific regulator of an adipocyte enhancer. Genes Dev 8: 1224–1234, 1994
32. Gao X, Huang L: A novel cationic liposome reagent for efficient transfection of mammalian cells. Biochem Biophys Res Commun 179: 280–285, 1991
33. Hunt MC, Nousiainen SE, Huttunen MK, Orii KE, Svensson LT, Alexson SE: Peroxisome proliferator-induced long chain acyl-CoA thioesterases comprise a highly conserved novel multi-gene family involved in lipid metabolism. J Biol Chem 274: 34317–34326, 1999
34. Forman BM, Chen J, Evans RM: Hypolipidemic drugs, polyunsaturated fatty acids, and eicosanoids are ligands for peroxisome proliferator-activated receptors α and δ. Proc Natl Acad Sci USA 94: 4312–4317, 1997
35. Hertz R, Berman I, Bar-Tana J: Transcriptional activation by amphipathic carboxylic peroxisomal proliferators is induced by the free acid rather than the acyl-CoA derivative. Eur J Biochem 221: 611–615, 1994
36. Dowell P, Peterson VJ, Zabriskie TM, Leid M: Ligand-induced peroxisome proliferator-activated receptor α conformational change. J Biol Chem 272: 2013–2020, 1997
37. Berger J, Bailey P, Biswas C, Cullinan CA, Doebber TW, Hayes NS, Saperstein R, Smith RG, Leibowitz MD: Thiazolidinediones produce a conformational change in peroxisomal proliferator-activated receptor-γ: Binding and activation correlate with antidiabetic actions in db/db mice. Endocrinology 137: 4189–4195, 1996
38. Allan GF, Leng X, Tsai SY, Weigel NL, Edwards DP, Tsai MJ, O'Malley BW: Hormone and antihormone induce distinct conformational changes which are central to steroid receptor activation. J Biol Chem 267: 19513–19520, 1992
39. Keidel S, LeMotte P, Apfel C: Different agonist- and antagonist-induced conformational changes in retinoic acid receptors analyzed by protease mapping. Mol Cell Biol 14: 287–298, 1994
40. Jørgensen C, Krogsdam AM, Kratchmarova I, Willson TM, Knudsen J, Mandrup S, Kristiansen K: Opposing effects of fatty acids and acyl-CoA esters on conformation and cofactor recruiment of peroxisome proliferator-activated receptors. Ann NY Acad Sci 967: 431–439, 2002
41. Motojima K: Differential effects of PPARα activators on induction of ectopic expression of tissue-specific fatty acid binding protein genes in the mouse liver. Int J Biochem Cell Biol 32: 1085–1092, 2000
42. Darimont C, Gradoux N, Cumin F, Baum HP, De Pover A: Differential regulation of intestinal and liver fatty acid-binding proteins in human intestinal cell line (Caco-2): Role of collagen. Exp Cell Res 244: 441–447, 1998
43. Poirier H, Niot I, Monnot MC, Braissant O, Meunier-Durmort C, Costet P, Pineau T, Wahli W, Willson TM, Besnard P: Differential involvement of peroxisome-proliferator-activated receptors α and δ in fibrate and fatty-acid-mediated inductions of the gene encoding liver fatty-acid-binding protein in the liver and the small intestine. Biochem J 355: 481–488, 2001
44. Helledie T, Grøntved L, Jensen SS, Kiilerich P, Rietveld L, Albrektsen T, Boysen MS, Nøhr J, Larsen LK, Fleckner J, Stunnenberg HG, Kristiansen K, Mandrup S: The gene encoding the acyl-CoA binding protein is activated by peroxisome proliferator activated receptor γ through an intronic response element functionally conserved between man and rodents. J Biol Chem 277: 26821–26830, 2002
45. Mandrup S, Sørensen RV, Helledie T, Nøhr J, Baldursson T, Gram C, Knudsen J, Kristiansen K: Inhibition of 3T3-L1 adipocyte differentiation by expression of acyl-CoA-binding protein antisense RNA. J Biol Chem 273: 23897–23903, 1998

Molecular and Cellular Biochemistry **239**: 165–172, 2002.

Analysis of tissue-specific and PPARα-dependent induction of FABP gene expression in the mouse liver by an *in vivo* DNA electroporation method

Kensei Fujishiro,[1] Yuka Fukui,[1,2] Osamu Sato,[1,2] Kohei Kawabe,[1,2] Koichi Seto[2] and Kiyoto Motojima[1,2]

[1]*Department of Biochemistry, School of Pharmaceutical Sciences, Toho University, Funabashi, Chiba;* [2]*Department of Biochemistry, Meiji Pharmaceutical University, Kiyose, Tokyo, Japan*

Abstract

Peroxisome proliferator (PPAR)α ligand Wy14,643 induces liver-fatty acid binding protein (FABP) spontaneously and heart-FABP gradually, but not intestine-FABP mRNA expression in the mouse liver. These strict regulations have not been reproduced in cultured cell systems. We applied a DNA electroporation method to directly introduce reporter gene constructs into the livers of mice. This system reproduced the *in vivo* responses of the above three FABP gene promoters to the PPARα ligand but not that of a promoter containing the typical three PPAR binding sites in tandem. Deletion and mutation analyses of the mouse L-FABP gene suggested that, in addition to the binding site for PPARα, a far upstream sequence is required for PPAR-dependent transactivation in the liver. In contrast to the cultured cell systems, our *in vivo* DNA electroporation method showed that PPARα binding to the promoter is necessary but not sufficient for PPARα ligand-dependent transcriptional activation of the L-FABP gene *in vivo*. (Mol Cell Biochem **239**: 165–172, 2002)

Key words: fatty acid binding protein, peroxisome proliferator-activated receptor, transcription, promoter

Abbreviations: FABP – fatty acid binding protein; PPAR – peroxisome proliferator-activated receptor; PPRE – peroxisome proliferator-activated receptor response element; RXR – retinoid X receptor; Wy 14,643 – 4-chloro-6-(2,3-xylidino)-2-pyrimidinyl-thio)acetic acid

Introduction

As key transcription factors of different genes participating in lipid homeostasis, three types of peroxisome proliferator-activated receptors (PPARs), α, β(δ) and γ, are known to regulate gene expression by sharing overlapping the roles in a cell- and tissue-specific manner. Their roles are diverse but those of PPARα are summarized as in fatty acid catabolism and those of PPARγ are in adipogenesis [1, 2]. PPARs, ligand dependent transcriptional factors, binds as a heterodimer with another nuclear receptor retinoid X receptor (RXR) to a direct repeat, the peroxisome proliferator-activated receptor response element (PPRE), located in the promoter sequences of several lipid metabolism-related enzyme genes [3]. A group of hypolipidemic drugs have been shown to bind PPARα and activate transcription of the target genes. The essential role of PPARα in hypolipidemic drug action has been demonstrated using PPARα null mice [4]. However, it has not been established whether PPARα, RXR and PPRE alone are not only necessary but also sufficient for *in vivo* regulated expression of the target genes. Our research lies in the possible involvement of other essential factors in PPAR-mediated *in vivo* responses.

To address these problems, we have focused on the mechanism of tissue-specific and PPARα-dependent induction of fatty acid binding protein (FABP) genes. FABPs bind long

Address for offprints: K. Motojima, Department of Biochemistry, Meiji Pharmaceutical University, 2-522-1 Noshio, Kiyose, Tokyo 204-8588, Japan (E-mail: motojima@my-pharm.ac.jp)

chain fatty acids in the cell and play important roles including cytosolic transport of fatty acids, regulation of fatty acid-responsive genes, and modulation of specific enzymes of lipid metabolic pathways [5]. Each FABP has a different substrate specificity and more interestingly each is encoded by a distinct gene and expression of many FABPs is strictly regulated in a tissue-specific manner under normal conditions [6]. However, we found that PPARα and/or γ ligands induced the mRNAs for liver-type (L-), intestine-type (I-), heart-type (H-), and adipose-type (A-)FABPs in the respective tissues of mice and that a potent PPARα ligand, Wy14,643, gradually induced ectopic expression of H-FABP and A-FABP mRNAs in the liver [7].

To study the structural basis of the gene promoters for responsiveness to the PPAR ligands in a tissue-specific manner, we cloned the mouse genomic sequences of the L-, I-, and H-FABP genes. We first tried to replicate the tissue-specific and PPAR ligand-dependent promoter activities in various cell lines. Weak transcriptional activation by a PPAR ligand was observed but the tissue-specific expression had almost disappeared in these systems. To overcome these problems, we developed an *in vivo* system to introduce the reporter genes by electroporation into the liver or muscle of living mice.

Materials and methods

Materials

Wy14,643, (4-chloro-6-(2,3-xylidino)-2-pyrimidinyl-thio)-acetic acid was purchased from Tokyo-Kasei (Tokyo, Japan). Dual-Luciferase Reporter Assay System and pGL reporter plasmids were purchased from Promega (Wisconsin, USA). pCAG-Luc plasmid was kindly provided by Jun-ichi Miyazaki (Osaka University, Japan). KOD polymerase and restriction enzymes were obtained from Toyobo (Osaka, Japan). TransIT™ *in vivo* gene delivery system was purchased from Mirus (Wisconsin, USA).

Cloning of mouse FABP promoters

Promoter sequences of mouse L-FABP and I-FABP genes were amplified from mouse genomic DNA using degenerate PCR primers. We synthesized specific primers for each gene and used them to carry out genomic walking using a mouse Genome Walker kit (Clontech) according to the manufacturer's protocol. Primers of the following sequences corresponding the 5′-ends of the mRNAs [8, 9] were used:

L-FABP R1, 5′-ATTGGTACTTGCCGGAGAAGTTCATGGT
R2, 5′-GGCTTCCTTTCCACAGGTGACCACAAC

I-FABP R1, 5′-CCAAGCTTCCTCTTCATCACATTAATGCCCATTTT
R2, 5′-TGTGTGTCTCTAGGAAAGCAGACCTCCA

The mouse H-FABP gene promoter sequence was obtained by PCR of mouse genomic DNA using the following primers [10]:

H-FABP Right, 5′-GATGGTAGTAGGCTTGGTCATGCTACCGACCTG
Left, 5′-TTATGTCCCTCTTCCTCTGAATCCCCAA

The PCR fragments were blunt-ended using KOD polymerase and cloned into the HincII site of pUC18. Sequencing and comparison with the published sequences of rat L-FABP [8], mouse I-FABP [9] and mouse H-FABP [10] genes confirmed that the cloned mouse L-FABP gene promoter was 2.3 kb in length, the I-FABP promoter was 2.4 kb and the H-FABP promoter was 1.6 kb.

Construction of reporter genes

The luciferase reporter plasmids were constructed by cloning the restriction fragments from the plasmids containing the promoter sequences into the restriction sites of the pGL3 promoter vector (Promega). To generate L2 (–91) and L3 (–53) truncated promoters of the L-FABP gene, PCR primers were designed to hybridize at the corresponding position and were coupled with the common downstream primer in the luciferase-coding region. Point mutations were introduced in the PPRE of the promoter (from TCACTGACCTATGG to TCAC*A*GACCTA*A*GG (–72 to –53)) using the GeneEditor™ *in vivo* site-directed mutagenesis system (Promega) and the corresponding mutated oligonucleotide.

Cell culture and DNA transfection

Fao (a subclone of rat hepatoma HIIE), IEC-6 (ATCC, rat small intestine epithelial cell line), C2C12 (ATCC, a subclone of the C2 mouse muscle cell line), and CV-1 (African green monkey kidney cell line) cells were cultured and differentiated under the conditions reported previously [11–13]. Fetal bovine serum was used after three cycles of charcoal treatment to remove possible endogenous PPAR ligands. For DNA transfection, an optimized method for each cell line was used; Fao, SuperFect™ (Qiagen); IEC-6 and C2C12, LipofectAMINE PLUS™ (GibcoBRL); CV-1, LipofectAMINE™ (GibcoBRL).

In vivo *direct gene transfer*

Normal male NZB mice (5–6 weeks of age) kept under a 12-h light–dark cycle with free access to water and food were used for the experiments. For *in vivo* electroporation, the indicated amounts of plasmid DNA in 15 μl of phosphate-buffered saline were injected with a 100 μl micro syringe into the exposed left liver lobe by cutting the skin. Immediately after the injection, the area was nipped with a needle-type or a tweezers-type electrode and square electric pulses were applied 10 times at 99 msec intervals at 25 V using an Electro-SquarePorator T820 (BTX, a Division Genetronics Inc., San Diego, USA). The incisions made on the skin were stitched immediately after electroporation. After injection, mice were fed a control diet (CE7, Clea Japan) or a diet containing Wy14,643 (0.05%) for the indicated periods.

Reporter gene assay

On the indicated days after injection, the transfected area of the liver was excised and weighed. The tissue was minced and homogenized with Passive Buffer (Promega) and lysed by repeated freezing and thawing. After centrifugation at 13,000 g for 10 min, the supernatant was monitored for firefly luciferase activity with a luminometer (Berthold, Germany). As the activity of *Renilla* luciferase was enhanced by the extract of Wy14,643-fed mouse livers for unknown reasons, the firefly luciferase activities could not be normalized by the *Renilla* luciferase activity as in the Double Luciferase System (Promega). Then, the firefly luciferase activity in relative light units was divided by the amounts of proteins used for the assay to give normalized values.

RNA preparation and Northern blotting analysis

Total RNA was prepared from the liver by the acid guanidinium isothiocyanate-phenol-chloroform extraction method [14]. Northern blotting analysis was carried out essentially as previously described [15]. After hybridization, the membranes were washed and autoradiographed with an intensifying screen at –80°C. For quantitative analysis, a BAS2000 image analyzer (Fujix, Japan) was used.

Results

Tissue-specific and PPARα-regulated expression of FABP mRNAs

L-FABP, I-FABP and H-FABP mRNA were induced by a PPARα ligand in the respective tissues of mice [7]. To directly examine the role of PPARα in the induction, we utilized PPARα-null mice [4]. The results of Northern blotting analyses of liver, intestine, skeletal muscle and fat RNA from (+/+) and (–/–) mice fed either a control diet or that containing 0.1%

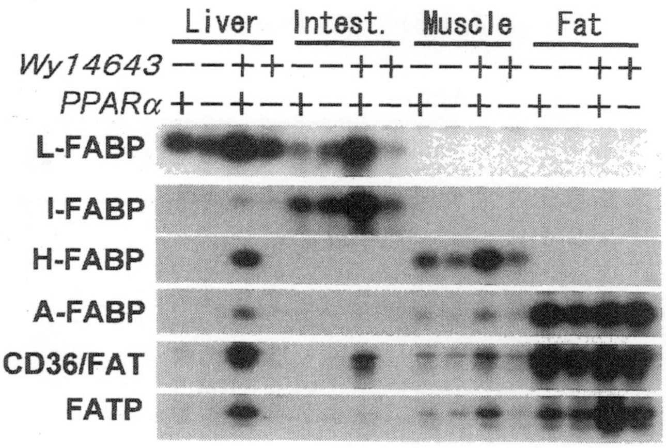

Fig. 1. Influence of the PPARα on Wy14,643-induction of FABP mRNAs in mouse tissues. Wild-type (+) and PPARα-null mice (–) were fed either control (–) diet or that containing 0.1% Wy14,643 (+). Total RNAs from the liver, intestine, skeletal muscle and adipose tissue were analyzed by Northern blotting using the cDNAs for L-FABP, I-FABP, H-FABP and A-FABP.

Wy14,643 are shown in Fig. 1. Each FABP mRNA was expressed specifically in the respective tissues of mice fed the control diet and were increased in (+/+) mice fed Wy14,643 in a tissue-specific manner. These tissue-specific increases in L-FABP, I-FABP and H-FABP mRNA levels and also ectopic induction of H-FABP and A-FABP mRNAs in the liver [7] were not observed in (–/–) mice, indicating an obligatory role of PPARα in Wy14,643 induction of FABP mRNAs in the respective tissues.

Promoter structures of the mouse L-FABP, I-FABP and H-FABP genes

To study the molecular basis of tissue-specific and PPARα-regulated expression of FABP genes, we cloned the promoter regions of the three mouse genes. Sequencing and comparison with the published sequence data including those of other species confirmed that the conserved PPAR binding sites (PPREs) were present in each promoter as summarized in Fig. 2.

Reporter analysis of the FABP gene promoters using the cultured cells

To examine whether the cloned promoter sequences are sufficient for tissue-specific and PPARα-regulated expression, we first performed a reporter assay using cultured cells, including not only CV-1 cells [16] that commonly used in reporter assays but also much more differentiated rat liver-derived Fao cells [11], mouse muscle-derived C2C12 cells [13] and rat intestine-derived IEC-6 cells [12]. The results of several ex-

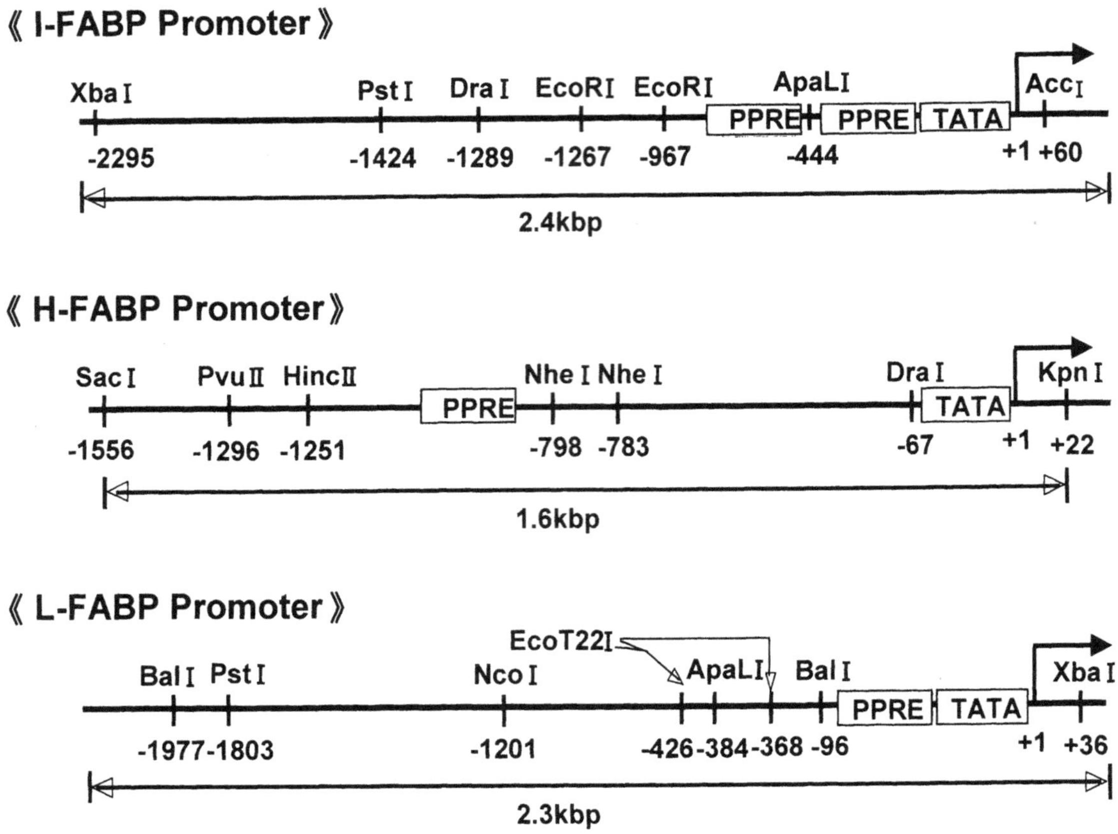

Fig. 2. Summary of the promoter structures of three mouse FABP genes analyzed in this study.

periments are summarized in Table 1. The I-FABP promoter did not respond to a PPARα ligand in any of these cells although the artificial tandem PPRE promoter responded in all cell types. The H-FABP promoter responded poorly in C2C12 cells. The L-FABP promoter but not the H-FABP or I-FABP promoter responded in Fao cells, but the basal expression levels of the L-FABP promoter in IEC-6 and C2C12 cells were much higher than those of the two other FABP promoters. Thus, the cultured cell system could not reproduce the *in vivo* responses and we could not conclude whether the cloned promoter sequences were or were not sufficient for tissue-specific and PPARα-regulated expression.

Table 1. Reporter assay of three FABP genes in the cultured cells (comparison of inducibility by Wy14,643)

PPAR/RXR	CV-1 +	IEC-6 +	C2C12 –	Fao –
PPRE	3.0 (1.0)	3.4 (1.0)	3.5 (1.0)	4.8 (1.0)
L-FABP	3.0 (1.9)	0.3 (9.5)	1.0 (9.6)	1.9 (3.2)
H-FABP	1.0 (0.3)	0.3 (3.3)	1.5 (4.4)	0.7 (0.2)
I-FABP	0.6 (0.1)	0.5 (0.1)	1.0 (0.1)	0.6 (2.0)

() – relative expression levels.

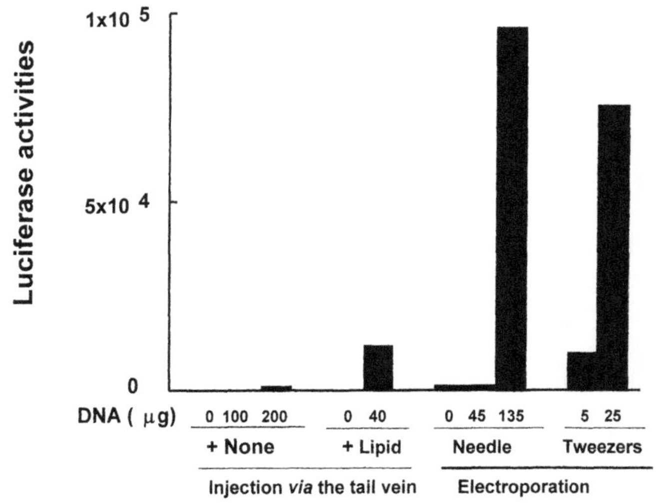

Fig. 3. Effects of *in vivo* gene transfer methods on the levels of reporter gene expression in the liver. Two methods were compared with modifications. The mice were injected via the tail vein with a solution of plasmid DNA encoding firefly luciferase directed by the CAGS promoter with or without cationic lipid, or the plasmid DNA was introduced into the liver by electroporation with a needle-type or a tweezers-type electrode. The luciferase activities were normalized by the amounts of protein used for the assay.

Comparison of the methods to introduce DNA into the mouse liver

To reproduce the *in vivo* responses of the FABP promoters, we developed an *in vivo* method for gene transfer into the mouse liver. To introduce plasmid DNA efficiently enough to perform quantitative reporter analysis of the FABP gene promoters, we first examined several published methods [17, 18] using a control reporter plasmid containing a strong CAGS promoter [19]. Introduction of plasmid DNA, naked or with cationic lipid, *via* the tail vein was not efficient enough for the promoter deletion studies, although the DNA might have been introduced into some proportion of the cells all over the liver. In contrast, the electroporation method resulted in highly efficient introduction of naked DNA into the localized cells. Furthermore, using the tweezers-type electrode gave more reproducible data than the needle-type electrode. During preparation of this manuscript, a similar method for *in vivo* gene electroporation was published elsewhere [20].

Sustained expression of the transferred reporter gene in the liver

We next examined the changes in the expression levels of the reporter gene in the mouse liver by measuring the liver luciferase activities at several days following *in vivo* electroporation. Luciferase activity levels gradually declined, but significant activities were maintained throughout the period

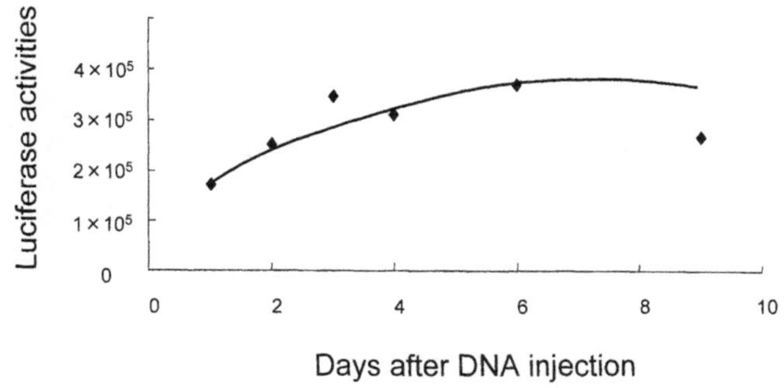

Fig. 4. Time course of changes in the reporter luciferase expression levels in the liver following a single introduction of the plasmid DNA. The reporter plasmid DNA was introduced into several mice on day 0, and the liver extracts were prepared at different time points as indicated for the reporter assay. The luciferase activities were normalized by the amounts of protein used for the assay.

examined (Fig. 4). The sustained expression of the transferred gene made it possible to study the responsiveness of the gene to a PPAR ligand.

Regulated expression of the FABP-promoter reporter genes

To examine whether *in vivo* transferred genes reproduce tissue-specific and PPARα ligand-activated expression in the liver, we introduced reporter genes containing the L-FABP, I-FABP or H-FABP gene promoters. Expression of lucifer-

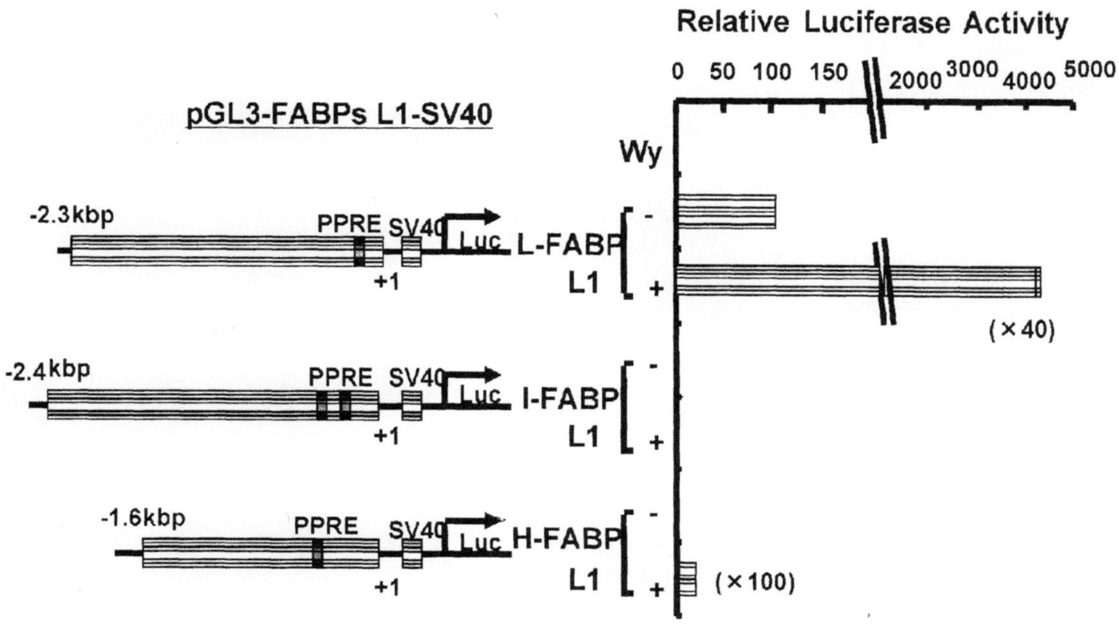

Fig. 5. In vivo reporter gene assay of the three mouse FABP gene promoters. Luciferase reporter plasmids containing the L-FABP, H-FABP or I-FABP gene promoter were introduced into the mouse liver by an *in vivo* DNA electroporation method. After feeding with a control diet or that containing Wy14,643 for 7 days, the luciferase activities in the liver extracts were assayed.

ase driven by the L-FABP promoter was detected in the livers of mouse fed the normal diet, and largely enhanced by feeding a diet containing Wy14,643. In contrast, I-FABP promoter did not respond at all, and the H-FABP promoter showed a slight response (Fig. 5). These results were consistent with those of the endogenous genes as shown by Northern blotting (Fig. 1). Thus, the *in vivo* gene transfer method makes it possible to analyze the mechanism of tissue-specific and PPARα-dependent induction of the L-FABP gene in the liver.

Deletion analysis of the L-FABP promoter

To examine the necessity and sufficiency of PPARα binding to the promoter in PPARα ligand-induced expression of the L-FABP gene, we next prepared reporter plasmids containing a series of deletions or an artificial fragment containing three copies of the acyl-CoA oxidase PPRE linked to the TK promoter [21]. These plasmids were transferred into the mouse liver *in vivo* and their responsiveness to Wy14,643 was determined. As shown in Fig. 6, the tandem PPREs and the short natural promoter containing sequence just up to the PPRE did not show transcriptional activation by a PPARα ligand. PPARα ligand-dependent transcriptional activation was observed only with the long natural promoter. The necessity of PPARα binding for the activation was confirmed by very low basal activity and little responsiveness of the long promoter containing the mutated PPRE. Thus, our *in vivo* gene transfer and deletion analysis showed that PPARα binding to the promoter is necessary but not sufficient for PPARα ligand-dependent transcriptional activation of the L-FABP gene *in vivo*.

Discussion

FABP gene expression is strictly regulated in a tissue-specific manner. Feeding mice with a PPARα ligand induced respective FABP mRNAs in the liver, intestine, and muscle in a PPARα-dependent manner (Fig. 1). Thus, the expression levels of PPARα in each tissue are high enough to promote transcriptional activation of respective genes and abundance of PPARα alone is not sufficient to induce all of the genes in one tissue. Some other tissue-specific factors are necessary for the action of PPARα in FABP gene expression in various tissues.

Experimental systems employed to date to analyze the relationship between PPAR action and promoter structure have been mostly limited to cultured cells and DNA transient transfection methods. In these systems, accessibility of transcription factors to the gene is easier and the concentrations of the translation products of the infected genes may be higher than those in the tissues in the body, making the transcriptional specificity low and thus the gene regulation in the transfected cells will not reflect the situation in the tissue [22]. Well-differentiated cells are difficult to transfect with DNA

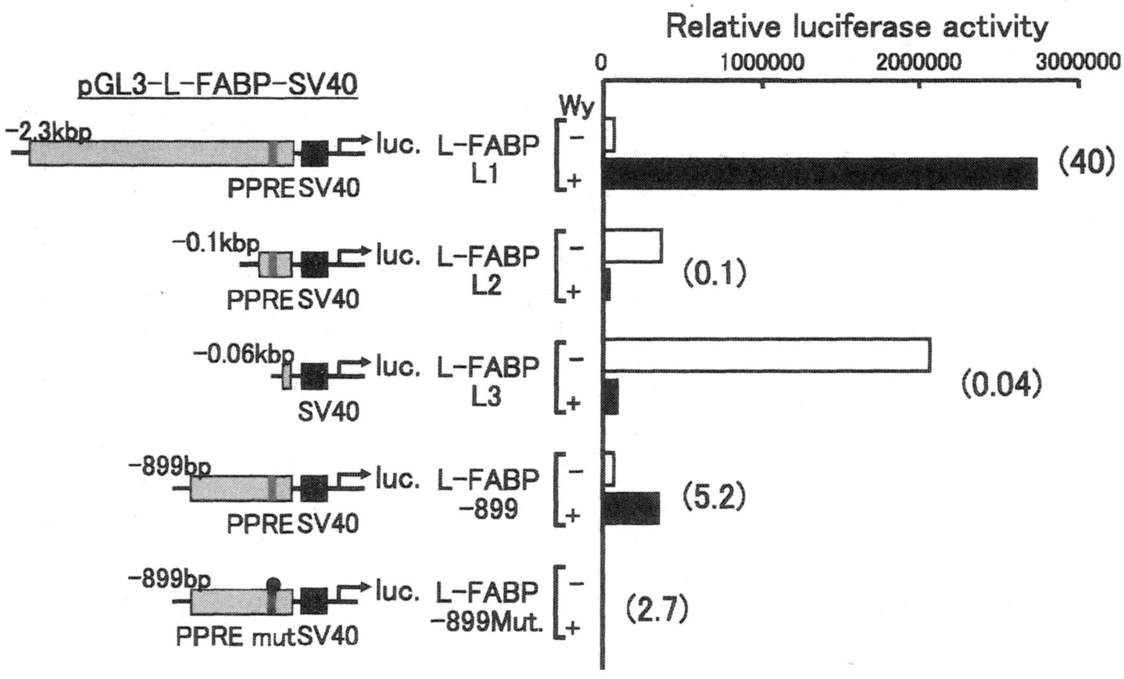

Fig. 6. Deletion and mutation analyses of the mouse L-FABP gene promoter. The luciferase reporter plasmids containing the deleted or mutated mouse L-FABP gene promoter as indicated were introduced into the mouse liver, and assayed for the reporter activities as described in the legend for Fig. 5.

and model cell lines such as CV-1 have been widely used for transfection studies. Previous systems were not intended to analyze the tissue-specific regulation of gene expression and few studies have been attempted from this point of view [16]. Our studies using cultured cells confirmed that the cell culture system is not suited to study the tissue-specific gene regulation (Table 1).

The *in vivo* gene transfer system described in this paper represents a new method to study the mechanism of tissue-specifically regulated gene expression, although the method should be improved in its efficiency and ease of operation. Deletion analysis of the mouse L-FABP gene promoter using the *in vivo* gene transfer method showed results clearly distinct from those obtained in similar experiments using cultured cells (Fig. 6). The PPRE and its surrounding sequence of the L-FBAP gene promoter alone were necessary but not sufficient for promotion of transcriptional activation by the PPARα ligand, Wy14,643. Artificial tandem PPRE promoter did not respond at all in the liver with or without the ligand. These observations suggested that PPARα is not a 'master regulator' in transcription of the L-FABP gene. Instead, several factors, some of which confer liver-specific expression and some of which confer physiological responses, coordinately control the gene expression. Supporting this model, an essential and direct role of hepatocyte nuclear factor 1α (HNF1α) in the transcriptional regulation of the L-FABP gene expression was suggested by analyzing HNF1α-null mice [23]. Furthermore, it was also reported that elevated expression of PPARα alone in HepG2 cells, in which the concentration of the expressed PPARα is quite low, could not promote transcriptional activation of the L-FABP gene although it induced the mRNAs of some of other PPARα-target genes, such as the long chain fatty acyl-CoA synthetase and carnitine palmitoyl acyl-CoA transferase [24].

Deletion analysis of the L-FABP gene promoter using the *in vivo* gene transfer method is currently in progress in our laboratory, and our preliminary results suggests that several negative and positive *cis* elements are present in the wide promoter region of the gene. Definite identification of these *cis* and then *trans* factors will require refinement of the *in vivo* gene transfer method. However, the results presented here first indicated that PPARα is required but not sufficient for liver-specific and PPARα-dependent induction of L-FABP gene expression.

Acknowledgements

The authors wish to thank Drs. J.M. Peters and F.J. Gonzalez for cooperation in preparing RNA samples from PPARα-null mice and Ms. M. Murakami and M. Hikita for technical assistance.

References

1. Kliewer SA, Xu HE, Lambert MH, Willson TM: Peroxisome proliferator-activated receptors: From genes to physiology. Recent Prog Horm Res 56: 239–263, 2001
2. Kersten S, Desvergne B, Wahli W: Roles of PPARs in health and disease. Nature 405: 421–424, 2000
3. Juge-Aubry C, Pernin A, Favez T, Burger AG, Wahli W, Meier CA, Desvergne B: DNA binding properties of peroxisome proliferator-activated receptor subtypes on various natural peroxisome proliferator response elements. Importance of the 5′-flanking region. J Biol Chem 272: 25252–25259, Date?
4. Lee SS, Pineau T, Drago J, Lee EJ, Owens JW, Kroetz DL, Fernandez-Salguero PM, Westphal H, Gonzalez FJ: Targeted disruption of the alpha isoform of the peroxisome proliferator-activated receptor gene in mice results in abolishment of the pleiotropic effects of peroxisome proliferators. Mol Cell Biol 15: 3012–3122, 1995
5. Glatz JFC, Storch J: Unrevelling the significance of cellular fatty acid-binding proteins. Curr Opin Lipid 12: 267–274, 2001
6. Heuckeroth RO, Birkenmeier EH, Levin MS, Gordon JI: Analysis of the tissue-specific expression, developmental regulation, and linkage relationship of a rodent gene encoding heart fatty acid binding protein. J Biol Chem 262: 9709–9717, 1987
7. Motojima K: Differential effects of PPARα activators on induction of ectopic expression of tissue-specific fatty acid binding protein genes in the mouse liver. Int J Biochem Cell Biol 32: 1085–1092, 2000
8. Sweetser DA, Lowe JB, Gordon JI: The nucleotide sequence of the rat liver fatty acid-binding protein gene. Evidence that exon 1 encodes an oligopeptide domain shared by a family of proteins which bind hydrophobic ligands. J Biol Chem 261: 5553–5561, 1986
9. Green RP, Cohn SM, Sacchettini JC, Jackson KE, Gordon JI: The mouse intestinal fatty acid binding protein gene: Nucleotide sequence, pattern of developmental and regional expression, and proposed structure of its protein product. DNA Cell Biol 11: 31–41, 1991
10. Treuner M, Kozak CA, Gallahan D, Grosse R, Muller T: Cloning and characterization of the mouse gene encoding mammary-derived growth inhibitor/heart-fatty acid-binding protein. Gene 147: 237–242, 1994
11. Motojima K, Passilly P, Peters JM, Gonzalez FJ, Latruffe N: Expression of putative fatty acid transporter genes are regulated by peroxisome proliferator-activated receptor α- and γ-activators in a tissue- and inducer-specific manner. J Biol Chem 273: 16710–16714, 1998
12. Dignass AU, Podolsky DK: Interleukin 2 modulates intestinal epitherial cell function *in vitro*. Exp Cell Res 225: 422–429, 1996
13. Saitoh O, Periasamy M, Kan M, Matsuda R: *Cis*-4-hydroxy-L-proline and ethyl-3,4-dihydroxybenzoate prevent myogenesis of C2C12 muscle cells and block MyoD1 and myogenin expression. Exp Cell Res 200: 70–76, 1992
14. Chomczynski P, Sacchi N: Single-step method of RNA isolation by acid guanidinium thiocyanate-phenol-chloroform extraction. Anal Biochem 162: 156–159, 1987
15. Motojima K, Peters JM, Gonzalez FJ: PPARα mediates peroxisome proliferator-induced transcriptional repression of non peroxisomal gene expression in mouse. Biochem Biophys Res Commun 2130: 155–158, 1997
16. Galli A, Pinaire J, Fischer M, Dorris R, Crabb DW: The transcriptional and DNA binding activity of peroxisome proliferator-activated receptor alpha is inhibited by ethanol metabolism. A novel mechanism for the development of ethanol-induced fatty liver. J Biol Chem 276: 68–75, 2001
17. Nomura T, Nakajima S, Kawabata K, Yamashita F, Takakura Y, Hashida M: Intratumoral pharmacokinetics and *in vivo* gene expression of naked plasmid DNA and its cationic liposome complexes after direct gene transfer. Cancer Res 57: 2681–2686, 1997

18. Nomura T, Yasuda K, Yamada T, Okamoto S, Mahato RI, Watanabe Y, Takakura Y, Hashida M: Gene expression and antitumor effects following direct interferon (IFN)-γ gene transfer with naked plasmid DNA and DC-chol liposome complexes in mice. Gene Ther 6: 121–129, 1999
19. Hisayasu S, Miyauchi M, Akiyama K, Gotoh T, Satoh S, Shimada T: *In vivo* targeted gene transfer into liver cells mediated by a novel galactosyl-D-lysine/D-serine copolymer. Gene Ther 6: 689–693, 1999
20. Muramatsu T, Ito N, Tamaoki N, Oda H, Park H-M: *In vivo* gene electroporation confers nutritionally-regulated foreign gene expression in the liver. Int J Mol Med 7: 61–66, 2001
21. Ricote M, Li AC, Willson TM, Kelly CJ, Glass CK: The peroxisome proliferator-activated receptor-γ is a negative regulator of macrophage activation. Nature 391: 79–82, 1998
22. Motojima K: Peroxisome proloferator-activated receptor (PPAR)-dependent and -independent transcriptional modulation of several non-peroxisomal genes by peroxisome proliferators. Biochimie 79: 101–106, 1997
23. Akiyama TE, Ward JM, Gonzalez FJ: Regulation of the liver fatty acid-binding protein gene by hepatocyte nuclear factor 1α (HNF1α). Alteration in fatty acid homeostasis in *HNF1α*-deficient mice. J Biol Chem 275: 27117–27122, 2000
24. Hsu M-H, Savas Ü, Griffin KJ, Johnson EF: Identification of peroxisome proliferator-responsive human genes by elevated expression of the peroxisome proliferator-activated receptor in HepG2 cells. J Biol Chem 276: 27950–27958, 2001

Molecular and Cellular Biochemistry **239**: 173–180, 2002.

Characterization of a new fatty acid response element that controls the expression of the locust muscle FABP gene

Qiwei Wu, Weihua Chang, Jutta Rickers-Haunerland, Tobi Higo and Norbert H. Haunerland

Department of Biological Sciences, Simon Fraser University, Burnaby, British Columbia, Canada

Abstract

In vertebrate and invertebrate muscles, the expression of fatty acid binding proteins (FABP) is induced by long chain fatty acids. To identify the fatty acid response elements that mediate this up-regulation, the gene of the FABP expressed in locust flight muscle was cloned, and its upstream sequences analyzed for potential regulatory elements. Comparison with other muscle FABP promoters revealed the presence of a 19-bp imperfect inverted repeat sequence that contains two hexanucleotide half sites (AGTGGT and ATGGGA), interspersed by 3 nucleotides. The promoter activity was studied with reporter gene constructs in L6 myoblasts, in which H-FABP expression is stimulated by long-chain fatty acids in a similar manner as in adult cardiomyocytes. The 19 bp element, located 180 bp upstream of the transcription start site, was found to be essential for the fatty acid induction of gene expression, and gel shift analysis confirmed that this fatty acid response element is capable of binding nuclear proteins both from rat myoblasts and locust muscle in the presence of fatty acids. A similar, but reverse sequence that is present upstream of all mammalian H-FABP promoters may modulate the expression of the rat H-FABP gene. (Mol Cell Biochem **239**: 173–180, 2002)

Key words: fatty acid binding protein, fatty acid response element, FARE, *Schistocerca gregaria*, inverted repeat, everted repeat, gene regulation, gel shift

Abbreviations: FABP – fatty acid binding protein; L-, A- and H-FABP – liver-, adipocyte- and heart-FABP; FARE – fatty acid response element; PPAR – peroxisome proliferator activated receptor; PPRE – peroxisome proliferator response element; DR-1 – direct repeat separated by 1 nucleotide; IR-3 – inverted repeat separated by 3 nucleotides; ER-3 – everted repeat separated by 3 nucleotides

Introduction

Fatty acid binding proteins are ubiquitous intracellular proteins that participate in fatty acid transport and metabolism. They belong to a conserved multi-gene family that also contains transport proteins for other hydrophobic ligands, such as retinoid acid [1]. In different tissues, distinct FABPs are expressed, reflecting the different metabolic roles of fatty acids. Generally, high levels of FABPs are found in tissues that utilize fatty acids at high rates, and the expression of FABP appears to increase when fatty acid demand increases [2, 3]. Evidence for such an up-regulation of FABP gene expression upon fatty acid exposure has been observed especially for the adipocyte (A-FABP), liver (L-FABP), and heart isoform of the FABP gene (H-FABP).

The concentration of H-FABP is directly proportional to the fatty acid dependent metabolic rate over the wide range found in various muscles. Fast twitch skeletal muscles that rely mostly on glucose possess relatively small amounts of the protein, while in mammalian heart FABP comprises up

Address for offprints: N.H. Haunerland, Department of Biological Sciences, Simon Fraser University, Burnaby, B.C. V5A 1S6, Canada (E-mail: haunerla@sfu.ca)

to 6% of all cytosolic proteins [2]. Flight muscles from vertebrates and invertebrates that encounter fatty acid metabolic rates 2 or 3 times higher have been found to contain FABP levels that are elevated by similar factors [4, 5]. Moreover, chronically increased lipid utilization has been found to lead to marked increases in FABP expression in various muscles, both in vertebrates and invertebrates [3–6] . Van Bilsen *et al.* [7, 8] demonstrated *in vitro* that prolonged exposure of cultured neonatal ventricular myocytes to exogenous fatty acids leads to a nearly 4-fold increase in FABP mRNA. All these observations suggest that an increased intracellular fatty acid concentration, maintained over certain time periods, serves as the signal to stimulate FABP expression. Thus, fatty acid itself may induce the expression of its own transport protein in muscle cells.

Fatty acids or their metabolites can modulate the expression of other members of the FABP gene family at the level of transcription initiation [9], by mechanisms similar to lipophilic hormones such as steroids, retinoids, and thyroxins. In these cases, nuclear receptors exist that bind, when complexed with the lipophilic ligand, alone or in concert with other proteins to cis-acting sequences upstream of the genes [10]. It has been demonstrated that both, the liver and adipocyte FABP genes are under the control of peroxisome proliferator activated receptors (PPAR), so called because of their activation by fibrate drugs known to stimulate the formation of peroxisomes [11, 12]. To date, three major forms of PPARs have been characterized; of these, PPARα is expressed in liver, kidney, and cardiac tissues, while PPARβ is ubiquitously distributed, including in the heart. PPARγ can be found in adipose tissue, lymphatic tissue, and the intestine [13].

Although PPARs are involved in the regulation of several muscle specific genes, they may not play a role in the fatty acid mediated up-regulation of the H-FABP gene. Several mammalian H-FABP genes lack peroxisome proliferator response elements (PPREs), and the PPRE-like sequence found upstream of the rodent genes [14] appears to be non-functional (Spener, personal communication). Thus, the involvement of a different, hitherto unknown fatty acid response element was suspected. The conserved gene structure of FABP genes, coupled with similar physiological responses to elevated fatty acid levels, suggests that analogous mechanisms control the expression of the mammalian and insect muscle FABP. Therefore, we reasoned that the suspected fatty acid response elements should be conserved as well, and that it may be possible to identify candidate sequences by comparing sequences upstream of the invertebrate and vertebrate muscle FABP promoters. If such sequences were found, they could be linked to reporter genes, and their expression studied in a suitable cell line. This paper summarizes our studies that led to the discovery of a novel fatty acid response elements in the insect muscle FABP gene, and shows additional experiments that point to a similar function of a related element found upstream of the mammalian FABP genes.

Materials and methods

Cloning and sequencing

The cloning of the locust muscle FABP gene is described in [15]. Sequencing was carried out on carried an ABI sequencer or with ABI's AmpliTaq DyeDeoxy terminator Cycle sequencing chemistry (PE Biosystems, Foster City, CA, USA) on the automatic ABI Model 373 Stretch DNA sequencer at Biotechnology Laboratory/Nucleic Acid Service Unit of the University of British Columbia.

Cell culturing and isolation

Male Sprague–Dawley rats were housed in the Animal Care facility of Simon Fraser University. Prior to heart dissection, rats were intra-peritoneally anesthetized with sodium pentobarbital. The heart was rapidly excised, placed into ice cold dissecting solution, and via the aorta attached to a water-jacketed Langendorff perfusion apparatus. Following perfusion first with aerated dissecting solution and then with collagenase solution, the digested tissue was gently disintegrated, and cells were transferred to culture dishes [16]. The fresh cardiomyocytes were cultured at approximately 60–80% confluency in Delbecco's Modified Eagle Medium (DMEM) with 10% fetal bovine serum (FBS), 150 μg/ml penicillin and 150 μg/ml streptomycin. L6 myoblasts were grown in the same medium for 3 days to approximately 80% confluency.

Fatty acid treatment

Fatty acid (600 μM)/BSA (200 μM) solutions were prepared by dissolving fatty acid free BSA (Sigma, Oakville, ON, Canada) and sodium salts of fatty acids (Sigma, Oakville, ON, Canada) in 0.9% NaCl. The filter-sterilized fatty acids–BSA complex (0.5 ml) was added to 4.5 ml fresh DMEM medium supplemented with 1% fetal bovine serum (final concentration of fatty acids 60 μM). Cells incubated with DMEM supplemented with 1% fetal bovine serum but without fatty acids were used as control samples.

Quantification of rat H-FABP expression

Prior to RNA extraction, cells were detached by 0.25% trypsin in phosphate buffered saline (0.01 M phosphate, 0.9% NaCl, pH 7.4) and collected by centrifugation. Total RNA was isolated by Totally-RNA isolation kit from Ambion (Austin, TX, USA). RNA was stored in ethanol at –80°C until used for RT-PCR. Ready-To-Go RT-PCR beads (Amersham Pharmacia Biotech, Piscataway, NJ, USA) were brought to final volume of 50 µl and 500–800 ng of total RNA were added. The reverse transcription reaction was carried out for 10 min at 25°C, 10 min at 60°C, and 15 min at 42°C, with random hexanucleotides as primers. The cDNA was denatured at 95°C for 1 min and amplified for 31 cycles of 30 sec at 95°C, 30 sec at 58°C, 1 min at 72°C. PCR-primers specific for exon 1 (upper primer R1, 5′-TAGCATGACCAAGCCGACCACAATC-3′) and exon 3 (lower primer R4, 5′-GTTCCCGTGTAAGCTTAGTCT CCTG-3′) of the rat H-FABP gene were used to amplify a 224 bp fragment of H-FABP mRNA. For the simultaneous amplification of 18 S internal standard RNA, a mix of natural and inactivated 18 S primers was added, yielding a 324 bp PCR product (18 S:competimer ratio 2:8 for myoblasts, 6:4 for cardiomyocytes; QuantumRNA 18s Internal Standard kit, Ambion, Austin, TX, USA).

Construction of the promoter-luciferase fusion plasmids

Different regions of the locust muscle FABP promoter were amplified from a genomic clone with forward primers (F), modified to contain a *Kpn*II restriction site, and reverse primers (R) containing a *Nhe*I restriction site (F–1135/–1115 5′-ATGGTACCTGCTAATAACTCCAATTGTC-3′; F–620/–600 5′-ATGGTACCCATA ATCACACTCATGTTA 3′; F–510/–490 5′-ATGGTACCTGACCATTGCAATAAGAT TT-3′; F–280/–260 5′-ATGGTACCACTGCGAAACAGATGAATGT-3′; F–130/–100 5′-ATGGTACCTAGCAGTCATGAAACAGAAT-3′; F–23/+5 5′-ATGGTACCGGCCG CCACCGGACGAGC-3′; R+52/+34 5′-AAGCTAGCGCTGCTGTGGTGGCGGT-3′; R+85/+63 5′-ATGCTAGCTACTTGATGCCTGCGAAT-3′; R–562/–544 5′-ATGCT AGCCTCAACAAGGAGATTCCG-3′). The PCR products were double digested with *Kpn*II and *Nhe*I, gel purified and cloned into the pGL3-basic luciferase vector (Promega, Madison, WI, USA). A 19-bp deletion mutant (Δ–180/–162) was constructed by amplifying the 5′ and 3′ flanking DNA, using PCR primers slightly modified to contain an *Afl*II restriction site (F–1135/–1115, R–280/–260 5′-ATGGTACCACTGCGAAACACAG ATGAATGT-3′; F–161/–143 5′-ATGGTACCACTGCGAAACAGATGAATG-T-3′, R+52/+34). Following digestion of the two PCR products with *Afl*II and ligation, the product was used as template for the PCR-based construction of deletion mutant vectors, as described above. All reporter constructs were sequenced to confirm the correct sequence and orientation. The various reporter gene constructs are depicted in Fig. 3.

Luciferase reporter gene assay

L6 were transiently transfected with the various reporter gene constructs, using LipofectAMINE 2000 reagent (Life Technologies, Burlington, ON, Canada) according to the manufacturer's instructions. For each dish, 5 µg luciferase reporter gene construct, 500 ng internal control vector pRL-TK (Promega, Madison, WI, USA) and 12 µl Lipofect reagent were gently mixed in 400 µl Opti-MEM and left at room temperature for 15 min. Subsequently, the transfection mixture was

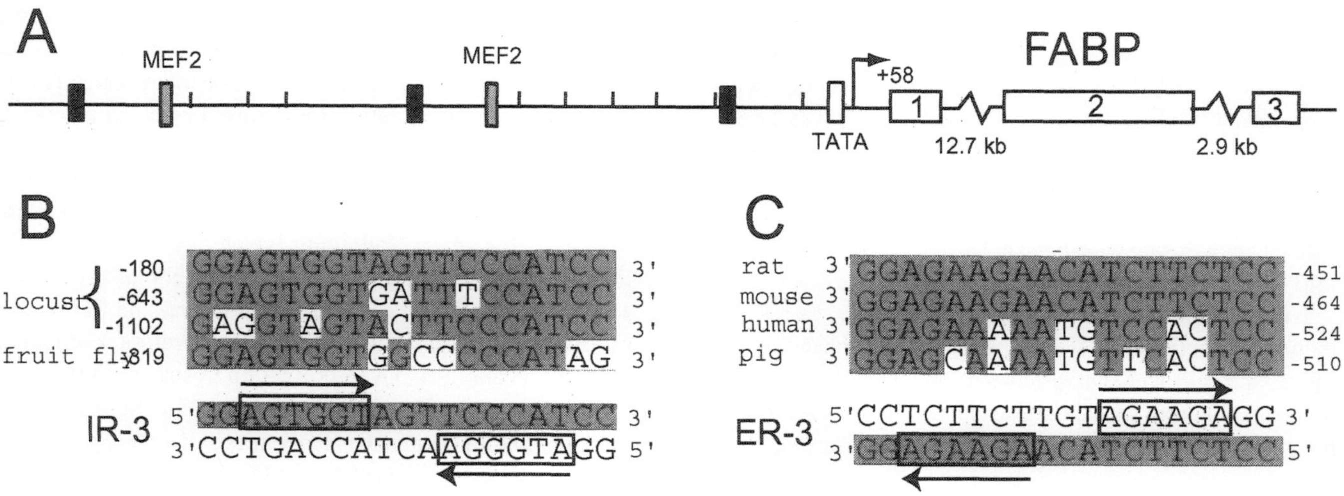

Fig. 1. The locust FABP gene and the conserved elements. (A) A schematic drawing of the muscle FABP gene from the desert locust, *Schistocerca gregaria*. The numbered open boxes depict the three introns of the gene. The arrow indicates the transcription start site. Shown is also the TATA box (–23 bp), 2 sequences with similarity to binding sites for MEF2, and the inverted repeat sequences (black boxes at –180, –643, –1102). The individual inverted repeat sequences, their analogue in the *Drosophila* FABP gene, and the consensus sequence with the two inward facing hexanucleotide half sites (IR-3) are shown in (B). (C) shows the elements found upstream of the cloned mammalian H-FABP genes, as well as the consensus sequence with two outward facing hexanucleotide half sites (ER-3).

added to the cells. After 6 h incubation at 37°C, 5% CO_2, the medium was changed back to DMEM. Following the 6 h transfection period, the fatty acid/BSA complex solution was added to cell culture media at final concentrations of 60 µM fatty acid. As control, the same volume of a 200 µM BSA solution was used. Incubations were carried out for 18 h. Cells were harvested and lysed with 200 µl passive lysis buffer (Promega, Madison, WI, USA), and cell debris was removed by centrifugation. The dual-luciferase reporter assay system (Promega, Madison, WI, USA) was employed to evaluate the relative luciferase activity, using a TD-20/20 luminometer (Turner Designs, Sunnyvale, CA, USA).

Preparation of nuclear extracts

Flight muscle was homogenized under liquid nitrogen, and the frozen muscle powder suspended in 4 volumes of homogenization buffer (10 mM Hepes-NaOH, pH 7.9, 0.35 M sucrose, 10 mM KCl, 1.5 mM $MgCl_2$, 0.1 mM EGTA, 0.5 mM dithiothreitol, 0.5 mM phenylmethylsulfonyl fluoride, 0.15 mM spermine, 0.5 mM spermidine, 2 µg/ml leupeptin and 2 µg/ml aprotinin; all from Sigma, Oakville, ON, Canada). The tissue was homogenized by 15–20 strokes in a Potter–Elvehjem glass homogenizer with a Teflon pestle. After addition of 0.1% Nonident P-40 the homogenate was filtered through two layers of cheesecloth and centrifuged for 10 min at 1300 × g. The pelleted nuclei were washed 3 times with homogenization buffer.

Nuclear extracts were prepared using the Nu-CLEAR extraction kit (Sigma, Oakville, ON, Canada). The nuclei were suspended in 3 volumes of 10 mM Hepes buffer (10 mM Hepes-NaOH, pH 7.9, 1.5 mM $MgCl_2$, 0.1 mM EGTA, 0.5 mM dithiothreitol, 5% glycerol, 1 mM phenylmethylsulfonyl fluoride, 2 µg/ml leupeptin and 2 µg/ml aprotinin). After addition of 400 mM NaCl, the suspension was stirred 30 min and centrifuged for 15 min at 13000 × g. The supernatant was dialyzed for 3 h against 100 volumes of 20 mM Hepes buffer (20 mM Hepes-NaOH, pH 7.9, 75 mM KCl, 0.1 mM EGTA, 0.5 mM dithiothreitol, 5% glycerol, 1 mM phenylmethylsulfonyl fluoride). The dialysate was then centrifuged at 13000 × g for 15 min, divided into aliquots and stored at –80°C.

Nuclei from cultured L6 cells were prepared in a similar manner. Culture dishes were rinsed with phosphate buffered saline, and cells were harvested. The pooled cells were centrifuged at 3000 rpm for 10 min, suspended in 3 volumes of homogenization buffer, and processed further as described above.

Electrophoretic mobility shift assay

DNA fragments of the locust FABP promoter for electrophoretic mobility-shift assays were amplified by PCR from the cloned reporter gene constructs –280/+53 (wild-type) and –280/+53 Δ–180/–162 (deletion mutant), with primers annealing at –280/–260 (5′-ACTGCGAAACAGATGAATGT-3′) and at –135/–154 (5′-GTCCAAACATCGAG TGTGA-3′). The PCR products (145 bp for wild-type, 126 bp for mutant) were end-labeled with (γ-^{32}P)-ATP (3000 Ci/mmol, NEN, Markham, ON, Canada) and T4 polynucleotide kinase. Nuclear extracts (~ 2 µg of protein) were pre-incubated in 20 µl of binding buffer (10 mM Tris-HCl, pH 7.5, 50 mM NaCl, 5 mM $MgCl_2$, 1 mM EDTA, 1 mM dithiothreitol, 50 µg/ml poly(dI-dC)/poly (dI-dC) and 5% glycerol) for 10 min at 24°C. The labelled probe (~ 15 ng) was added and incubated at 24°C for 30 min. Subsequently, the reaction mixture was loaded onto non-denaturing polyacrylamide gels (5% T, 3.3% C) and electrophoresed at 20 V/cm for 30 min in 0.5 × TBE (45 mM Tris, 45 mM boric acid, 1 mM EDTA). The gels were dried and analyzed by autoradiography. For the competition experiments, the pre-incubation was performed in the presence of unlabelled competitor DNA at the molar excess indicated in the figure legends. Rat H-FABP (gift from Dr. F. Spener, University of Münster) was added in one experiment, as indicated. Gelshifts with the rat H-FABP element were carried out in the same manner, except that the double stranded synthetic oligonucleotide (5′ CCTCTTCTGTCAGAAGAG-G 3′, –510 → –529) was used as probe.

Results and discussion

The locust muscle FABP gene

The gene coding for locust muscle FABP was amplified by PCR and cloned, together with 1.2 kb of upstream sequence [15]. The sequence coding for the 607 bp cDNA is interrupted by two introns of 12.7 and 2.9 kb, inserted in analogous positions as the first and third intron of the mammalian homologues: one intron is found following the sequence coding for glycine 26, which is located in the turn of the helix-turn helix motif that shields the binding cavity, the other intron before threonine 117. An additional intron found in all mammalian FABP genes (following lysine 84), is absent in the locust FABP gene (Fig. 1). Both introns contain repetitive sequences also found in other locust genes, and the second intron contains a GT-microsatellite. The promoter sequence includes a canonical TATA box 24 bp upstream of the transcription start site. The upstream sequence contains various potential myocyte enhancer sequences and a 160 bp segment that is repeated 3 times. This finding raised the possibility that the repeats contain regulatory elements which could act more efficiently in multiple copies. A comprehensive sequence analysis did not reveal any known transcription factor binding sites within these repeats. Noteworthy, however, is the presence of a 19 bp inverted repeat sequence, 5’-GGAGTGGTA N TTCCCATCC-3’. A similar, partially palindromic sequence is also found upstream of the putative *Drosophila* FABP promoter (Fig. 1b). While this sequence is not recognized as a transcription factor binding site, it contains inverted repeats of two hexanucleotide half-sites: 5′ gg AGTGGT nnn TCCCAT cc 3′. This sequence is similar to the response elements for steroid hormone receptors, which generally bind to inverted repeat element separated by 3 nucleotides (IR-3) with the half-site sequences 5′ AGAACA 3′ or 5′ AGGTCA 3′.

Interestingly, a strikingly similar, but reversed palindromic sequence is found within 600 bp upstream of the promoter of all mammalian heart FABP genes (Fig. 1c), however with consensus half-sites facing outward (everted repeat, ER-3).

The fact that this conserved 19 bp sequence is found within the promoter region of all heart and muscle FABP genes cloned so far, which otherwise does not show obvious areas of sequence conservation, suggested its involvement in the regulation of the FABP gene. Given its similarity to previously characterized response elements, it appeared possible that this sequence is involved in the induction of gene expression by fatty acids. In order to investigate this, a cell system was needed that is suitable to measure changes in the expression of reporter gene constructs of the FABP promoter. Suitable cells should be easily transfected with reporter gene constructs, and respond to fatty acid treatment in a similar manner as differentiated muscle cells. Since muscle cells are terminally differentiated and hence not suitable for protocols involving transfection with reporter gene constructs and lengthy fatty acid incubation periods, we decided to investigate whether undifferentiated myoblasts can be used for these studies.

Rat myoblasts are suitable for expression studies

In order to compare the expression of the H-FABP gene in undifferentiated and differentiated myocytes, H-FABP mRNA was measured by reverse-transcription PCR, yielding PCR products of 224 bp. To account for variations in template amount, 18 S RNA sequence was used as internal standard. A mixture of regular and blocked primers (competimers) was necessary to achieve similar intensities between the gene specific PCR products and the 324 bp 18S PCR product. H-FABP mRNA can be found in myoblasts and adult cardiomyoctes, albeit at different levels (Fig. 2). Quantitative comparison with the PCR products from myoblasts indicate far higher mRNA levels in cardiomyocytes. Because of the relative abundance of H-FABP mRNA in the heart, 18S RNA could only be seen when the concentration of the unblocked ribosomal RNA primers was increased. Thus, we established that the H-FABP is expressed in myoblasts, but at a lower level then in cardiomyocytes. To study the induction of H-FABP expression by fatty acids, cultured myoblasts and isolated cardiomyocytes were treated with linoleic acid. H-FABP

Table 1. Induction of H-FABP expression by long chain fatty acids

Fatty acid	mRNA levels (× control)
Myoblasts	
16:0	3.0 ± 0.7
18:1	1.9 ± 0.2
18:2	2.0 ± 0.5
18:3	2.7 ± 0.5
20:4	2.2 ± 0.8
Cardiomyocytes	
18:2	2.4 ± 0.3

Values are the mean of 3–6 independent determinations ± S.D.

mRNA increased in myoblasts and cardiomyocytes following 10 h of incubation. Quantitative analysis revealed a similar increase in myoblasts and cardiomyocytes. Saturated and unsaturated long chain-fatty acids all stimulated H-FABP gene expression to similar degrees (Table 1). Induction of gene expression was visible already after 30 min incubation in both cell types, when the primary transcript was quantified in an analogous manner [17]. These findings show that fatty acid treatment affects gene expression similarly in L6 cells and fully differentiated myocytes; thus, myoblasts can be useful to study the activity of the FABP promoter *in vitro*.

Analysis of the locust FABP promoter

We used a dual luciferase reporter gene assay for the sensitive detection of the expression under the control of the insect promoter [18]. Transfection of rat myoblasts with a reporter gene construct containing the full locust FABP promoter resulted in strong expression of luciferase. The enzyme activity detected 18 h post transfection was more than 12-fold higher than found for a promoter-less luciferase vector, and nearly half as strong as for a control vector containing the strong, universal SV 40 promoter, demonstrating that the locust muscle FABP promoter is effective in a heterologous system. Reporter gene expression was measured with various constructs of decreasing length, gradually eliminating potential regulatory elements such as MEF2 and an inverted repeat that is present 3 times within the first thousand basepairs upstream of the TATA box. Similar levels of luciferase activity were measured for all constructs that contained at least 280 bp of upstream sequence (Fig. 3). A shorter construct that started just downstream of the inverted repeat sequence also expressed luciferase, however much weaker (< 50%) than the longer constructs. Constructs that did not contain the TATA box (–1153/–562, –23/+52), or expressed luciferase out of frame (–1135/+85, luciferase cDNA inserted 29 bp downstream of the FABP start codon) led to drastically reduced levels of luciferase activity. To further determine the impact

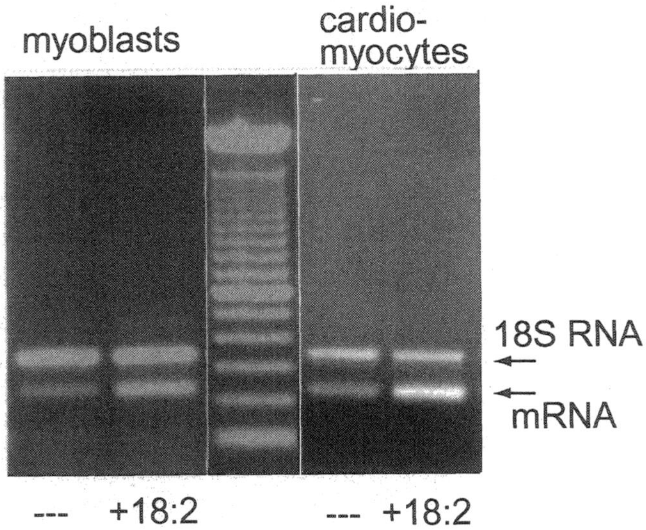

Fig. 2. RT-PCR of rat H-FABP mRNA. Cells were incubated for 10 h with culture media containing 20 μM BSA or 60 μM linoleic acid complexed to 20 μM BSA, as described in Materials and methods. Total RNA (500 ng) was used as template for multiplex PCR, with primers specific for 18 S RNA and H-FABP mRNA. The specific primer: competimer ratio was 2:8 for myotubes, and 4:6 for cardiomyocytes.

of the inverted repeat element, a deletion mutant was constructed that contained the entire promoter (–280/+52), but omitted the 19-bp repeat (Δ–180/–162). The deletion mutant promoter continued to strongly stimulate reporter gene expression; luciferase activity was approximately one third lower than for the wild-type promoter, but remained still higher than for the minimal active promoter construct (–130/+52). When transformed cells were treated with fatty acids, expression rates were unaltered for control constructs, but markedly increased for constructs containing the FABP promoter with the inverted repeat elements. The increase was highest (more than 2-fold) for the –280/+52 construct that contained one copy of the inverted repeat element, and somewhat lower (~ 1.5-fold) for longer constructs. Treatment with saturated and mono-unsaturated fatty acids also led to a clear increase in luciferase expression (~ 1.5-fold, [18]). However, neither the minimal promoter (–130/+52) nor the construct in which the inverted repeat had been deleted (Δ–180/–162) responded to fatty acid treatment, indicating that the inverted repeat is indeed involved in the fatty acid response.

Characterization of the locust fatty acid response element

To see whether the inverted repeat element does bind a nuclear transcription factor, electrophoretic mobility shift assays were carried out. A 145-bp fragment of the upstream region containing the repeat element (–280/–135) was shifted to lower mobility in the presence of nuclear proteins isolated from rat myoblasts (Fig. 4). The shift, however, did not occur with an analogous fragment without the 19-bp sequence (–280/–135 Δ-180/–162). Increasing amounts of the double-stranded

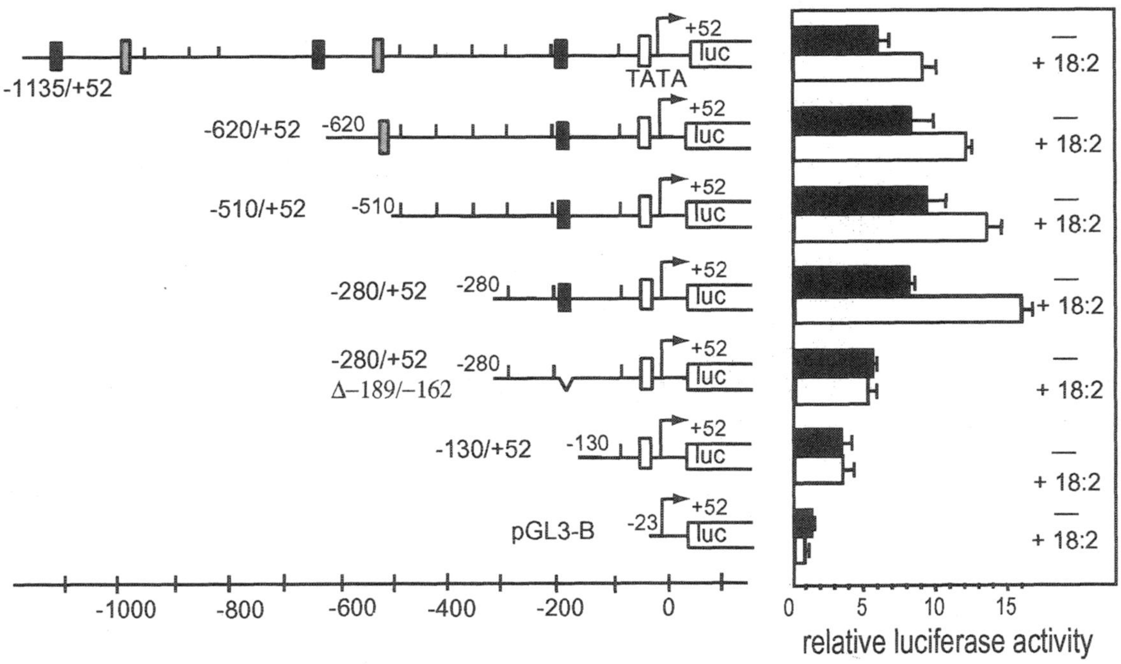

Fig. 3. Luciferase expresssion from the locust muscle FABP promoter. Myoblasts were transfected with reporter gene constructs, as indicated. The transfected cells were incubated for 18 h with BSA alone (black bars), or with 60 μM linoleic acid/BSA (white bars) as described in Experimental procedures. Each result represents the average of 3–6 independent measurements ± S.D.

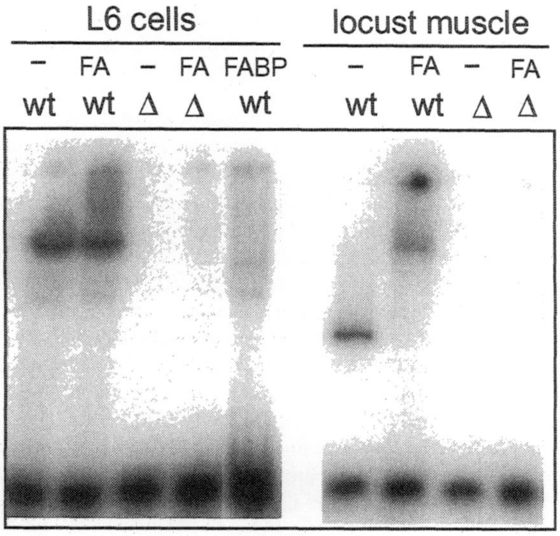

Fig. 4. Binding of nuclear proteins to the locust inverted repeat element. Electrophoretic mobility shift assays were carried out with nuclear extracts from L6 myoblasts and locust flight muscle. Each extract was treated with a radiolabeled 145-bp probe (representing the sequence –280/–135) that included the inverted repeat element (native probe) or an identical 126-bp probe where the element had been deleted (Δ–180/–162 probe), as shown above the lanes. BSA (–, final concentration 0.1 μM), fatty acid (FA, final concentration 0.3 μM linoleic acid/0.1 μM BSA solution (FA), or FABP (FABP, ~ 1 μg) was added to the nuclear extracts, as indicated above the lanes.

19-bp oligonucleotide competed with the labeled nucleotide, and gradually eliminated the gel-shift band. An excess of the unlabeled full fragment used in the gel-shift experiments also eliminated the band, but an otherwise identical fragment that lacked the element did not reduce binding. Nuclear proteins from locust flight muscle also interacted with the nucleotide fragments. A strong gel-shift band was seen when the labeled 145-bp probe was mixed with locust nuclear proteins; the band, however, appeared at a different location. Similar as in the rat extract, the binding was eliminated by an excess of the unlabeled 19-bp element, but not affected by the fragment devoid of the element, clearly indicating specific interactions. The addition of fatty acids to the nuclear extracts prior to gel-shift analysis led to a shift of the band seen for locust proteins. After treatment with 0.3 μM linoleic acid complexed to BSA, the labeled band moved to the same location as seen with rat extract, possibly indicating a change in protein binding induced by fatty acids. BSA alone, without fatty acids, did not alter the gel-shift pattern. Fatty acids also did not induce any binding to the deletion mutant nucleotide. The addition of FABP to the gel shift mixture, however, eliminated either band, indicating that fatty acid is essential for the binding of nuclear proteins to the fatty acid response element (Fig. 4).

Model for the regulation of the locust muscle FABP gene

From the results presented above, it is likely that two different factors bind to the fatty acid response element. At least one of these, but possibly both are activated by the binding of long chain fatty acids. This behavior is similar to the other known fatty acid response elements, which bind heterodimers of the transcription factors PPAR and RXR. However, in contrast to PPREs which are direct repeat elements (DR-1), the fatty acid response element in locust flight muscle is an inverted repeat (IR-3), similar to steroid hormone response elements. Fatty acids are required for full activity, and their availability to nuclear proteins depends on the intracellular concentration of FABP. FABP appears to have free access to the nuclear lumen [19], and thus this protein's contribution must be considered. An analogous participation of L-FABP in the PPAR mediated control of expression of the L-FABP gene has recently been shown by Wolfrum *et al.* [20].

Evidence for a fatty acid response element in the rat H-FABP promoter

The presence of a related sequence within 600 bp of upstream sequence of all mammalian heart FABP genes suggests an analogous function. Indeed, preliminary data confirm that fatty acids stimulate the expression of a reporter gene under the control of the rat promoter that contains the elements,

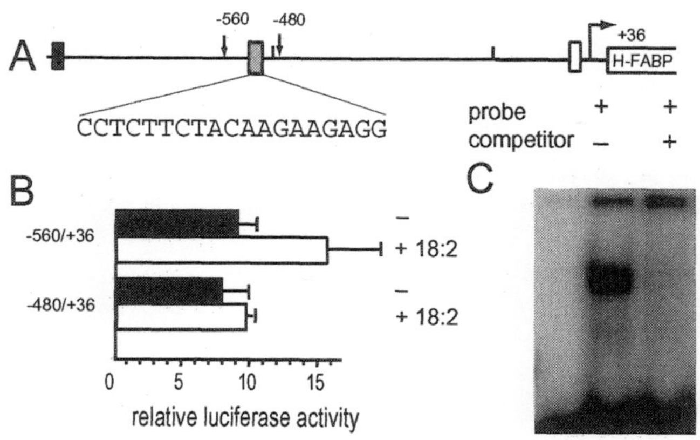

Fig. 5. Characterization of the rat H-FABP element. The promoter region of the rat H-FABP gene is shown in (A). White box: TATA box; black box: PPRE-like element; grey box: 19 bp everted repeat element. The 5′ end of the two reporter gene constructs are marked with an arrow. (B) Myoblasts were transfected with the luciferase reporter gene constructs, as described in Materials and methods. Luciferase activity was measured after incubation for 18 h with BSA (black bars) or the BSA/fatty acid complex (open bars). Each value is the mean of 3 determinations ± S.D. (C) The double-stranded 19 bp everted repeat element was radiolabeled and incubated with nuclear extracts from L6 cells, in the absence and presence of a 10-fold molar excess of the unlabelled 19-bp element, as indicated above the lanes.

while the response is lost in shorter constructs (Fig. 5b). Moreover, in gel shift assays the mammalian 19 bp element binds strongly to proteins from an L6 nuclear extract (Fig. 5c). However, it remains to be seen whether this element serves any regulatory function *in vivo*.

Acknowledgements

This research was supported by grants from the Natural Science and Engineering Research Council of Canada and the Heart and Stroke Foundation of BC and Yukon. We thank Ian Dawe for cloning two of the reporter gene constructs.

References

1. Vogel Hertzel A, Bernlohr DA: The mammalian fatty acid-binding protein multigene family: Molecular and genetic insights into function. Trends Endocrinol Metab 11: 175–180, 2000
2. Veerkamp JH, van Moerkerk HTB: Fatty acid binding protein and its relation to fatty acid oxidation. Mol Cell Biochem 123: 101–106, 1993
3. van Breda E, Keizer HA, Vork MM, Surtel DAM, de Jong YF, van der Vusse GJK, Glatz JFC: Modulation of fatty-acid-binding protein content of rat heart and skeletal muscle by endurance training and testosterone treatment. Pflügers Arch 421: 274–279, 1992
4. Guglielmo CG, Haunerland NH, Williams TD: Fatty acid binding protein, a major protein in the flight muscle of migrating Western Sandpipers. Comp Biochem Physiol 119B: 549–555, 1998
5. Chen X, Haunerland NH : Fatty acid binding protein expression in locust flight muscle. Induction by flight, adipokinetic hormone, and low density lipophorin. Insect Biochem 24: 573–579, 1994
6. Glatz JF, van Breda E, Keizer HA, de Jong YF, Lakey JR, Rajotte RV, Thompson A, van der Vusse GJ, Lopaschuk GD: Rat heart fatty acid-binding protein content is increased in experimental diabetes. Biochem Biophys Res Commun 199: 639–646, 1994
7. Van Bilsen M, de Vries JE, van der Vusse GJ : Long-term effects of fatty acids on cell viability and gene expression of neonatal cardiac myocytes. Prostaglandins Leukotrienes Essent Fatty Acids 57: 39–45, 1997
8. van der Lee KA, Vork MM, De Vries JE, Willemsen PH, Glatz JF, Reneman RS, van der Vusse GJ, van Bilsen M: Long-chain fatty acid-induced changes in gene expression in neonatal cardiac myocytes. J Lipid Res 41: 41–47, 2000
9. van Bilsen M, van der Vusse GJ, Reneman RS: Transcriptional regulation of metabolic processes: Implications for cardiac metabolism. Pflügers Archs 437: 2–14, 1998
10. Mangelsdorf DJ, Thummel C, Beato M, Herrlich P, Schutz G, Umesono K, Blumberg B, Kastner P, Mark M, Chambon P *et al.*: The nuclear receptor superfamily: The second decade. Cell 83: 835–839, 1995
11. Wolfrum C, Ellinghaus P, Fobker M, Seedorf U, Assmann G, Börchers T, Spener F: Phytanic acid is ligand and transcriptional activator of murine liver fatty acid binding protein. J Lipid Res 40: 708–714, 1999
12. Frohnert BI, Hui TY, Bernlohr DA: Identification of a functional peroxisome proliferator-responsive element in the murine fatty acid transport protein gene. J Biol Chem 274: 3970–3977, 1999
13. Braissant O, Foufelle F, Scotto C, Dauca M, Wahli W: Differential expression of peroxisome proliferator-activated receptors (PPARs): Tissue distribution of PPAR-alpha, -beta, and -gamma in the adult rat. Endocrinology 137: 354–366, 1996
14. Zhang J, Rickers-Haunerland J, Dawe I, Haunerland, NH: Structure and chromosomal location of the rat gene encoding the heart fatty acid-binding protein. Eur J Biochem 266: 347–351, 1999
15. Wu Q, Andolfatto P, Haunerland NH : Cloning and sequence of the gene encoding the muscle fatty acid binding protein from the desert locust, *Schistocerca gregaria*. Insect Biochem Mol Biol 31: 553–562, 2001
16. Rodriguez B, Severson D: Preparation of cardiomyocytes. In: J.H. McNeill (ed). Biochemical Techniques in the Heart. CRC Press, Boca Raton, FL, 1997, pp 101–115
17. Chang W, Rickers-Haunerland J, Haunerland NH: Induction of cardiac FABP gene expression by long chain fatty acids in cultured rat muscle cells. Mol Cell Biochem 221: 127–132, 2001
18. Wu Q, Haunerland NH: A novel fatty acid response element controls the expression of the flight muscle FABP gene of the desert locust, *Schistocerca gregaria*. Eur J Biochem 268: 5894–5900, 2001
19. Haunerland NH, Andolfatto P, Chisholm JM, Wang Z, Chen X: Fatty acid binding protein in locust flight muscle. Developmental changes of expression, concentration, and intracellular distribution. Eur J Biochem 210: 1045–1051, 1992
20. Wolfrum C, Borrmann CM, Börchers T, Spener F: Fatty acids and hypolipidemic drugs regulate peroxisome proliferator-activated receptors alpha – and gamma-mediated gene expression via liver fatty acid binding protein: A signaling path to the nucleus. Proc Natl Acad Sci USA 98: 2323–2328, 2001

Molecular and Cellular Biochemistry **239:** 181–192, 2002.

Regulation of fatty acid transport and membrane transporters in health and disease

Arend Bonen,[1] Joost J.F.P. Luiken[2] and Jan F.C. Glatz[2]

[1]*Department of Kinesiology, University of Waterloo, Waterloo, ON, Canada;* [2]*Department of Physiology, Maastricht University, Maastricht, The Netherlands*

Abstract

Long chain fatty acid uptake across the plasma membrane occurs, in part, via a protein-mediated process involving a number of fatty acid binding proteins known as fatty acid transporters. A critical step in furthering the understandings of fatty acid transport was the discovery that giant vesicles, prepared from tissues such as muscle and heart, provided a suitable system for measuring fatty acid uptake. These vesicles are large (10–15 µm diameter), are oriented fully right side out, and contain cytosolic FABP in the lumen, which acts as a fatty acid sink, while none of the fatty acid taken up is metabolized or associated with the plasma membrane. The key fatty acid transporters FAT/CD36 and FABPpm are expressed in muscle and heart and their plasma membrane content is positively correlated with rates of fatty acid transport. These transporters are regulated acutely (within minutes) and chronically (days). For instance, both muscle contraction and insulin can translocate FAT/CD36 from an intracellular pool to the plasma membrane, thereby increasing fatty acid transport. With obesity, fatty acid transport is increased along with a concomitant increase in plasmalemmal FAT/CD36 (heart, muscle) and FABPpm (heart only), but without change in the expression of these transporters. This latter observation suggests that some of the fatty acid transporters are permanently relocated to the plasma membrane. In other studies it also appears that fatty acid transport rates are altered in a reciprocal manner to glucose transport. Since disorders in lipid metabolism appear to be an important factor contributing to the etiology of a number of common human diseases such as diabetes and obesity, our evidence that protein-mediated fatty acid transport is a key step in lipid metabolism allows the speculation that malfunctioning of the fatty acid transport process could be a common critical factor in the pathogenesis of these diseases. (Mol Cell Biochem **239**: 181–192, 2002)

Key words: fatty acid-binding protein, FAT/CD36, giant vesicles, skeletal muscle, insulin resistance

Introduction

In the past decade considerable evidence has accumulated that long chain fatty acids enter cells via a protein mediated mechanism involving a number of fatty acid binding proteins (cf. [1, 2]). For convenience of terminology, this system has been termed a fatty acid transport system, although it is unlikely that the process of moving fatty acids across the plasma membrane involves pore formation as occurs with glucose transporters. Moreover, the presence of a protein-mediated mechanism does not negate the fact that a proportion of fatty acids also diffuses across the plasma membrane [3].

Recent work from our laboratories has focussed on the examination of the regulation of fatty acid transport in heart and skeletal muscle, which are metabolically important tissues. The heart derives over 70% of its energy requirement from the oxidation of fatty acids. Skeletal muscle, because of its mass (40% of body weight) and variable metabolic rate, is also a key tissue involved in regulating fatty acid homeostasis, although its role is often overlooked. The focus of this review, based on our work, is to (a) describe the system that we have developed for assessing fatty acid transport in heart and muscle, (b) to compare fatty acid transport rates in heart and muscle, and relate this to the presence of fatty acid transporters, (c) to examine how fatty acid transport is regulated by fatty acid transporters, both acutely (within minutes) and chronically (days) and (d) to examine how fatty acid transport is upregulated in obesity and

Address for offprints: A. Bonen, Department of Kinesiology, University of Waterloo, Ontario, N2L 3G1, Canada (E-mail: abonen@healthy.uwaterloo.ca)

diabetes. These studies have begun to demonstrate that the transport of fatty acids and its regulation by plasma membrane fatty acid transporters is complex and resembles in some ways the regulation of glucose transport by the glucose transporter GLUT4.

Assessing fatty acid transport in heart and muscle

Determining fatty acid transport

In a number of studies fatty acid transport had been assessed in respiring preparations (hepatocytes, adipocytes, cardiac myocytes) [4–6]. These preparations can provide an estimate of fatty acid uptake across the plasma membrane, providing that determinations are made over a short period of time, so as to minimize fatty acid metabolism (esterification and oxidation). In such studies it has been necessary to adjust the fatty acid uptake rates for the confounding effects of fatty acid metabolism, but it has not been possible to determine the adequacy of such corrections. However, ideally, transport studies should be performed in the absence of any metabolism of the substrate, so as to not confound the rate of transport across the plasma membrane. There are two ways in which to effectively eliminate the confounding effects of substrate metabolism: (a) by using a non-metabolizable analog or (b) by using membrane preparations devoid of organelles and enzymes which facilitate the activation, oxidation and/or storage of the substrate. In the area of glucose transport many investigators have used one of two glucose analogs to examine glucose transport in respiring biologic tissues such as the heart, muscle and adipocytes. These analogs are 3-O-methyl glucose which is not metabolized in the cell and 2-deoxy-D-glucose, which is not metabolized beyond glucose-6-phosphate, although the possible rate limiting effects of hexokinase remain a concern with this analog. Alternatively, glucose transport has been determined in small vesicle preparations made from biologic tissues such as skeletal muscle [7] in which the capacity to metabolize glucose is absent.

Fatty acid analogs

Iodinated fatty acid analogs such as β-methyliodophenyl-pentadecanoic acid (BMIPP) and iodo-heptadecanoic acid (IHDA) have been used as an index of fatty acid transport [8, 9]. This may not be appropriate, since BMIPP is metabolized once it has been taken up into the cell [10–12]. In the heart, and presumably other tissues, the uptake of the radioiodinated 3(R)-isomer of BMIPP is taken up more rapidly than the 3(S)-isomer of BMIPP, and the it was suggested that this may be indicative of relative differences in the membrane transport of the two isomers [11]. Direct comparisons of the rates of vesicular transport of fatty acid analogs with naturally occurring fatty acids are not available. Such experiments are necessary if BMIPP is to be used as an index of fatty acid transport rates *in vivo*.

Small vesicles

Some preparations, derived from fractionated tissue in which the plasma membranes spontaneously form small (< 1 μm), resealed vesicles, are not appropriate for the study of fatty acid transport, and possibly other substrates. First, approximately 40% of these small vesicles are oriented inside out [13], which may be problematical depending how the transport proteins are organized at the plasma membrane. Second, the preparatory steps in small vesicle generation involve disruption of the cell membrane, which will predictably result in the almost complete loss of the 15 kDa cytosolic fatty acid binding protein (FABPc). This latter protein binds fatty acids in the cytosol after they have crossed the plasma membrane. Thus, the loss of FABPc from the interior of the small vesicles make this preparation virtually unsuitable for fatty acid transport studies. However, another vesicle preparation, known as giant vesicles, had been used in our laboratory [14] and others [15–17] to examine glucose and lactate transport, and this preparation seemed to offer a potential model system for examining fatty acid transport.

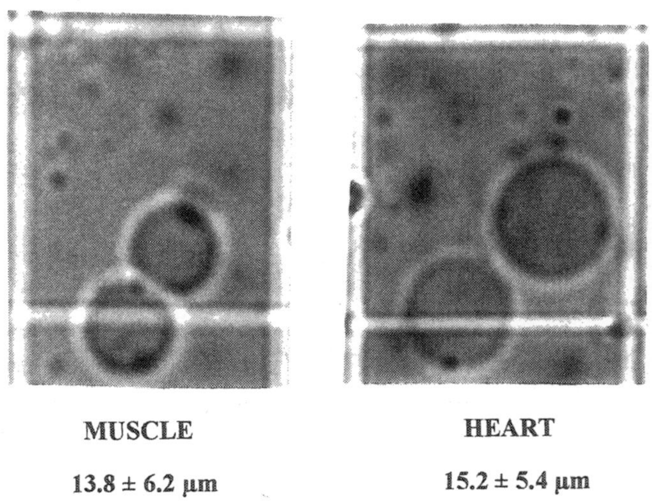

Fig. 1. Giant sarcolemmal vesicles derived from mixed skeletal muscles and heart. Diameters of vesicles (μm) are not different in heart and muscle.

Giant vesicles

When we began to examine fatty acid transport a key consideration was to have available a vesicle preparation that could be obtained from biologically important tissues. But, such a preparation also had to satisfy a number of other key criteria. These were as follows: (1) fatty acid metabolism must be absent, (2) the vesicles must be oriented right side out, (3) the vesicles must contain significant quantities of FABPc to act as a sink for the fatty acids that have crossed the plasma membrane, and (4) if possible, the vesicles should also be obtainable from metabolically important tissues such as heart, muscle, liver and adipose tissue, so as to be able to compare the responses of different tissues to metabolic perturbations that occur during disease states such as obesity and diabetes.

It seemed that giant vesicles obtained from skeletal muscle [14–17] might satisfy some of the foregoing criteria. It was already known that the giant sarcolemmal vesicles were oriented right side out, as these had been used previously in patch clamp studies (see [18]). Therefore, we began using the giant vesicles to examine whether these were suitable for fatty acid transport.

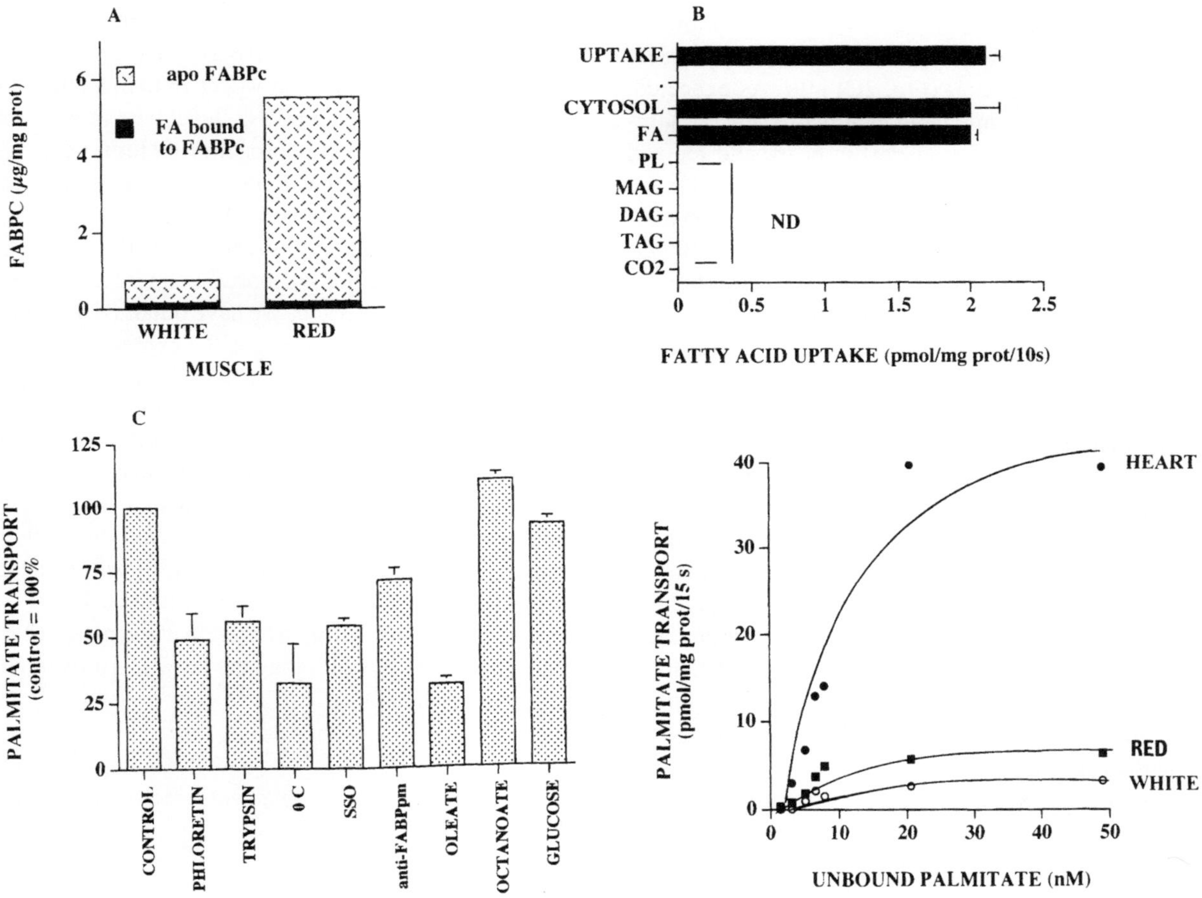

Fig. 2. Characterization of fatty acid uptake by giant vesicles derived from skeletal muscle. (A) Content of FABPc (μg/mg protein) in white and red muscle giant vesicles. The black portion of each bar graph reflects the maximal fatty acid concentrations used in the transport assays that could be bound to FABPc, the stippled section of each bar graph reflects the apo FABPc. These data indicate that there is a great excess of FABPc (31 and 5 fold excess in red and white muscles respectively) which acts as a fatty acid sink in the giant vesicles. This excess means it is not limiting to fatty acid uptake (data redrawn from [19]). (B) Fatty acid taken up was fully recovered in the lumen of the vesicles (cytosol) and was present as fatty acids (FA) as determined by thin layer chromatography. This indicated that fatty acids were not associated with the plasma membrane. Also, fatty acids were not metabolized, as none of the fatty acid was present as phospholipids (PL), mono-(MAG), di-(DAG), and triacylglycerols (TAG), oxidation (CO_2 production) was not occurring (data redrawn from [19]). (C) Protein mediated mechanism of fatty acid uptake is demonstrated in these experiments protein modifying agents (phloretin, trypsin), assay temperature reduction, and inhibitors of fatty acid transporters (anti-FABPpm and SSO) reduced fatty acid transport. Competitive inhibition was demonstrated by the displacement of palmitate uptake by excess oleate. Specificity of the transport system was shown by the lack of effect of octanoate and glucose on fatty acid transport (data redrawn from [19]). (D) Kinetics of palmitate transport in giant vesicles obtained from heart, and red and white muscles. V_{max} in heart is approximately 5–6 fold greater than in red muscle, and V_{max} in red muscle is approximately 2-fold greater than in white muscle. In all tissues $K_{1/2}$ was similar (~ 6–9 nM) (data redrawn from [19, 21]).

Characterization of fatty acid transport by giant vesicles

Fortuitously, the giant vesicles met all of the criteria that we required (to use them as a model system). The giant sarcolemmal vesicles contained excess FABPc, so that its concentration could not be seen as a rate limiting step in fatty acid transport (Fig. 2A). There was no evidence of any fatty acid metabolism, either oxidation or esterification, and all of the fatty acid taken up into the giant vesicle was recovered as fatty acids in the lumen of the vesicle, none was associated with the plasma membrane (Fig. 2B). We also demonstrated that vesicular palmitate transport was inhibitable by protein modifying agents (phloretin, trypsin), by inhibitors of fatty acid transport proteins (FABPpm antisera, sulfo-*N*-succinimidyl oleate (SSO)), by reduced assay temperatures, and by excess long chain fatty acid (oleate) (Fig. 2C) [19]. As expected, palmitate transport was not altered by glucose (Fig. 2C). Finally, the rates of fatty acid transport in heart, and red and white skeletal muscles (heart >>red>white) reflect the known differences in the rates of fatty acid metabolism in these tissues (Fig. 2D). Collectively, these studies indicated that giant sarcolemmal vesicles provided a suitable preparation for the study of fatty acid uptake, and that in skeletal muscle, fatty acids are taken across the plasma membrane largely via a protein mediated mechanism.

Subsequently, we have succeeded, through slight adaptations of the method used for muscle, to also prepare giant vesicles from heart, liver and adipose tissue [20]. Studies with giant vesicles obtained from the heart, also revealed a fatty acid transport system with similar characteristics to those observed in skeletal muscle [21]. Thus, we have successfully developed and characterized the giant vesicles as a suitable preparation for the examination and regulation of fatty acid transport. Due to the absence of any fatty acid metabolism in the giant vesicles, this preparation permits us to use naturally occurring fatty acids, thereby avoiding the potential problems of fatty acid analogs discussed above. A major advantage of these giant vesicles is that they can be obtained from a number of metabolically important tissues, thereby facilitating the study of fatty acid transport and its regulation in health and disease in these different tissues simultaneously.

Fatty acid transport and fatty acid transporters

Identification of fatty acid transporters

A number of fatty acid binding proteins, known as fatty acid transporters, have been identified. Among these are the 40 kDa peripheral membrane fatty acid binding protein (FABPpm) [22], the 63 kDa fatty acid transport proteins, FATP1-5 [23, 24], and the 53 kDa native, non-glycosylated fatty acid translocase (FAT/CD36, the rat analog of the human CD36), which, when fully glycosylated, is typically observed at 88 kDa [25, 26]. For each of these transporters there is evidence that they can independently increase fatty acid transport when overexpressed in a cell line [22, 24, 26]. However, their precise functioning in the transmembrane process and their exact roles in biological tissues are not yet fully understood.

Linking fatty acid transport to fatty acid transporters

Clarifying the roles of fatty acid transporters has not been simple. Some studies that have attempted to either infer changes in fatty acid transport based on measurement of fatty acid transporter mRNAs only [27], or to link fatty acid transporter expression at the mRNA level with rates of fatty acid transport [5, 28]. These approaches are limiting, since the inherent assumption is that there is a direct correlation between the abundance of the fatty acid transporter mRNA, its protein product, including post-translational modification, and the rates of fatty acid transport. Such direct linkages do not necessarily have to occur. We have obtained definitive proof that under some circumstances the expression of one of the fatty acid transporters, FAT/CD36, is regulated at the post-transcriptional level, indicating that there is no direct relationship between the fatty acid transporter transcript and its protein product [20]. As outlined in detail below, we also have evidence that there is not necessarily a strict relationship between rates of fatty acid transport and the total available pool of fatty acid transport proteins [20], since such proteins can exist in discrete locations within the cell. It is only the fatty acid transport proteins present at the plasma membrane that participate in facilitating the fatty acid movement across the plasma membrane.

Fatty acid transport and plasma membrane fatty acid transporters

In our studies we have compared the rates of fatty acid transport with the content of fatty acid transporters at the plasma membrane. Among muscle tissues, the maximal fatty acid transport rates, as assessed in giant vesicles derived from these tissues, differed substantially (heart >>red>white muscle) (Fig. 3). Importantly, these maximal rates of transport in the three muscle types paralleled the well known differences in fatty acid metabolism among these tissues. There was also a good relationship between either plasma membrane FAT/CD36 or plasma membrane FABPpm and fatty acid transport (Fig. 3). Interestingly, there was an inverse relationship between FATP1 and fatty acid transport (Fig. 3). However, we cannot exclude the possibility that these observations for

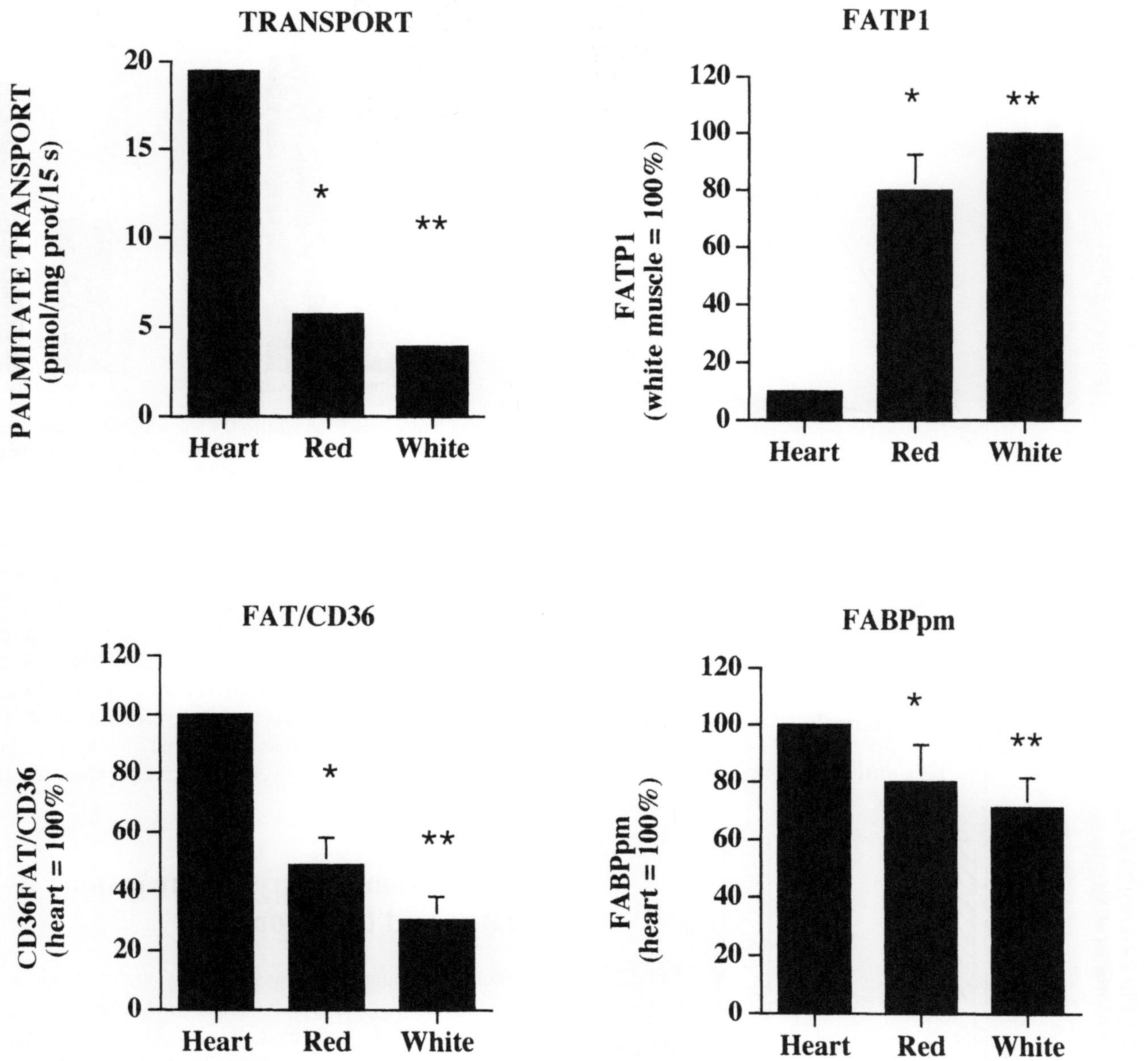

Fig. 3. Comparison of palmitate transport in vesicles from heart and red and white muscles with the plasma membrane content of fatty acid transporters, FAT/CD36, FATP1, and FABPpm (data redrawn from [21]). *$p < 0.05$ Red vs. heart; **$p < 0.05$ White vs. red.

FATP1 may have been due in part to the problems encountered, in that study the FATP1 antibody, which did not work well in biologic tissues, although it was adequate in some cell lines (Bonen, unpublished data). Others have questioned the transport function of FATP1, believing it to be a very long chain fatty acyl-CoA synthetase [29, 30]. Based on our studies, it seems that in heart and muscle, both FAT/CD36 and FABPpm are physiologically important transporters in moving fatty acids across the plasma membrane, with FATP1 perhaps playing a permissive role. Therefore, in subsequent studies we have focussed on FAT/CD36 and FABPpm.

Possible interaction of FABPpm and FAT/CD36 to transport fatty acids

It has been proposed previously that FAT/CD36 and FABPpm may collaborate to transport fatty acids across the plasma membrane [31, 32]. Some indirect evidence for this comes from studies in which we were able to inhibit vesicular fatty acid uptake in heart and muscle by the FAT/CD36 inhibitor SSO and by antisera to FABPpm (Fig. 4) [21, 33]. Notably, when SSO and the FABPpm antisera were combined, in concentrations that would each elicit only partial inhibition, we

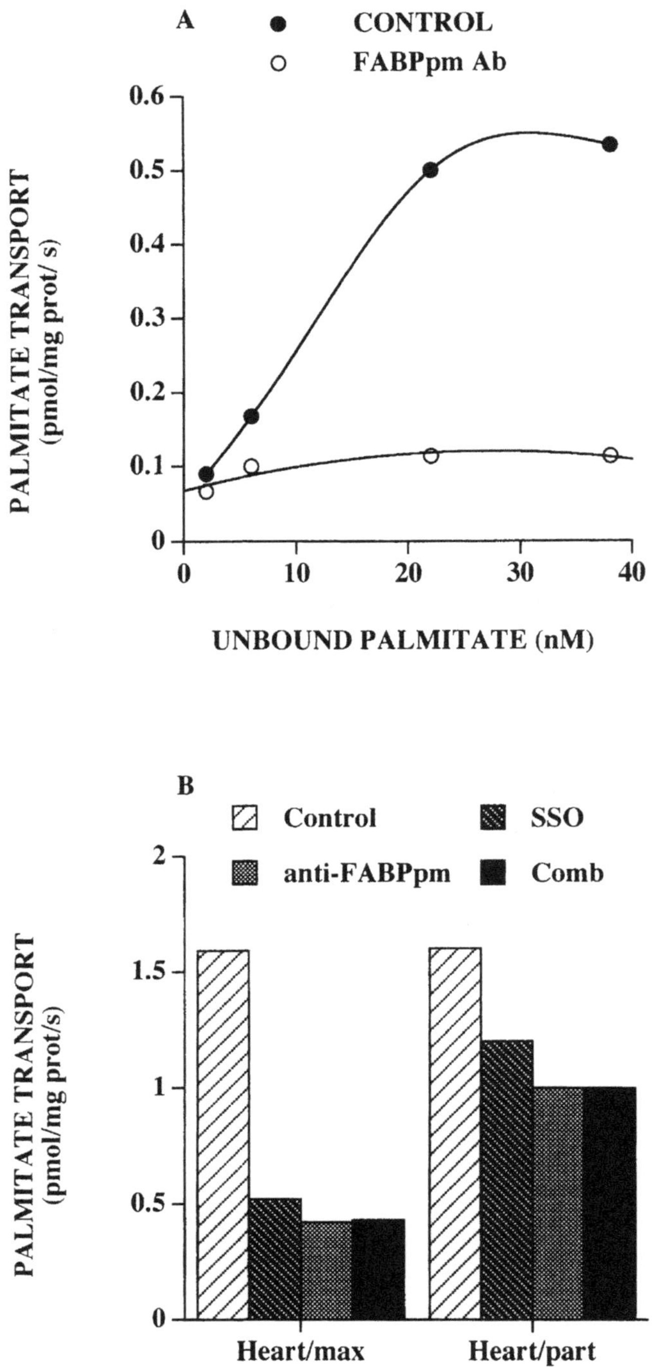

Fig. 4. Inhibition of vesicular fatty acid transport (A) by FABPpm antisera (inhibitor of FABPpm) in giant vesicles obtained from skeletal muscles (redrawn from (33)), and (B) by SSO (inhibitor of FAT/CD36), and FABPpm antisera and the combination (Comb) of SSO + FABPpm antisera in giant vesicles obtained from heart. Experiments in the heart (B) were conducted with maximal (max) and submaximal (50%) part quantities of the inhibitors. Since the independent effects of SSO and FABPpm antisera produced similar effects, and because their effects were not additive, either under maximal or submaximal inhibitor conditions, it suggests that the fatty acid transporters FAT/CD36 and FABppm may cooperate to facilitate the transmembrane movement of fatty acids (data redrawn from [21]).

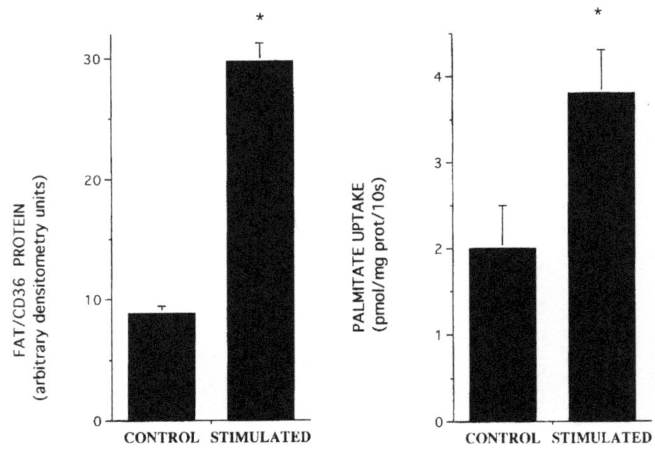

Fig. 5. Chronic muscle stimulation (24 h/day for 7 days) designed to increase the oxidative capacity of skeletal muscle also increased the plasma membrane content of FAT/CD36, and this resulted in an increased rate of vesicular fatty acid transport (data redrawn from [35]).

did not observe a further decrease in fatty acid transport [21]. Thus, inhibiting either FAT/CD36 or FABPpm is sufficient to inhibit fatty acid transport in muscle and heart, while there is no additive effect of SSO and FABPpm antisera on reducing fatty acid transport. Collectively, these observations begin to suggest that in some manner these proteins collaborate to transport fatty acids across the plasma membrane. The molecular mechanism involved in this cooperation remains to be determined.

Regulation of fatty acid transport by fatty acid transporters

There are a number of ways that fatty acid transport is regulated. Assuming that passive diffusional uptake of fatty acids is not altered and knowing that in muscle the transmembrane gradient is of fatty acids is high [34], then chronically altered rates of fatty acid transport are most likely due to (a) the expression of the fatty acid transporters, (b) their subcellular distribution, and (c) possibly, their altered activity at the plasma membrane. The acute regulation of fatty acid transport occurs through either (b) or (c), since protein expression levels are likely not altered within a very short period of time (< 30 min).

Altering the FAT/CD36 protein expression alters fatty acid transport

In rat skeletal muscle, induced to contract chronically (7 days) via low frequency electrical stimulation, FAT mRNA and FAT/CD36 protein were increased. This resulted in an in-

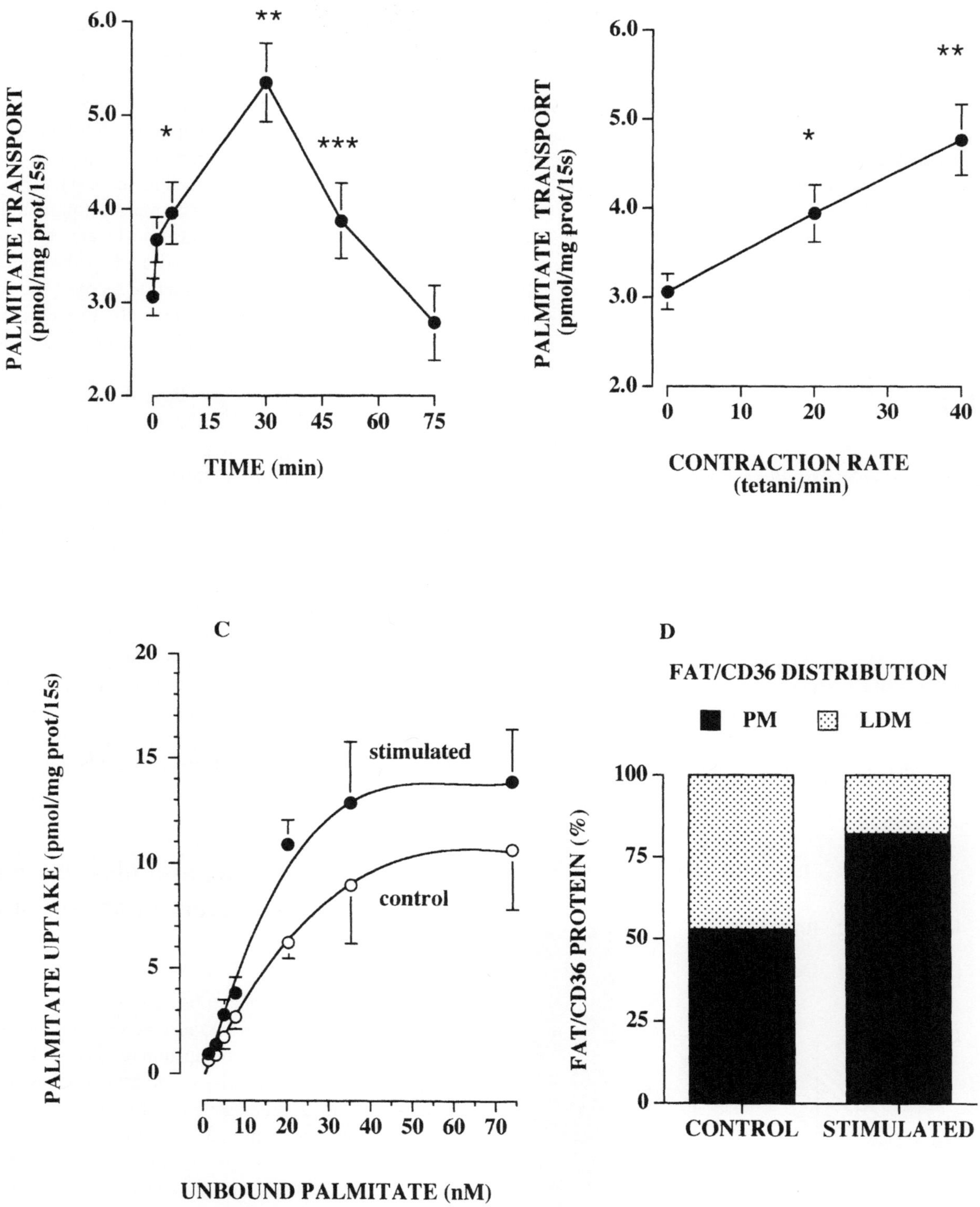

Fig. 6. Muscle contractility (30 min) increases vesicular fatty acid transport due to the translocation of FAT/CD36 from a low density microsomal (LDM) to the plasma membrane (PM) (all data redrawn from [42]). (A) Muscle contraction (0–30 min) rapidly increases vesicular fatty acid transport, and vesicular fatty acid transport is reduced to rates observed in resting muscle during the post-contraction (min 30–75 min) period. (B) Vesicular fatty acid transport is increased in proportion to the rate of muscle contraction. (C) Kinetics of vesicular fatty acid transport in resting and contracting (30 min) muscles. V_{max} was increased while $K_{1/2}$ remained unchanged. (D) Redistribution of FAT/CD36. At rest the proportions of FAT/CD36 in the low density microsomal (LDM) and the plasma membrane (PM) fractions were similar, but with contraction the FAT/CD36 content in LDM was reduced, while concomitantly the PM content of FAT/CD36 was increased (i.e. FAT/CD36 is translocated).

crease in plasma membrane FAT/CD36 and in fatty acid transport, as assessed in giant sarcolemmal vesicles [35] (Fig. 5). In transgenic mice that overexpress FAT/CD36 in muscle, circulating fatty acid levels were reduced and fatty acid oxidation was increased, although only during muscle contraction [36], suggesting that FA uptake was increased only when the muscle's energetic demands were increased. Conversely, in FAT/CD36 null mice, circulating fatty acid levels are increased [37] while the tissue uptake of BMIPP, an indirect measure of fatty acid transport, is reduced in heart, muscle and adipose tissue [8]. Thus, collectively, these studies indicate that altering the expression of the fatty acid transport protein FAT/CD36 results in parallel changes in muscle fatty acid uptake rates. Similar studies are not available for FABPpm.

Muscle contraction and insulin rapidly increases fatty acid transport due to the translocation of FAT/CD36

Acute regulation of fatty acid transport was suggested by studies in skeletal muscle, which indicated that total fatty acid taken up was increased during muscle contraction [38, 39] and by insulin, the latter involving a PI3-kinase mediated mechanism [40, 41]. To examine the contraction effect further we electrically stimulated the sciatic nerve in order to cause all rat hind limb, skeletal muscle fibers to contract, without inducing fatigue. Subsequently, we prepared giant vesicles and performed transport studies [42]. This work showed that with muscle contraction fatty acid transport was increased rapidly (Fig. 6A) and in relation to muscle contraction intensity (Fig. 6B). Kinetic studies of fatty acid uptake showed that V_{max} was increased while there was no change in K_m (Fig. 6C). From fractionation studies of muscle tissue, using a continuous Percoll gradient, we showed that FAT/CD36 is present in both an intracellular pool and in the plasma membrane [42]. Importantly, with muscle contraction a portion of this intracellular pool was translocated to the plasma membrane, thereby increasing fatty acid uptake acutely (Fig. 6D) [42]. During the post-contraction recovery period fatty acid transport returned to control levels (Fig. 6A), and presumably FAT/CD36 was internalized from the plasma membrane.

In more recent work we [43] have shown that insulin also translocates FAT/CD36, (Fig. 7) and this appears to involve the post-receptor signalling protein PI3-kinase. The effects of insulin and muscle contraction on fatty acid uptake by intact muscle are additive [40], and this may suggest that there are separate intracellular pools of FAT/CD36 that are sensitive to insulin- or contraction-activated signals. Such insulin and contraction-sensitive intracellular transport protein pools have already been identified for GLUT4 [44]. Thus, it is clear that fatty acid transport is regulated acutely within minutes by muscle contraction and by insulin, and that the regulation involves the cellular redistribution of FAT/CD36. However, whether FABPpm is redistributed by these physiologic stimuli is not known, although we have observed that this protein also is present in an intracellular pool (Bonen *et al.*, unpublished data).

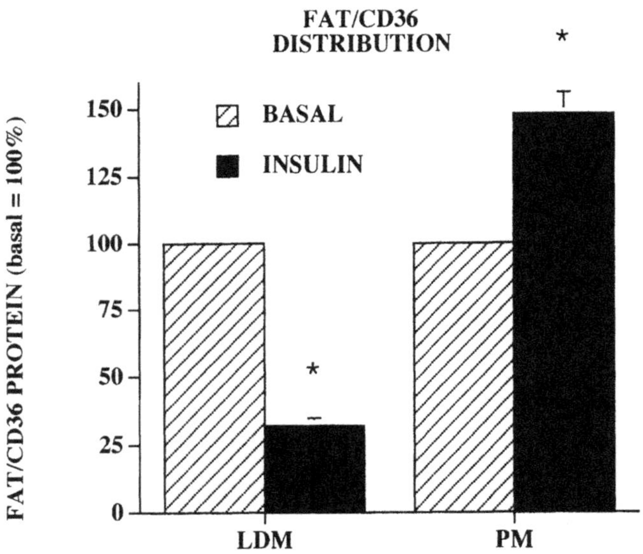

Fig. 7. Insulin induces the translocation of FAT/CD36 in skeletal muscle. In the presence of insulin the FAT/CD36 content in the LDM fraction is reduced, while concomitantly the PM content of FAT/CD36 was increased (i.e. FAT/CD36 is translocated) (data redrawn from [43]).

Fatty acid transport and plasmalemmal fatty acid transporters in obesity and diabetes

In recent years it has been observed that there is a strong correlational relationship between the accumulation of intramuscular triacylglycerols and insulin resistance [45]. This intramuscular triacylglycerol accumulation in obesity may be associated with an increased rate of fatty acid transport. In obese Zucker rats, a model of insulin resistance, and in streptozotocin-induced diabetic animals, with known increments in fatty acid metabolism we have examined vesicular fatty acid transport and fatty acid transporters.

In obese Zucker rats, fatty acid transport in muscle and heart was markedly increased (Fig. 9). However, this was not attributable to changes in the expression of fatty acid transporters, either at the mRNA or protein levels. Rather, the increased fatty acid transport was attributable to increased fatty acid transporter, FAT/CD36, at the plasma membrane (Fig. 9). Because

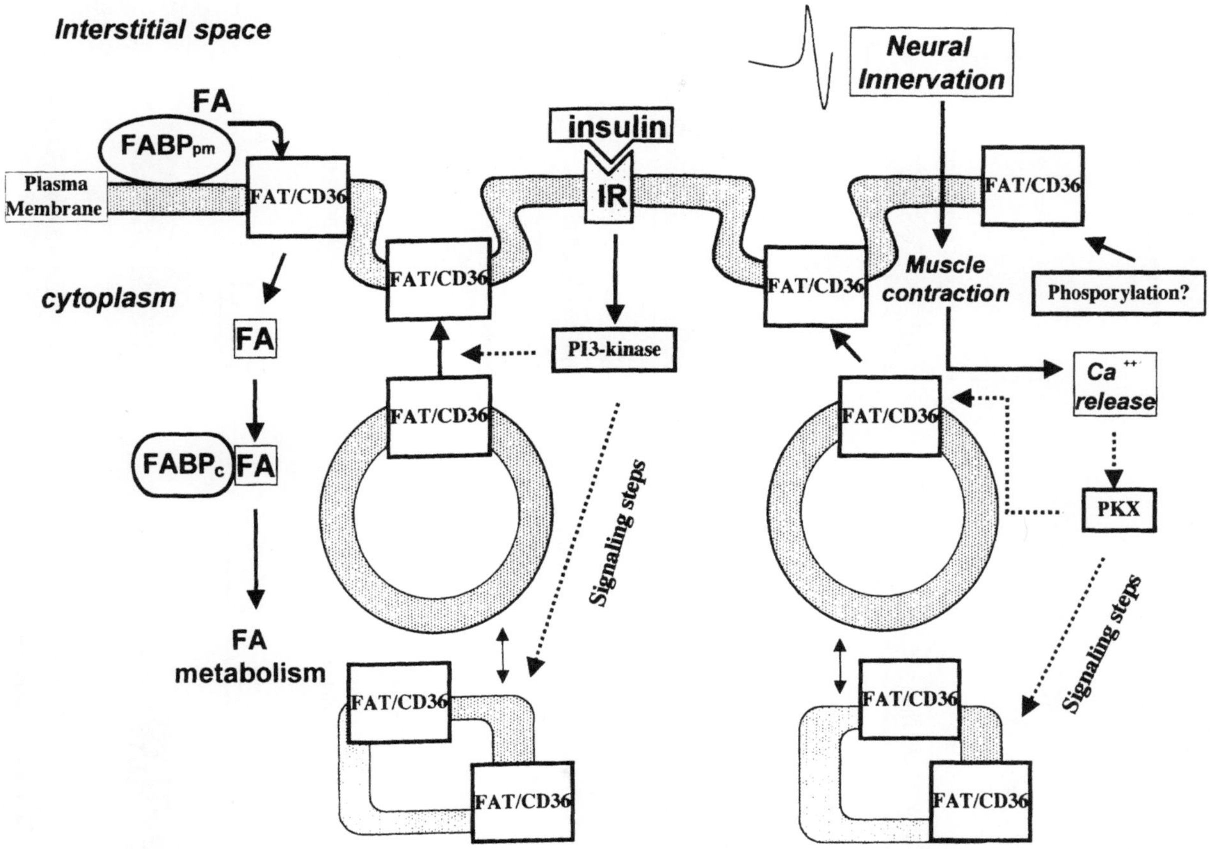

Fig. 8. Schematic representation of fatty acid transport in skeletal muscle. This scheme demonstrates that fatty acids (FA) cross the plasma membrane via fatty acid transport or binding proteins. We show that may be trapped by the peripheral membrane protein FABPpm to be donated to FAT/CD36 located at the plasma membrane (this is speculative, see also Fig. 4, for some indirect support for this idea). Once FA is taken across the plasma membrane it is complexed to FABPc and is then metabolized. Experimental work (see [42, 43]) has shown that FAT/CD36 can be recruited to the plasma membrane from an intracellular pool by either muscle contraction [42] or by insulin involving the PI3-kinase signalling system [43]. We have shown that there are likely to be separate insulin- and contraction-sensitive intracellular depots of FAT/CD36, since insulin and contraction stimulate FA uptake additively in isolated muscles [40]. The contraction-induced translocation may involve a Ca^{2+}-activated signalling system, and it is possible that the phosphorylation of surface FAT/CD36 is a mechanism to increase the transport capacity of plasma membrane associated FAT/CD36.

the total pool of fatty acid transporters was not changed when the plasma membrane FAT/CD36 was increased, it appears that in muscle and heart of obese Zucker rats a larger part of the intracellular pool of fatty acid transporters are permanently located at the plasma membrane.

Alterations in transporter redistribution associated with obesity have also been observed for GLUT4 in both muscle [46] and heart [47]. In healthy muscles [46] and heart [47] GLUT4 is mobilized by insulin from intracellular stores to the sarcolemma, whereas in obese heart and muscles intracellular GLUT4 translocation by insulin is inhibited [46, 47], indicating that the GLUT4 translocation mechanism is impaired. In contrast, in obese rats more FAT/CD36 is located at the plasma membrane, when compared to lean animals (Fig. 9). Taken all together, we speculate that the patterns of subcellular distribution of FAT/CD36 and GLUT4 at the plasma membrane and the intracellular compartments appear to be diametrically opposed in obesity. This may point towards a linkage between insulin resistance and the augmented uptake of fatty acids in obesity. An increased rate of fatty acid transport has recently also been observed in obese individuals (Bonen *et al.*, unpublished data).

Finally, this linkage between insulin sensitivity and fatty acid transport was also demonstrated in several other recent studies in our laboratories. Specifically, we have found that with streptozotocin-induced diabetes fatty acid transport is increased in heart and muscle due to an increased expression of FAT/CD36 and its associated increase at the plasma membrane ([48] and Luiken *et al.*, unpublished data). In type 2 diabetic patients we have also observed an increased rate of fatty acid transport in skeletal muscle (Bonen *et al.*, unpublished data). Conversely, chronic leptin treatment (2 weeks) which

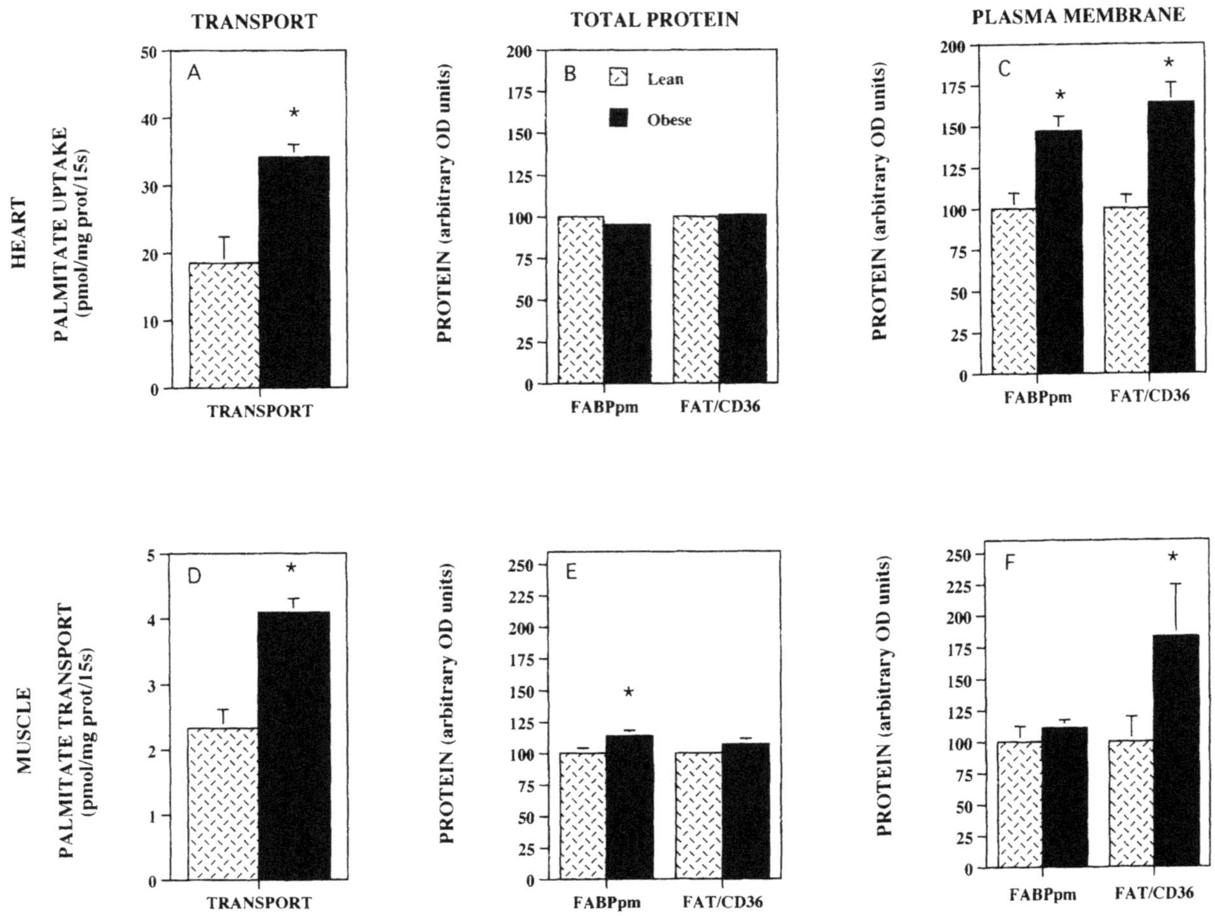

Fig. 9. Increased rates of fatty acid transport in heart (A) and skeletal muscle (D) of obese Zucker rats. This increase was not associated with altered protein expression (total protein) in heart (B) or muscle (E), instead the transport increases were associated with increases in plasma membrane FAT/CD36 and FABPpm in the heart (C), and FAT/CD36 but not FABPpm in muscle (F). These data indicate that in muscle and heart of obese Zucker rats a larger part of the intracellular pool of fatty acid transporters are permanently located at the plasma membrane (note that both FAT/CD36 [42, 43] and FABPpm (unpublished data) are present in an intracellular depots in muscle.

improves insulin sensitivity in muscle [49] was associated with a reduction in fatty acid transport and FAT/CD36 and FABPpm proteins (Fig. 10) [50]. Thus, a number of studies have begun to demonstrate that fatty acid transport rates are altered in a reciprocal manner to glucose transport.

Concluding remarks

It is becoming increasingly evident that disorders in lipid metabolism are an important factor contributing to the etiology of a number of common human diseases such as diabetes and obesity (cf. [51]). Our evidence that protein-mediated fatty acid transport plays a key role in the regulation of muscular fatty acid metabolism suggests that malfunctioning of the fatty acid transport process could be another critical factor in the pathogenesis of these diseases. Therefore, it is essential to develop a thorough understanding of the factors regulating fatty acid transport, the mechanisms accounting for alterations in the expression of fatty acid transporters, and the signaling processes involved in their subcellular distribution. The fatty acid transport system is therefore a potential target for therapeutic intervention in obesity and diabetes.

Acknowledgements

These studies were supported by grants from the Canadian Institutes of Health Research (A.B), the Heart and Stroke Foundation of Ontario (A.B.), and the Netherlands Heart Foundation (grant D98.012: J.L. and J.G.). Joost J.F.P. Luiken is a Dekker post-doctoral fellow of the Netherlands Heart Foundation.

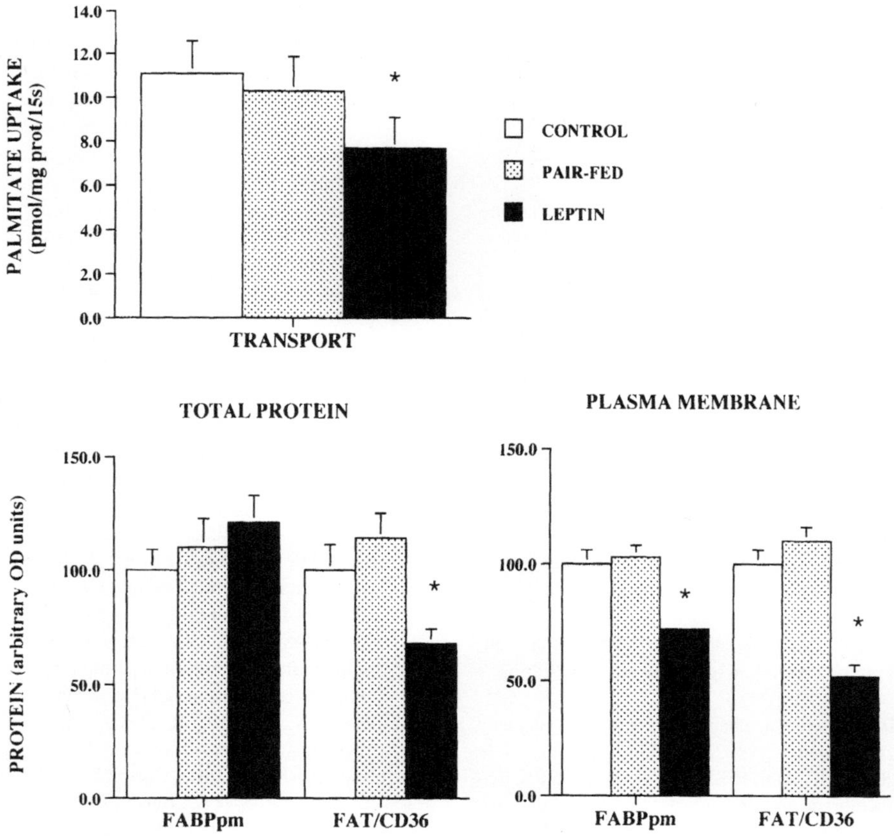

Fig. 10. Chronic (2 week) administration of leptin infused via an osmotic minipump resulted in a reduction in skeletal muscle palmitate transport, FAT/CD36 repression, and reduced plasmalemmal content of FABPpm and FAT/CD36 (redrawn from [50]).

References

1. Abumrad NA, Harmon C, Ibrahimi A: Membrane transport of long-chain fatty acids: Evidence for a facilitated process. J Lipid Res 39: 2309–2318, 1998
2. Luiken JJFP, Schaap FG, van Nieuwenhoven FA, van der Vusse GJ, Bonen A, Glatz JFC: Cellular fatty acid transport in heart and muscle as facilitated by proteins. Lipids 34: S169–S175, 1999
3. Hamilton JA, Kamp F: How are free fatty acids transported in membranes? Is it by proteins or by free diffusion through the lipids? Diabetes 48: 2255–2269, 1999
4. Abumrad NA, Park JH, Park CR: Permeation of long-chain fatty acid into adipocytes. Kinetics, specificity, and evidence for involvement of a membrane protein. J Biol Chem 259: 8945–8953, 1984
5. Berk PD, Zhou S-L, Kiang C-L, Stump D, Bradbury M, Isola L: Uptake of long chain fatty acids is selectively up-regulated in adipocytes of Zucker rats with genetic obesity and non-insulin-dependent diabetes mellitus. J Biol Chem 272: 8830–8835, 1997
6. Luiken JJFP, van Nieuwenhoven FA, America G, van der Vusse GJ, Glatz JFC: Uptake and metabolism of palmitate by isolated cardiac myocytes from adult rats: Involvement of sarcolemmal proteins. J Lipid Res 38: 725–758, 1997
7. King PA, Hirschman MF, Horton ED, Horton ES: Glucose transport in skeletal muscle membrane vesicles from control and exercised rats. Am J Physiol 257 (Cell Physiol): C1128–C1134, 1989
8. Coburn CT, Knapp FF Jr, Febbraio M, Beets AL, Silverstein RL, Abumrad NA: Defective uptake and utilization of long chain fatty acids in muscle and adipose tissue of CD36 knockout mice. J Biol Chem 275: 32523–32529, 2000
9. Hajri T, Ibrahimi A, Coburn CT, Knapp FF Jr, Kurtz T, Pravenec M, Abumrad NA: Defective fatty acid uptake in the spontaneously hypertensive rat is a primary determinant of altered glucose metabolism, hyperinsulinemia, and myocardial hypertrophy. J Biol Chem 276: 23661–23666, 2001
10. Kropp J, Eisenhut M, Ambrose KR, Knapp FF Jr, Franke WG: Pharmacokinetics and metabolism of the methyl-branched fatty acid (BMIPP) in animals and humans. J Nucl Med 40: 1484–1491, 1999
11. Mokler FT, Lin Q, Luo H, McPherson DW, Beets AL, Bockisch A, Kropp J ,Knapp FF Jr: Dual-label studies with [125I]-3(R)/[131I]-3(S)-BMIPP show similar metabolism in rat tissues. J Nucl Med 40: 1918–1927, 1999
12. Lin Q, Luo H, Mokler F, Beets AL, Ambrose KR, McPherson DW, Kropp J, Knapp FF Jr: Effects of configuration on the myocardial uptake of radioiodinated 3(R)-BMIPP and 3(S)-BMIPP in rats. J Nucl Med 38: 1434–1441, 1997
13. McDermott JC, Bonen A: Lactate transport by skeletal muscle sarcolemmal vesicles. Mol Cell Biochem 122: 113–121, 1993
14. McCullagh KJA, Juel C, O'Brien M, Bonen A: Chronic muscle stimulation increases lactate transport in rat skeletal muscle. Mol Cell Biochem 156: 51–57, 1996
15. Juel C: Muscle lactate transport studied in sarcolemmal giant vesicles. Biochim Biophys Acta 1065: 15–20, 1991
16. Kristiansen S, Wojtaszewski JF, Juel C, Richter EA: Effect of glucose-

6-phosphate and pH on glucose transport in skeletal muscle plasma membrane giant vesicles. Acta Physiol Scand 150: 227–233, 1994
17. Kristiansen S, Hargreaves M, Richter EA: Progressive increase in glucose transport and GLUT-4 in human sarcolemmal vesicles during moderate exercise. Am J Physiol 272: E385–E389, 1997
18. Juel C: Lactate – proton cotransport in skeletal muscle. Physiol Rev 77: 1–37, 1997
19. Bonen A, Luiken JJFP, Lui S, Dyck DJ, Kiens B, Kristiansen S, Turcotte L, van der Vusse GJ, Glatz JFC: Palmitate transport and fatty acid transporters in red and white muscles. Am J Physiol Endocrinol Metab 275: E471–E478, 1998
20. Luiken JJFP, Arumugam Y, Dyck DJ, Bell RC, Pelsers ML, Turcotte LP, Tandon NN, Glatz JFC, Bonen A: Increased rates of fatty acid uptake and plasmalemmal fatty acid transporters in obese Zucker rats. J Biol Chem 2001 (in press)
21. Luiken JJFP, Turcotte LP, Bonen A: Protein-mediated palmitate uptake and expression of fatty acid transport proteins in heart giant vesicles. J Lipid Res 40: 1007–1016, 1999
22. Isola LM, Zhou SL, Kiang CL, Stump DD, Bradbury MW, Berk PD: 3T3 fibroblasts transfected with a cDNA for mitochondrial aspartate aminotransferase express plasma menbrane fatty acid-binding protein and saturable fatty acid uptake. Proc Natl Acad Sci USA 92: 9866–9870, 1995
23. Hirsch D, Stahl A, Lodish HF: A family of fatty acid transporters conserved from mycobacterium to man. Proc Natl Acad Sci USA 95: 8625–8629, 1998
24. Schaffer JE, Lodish HF: Molecular mechanism of long-chain fatty acid uptake. Trends Cardiovasc Med 5: 218–224, 1995
25. Abumrad NA, El-Maghrabi MR, Amri E-Z, Lopez E, Grimaldi P: Cloning of a rat adipocyte membrane protein implicated in binding or transport of long chain fatty acids that is induced during preadipocyte differentiation. Homology with human CD36. J Biol Chem 268: 17665–17668, 1993
26. Ibrahimi A, Sfeir Z, Magharaine H, Amri EZ, Grimaldi P, Abumrad NA: Expression of the CD36 homolog (FAT) in fibroblast cells: Effects on fatty acid transport. Proc Natl Acad Sci USA 93: 2646–2651, 1996
27. Memon RA, Fuller J, Moser AH, Smith PJ, Grunfeld C, Feingold KR: Regulation of putative fatty acid transporters and Acyl-CoA synthetase in liver and adipose tissue in ob/ob mice. Diabetes 48: 121–127, 1999
28. Berk PD, Zhou S-L, Kiang C-L, Stump D, Fan X, Bradbury M: Selective upregulation of fatty acid uptake by adipocytes characterizes both genetic and diet-induced obesity in rodents. J Biol Chem 274: 28626–28631, 1999
29. Coe R, Johnston-Smith A, Frohnert BI, Watkins PA, Bernlohr DA: The fatty acid transport protein (FATP1) is a very long chain acyl-CoA synthetase. J Biol Chem 274: 36300–36304, 1999
30. Watkins PA, Lu JF, Steinberg SJ, Gould SJ, Smith KD, Braiterman LT: Disruption of the *Saccharomyces cerevisiae* FAT1 gene decreases very long-chain fatty acyl-CoA synthetase activity and elevates intracellular very long-chain fatty acid concentrations. J Biol Chem 273: 18210–18219, 1998
31. Glatz JFC, van Nieuwenhoven FA, Luiken JJFP, Schaap FG, van der Vusse GJ: Role of membrane -associated and cytoplasmic fatty acid binding proteins and cellular fatty acid metabolism. Prostagland Leuk Essent Fatty Acids 4–5: 373–378, 1997
32. Glatz JFC, Storch J: Unravelling the significance of cellular fatty acid binding proteins. Curr Opin Lipidol 12: 276–274, 2001
33. Turcotte LP, Swenberger JR, Tucker MZ, Trump G, Yee AJ, Luiken JJFP, Bonen A: Muscle palmitate transport and binding are saturable and inhibited by antibodies to FABPpm. Mol Cell Biochem 210: 53–63, 2000
34. Van der Vusse GJ, Roemen THM: Gradient of fatty acids from blood plasma to skeletal muscle in dogs. J Appl Physiol 78: 1839–1843, 1995
35. Bonen A, Dyck DJ, Ibrahimi A, Abumrad NA: Muscle contractile activity increases fatty acid metabolism and transport and FAT/CD36. Am J Physiol (Endocrinol Metab) 276: E642–E649, 1999
36. Ibrahimi A, Bonen A, Blinn WD, Hajri T, Li X, Zhong K, Cameron R, Abumrad NA: Muscle-specific overexpression of FAT/CD36 enhances fatty acid oxidation by contracting muscles, reduces plasma triglycerides and fatty acids, and increases plasma glucose and insulin. J Biol Chem 274: 26761–26766, 1999
37. Febbraio M, Abumrad NA, Hajjar DP, Sharma K, Cheng W, Frieda S, Pearce A, Silverstein RL: A null mutation in murine CD36 reveals an important role in fatty acid and lipoprotein metabolism. J Biol Chem 274: 19055–19062, 1999
38. Dyck DJ, Bonen A: Muscle contraction increases palmitate esterification and oxidation, and triacylglycerol oxidation. Am J Physiol (Endocrinol Metab) 275: E888–E896, 1998
39. Turcotte LP, Richter EA, Kiens B: Increased plasma FFA uptake and oxidation during prolonged exercise in trained vs. untrained humans. Am J Physiol 262: E791–E799, 1992
40. Dyck DJ, Steinberg G, Bonen A: Insulin increases FFA uptake and esterification but reduces lipid utilization in isolated contracting muscles. Am J Physiol Endocrinol Metab 281: E600–E607, 2001
41. Muoio DM, Dohm GL, Tapscott EB, Coleman RA: Leptin opposes insulin's effects on fatty acid partitioning in muscles isolated from obese ob/ob mice. Am J Physiol (Endocrinol Metab) 276: E913–E921, 1999
42. Bonen A, Luiken JJFP, Arumugam Y, Glatz JFC, Tandon NN: Acute regulation of fatty acid uptake involves the cellular redistribution of fatty acid translocase. J Biol Chem 275: 14501–14508, 2000
43. Luiken JJFP, Dyck DJ, Han X-X, Tandon NN, Arumugam Y, Glatz JFC, Bonen A: Insulin induces the translocation of the fatty acid transporter FAT/CD36 to the plasma membrane. Am J Physiol Endocrinol Metab 2001 (in press)
44. Lemieux K, Han X-X, Dombrowski L, Bonen A, Marette A: The transferrin receptor defines two distinct contraction-responsive GLUT4 vesicle populations. Diabetes 49: 183–189, 2000
45. Pan DA, Lillioja S, Kriketos AD, Milner MR, Baur LA, Jenkins AB, Storlien LH: Skeletal muscle triglyceride levels are inversely related to insulin action. Diabetes 46: 983–988, 1997
46. Brozinick JT Jr, Etgen GJ, Yaspelkis BB III, Kang HY, Ivy JL: Effects of exercise training on muscle GLUT-4 protein content and translocation in obese Zucker rats. Am J Physiol (Endocrinol Metab) 265: E419–E427, 1993
47. Uphues I, Kolter T, Goud B, Eckel J: Failure of insulin-regulated recruitment of the glucose transporter GLUT4 in cardiac muscle of obese Zucker rats is associated with alterations of small-molecular-mass GTP-binding proteins. Biochem J 311: 161–166, 1995
48. Pelsers ML, Lutgerink JT, van Nieuwenhoven FA, Tandon NN, van der Vusse GJ, Arends JW, Hoogenboom HR, Glatz JFC: A sensitive immunoassay for rat fatty acid translocase (CD36) using phage antibodies selected on cell transfectants: Abundant presence of fatty acid translocase/CD36 in cardiac and red skeletal muscle and up- regulation in diabetes. Biochem J 337: 407–414, 1999
49. Yaskelpis III BB, Ansari L, Ramey EA, Loy SF: Chronic leptin administration increases insulin-stimulated skeletal muscle glucose uptake. Metabolism 48: 671–676, 1999
50. Steinberg GR, Dyck DJ, Calles-Escandon J, Tandon NN, Luiken JJFP, Glatz JF, Bonen A: Chronic leptin administration decreases fatty acid uptake and fatty acid transporters in rat skeletal muscle. J Biol Chem 2002 (in press)
51. Kelley DA, Mandarino LJ: Fuel selection in human skeletal muscle in insulin resistance. A reexamination. Diabetes 49: 677–683, 2000

Molecular and Cellular Biochemistry **239:** 193–197, 2002.

The impact of overexpression and deficiency of fatty acid translocase (FAT)/CD36

M. Febbraio,[1] E. Guy,[1] C. Coburn,[2] F.F. Knapp Jr,[3] A.L. Beets,[3] N.A. Abumrad[2] and R.L. Silverstein[1]

[1]*Department of Medicine, Weill Medical College of Cornell University, New York, NY;* [2]*Department of Physiology, State University of New York – Stony Brook, Stony Brook, NY;* [3]*Nuclear Medicine Group, Health Science Research Center, Oak Ridge National Laboratory, Oak Ridge, TN, USA*

Abstract

Fatty acid translocase (FAT)/CD36 has been associated with diverse normal and pathologic processes. These include scavenger receptor functions (uptake of apoptotic cells and modified lipid), lipid metabolism and fatty acid transport, adhesion, angiogenesis, modulation of inflammation, transforming growth factor-β activation, atherosclerosis, diabetes and cardiomyopathy. Although CD36 was identified more than 25 years ago, it is only with the advent of recent genetic technology that *in vivo* evidence has emerged for its physiologic and pathologic relevance. As these *in vivo* studies are expanded, we will gain further insight into the mechanism(s) by which CD36 transmits a cellular signal, and this will allow the design of specific therapeutics that impact on a particular function of CD36. (Mol Cell Biochem **239**: 193–197, 2002)

Key words: fatty acid translocase, CD36, CD36 null, fatty acid uptake

Introduction

Protein facilitated transport of long chain fatty acids (LCFAs) has been a subject of controversy because data suggest that LCFAs can freely diffuse across cell membranes [1–5]. In a series of careful and elegant studies of LCFA permeation across rat adipocyte membranes, Abumrad *et al.* identified a saturable component that was abolished by treatments that disrupted proteins, showed specificity for chain length and saturation of fatty acids and was inhibitable, all characteristics of a protein mediated phenomenon [6, 7]. This group purified the specific protein, made tryptic fragments, and constructed oligonucleotide probes with which to screen a rat cDNA library [8]. The protein they cloned was the homolog of human CD36, which did not have structural characteristics of classic transporters [9]. In addition to CD36, other plasma membrane LCFA transporters have now been identified, and what is perhaps most interesting about these proteins is their lack of homology [10–11]. The contentiousness relating to this function of CD36 and these other proteins may be partly attributed to the use of the word 'transporter' or 'translocator', which may imply a classic energy-dependent receptor mediated process. Although a mechanism has not been defined, and it has not been clarified whether CD36 affects LCFA transport alone or with other proteins in a complex, the experimental data support a role for CD36 in regulated LCFA uptake in a variety of tissues [12–25]. Studies in humans and mice deficient in CD36 indicate that CD36 determines LCFA uptake in the heart, and recent studies of CD36 transgenic and null mice have confirmed a physiologic role for fatty acid transporters [14, 16, 19, 20, 22, 25]. In humans, whether absence of CD36 leads to diabetes or cardiac pathology remains controversial [22, 26–31]. Further investigation in animal models promises to elucidate the possible connection between CD36 and these diseases.

Address for offprints: M.L. Febbraio, Department of Hematology and Oncology, Weill Medical College of Cornell University, 1300 York Avenue, New York, NY 10021-4896, USA (E-mail: mjfebbra@med.cornell.edu)

Earlier studies on CD36

CD36 is an 88 kD ditopic protein remarkable for its numerous complex N-linked sugar residues and for its two short (10–13 amino acids) cytoplasmic domains [32, 33]. Both these cellular domains contain two cysteine residues and can be palmitoylated [34]. CD36 is found on a variety of cell types, including platelets, monocytes/macrophages, adipocytes, capillary endothelial cells, skeletal muscle and specialized epithelium of the breast and retina [12, 35–39]. Prior to its identity with FAT, CD36 was primarily recognized to be a receptor for thrombospondin-1 (TSP-1) and a class B scavenger receptor capable of binding native and modified lipoproteins and anionic phospholipids [39–41]. CD36 has more recently been shown to co-fractionate with components of caveolae, which are specialized plasma membrane domains that may function in aggregation of receptors and facilitate signal transduction in cells [42–44]. CD36 cooperates with TSP-1 and the integrin $''_{v}\$_{3}$ to effect the uptake of apoptotic cells, and with the integrin $''_{v}\$_{5}$ to effect removal of shed photoreceptor rod outer segments by retinal pigment epithelial cells [38, 45]. In endothelial cells, interaction of CD36 with TSP-1 can lead to inhibition of angiogenesis promoted by vascular endothelial cell growth factor (VEGF) or basic fibroblast growth factor (FGF-2), by activation of a pro-apoptotic signaling cascade [46, 47].

CD36 and fatty acid metabolism

In vitro work had implicated CD36 in LCFA uptake in fat and muscle. In addition to transport studies in adipocytes, it was shown that mRNA expression of CD36 was strongly induced during differentiation of adipocytes, and that LCFA uptake during differentiation coincided with CD36 expression [48]. CD36 expression in muscle correlates with oxidative potential: it is highly expressed in muscles rich in oxidative fibers and low in muscles that are predominantly glycolytic [17–19]. Stimulation of isolated sarcolemmal vesicles results in increased CD36 expression and increased FA uptake [49, 50]. Transfection of CD36 into non-expressing cells significantly increased the high affinity binding and uptake of LCFA, and this was abolished in the presence of an inhibitor of protein mediated transport [13]. Transfection and inhibitor treatments did not alter the low affinity diffusable component.

In vitro studies supported a role for CD36 in effecting LCFA transport in multiple cell types but it remained unclear if CD36 played a physiologic role in whole body LCFA homeostasis. In 1999, Binas *et al.*, showed that absence of cardiac cytoplasmic fatty acid binding protein gave rise to a poorly defined cardiomyopathy [51]. Cytoplasmic fatty acid binding proteins may participate in the trafficking of fatty acids within the cell, such that, depending upon the metabolic state, fatty acids can be utilized in $-oxidation, synthesis of cholesterol esters and phospholipids, or converted to triacylglycerol for storage [52]. Cytoplasmic fatty acid binding proteins are downstream of plasma membrane transporters such as CD36, and absence of these proteins may result in improper fatty acid targeting and/or the efflux of free (non-esterified) fatty acids from the cell.

In the past several years, several animal models have become available that have allowed exploration of the question of the *in vivo* relevance of plasma membrane LCFA transporters such as CD36. Our laboratory created a CD36 null mouse, and Abumrad has made a muscle creatine kinase (MCK)-CD36 transgenic mouse which over expresses CD36 in muscle [14, 25]. Aitman *et al.* have identified genetic defects in CD36 leading to a null phenotype in at least one strain of the spontaneously hypertensive rat (SHR), and he and colleagues have created a SHR congenic, while Pravenec *et al.* generated a transgenic [53, 54]. These animal models link CD36 expression with tissue competence to take up LCFA, substrate choice and other changes in whole body metabolism and homeostasis.

Adipocytes isolated from age and sex matched wild type and CD36 null mice (both lines were established from litter mates after 4 backcrosses to C57Bl/6) showed a significant decrease in cell associated radiolabeled oleate or palmitate in transport studies, and kinetic analysis using an Hanes plot determined that the high affinity component of LCFA was lacking, while the low affinity, diffusable component was unaltered [25].

In addition to adipocytes, CD36 facilitates fatty acid translocation in heart and skeletal muscle, where LCFAs are important substrates for energy production. MCK-CD36 transgenic mice showed enhanced muscle fatty acid oxidation, decreased plasma levels of circulating fatty acids and triacylglycerols, and decreased fat deposition [14]. CD36 null mice had increased plasma levels of cholesterol, triacylglycerol and fatty acids, and fasting hypoglycemia [25]. These changes reflect an impact on energy utilization from LCFAs and suggest an important role for CD36 in lipoprotein and lipid homeostasis. SHR have defective adipocyte transport of LCFAs, and a type II diabetes syndrome [53]. Replacement of the mutated CD36 allele with a wild type allele using a congenic or transgenic strategy reversed the diabetic phenotype and lowered plasma triacylglycerol levels [53, 54]. Studies in humans have also provided support for a physiological role for CD36 in LCFA translocation. Lack of uptake of a LCFA analog in heart correlated with the CD36 phenotype, and some studies showed a correlation with diabetes or cardiomyopathy [22, 26–31]. These latter reports remain controversial because of the small cohort sizes.

BMIPP and IPPA are LCFA analogs that can be used to quantitate LCFA uptake in tissues (55). Because of its branched methyl chain, BMIPP cannot be $-oxidized, and thus has prolonged retention in tissues. In contrast, IPPA is metabolized and oxidized essentially as palmitic or oleic acid. In studies with wild type and CD36 null mice, both of these tracers gave essentially identical results. Age and sex-matched wild type and CD36 null mice were injected via the lateral tail vein with 200:1 of the BMIPP/BSA or IPPA/BSA solution (14–42:Ci) [19]. After 30 min (BMIPP uptake plateaus by this time, as determined in previous experiments), animals were sacrificed and tissues removed, rinsed with saline and blotted dry. BMIPP/IPPA incorporation was determined by (-scintillation counting. A sample of the injected dose was counted to determine the maximum number of counts. We found a profound decrease in uptake of both LCFA analogs in heart (50–80%), skeletal muscle (40–75%) and adipose tissue (60–70%) from null mice as compared with wild type controls [19]. The defect in skeletal muscle correlated with oxidative potential, in that it was greatest in the heart and diaphragm as compared with the external oblique or hip muscles [19]. In contrast, there was no difference in uptake of either tracer in liver, a tissue that takes in LCFAs independent of CD36.

Pathophysiology of CD36

CD36 null mice are not diabetic. This variance between mice and rats may reflect species differences in glucose/fatty acid metabolism, or a difference in genetic modulators, or compensatory mechanisms. However, CD36 null mice do have altered lipid and glucose metabolism, as previously outlined, and study of insulin sensitivity at the level of individual tissues is under way. The understanding of the interrelationship between CD36, fatty acid and glucose uptake and utilization may have major impact on treatment strategies. For example, changes in adipocyte CD36 expression have been noted in response to thiazolidinediones. Increased expression of adipocyte CD36 may promote lower plasma fatty acid levels by increasing fat storage, and be part of the mechanism by which these drugs improve insulin resistance in type II diabetics. The effect of thiazolidinediones on muscle CD36 has not yet been reported, but muscle fatty acid oxidation has been shown to increase, suggesting that CD36 is up-regulated in this tissue as well. Interestingly, Li *et al.* reported that treatment of LDL receptor deficient mice with thiazolidinediones improved insulin sensitivity in male (but not female) mice, and resulted in decreased atherosclerosis in male mice, in spite of the impact these drugs have on macrophage CD36 expression [56]. It may be that insulin resistance has the greatest pro-atherogenic effect in this model, and correction of this counters any negative effect due to potential up-regulation of macrophage CD36. Importantly, this study underscores the complexity of the regulation of CD36 in different tissues and in response to different environmental cues.

Conclusion

In summary, the newly developed mouse models, and the ability to pharmacologically modulate CD36, provide exciting new tools with which to explore the role of this protein in normal homeostasis and in disease states related to fatty acid, lipid and lipoprotein metabolism.

Acknowledgements

We gratefully acknowledge support from the National Institutes of Health (NHBLI, R29HL58559), and the generous gift from the Silbermann Foundation.

References

1. Hamilton JA, Johnson RA, Corkey B: Fatty acid transport: The diffusion mechanism in model and biological membranes. J Mol Neurosci 16: 99–108; discussion 151–157, 2001
2. Hamilton JA, Kamp F: How are free fatty acids transported in membranes? Is it by proteins or by free diffusion through the lipids? Diabetes 48: 2255–2269, 1999
3. Hamilton JA: Transport of fatty acids across membranes by the diffusion mechanism. Prostagland Leukot Essent Fatty Acids 60: 291–297, 1999
4. Hamilton JA: Fatty acid transport: Difficult or easy? J Lipid Res 39: 467–481, 1998
5. Zakim D: Fatty acids enter cells by simple diffusion. Proc Soc Exp Biol Med 212: 5–14, 1996
6. Abumrad NA, Park JH, Park CR: Permeation of long-chain fatty acid into adipocytes. Kinetics, specificity, and evidence for involvement of a membrane protein. J Biol Chem 259: 8945–8953, 1984
7. Abumrad NA, Perkins RC, Park JH: Mechanism of long chain fatty acid permeation in the isolated adipocyte. J Biol Chem 256: 9183–9191, 1981
8. Harmon CM, Abumrad NA: Binding of sulfosuccinimidyl fatty acids to adipocyte membrane proteins: Isolation and amino-terminal sequence of an 88-kD protein implicated in transport of long-chain fatty acids. J Membr Biol 133: 43–49, 1993
9. Abumrad NA, el-Maghrabi MR, Amri EZ: Cloning of a rat adipocyte membrane protein implicated in binding or transport of long-chain fatty acids that is induced during preadipocyte differentiation. Homology with human CD36. J Biol Chem 268: 17665–17668, 1993
10. Stahl A, Gimeno RE, Tartaglia LA: Fatty acid transport proteins: A current view of a growing family. Trends Endocrinol Metab 12: 266–273, 2001
11. Abumrad N, Coburn C, Ibrahimi A: Membrane proteins implicated in long-chain fatty acid uptake by mammalian cells: CD36, FATP and FABPm. Biochim Biophys Acta 1441: 4–13, 1999

12. Abumrad NA, el-Maghrabi MR, Amri EZ: Cloning of a rat adipocyte membrane protein implicated in binding or transport of long-chain fatty acids that is induced during preadipocyte differentiation. Homology with human CD36. J Biol Chem 268: 17665–17668, 1993
13. Ibrahimi A, Sfeir Z, Magharaie H: Expression of the CD36 homolog (FAT) in fibroblast cells: Effects on fatty acid transport. Proc Natl Acad Sci USA 93: 2646–2651, 1996
14. Ibrahimi A, Bonen A, Blinn WD: Muscle-specific overexpression of FAT/CD36 enhances fatty acid oxidation by contracting muscle, reduces plasma triglycerides and fatty acids, and increases plasma glucose and insulin. J Biol Chem 274: 26761–26766, 1999
15. Van Nieuwenhoven FA, Verstijnen CP, Abumrad NA: Putative membrane fatty acid translocase and cytoplasmic fatty acid-binding protein are co-expressed in rat heart and skeletal muscles. Biochem Biophys Res Commun 207: 747–752, 1995
16. Watanabe K, Ohta Y, Toba K: Myocardial CD36 expression and fatty acid accumulation in patients with type I and II CD36 deficiency. Ann Nucl Med 12: 261–266, 1998
17. Bonen A, Luiken JJ, Liu S: Palmitate transport and fatty acid transporters in red and white muscles. Am J Physiol 275: E471–478, 1998
18. Bonen A, Dyck DJ, Ibrahimi A: Muscle contractile activity increases fatty acid metabolism and transport and FAT/CD36. Am J Physiol 276: E642–649, 1999
19. Coburn CT, Knapp FF Jr, Febbraio M: Defective uptake and utilization of long chain fatty acids in muscle and adipose tissues of CD36 knockout mice. J Biol Chem 275: 32523–32529, 2000
20. Fukuchi K, Nozaki S, Yoshizumi T: Enhanced myocardial glucose use in patients with a deficiency in long-chain fatty acid transport (CD36 deficiency). J Nucl Med 40: 239–243, 1999
12. Hajri T, Ibrahimi A, Coburn CT: Defective fatty acid uptake in the spontaneously hypertensive rat is a primary determinant of altered glucose metabolism, hyperinsulinemia and myocardial hypertrophy. J Biol Chem 25: 25, 2001
22. Nozaki S, Tanaka T, Yamashita S: CD36 mediates long-chain fatty acid transport in human myocardium: Complete myocardial accumulation defect of radiolabeled long-chain fatty acid analog in subjects with CD36 deficiency. Mol Cell Biochem 192: 129–135, 1999
23. Pravenec M, Landa V, Zidek V: Transgenic rescue of defective Cd36 ameliorates insulin resistance in spontaneously hypertensive rats. Nat Genet 27: 156–158, 2001
24. Collison M, Glazier AM, Graham D: Cd36 and molecular mechanisms of insulin resistance in the stroke-prone spontaneously hypertensive rat. Diabetes 49: 2222–2226, 2000
25. Febbraio M, Abumrad NA, Hajjar DP: A null mutation in murine CD36 reveals an important role in fatty acid and lipoprotein metabolism. J Biol Chem 274: 19055–19062, 1999
26. Watanabe K, Toba K, Ogawa Y: Hypertrophic cardiomyopathy with type I CD36 deficiency. Jpn Circ J 62: 541–542, 1998
27. Tanaka T, Sohmiya K, Kawamura K: Is CD36 deficiency an etiology of hereditary hypertrophic cardiomyopathy? J Mol Cell Cardiol 29: 121–127, 1997
28. Watanabe K, Ohta Y, Toba K: Myocardial CD36 expression and fatty acid accumulation in patients with type I and II CD36 deficiency. Ann Nucl Med 12: 261–266, 1998
29. Gotoda T, Iizuka Y, Yamada N: Complex connection between CD36 and atherosclerosis, lipid metabolism, and insulin resistance syndromes. Curr Atheroscler Rep 2: 453–454, 2000
30. Miyaoka K, Kuwasako T, Hirano K: CD36 deficiency associated with insulin resistance. Lancet 357: 686–687, 2001
31. Yanai H, Chiba H, Morimoto M: Type I CD36 deficiency in humans is not associated with insulin resistance syndrome. Thromb Haemost 83: 786., 2000
32. Gruarin P, Thorne RF, Dorahy DJ: CD36 is a ditopic glycoprotein with the N-terminal domain implicated in intracellular transport. Biochem Biophys Res Commun 275: 446–454, 2000
33. Oquendo P, Hundt E, Lawler J: CD36 directly mediates cytoadherence of Plasmodium falciparum parasitized erythrocytes. Cell 58: 95–101, 1989
34. Tao N, Wagner SJ, Lublin DM: CD36 is palmitoylated on both N- and C-terminal cytoplasmic tails. J Biol Chem 271: 22315–22320, 1996
35. Greenwalt DE, Lipsky RH, Ockenhouse CF: Membrane glycoprotein CD36: A review of its roles in adherence, signal transduction, and transfusion medicine. Blood 80: 1105–1115, 1992
36. Greenwalt DE, Mather IH: Characterization of an apically derived epithelial membrane glycoprotein from bovine milk, which is expressed in capillary endothelia in diverse tissues. J Cell Biol 100: 397–408, 1985
37. Swerlick RA, Lee KH, Wick TM: Human dermal microvascular endothelial but not human umbilical vein endothelial cells express CD36 *in vivo* and *in vitro*. J Immunol 148: 78–83, 1992
38. Ryeom SW, Sparrow JR, Silverstein RL: CD36 participates in the phagocytosis of rod outer segments by retinal pigment epithelium. J Cell Sci 109: 387–395, 1996
39. Endemann G, Stanton LW, Madden KS, *et al.*: CD36 is a receptor for oxidized low density lipoprotein. J Biol Chem 268: 11811–11816, 1993
40. Ryeom SW, Silverstein RL, Scotto A: Binding of anionic phospholipids to retinal pigment epithelium may be mediated by the scavenger receptor CD36. J Biol Chem 271: 20536–20539, 1996
41. Silverstein RL, Asch AS, Nachman RL: Glycoprotein IV mediates thrombospondin-dependent platelet-monocyte and platelet-U937 cell adhesion. J Clin Invest 84: 546–552, 1989
42. Lisanti MP, Scherer PE, Vidugiriene J: Characterization of caveolin-rich membrane domains isolated from an endothelial-rich source: Implications for human disease. J Cell Biol 126: 111–126, 1994
43. Dorahy DJ, Lincz LF, Meldrum CJ: Biochemical isolation of a membrane microdomain from resting platelets highly enriched in the plasma membrane glycoprotein CD36. Biochem J 319: 67–72, 1996
44. Lee H, Woodman SE, Engelman JA: Palmitoylation of caveolin-1 at a single site (Cys-156) controls its coupling to the c-Src tyrosine kinase. J Biol Chem 12: 12, 2001
45. Savill J, Hogg N, Ren Y: Thrombospondin cooperates with CD36 and the vitronectin receptor in macrophage recognition of neutrophils undergoing apoptosis. J Clin Invest 90: 1513–1522, 1992
46. Dawson DW, Pearce SF, Zhong R: CD36 mediates the *in vitro* inhibitory effects of thrombospondin-1 on endothelial cells. J Cell Biol 138: 707–717, 1997
47. Jimenez B, Volpert OV, Crawford SE: Signals leading to apoptosis-dependent inhibition of neovascularization by thrombospondin-1. Nat Med 6: 41–48, 2000
48. Sfeir Z, Ibrahimi A, Amri E: Regulation of FAT/CD36 gene expression: Further evidence in support of a role of the protein in fatty acid binding/transport. Prostaglandins Leukot Essent Fatty Acids 57: 17–21, 1997
49. Luiken JJ, van Nieuwenhoven FA, America G: Uptake and metabolism of palmitate by isolated cardiac myocytes from adult rats: Involvement of sarcolemmal proteins. J Lipid Res 38: 745–758, 1997
50. Luiken JJ, Willems J, van der Vusse GJ: Electrostimulation enhances FAT/CD36-mediated long-chain fatty acid uptake by isolated rat cardiac myocytes. Am J Physiol Endocrinol Metab 281: E704–712, 2001
51. Binas B, Danneberg H, McWhir J: Requirement for the heart-type fatty acid binding protein in cardiac fatty acid utilization. Faseb J 13: 805–812, 1999

52. Glatz JF, van Nieuwenhoven FA, Luiken JJ: Role of membrane-associated and cytoplasmic fatty acid-binding proteins in cellular fatty acid metabolism. Prostaglandins Leukot Essent Fatty Acids 57: 373–378, 1997
53. Aitman TJ, Glazier AM, Wallace CA: Identification of Cd36 (Fat) as an insulin-resistance gene causing defective fatty acid and glucose metabolism in hypertensive rats. Nat Genet 21: 76–83, 1999
54. Pravenec M, Landa V, Zidek V: Transgenic rescue of defective Cd36 ameliorates insulin resistance in spontaneously hypertensive rats. Nat Genet 27: 156–158, 2001
55. Knapp FF, Jr., Kropp J, Franken PR: Pharmacokinetics of radioiodinated fatty acid myocardial imaging agents in animal models and human studies. Quart J Nucl Med 40: 252–269, 1996
56. Li AC, Brown KK, Silvestre MJ: Peroxisome proliferator-activated receptor gamma ligands inhibit development of atherosclerosis in LDL receptor-deficient mice. J Clin Invest 106: 523–531, 2000

Molecular and Cellular Biochemistry **239**: 199–202, 2002.

Intestinal lipid absorption is not affected in CD36 deficient mice

Jeltje R. Goudriaan,[1] Vivian E.H. Dahlmans,[1] Maria Febbraio,[2] Bas Teusink,[1] Johannes A. Romijn,[3] Louis M. Havekes[1,4] and Peter J. Voshol[1,3]

[1]*TNO-Prevention and Health, Gaubius Laboratory, Leiden, The Netherlands;* [2]*Division of Hematology/Oncology, Cornell University, New York, USA;* [3]*Department of Endocrinology and Diabetes and* [4]*Department of Cardiology and General Internal Medicine, Leiden University Medical Center, Leiden, The Netherlands*

Abstract

Increasing evidence has implicated the membrane protein CD36 (or fatty acid translocase, FAT) to be involved in high affinity fatty acid uptake. CD36 is expressed in tissues active in fatty acid metabolism, like adipose tissue and skeletal and cardiac muscle, but also in intestine. CD36 is localized in the intestine mainly in the jejunal villi, where it is confined to enterocyte apical membrane.

The aim was to determine the role of CD36 in intestinal lipid absorption. Lipid absorption was determined by administering ^{3}H-labeled triolein and ^{14}C-labeled palmitic acid as an olive oil bolus by intragastric gavage and determine appearance of ^{3}H and ^{14}C label in plasma, after blocking lipolysis by i.v. injections of Triton WR 1339. Surprisingly, no differences in plasma appearance of ^{3}H-label or ^{14}C-label were observed in CD36$^{-/-}$ mice compared to wild type controls. These results suggest that CD36 does not play a role in intestinal lipid absorption after an acute lipid load. (Mol Cell Biochem **239**: 199–202, 2002)

Key words: CD36, lipoproteins, fatty acid transport, triglyceride metabolism, transgenic mice

Introduction

Fatty acid translocase (FAT), a membrane fatty acid transporter, also called CD36, belongs to the scavenger receptor family class B and is involved in binding of several ligands, e.g. oxidated LDL, long-chain fatty acids (LCFA), and the proteins collagen and thrombospondin [1–5].

Findings by Abumrad *et al.* [2], showed that CD36 is abundant in peripheral tissues active in fatty acid (FA) metabolism like skeletal and cardiac muscle, adipose tissue but also in intestine, spleen and testis. Low expression of CD36 was found in liver [2].

In addition to its role as scavenger receptor, it was found that CD36 is involved in high affinity uptake of fatty acids in adipocytes, and cardiac and skeletal muscle [2, 6–8].

In line with this, Coburn *et al.* [9] showed that fatty acid uptake, as determined by the uptake of a slowly oxidized FA analog 15-(*p*-iodophenyl)-3-(*R,S*)-methyl pentadecanoic acid (BMIPP) is strongly impaired in muscle and adipose tissue of CD36 deficient mice. This impairment in peripheral FA uptake is reflected by the increased plasma FA concentrations measured in these CD36 deficient mice [10].

In the intestine CD36 is localized mainly in the jejunal villi, where it is confined to enterocyte apical membrane. This is the main site of FA (lipid) absorption and since CD36 acts as a FA transporter it was supposed to play a role in intestinal uptake of fatty acids [7].

The aim of the present study was to determine the impact of CD36 deficiency on food intake and intestinal lipid absorption. We clearly show that intestinal lipid absorption is not affected in CD36 deficient mice, although daily food intake seems to be reduced in these mice.

Address for offprints: P.J. Voshol, TNO-Prevention and Health, Gaubius Laboratory, Division VBO, Zernikedreef 9, 2333 CK Leiden, P.O. Box 2215, 2301 CE Leiden, The Netherlands (E-mail: pj.voshol@pg.tno.nl)

Materials and methods

Animals

CD36$^{-/-}$ mice were generated by targeted homologous recombination and back-crossed 4 times to C57Bl/6. Wild type control littermates (CD36$^{+/+}$) were bred from the same cross and were therefore of identical genetic background [10]. Mice (both males and females) used in the experiments were 4–6 months of age and were housed under standard conditions with free access to water and food (standard rat-mouse chow diet). Experiments were performed after overnight fast as indicated.

Food intake

Food intake was monitored per cage during a period of 3 days. Average food intake per mouse per 24 h was calculated.

Plasma lipid analysis

In all experiments, blood was collected by tail bleeding in chilled paraoxonized capillary tubes to prevent *in vitro* lipolysis [11]. These tubes were placed in ice and immediately centrifuged at 4°C. Levels of total plasma cholesterol, plasma triglycerides (TG) (without free glycerol) and free fatty acids (FFAs) were determined using enzymatically available kits #236691 (Boehringer Mannheim), #337-B (Sigma GPO-Trinder kit, St. Louis, MA, USA) and a Nefa-C kit (Wako Chemicals, Neuss, Germany), respectively.

Intestinal lipid absorption

To study the intestinal fatty acid and lipid uptake, CD36$^{-/-}$ and CD36$^{+/+}$ female mice, were fasted (overnight) and injected intravenously with 500 mg Triton WR 1339 per kg body weight as a 15% solution in 0.9% NaCl. Serum lipoprotein clearance is virtually completely inhibited under these circumstances [12].

Directly after the Triton injection, mice were given an intragastric 200 µl olive oil bolus with ^{3}H-labeled triglycerides (7 µCi glycerol-[^{3}H]-oleate (Amersham)) and ^{14}C-labeled fatty acids (2 µCi [9,10(n)-^{14}C] palmitic acid (Amersham). Blood samples were drawn 30, 60, 90, 120, 180 and 240 min after the bolus and plasma was collected after centrifugation. The amount of ^{3}H and ^{14}C radioactivity was determined in plasma and TLC analysis revealed that > 90 % of the label appeared in the TG fraction.

Statistical analysis

The Mann–Whitney nonparametric test for two independent samples was used to define differences between CD36$^{-/-}$ and CD36$^{+/+}$ mice. The criterion for significance was set at $p < 0.05$. All data are presented as means ± S.D.

Results and discussion

Since comparable data were obtained in both female and male mice, only data from male mice are presented. Table 1 shows the basic characteristics determined in CD36$^{-/-}$ and control mice. Plasma free fatty acids and triglyceride concentrations were found to be higher in CD36$^{-/-}$ compared to wild type mice. No significant difference was found in plasma cholesterol concentration. These data show that our mice as obtained from the original breeding colony [10] are comparable in their phenotype. Body weight of CD36$^{-/-}$ male mice was found to be lower compared to age matched wild type mice. Similar findings, although less pronounced were observed with females (results not shown). These data suggest altered food intake and/or lipid absorption in these CD36 deficient mice. Daily food intake showed a significant lower intake in CD36$^{-/-}$ mice compared to wild types. The difference was more pronounced in male mice, with no significant difference in female mice (data not shown). Decreased body weight and decreased food intake especially in male CD36$^{-/-}$ mice suggest alterations in energy consumption, physical activity and/or intestinal lipid absorption in these animals. Food

Table 1. Body weight, food intake and plasma lipid concentrations in CD36$^{-/-}$ and CD36$^{+/+}$ mice

Genotype	Body weight	Food intake	Plasma lipid levels (mmol/L)		
	(g)	(g/24 h/mouse)	TC	TG	FFA
CD36$^{+/+}$	29.7 ± 3.1	4.6 ± 0.7	1.6 ± 0.0	0.2 ± 0.0	0.8 ± 0.1
CD36$^{-/-}$	24.5 ± 1.5*	3.9 ± 0.3*	1.5 ± 0.1	0.4 ± 0.1*	1.7 ± 0.2*

At 16 weeks of age body weight was measured and food intake was monitored per cage. Plasma cholesterol (TC), triglyceride (TG) and free fatty acid (FFA) levels were measured after overnight fasting as described in the Materials and methods section. Values represent the mean ± S.D. of 7 CD36$^{+/+}$ and 8 CD36$^{-/-}$ mice (body wt. and food intake), 5 CD36$^{+/+}$ and 6 CD36$^{-/-}$ mice (plasma lipid levels). *$p < 0.05$, indicating the difference between CD36$^{-/-}$ and CD36$^{+/+}$ mice, using nonparametric Mann–Whitney tests.

intake is partly correlated with energy consumption which in return is related to body weight. This might suggest that the lower daily food intake in CD36 deficient mice could be partly explained by their decrease in body weight.

Since CD36 is highly expressed in the apical membrane of enterocytes [2, 7] and was suggested to play a role in intestinal lipid absorption, we questioned whether CD36$^{-/-}$ mice had an altered intestinal lipid absorption. Intestinal lipid absorption was determined by administering ^{3}H-labeled triolein and ^{14}C-labeled palmitic acid by intragastric gavage and determine appearance of ^{3}H and ^{14}C label in plasma, after blocking lipolysis by i.v. injections of Triton WR 1339. This double label test was performed to test the possibility that lipid malabsorption in CD36 deficient mice is due to impaired lipolysis of TG in the intestinal lumen [13]. And since CD36 transports free fatty acids, labeled palmitic acid was used to see effects of CD36 deficiency independent of possible intestinal lipolysis defects in these mice. Surprisingly, no differences in plasma appearance of ^{14}C-label was observed in CD36$^{-/-}$ compared to wild type mice (Fig. 1). These results indicate apparent normal absorption of free fatty acids from the intestinal lumen. Furthermore, lipolysis in the intestinal lumen appears to be normal since the ^{3}H-label appeared with equal rate in CD36$^{-/-}$ and wild type mice. These results further show that sequential effects in the enterocytes, uptake, esterification and chylomicron formation is also not altered in these CD36 deficient mice. Impaired chylomicron formation leads to strongly decreased plasma appearance rates of label after an acute lipid load in mice [14].

These data imply that either CD36 is not the main player in intestinal lipid absorption or that due to counter regulation of other fatty acid transporters, lipid absorption is not affected. Other transporters which are suggested to play a role in intestinal lipid absorption are liver-type fatty acid binding protein (L-FABP), intestinal FABP (I-FABP) [15] and ileal lipid binding protein (I-LBP) [16]. Further analysis of expression levels of these suggested intestinal lipid transporters in intestines of CD36 deficient mice would be necessary to elucidate this point.

Overall these data suggest possible alteration in body composition in CD36 deficient mice, possibly related to alterations in energy metabolism and/or physical activity in these CD36$^{-/-}$ mice. Preliminary data showed indeed decreased adipose tissue mass in CD36 mice compared to wild type animals (Personal observations, Febbraio 2001). Further research is needed to elucidate body composition and adipose tissue in these mice in combination with energy expenditure and physical activity to give an answer as to why CD36 deficient mice have lower body weight and food intake.

In conclusion we clearly show that CD36 although expressed at the apical membrane on enterocytes is not essential in intestinal lipid absorption.

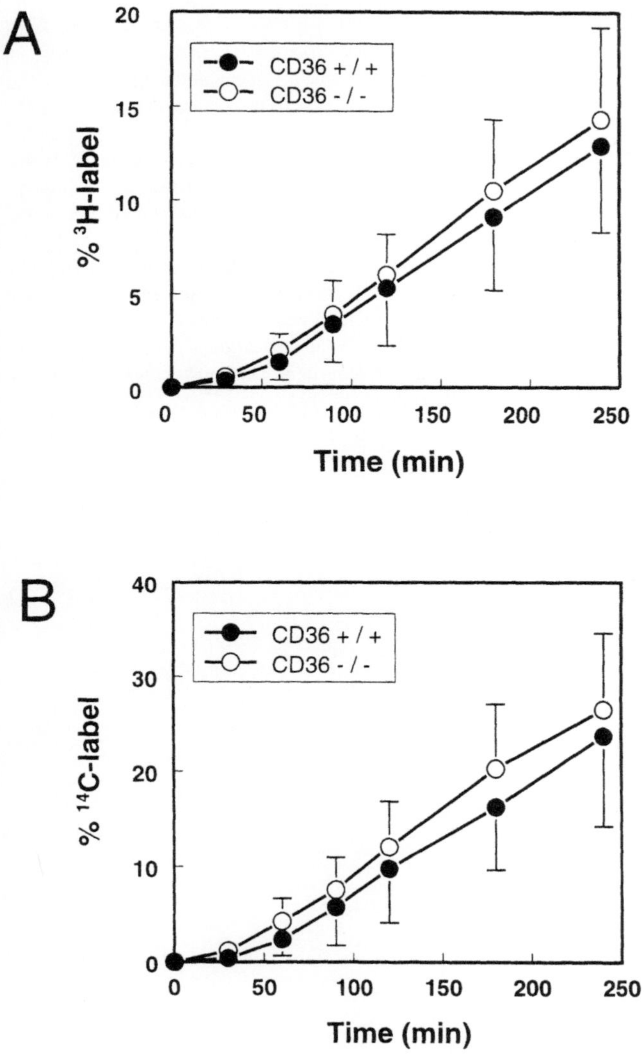

Fig. 1. Intestinal lipid absorption of ^{3}H-labeled triglycerides (A) and ^{14}C-labeled fatty acids (B) in CD36$^{-/-}$ and CD36$^{+/+}$ mice. Mice were fasted (overnight) and injected intravenously with Triton. Directly after the Triton injection, mice were given an olive oil bolus including ^{3}H-labeled triolein and ^{14}C-labeled palmitic acid by gavage. The amount of plasma radioactivity (^{3}H and ^{14}C) was determined and shown as a percentage of the given bolus. Values represent the mean ± S.D. of 4 CD36$^{-/-}$ and 5 CD36$^{+/+}$ mice.

Acknowledgments

We are grateful to Miek Jong and Martin Muurling for their technical support and discussions to the study. This research was supported by the Netherlands Foundation for Scientific Research (903-39-192 and 903-39-194) and by the Netherlands Heart Foundation (97067).

References

1. Endemann G, Stanton LW, Madden KS, Bryant CM, White RT, Protter AA: CD36 is a receptor for oxidized low density lipoprotein. J Biol Chem 268: 11811–11816, 1993
2. Abumrad NA, El-Maghrabi MR, Amri EZ, Lopez E, Grimaldi PA: Cloning of a rat adipocyte membrane protein implicated in binding or transport of long-chain fatty acids that is induced during preadipocyte differentiation. J Biol Chem 268: 17665–17668, 1993
3. Tandon NN, Kralisz U, Jamieson GA: Identification of glycoprotein IV (CD36) as a primary receptor for platelet-collagen adhesion. J Biol Chem 264: 7576–7583, 1989
4. Silverstein RL, Asch AS, Nachman RL: Glycoprotein IV mediates thrombospondin-dependent platelet-monocyte and platelet-U937 cell adhesion. J Clin Invest 84: 546–552, 1989
5. Greenwalt DE, Lipsky RH, Ockenhouse CF, Ikeda H, Tandon NN, Jamieson GA: Membrane glycoprotein CD36: A review of its role in adherence, signal transduction, and transfusion medicine. Blood 80: 1105–1115, 1992
6. van Nieuwenhoven FA, Verstijnen CPHJ, Abumrad NA, Willemsen PHM, van Eys GJJM, van der Vusse GJ, Glatz JFC: Putative membrane fatty acid translocase and cytoplasmic fatty acid-binding protein are co-expressed in rat heart and skeletal muscles. Biochem Biophys Res Commun 207: 747–752, 1995
7. Poirier H, Degrace P, Niot I, Bernard A, Besnard P: Localization and regulation of the putative membrane fatty-acid transporter (FAT) in the small intestine, comparison with fatty acid-binding proteins (FABP). Eur J Biochem 238: 368–373, 1996
8. Abumrad NA, Harmon C, Ibrahimi A: Membrane transport of long-chain fatty acids: Evidence for a facilitated process. J Lipid Res 39: 2309–2318, 1998
9. Coburn CT, Knapp FF Jr, Febbraio M, Beets AL, Silverstein RL, Abumrad NA: Defective uptake and utilization of long chain fatty acids in muscle and adipose tissues of CD36 knockout mice. J Biol Chem 275: 32523–32529, 2000
10. Febbraio M, Abumrad NA, Hajjar DP, Sharma K, Cheng W, Frieda S, Pearce A, Silverstein RL: A null mutation in murine CD36 reveals an important role in fatty acid and lipoprotein metabolism. J Biol Chem 274: 19055–19062, 1999
11. Zambon A, Hashimoto SI, Brunzell JD: Analysis of techniques to obtain plasma for measurement of levels of free fatty acids. J Lipid Res 34: 1021–1028, 1993
12. Aalto-Setala K, Fisher EA, Chen X, Chajek-Shaul T, Hayek T, Zechner R, Walsh A, Ramakrishnan R, Ginsberg HN, Breslow JL: Mechanism of hypertriglyceridemia in human apolipoprotein (Apo) CIII transgenic mice. J Clin Invest 90: 1889–1900, 1992
13. Kalivianakis M, Minich DM, Bijleveld CM, van Aalderen WM, Stellaard F, Laseur M, Vonk RJ, Verkade HJ: Fat malabsorption in cystic fibrosis patients receiving enzyme replacement therapy is due to impaired intestinal uptake of long-chain fatty acids. Am Clin Nutr 69: 127–134, 1999
14. Voshol PJ, Minich DM, Havinga R, Elferink RP, Verkade HJ, Groen AK, Kuipers F: Postprandial chylomicron formation and fat absorption in multidrug resistance gene 2P-glycoprotein-deficient mice. Gastroenterology 118: 173–182, 2000
15. Stremmel W, Lotz G, Strohmeyer G, Berk PD: Identification, isolation and partial characterization of a fatty acid binding protein from rat jejunal microvillous membranes. J Clin Invest 75: 1068–1076, 1985
16. Lin MC, Kramer W, Wilson FA: Identification of cytosolic and microsomal bile acid-binding proteins in rat ileal enterocytes. J Biol Chem 265: 14986–14995, 1990

Molecular and Cellular Biochemistry **239**: 203–211, 2002.

Surface expression of fatty acid translocase (FAT/CD36) on platelets in myeloproliferative disorders and non-insulin dependent diabetes mellitus: Effect on arachidonic acid uptake

Hasib Salah-Uddin,[1] Margaret J. Gordon,[1] Isobel Ford,[2] Narendra N. Tandon,[3] Mike Greaves[2] and Asim K. Duttaroy[1,4]

[1]*Rowett Research Institute, Aberdeen, Scotland, UK;* [2]*Department of Medicine and Therapeutics, University of Aberdeen, Scotland, UK;* [3]*Otsuka Pharmaceutical Inc., Maryland, MA, USA;* [4]*Institute for Nutrition Research, University of Oslo, Blindern, Oslo, Norway*

Abstract

Increased platelet reactivity has been implicated in the vascular complications of myeloproliferative diseases and diabetes mellitus. The mechanisms of platelet hyperresponsiveness have not been fully explained. Expression of CD36 or fatty acid translocase (FAT) and its role in arachidonic acid (AA) uptake by platelets were examined in subjects with myeloproliferative disorders (MPD), those with non-insulin-dependent diabetes mellitus (NIDDM), and in normal, healthy, age-matched controls. Surface expression of CD36 on platelet membranes was increased in MPD (10.94 ± 0.76 pmol/mg protein) compared with normal controls (6.94 ± 0.48 pmol/mg protein), $p < 0.001$. Total platelet content of CD36 was also significantly higher (32.1 ± 0.61 pmol/mg protein, $p < 0.01$) compared with those in sex and age matched normal controls (25.7 ± 1.09 pmol/mg protein). In contrast, platelet surface expression of CD36 in NIDDM (6.5 ± 0.56 pmol/mg protein) was not significantly different from those of normal controls despite higher total content of CD36 (32.8 ± 1.2, pmol/mg protein, $p < 0.01$). Intact MPD platelets bound significantly more arachidonic acid (AA) (1.53 ± 0.16 nmol/mg protein, $p < 0.05$), compared with controls (1.12 ± 0.07 nmol/mg protein) or NIDDM subjects (1.16 ± 0.16 nmol/mg protein). The capacity of MPD platelet membranes to bind ^{14}C-AA was also increased (1.72 ± 0.25 nmol/mg protein, $p < 0.05$) compared with that of controls (1.62 ± 0.05 nmol/mg protein) and of NIDDM (1.22 ± 0.08 nmol/mg protein). This is consistent with higher surface expression of CD36 in MPD platelets. Membrane fatty acid analysis indicated that the % of AA in platelet phospholipids was significantly lower in MPD (3.15 ± 0.81%) compared with the controls (5.62 ± 1.7%, $p < 0.05$. The AA content of diabetic platelets (4.82 ± 1.1%) was not significantly different from normal controls. In summary, both total and surface expression of CD36 are increased in MPD, consistent with an enhanced capacity for uptake of AA by platelets. Increased expression of CD36 in platelets may play a role in the vaso-occlusive manifestations of MPD. (Mol Cell Biochem **239**: 203–211, 2002)

Key words: FAT, CD36, fatty acid translocase, arachidonic acid, platelets, diabetes, myeloproliferative disorders, fatty acid uptake, thromboxane A2, membrane phospholipids

Introduction

Vascular occlusions are a major cause of morbidity and death in myeloproliferative disorders (MPD) [1] and diabetes mellitus [2]. In MPD, multiple platelet abnormalities result from clonally derived megakaryocytopoiesis [3]. The abnormalities leading to thrombosis have not been fully characterised, but may include an increased thromboxane A_2 (TxA_2) synthesis, increased granule secretion and spontaneous platelet aggregation [4, 5]. Abnormalities in platelet

Address for offprints: A.K. Duttaroy, Institute for Nutrition Research, University of Oslo, P.O. Box 1046, Blindern, N-0316 Oslo, Norway (E-mail: a.k.duttaroy@basalmed.uio.no)

membrane glycoproteins (GP) and altered platelet aggregation were also observed in MPD platelets [5–7]. Variable abnormalities in membrane GP content of non-stimulated platelets have been described in MPD [8], however the pattern of GP redistribution in these patients has rarely been addressed [9, 10]. Platelet abnormalities with increased platelet reactivity, are also a feature of diabetes mellitus. They may contribute to the pathogenesis of microvascular complications [11]. Platelet membrane GPs serve a variety of functions including as receptors for ligands necessary for platelet adhesion to leukocytes, fibrin, endothelial cells and extracellular matrix. For example, the GPIIb/IIIa complex possesses activation dependent receptors for fibrinogen, fibronectin and von Willebrand factor (vWf) essential for platelet aggregation and adherence [12]. In contrast, GPIV (CD36) which is present on human platelets and other cells including monocytes/macrophages, and placenta [13] mediates platelet adhesion to monocytes via bound thrombospondin [14–16]. CD36 is also known as fatty acid translocase (FAT) and is composed of a highly glycosylated polypeptide chain with an apparent molecular mass of 88 kDa [17]. It is a multifunctional protein and has a number of putative ligands including oxidised LDL, and fatty acids [13]. A series of studies have shown that CD36/FAT is involved in free fatty acid uptake [13]. It is highly induced during adipocyte differentiation and in diabetes, and the induction is paralleled with an increase in free fatty acid (FFA) uptake by the cells [18]. Furthermore, we have shown recently that CD36/FAT is involved in arachidonic acid uptake by human platelets [19]. This observation was later confirmed by others through lack of arachidonic acid-induced aggregation in the presence of anti-CD36 antibody [20]. Arachidonic acid is the substrate for synthesis of TxA_2, a potent platelet agonist and vasoconstrictor [21, 22]. To serve as the substrate for synthesis of TXA_2, arachidonic acid is liberated from membrane phospholipids by phospholipases [21]. However, exogenous arachidonic acid is readily taken up by platelets and consequently the need for release of arachidonic acid from membrane phospholipids may be bypassed. Absorbed arachidonic acid is available to platelets and is converted rapidly to TxA_2 [21]. The extent to which this process occurs depends largely on the concentration of exogenous arachidonic acid as well as the arachidonic acid uptake activity of the platelets. Arachidonic acid uptake has also been reported to be significantly higher in platelets from patients with diabetes mellitus [2, 23–25]. However very little information is available regarding platelet arachidonic acid uptake and metabolism in MPD. We postulated that altered expression of CD36/FAT could lead to hyperactivation of platelets in diabetes and MPD through increased uptake and metabolism of arachidonic acid. The aim of the present study was therefore, to investigate both CD36/FAT expression and arachidonic acid uptake by platelets from subjects with MPD and NIDDM.

Materials and methods

Materials

[^{14}C]Arachidonic acid (58 mCi/mmol) was obtained from Amersham, UK. Na^{125}I was obtained from ICN Biomedicals, UK. The monoclonal antibodies to CD36 (131.1, 131.2 and 131.7), supplied by Dr. N.N. Tandon (Otsuka America Pharmaceutical). These monoclonal antibodies were obtained from mouse ascites fluid and prepared using a hybridoma cell line and characterised [26–29]. Triton X-114, protease inhibitors, essentially fatty acid free bovine serum albumin and unlabelled arachidonic acid were obtained from Sigma, UK. All other reagents used were of analytical grade.

Subjects and study protocol

Blood samples were obtained from healthy donors and patients (MPD and NIDDM) upon informed consent and in accordance with the Declaration of Helsinki. The age range of patients suffering from MPD and NIDDM was between 65 and 86 years of age. Control subjects were comparable for age and sex. This study protocol was approved by the Joint Ethical Committee of the Grampian Health Board and the University of Aberdeen (Scotland, UK).

Purification of CD36 from human platelets

Outdated platelet concentrates were obtained from the North East Scotland Blood Transfusion Service, Aberdeen. Purification of CD36 was carried out according to the method previously described [30]. Platelets (55×10^9) were suspended in 5 mM EGTA and centrifuged at 3200 × g for 3.5 min at 22°C to sediment red blood cells. The supernatant was then removed and further centrifuged at 3200 × g for 15 min to pellet platelets which were then resuspended and washed twice with buffer (96.5 mM NaCl, 85.7 mM glucose, 10 mM Tris.HC1, 1 mM EDTA, and 5 mM EGTA, pH 7.4). The platelets were then suspended in the above buffer without EDTA and EGTA. CD36 was then purified to an electrophoretic homogeneity using these platelets as starting material. The purified protein was stored at –80°C for further use.

Electrophoretic separation and Western blot analysis of membrane proteins

Polyclonal antisera against the purified CD36 was used [31]. Western blot analysis of solubilised platelet membrane proteins and purified CD36 was carried out as described by Campbell and Dutta-Roy [32]. After electrophoresis, proteins were transferred onto PVDF membrane at 70°C for 2 h. The membrane was probed for the presence of CD36 by incubating with rabbit polyclonal antiserum to CD36. Antibody-antigen complexes were then detected with the secondary antibody (horseradish peroxidase-linked anti-rabbit IgG fraction of donkey polyclonal antiserum, Scottish Antibody Production Unit, UK), as described [33].

Protein assay

Protein concentration of samples was determined using the Bio-Rad Protein Microassay [34].

Iodination of monoclonal antibodies (131.1 and 131.7) against CD36. Antibodies were iodinated using the chloramine T procedure as described by Legrand *et al.* [35]. Typically, 500 μCi of stock $Na^{125}I$ was added to 250 μg of antibodies (131.7) in a total vol. of 625 μl of 0.5 M sodium phosphate (pH7.5) in a 1.5 ml tube, followed by the addition of 5 μl of chloramine T (10 mg/ml in phosphate buffered saline (PBS), pH7.4). After incubation at room temperature for 90 sec, 50 μl of chloramine T stop reagent (2.4 mg/ml $Na_2S_2O_5$, 10 mg/ml tyrosine (saturated), 10% glycerol, 0.1% xylene cylanol in PBS), kept on ice was added. Iodinated protein was then purified using a Sephadex G-25 (5 ml bed volume) column (Pharmacia-LKB Biotechnology, Sweden). If the incorporation exceeded 95% the antibody was stored in small aliquots 100–200 μl /vial at –80°C.

Platelet isolation from freshly drawn blood

Blood (60 ml) from normal, or patients with MPD and NIDDM was collected into acid citrate anticoagulant (final concentration 15 mM). Platelet-rich plasma (PRP) was obtained by centrifugation at 180 × g for 10 min at 20°C. Platelets were washed by repeated centrifugations at 670 × g for 10 min in a modified Tyrode's buffer (0.137 M NaCl, 2.68 mM KCl, 11.9 mM $NaHCO_3$, 0.146 mM NaH_2PO_4, 1 mM $MgCl_2$, pH 6.5) containing 3.5 mg/ml bovine serum albumin (BSA), 100 nM PGE_1 and 1 U/ml apyrase [35]. Platelets were resuspended (~1.25×10^8 platelets/ml) in modified Tyrode's buffer. pH 7.4, containing 2 mM $CaCl_2$, 1 mM $MgCl_2$ and 3.5 mg/ml of BSA [36]. Platelet suspensions contained less than 0.005% monocytes, assessed by microscopic examination after methanol/ methylene blue/ eosin staining (Bayer Diagnostics, USA).

Platelet lysate preparation

Platelets (~2×10^9 platelets/ml) as isolated above were resuspended in Tris-buffered saline, TBS (10 mM Tris-HCl, 150 mM NaCl, pH 7.4) containing protease inhibitors. Membrane proteins were solubilised in 1% (v/v) Triton X-100 by gentle agitation for 30 min at 4°C. Soluble proteins were separated by ultracentrifugation for 30 min at 100,000 × g and 4°C, and immediately frozen at –80°C.

Platelet membrane preparation

Platelets as isolated above, washed in a buffer containing 1 mM EDTA, 0.15 M NaCl and 20 mM Tris.HCl, pH 7.5 was used instead of modified Tyrode's buffer to wash platelets. Platelets were then resuspended in the above buffer (~1×10^9 platelets/ml). Crude membranes were prepared by lysing washed platelets by repeated freeze/thaw using liquid N_2 followed by centrifugation at 10,000 × g for 10 min at 4°C to remove cell debris and unbroken platelets and then at 100,000 × g for 30 min to sediment the platelet membrane fraction, as described [37].

Radioimmunoassay for total CD36

Radioimmunoassay for total CD36 in platelets was carried out as described [8]. The immunocapture of CD36 by a solid-phase adsorbed monoclonal antibody, MoAb 131.2 was carried out. Polystyrene tubes were coated overnight at room temperature with 100 μl of MoAb 131.2 at a concentration of 10 μg/ml in TBS containing 2 mM Ca^{2+}. After four washes with TBS containing 2 mM Ca^{2+}, 1 mM Mg^{2+}, 0.05% (v/v) Tween 20 and 0.02% (w/v) sodium azide (TBS-Tween), the tubes were incubated for 60 min with 100 μl of a 1.5% (w/v) solution of BSA in TBS in order to block unreacted sites. Tubes were then washed four times with TBS-Tween and incubated for 150 min with 100 μl of purified CD36 or platelet lysate, both diluted in TBS containing 0.1% (v/v) Triton X-100. After four washes, the tubes were incubated for 150 min with 100 μl of radioiodinated monoclonal antibody 131.1 (~500,000 cpm/well) at a concentration of 10 μg/ml in TBS-Tween. After four final washes, the radioactivity of each tube was counted in a Cobra 5003 gamma counter (Canberra-Packard, UK). In control experiments, tubes were coated with monoclonal antibody, 131.2 and incubated with TBS-Tween lacking platelet lysate. Background radioactivity (less than 10%) was subtracted from the total radioactivity and the amount of monoclonal antibody, 131.1 bound was expressed in nanograms of ^{125}I-monoclonal antibody, 131.1 bound per well. Results obtained with tested samples were related to the standard curve obtained on the same day with purified CD36 and expressed as micrograms of CD36 per 100 μg of platelet proteins. Identical standard curves were obtained when purified CD36 was diluted in platelet lysate from a Nak^a-negative donor [38], showing that under our experimental conditions, no GPIV alteration occurred in the platelet lysate.

Binding of radioiodinated monoclonal antibody (131.7) to platelets

In order to investigate the expression of CD36 on platelet surface, direct binding experiments using ^{125}I-monoclonal antibody (MAB), 131.7, were carried out. Typically, platelet suspensions were incubated with various concentrations of the ^{125}I-monoclonal antibody, 131.7 (1, 2, 4, 6, 8 and 10 µg/ml) for 30 min at room temperature, in order to determine the saturating concentration of antibody in patients' and normal platelets. Aliquots (100 µl) were then layered over 500 µl of silicone oil and centrifuged at 12,000 × g for 5 min. The radioactivity associated with the pellets was counted in a Cobra 5003 gamma counter (Canberra-Packard, UK). Background radioactivity, determined by centrifuging each concentration of antibody in buffer without platelets, was subtracted from the radioactivity in the platelet pellets. Platelet suspensions (10^9 cells/ml) incubated with saturating concentrations of the ^{125}I-MoAb 131.7 were carried out to determine the surface expression of CD36 in platelets of individual group of patients and normal controls. Results are expressed as CD36/mg of platelet protein. Saturating concentrations of monoclonal antibody for control and NIDDM platelets was 4 µg/ml whereas the value for MPD was 6 µg/ml.

Non-specific binding at respective saturating binding concentrations was assessed through competition binding between radiolabelled antibody and a 500-fold excess of unlabelled antibody.

[14C] Arachidonic acid binding assay

Binding [^{14}C]arachidonic acid to platelets was determined as described by Dutta-Roy *et al.* [19]. [^{14}C]Arachidonic acid was mixed with BSA (fatty acid free) to obtain molar ratio of [^{14}C]arachidonic acid to BSA of 1:1 in PBS pH 7.4. Platelet membranes (75 µg) were incubated with [^{14}C]arachidonic acid bound to BSA in a 1:1 molar ratio for 20 min at 37°C in PBS pH 7.4, in a total volume of 500 µl. The non-specific binding was determined from parallel experiments adding a 500-fold excess of unlabelled arachidonic acid to the incubations. At the end of the incubations the unbound ligand was separated from the bound ligand by rapid vacuum filtration of 400 µl aliquots of the mixtures through Whatman GF/C glass fibre microfilters (Whatman, UK) which had been pre-soaked in ice-cold PBS. Each filter was washed with 3 × 5 ml of ice-cold 2 µM BSA in PBS. The filters were then dried and suspended in 5 ml of scintillation fluid (Ultima Gold, Packard, UK) before scintillation counting (Tricarb 1900CA, Packard, UK).

The saturating concentration of AA was determined using various concentrations (0.4, 0.6, 0.8, 1.0, 1.2 and 1.4 µM) of radiolabelled AA. Saturating concentration of radiolabelled AA for patients and normal controls was 1 µM.

Lipid analysis

Platelet and erythrocyte lipids were extracted by the method of Bligh and Dyer [39]. Lipids fractions were then separated by thin layer chromatography (TLC) using thin layer silica plates (Silica Gel, LK5), into cholesterol ester, triacylglycerol, FFA, diacylglycerol, and phospholipids. Total phospholipids were scraped off the plate, extracted 3 times with chloroform:methanol (2:1, v/v) and separated by TLC (LK5 plates) into phosphatidylcholine, phosphatidylserine, phosphatidylethanolamine, phosphatidylinositol using the solvent system chloroform/ethanol/water/triethylamine (30:34:8:35, v/v/v/v) [40]. Lipid fractions were identified by co-chromatography with suitable standards and visualised under UV light after spraying with 0.2% dichlorofluorecene in ethanol. Appropriate bands were scrapped off and fatty acid compositional analysis was performed.

Fatty acid composition was then determined after their derivatisation with pentafluorobenzyl bromide [41]. The derivatised fatty acids were then analysed by gas liquid chromatography using electron-capture detection in a Hewlett Packard 6890 gas chromatography system. Data acquisition and analysis were performed using Hewlett Packard Chem Station software. Pentaflurobenzyl esters (PFBE) of the fatty acid samples were analysed on a 30 m × 0.258 mm, 0.25 µm DB-23 column (Jones Chromatography, UK) using high purity helium as the carrier gas. The ^{63}Ni electron-capture detector used nitrogen gas and was maintained at a temperature of 300°C throughout. The oven temperature was held at 85°C for 2 min, increased to 165°C at a rate of 40°C/ min, then increased to 250°C, at a rate of 3.5°C / min and held for 1 min at 250°C. The PFBE of the fatty acids were quantified relative to the internal standard heptadecanoylpentaflurobenzyl ester.

Statistical analysis

Results were analysed using the Student's *t*-test. The results are presented as the mean ± S.D.

Results

The total CD36/FAT content in platelets isolated from age and sex matched normal controls as well as from MPD and NIDDM patients were measured using a radioimmuno assay. The platelets of MPD patients exhibited a mean level of 30.13 ± 0.61 pmol of CD36/mg protein, whereas NIDDM platelets had 32.88 ± 1.24 pmol of CD36/mg protein compared with that of age and sex matched controls (25.70 ± 1.09 pmol of CD36/mg protein). Despite a broad inter individual variability (Fig. 1), there was a significant difference between the val-

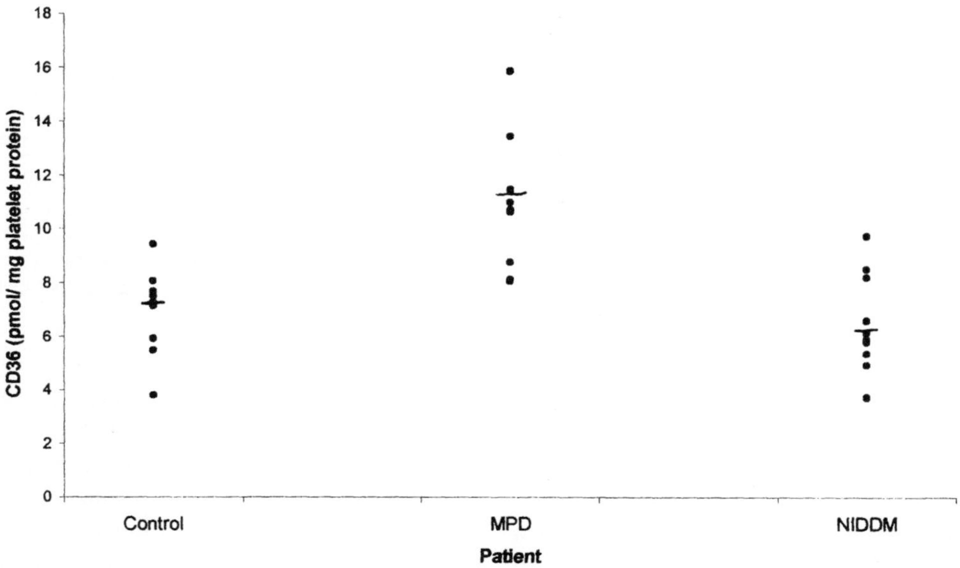

Fig. 1. Surface expression of CD36 on platelets. Each point represents a different subjects, and the mean value of each group is indicated (–). For details please see the Materials and methods section.

ues of patients (MPD and NIDDM) and normal platelets. There was no significant difference between the MPD and NIDDM platelets.

To determine whether this increase in total CD36/FAT content in patients' platelets was correlated with an enhanced surface expression of the molecule, we measured the binding of ^{125}I labelled monoclonal antibody (131.7) to intact platelets from MPD and NIDDM patients. Expression of CD36 on the platelet membrane was increased in MPD (10.94 ± 0.76 pmol/mg protein) compared with normal controls (6.94 ± 0.48 pmol/mg protein), $p < 0.001$. However, despite increase in total CD36 molecules in NIDDM, surface expression of CD36 in platelets of NIDDM patients (6.5 ± 0.56 pmol/mg protein) was not significantly different from those of normal controls (Fig. 2).

The fatty acid composition of platelet phospholipids was given in Table 1. The level of arachidonic acid in MPD platelets (3.1 ± 0.1% of total fatty acids present in platelet membrane phospholipids) appeared to be significantly lower than in age-sex matched controls (5.6 ± 1.7% of total fatty acids present in platelet membrane phospholipids, $p < 0.001$). Whereas no significant difference was observed in arachi-

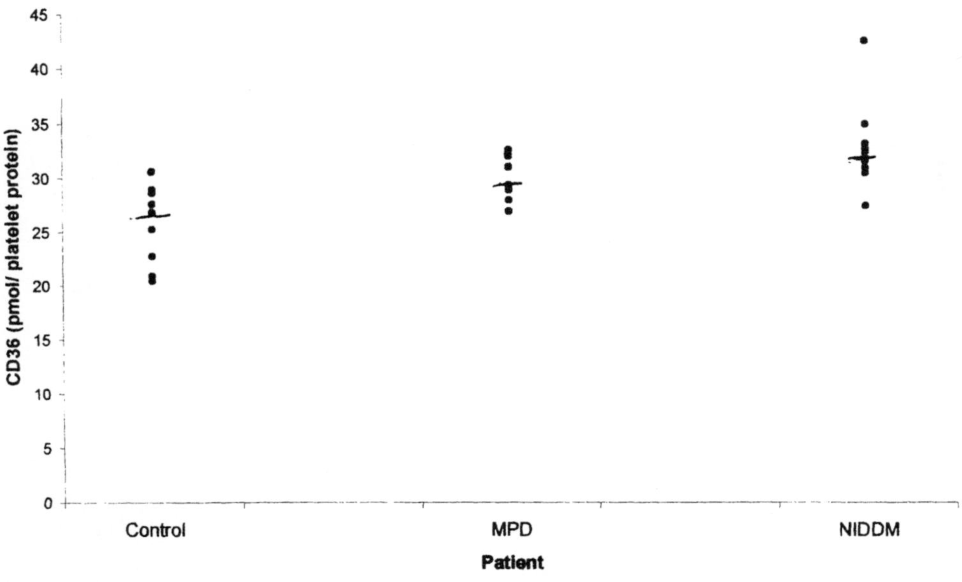

Fig. 2. Total CD36 content of platelets. Each point represents a different subjects and the mean value of each group is indicated (–). For details please see the Materials and methods section.

Table 1. Distribution of fatty acids in platelet phospholipids

% Fatty acid (FA)	Control Mean ± S.E.	MPD Mean ± S.E.	NIDDM Mean± S.E.
14:0	12.1 ± 1.2	10.2 ± 1.0†	10.6 ± 1.4
16:0	26.7 ± 2.1	32.3 ± 2.4‡	30.4 ± 2.3
18:0	14.2 ± 1.7	13.7 ± 1.5	13.2 ± 1.7
18:1 (n-9)	23.8 ± 1.5	23.1 ± 2.3	21.7 ± 2.1
18:2 (n-6)	9.5 ± 1.1	6.4 ± 0.8	7.8 ± 0.8†
18:3 (n-6)	0.9 ± 0.9	0.0 ± 0.0	0.2 ± 0.0
20:4 (n-6)	5.6 ± 1.7	3.1 ± 0.1•	4.8 ± 1.1
20:5 (n-3)	0.7 ± 1.3	0.3 ± 0.2	1.0 ± 0.7
22:6 (n-3)	0.6 ± 0.2	0.4 ± 0.1	0.8 ± 0.4

Values are expressed as mean ± S.E.M. *$p < 0.05$; †$p < 0.01$; ‡$p < 0.005$; •$p < 0.001$ vs. control.

donic acid content between control and NIDDM (4.8 ± 1.1%). Similarly, myristic (14:0) and linoleic (18:2 n-6) acids were lower in MPD compared with control. Whereas palimitic (16:0) and arachidic (20:0) acids were both increased in MPD platelets ($p < 0.005$; $p < 0.001$, respectively) compared with control. No such difference was found in levels of arachidonic acid in erythrocytes between patients with MPD and controls (data not shown). The fatty acid composition of platelet membrane phospholipids is given in Table 2. Distribution of arachidonic acid in platelet membrane phospholipid classes indicated that arachidonic acid content in combined phosphatidylinositol and phosphatidylserine classes were found to be significantly lower in MPD than in controls ($p < 0.001$).

Arachidonic acid uptake was then determined in platelet membranes and intact cells using radiolabelled arachidonic

Table 2. Levels of arachidonic acid in platelet phospholipid fractions

	ARA (ng/mg total lipid)		
	PC	PE	PI + PS
Control	0.24 ± 0.14	0.94 ± 0.78	0.83 ± 1.03
MPD	0.17 ± 0.17	0.38 ± 0.31	0.26 ± 0.36**
NIDDM	0.40 ± 0.61	0.79 ± 0.76	0.84 ± 1.07

PC – phosphatidylcholine; PI + PS – phosphatidylinositol and phosphatidylserine; PE – phosphatidylethanolamine. Values are expressed as mean ± S.D. **$p < 0.001$ vs. control.

acid. Consistent with the increased surface expression of CD36, intact platelets from MPD patients bound significantly more arachidonic acid (1.53 ± 0.16 nmol/mg protein), compared with controls (1.12 ± 0.07 nmol/mg protein) or NIDDM subjects (1.16 ± 0.16 nmol/mg protein) $p < 0.05$. The capacity of MPD platelet membranes to bind ^{14}C-arachidonic acid was also increased (1.72 ± 0.25 nmol of ^{14}C-arachidonic acid bound per mg of protein in MPD ($p < 0.05$); 1.62 ± 0.05 nmol/mg protein in controls; 1.22 ± 0.08 nmol/mg protein in NIDDM) (Fig. 3).

Scatchard analysis of binding data indicate that Kd values for arachidonic acid binding to these platelets did not differ between the normal and patients platelets (0.47 ± 0.09 μM for normal platelets; 0.43 ± 0.09 μM for NIDDM platelets, and 0.57 ± 0.09 μM for MPD platelets, $p > 0.5$). However, B_{max} value for MPD platelets (3.08 ± 0.20 nmol/mg protein) was much higher than those of normal platelets (1.89 ± 0.13 nmol/mg protein) or NIDDM platelets (1.87 ± 0.13 nmol/mg protein, $p < 0.005$).

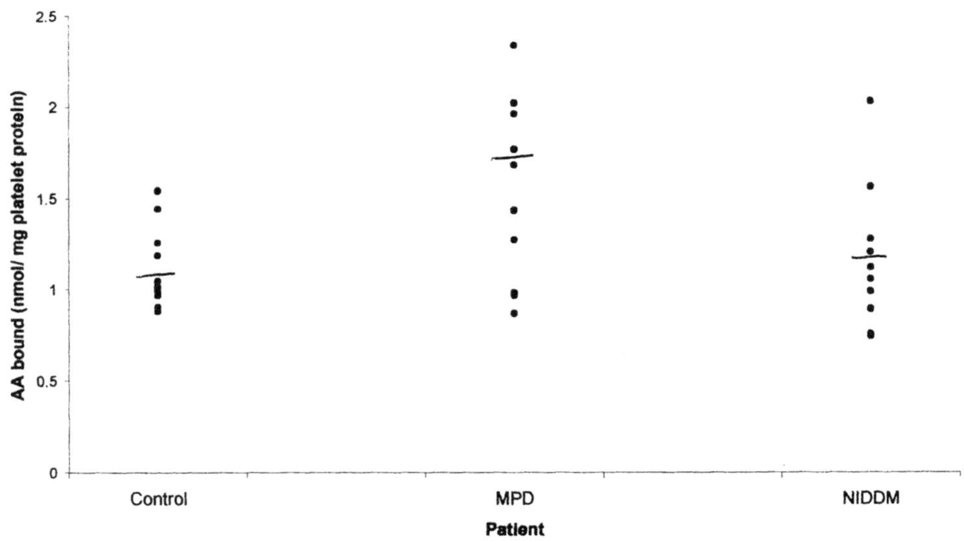

Fig. 3. Arachidonic acid uptake by platelets. Each point represents a different subjects and the mean value of each group is indicated (–). For details please see the Materials and methods section.

Discussion

Here we report that both surface expression and total content of CD36 are increased in MPD platelets compared with those of age and sex matched normal controls. In contrast total, but not surface CD36 is increased in platelets from subjects with type II diabetes. Furthermore, arachidonic acid uptake is also increased in MPD platelets, but not in diabetic platelets, in keeping with the hypothesis that the increased surface expression of CD36/FAT may be responsible for the increased uptake. The increase in CD36 surface expression in MPD platelets previously been reported by Thiebert *et al.* [8]. The extent of surface expression of CD36 may be called into question in the aforementioned study as the antibody, FA6-152, may induce platelet activation (Tandon, 1999, personal communication), which may lead to increased CD36 surface expression from internal pools [10]. The antibody used in this study (MoAb 131.7) does not activate platelets but inhibits platelet adhesion to type 1 and insoluble collagen [28, 29]. The use of MoAb 131.7 in this study, allows a more accurate measurement of CD36. Cellular fatty acid uptake is mediated in part by multiple membrane-associated transport or binding proteins (FABPpm, FAT/CD36, FATP) [13]. Some cell types such as cardiomyocyte, adipocyte or placental trophoblast have multiple membrane-associated transport/binding proteins, and these membrane proteins are proposed to work in tandem for effective extraction of free fatty acids from plasma [13]. In contrast, human platelets contain only one type of membrane fatty acid transporter (CD36/FAT), therefore any alteration in the expression of CD36/FAT in human platelets may have significant effect on fatty acid uptake. We have shown earlier that CD36/FAT is involved, at least in part, in the uptake of exogenously added arachidonic acid by human platelets [19]. Recent studies by Kerkhoff *et al.* [42] have also provided some evidence that the uptake of exogenous arachidonic acid by human umbilical vein endothelial cells was predominantly mediated by FAT/CD36. Irreversible aggregation of platelets depends on intake of arachidonic acid once liberated from platelets and production of its metabolite TxA_2 [20]. Therefore, the involvement of CD36 in arachidonic acid-induced platelet aggregation may be highly significant in diseases characterized by enhanced expression of CD36/FAT as well as increased arachidonic acid uptake by platelets. In a calcium-deficient state, CD36-deficient platelets exhibited a delay and decline of irreversible aggregation following agonist stimulation. Defective aggregation of CD36-deficient platelets in a calcium-deficient state seemed to be due to defective uptake of arachidonic acid [20]. In addition, uptake of oxidised LDL by platelets is also mediated by CD36 [43], and this may enhance TxA_2 synthesis by these platelets [44], as generally observed in MPD platelets. CD36 has been shown to be up-regulated in various cells via induction of a nuclear transcription factor, peroxisome proliferator activated receptor (PPAR) gamma [45]. The mechanisms involved in increased expression of CD36/FAT in platelets (de-nucleated cells) of MPD is not known at present. However, it is possible that this up-regulation of CD36 may occur during megakaryocytopoiesis. Although increased arachidonic acid uptake has been reported in diabetic platelets by several workers [23–25], we did not observe any such increase in arachidonic acid uptake, nor in the surface expression of CD36. This situation is in clear contrast to that of MPD. Whether the degree of glycaemic control could influence the results obtained for arachidonic acid uptake in our diabetic population is unknown, although intake of antioxidants and certain drugs decrease CD36/FAT levels in human macrophages [45, 46]. In NDDM platelet membrane phospholipids , no changes in the levels of arachidonic acid were observed compared with that of controls. It still remains unclear whether there is any alteration of incorporation of endogenous arachidonic acid into platelet membrane phospholipids of NIDDM. Increased [47], reduced [48] and unaltered [49] arachidonic acid levels in membrane phospholipids in NIDDM subjects have been reported.

Despite the increased uptake by MPD platelets, arachidonic acid levels in these platelets are lower. Increased consumption of arachidonic acid to support TxA_2 synthesis in MPD may have contributed to the depression of arachidonic acid in MPD platelets. Our findings of normal arachidonic acid content of corresponding erythrocyte membranes is consistent with this, as our earlier observation that normal aggregation to exogenous arachidonic acid may be accompanied by reduced aggregability and thromboxane synthesis in response to collagen in MPD [50].

Increased levels of CD36 and with increased uptake of arachidonic acid could lead to overproduction of TxA_2 in MPD. Such a mechanism might contribute to the vaso-occlusive manifestations which form part of the natural history. Increased platelet thromboxane synthesis *in vivo* in polycythemia vera and thromboxane dependent platelet activation preceding thrombosis in thrombocythaemia are consistent with putative pathogenic mechanism [4, 5]. Since CD36 is also involved in platelet aggregation and adhesion to collagen via several mechanisms in addition to its role as uptake protein for arachidonic acid increased surface expression and enhance platelet reactivity through thromboxane-independent pathways. These findings could have implications in relation to choice of platelet inhibitory medications in MPD.

Acknowledgements

A preliminary account of this work was presented at the Multiple Risk factors in Cardiovascular Disease meeting, Venice, Italy, October 1999.

References

1. Cortelazzo S, Viero P, Barbui T: Platelets activation in myeloproliferative disorders. Thromb Haemost 45: 211–213, 1981
2. Duttaroy AK: Insulin mediated processes in platelets, monocytes/ macrophages and erythrocytes: Effects of essential fatty acid metabolism. Prost Leuk Essntl Fatty Acids 51: 385–389, 1994
3. Wehmeier A, Sudhoff T. Meierkord F: Relation of platelet abnormalities to thrombosis and hemorrhage in chronic course of myeloproliferative disorders. Eur J Haematol 45: 191–197, 1997
4. Landolfi R, Ciabattoni G, Patrignani P, Castellana MAL, Pogliani E, Bizzi B, Patrono C: Increased thromboxane biosynthesis in patients with polycythemia vera: Evidence for aspirin-suppressible platelet activation *in vivo*. Blood 80: 1965–1971, 1992
5. Van Genderen PJJ, Prins FJ, Michiels JJ, Schror K: Thromboxane-dependent platelet activation *in vivo* precedes arterial thrombosis in thrombocythaemia: As an antithrombotic agent. Br J Haematol 104: 438–441, 1999
6. Balduini CL, Bertolino G, Noris P, Piletta GC: Platelet aggregation in platelet rich plasma and whole blood in 120 patients with myeloproliferative disorders. Am J Clin Pathol 95: 82–86, 1991
7. Kaywin P, McDonough M, Insel PA, Shattil SJ: Platelet function in essential thrombocythemia. Decreased epinephrine responsiveness associated with a deficiency of platelet alfa-adrenergic receptors. N Engl J Med 299: 505–509, 1978
8. Thibert V, Bellucci S, Cristofari M, Gluckman E, Legrand C: Increased platelet CD36 constitutes a common marker in myleoproliferative disorders. Br J Haematol 91: 618–624, 1995
9. Le Blanc K, Lindahl T, Rosendahl K, Samuelsson J: Impaired platelet binding of fibrinogen due to a lower number of GPIIb/IIIa receptors in polycythemia vera. Thromb Res 91: 287–295, 1998
10. Michelson AD, Wencel-Drake JD, Kestin AS, Barnard MR: Platelet activation results in a redistribution of glycoprotein IV (CD36). Arterioscler Thromb 14: 1193–1201, 1994
11. Ford I, Malik RA, Newrick PG, Preston FE, Ward JD, Greaves M: Thromb Haemost 68: 628–633, 1992
12. Phillips DR, Charo IF, Parise LV, Fitzerglad LA:The platelet membrane glycoprotein IIb-IIIa complex. Blood 71: 831–843, 1988
13. Duttaroy AK: Cellular uptake of long-chain fatty acids: Role of membrane-associated fatty acid-binding/transport proteins. Cell Mol Life Sci 57: 1360–1372, 2000
14. Ruggeri ZM: The platelet glycoprotein Ib-IX complex. Prog Hemost Thromb 10: 35–68, 1991
15. Silverstein RL, Asch AS, Nachman RL: Glycoprotein IV mediates thrombospondin-dependent platelet-monocyte and platelet-U937 cell adhesion. J Clin Invest 84: 546–552, 1989
16. Greenwalt DE, Lipskey RH, Ockenhouse CF, Ikeda H, Tandon NN, Jamieson GA: Membrane glycoprotein CD36: A review of its role in adherence, signal transduction, and transfusion medicine. Blood 80: 1105–1115, 1992
17. Abumrad NA, El-Maghrabi MR, Amri EZ, Lopez E, Grimaldi PA: Cloning of a rat adipocyte membrane-protein implicated in binding or transport of long chain fatty acids that is induced during preadipocyte differentiation: Homology with human CD36. J Biol Chem 268: 17665–17668, 1993
18. Greenwalt DE, Scheck SH, Rhinehart T: Heart CD36 expression is increased in murine models of diabetes and in mice fed a high fat diet. J Clin Invest 96: 1382–1388, 1995
19. Duttaroy AK, Gordon MJ, Campbell FM, Crosbie L: Arachidonic acid uptake by human platelets is mediated by CD36. Platelets 7: 291–295, 1996
20. Ikeda H: Platelet membrane CD36. Hokkaido Igaku Zasshi 74: 99–104, 1999
21. Arita H, Nakano T, Hanasaki K: Thromboxane A_2 – its generation and role in platelet activation. Prog Lipid Res 28: 273–301, 1989
22. Duttaroy AK, Ray TK, Sinha AK: Prostacyclin stimulation of the activation of blood coagulation factor X by platelets. Science 231: 385–388, 1986
23. Perez C, Ramiro JM, Campillo JE: Increased arachidonic acid uptake by platelets from insulin-dependent diabetics and diabetic rats. Diabet Res Clin Prac 7: 69–77, 1989
24. Mikhailidis DP, Jeremy JY, Barradas MA, Dandona P: Increased arachidonic acid incorporation into platelet phospholipids in type-2 (non-insulin-dependent) diabetes. Diabetologia 27: 242–248, 1984
25. Prisco D, Rogasi PG, Pannicia R, Abbate R, Gensini GF, Pinto S, Vanni D, Neri Seneri GG: Altered membrane fatty acid composition and increased thromboxane A_2 generation in platelets from patients with diabetes. Prost Leuk Essen Fatty Acids 35: 15–33, 1989
26. Kieffer N, Bettaieb A, Legrand C, Coulombel L, Vainchenker W, Edelman L, Breton-Gorius J: Developmentally regulated expression of a 78 Kda erythoblast membrane glycoprotein immunologically related to the platelet thrombospondin receptor. Biochem J 262: 835–842, 1988
27. Beiso P, Pidard D, Fournier D, Dubernard V, Legrand C: Studies on the interaction of platelet glycoprotein IIb-IIIa and glycoprotein IV with fibrinogen and thrombospondin; a new immunochemical approach. Biochim Biophys Acta 1033: 7–12, 1990
28. Matsuno K, Diaz-Ricart M, Montgomery RR, Aster RH, Jamieson GA, Tandon NN: Inhibition of platelet adhesion to collagen by monoclonal anti-CD36 antibodies. Br J Haematol 92: 960–967, 1996
29. Diaz-Ricart M, Tandon NN, Gomezoritz G, Carretero M, Escolar G, Ordinas A, Jamieson GA : Antibodies to CD36 (GPIV) inhibit platelet-adhesion to sub-endothelial surfaces under flow conditions. Arterioscler Thromb Vasc Biol 16: 883–888, 1996
30. Tandon NN, Kralisz U, Jamieson GA: Identification of glycoprotein IV (CD36) as a primary receptor for platelet collagen adhesion. J Biol Chem 264: 7576–7583, 1989
31. Campbell FM, Bush P, Veerkamp JH, Duttaroy AK: Distribution of membrane associated and cytoplasmic fatty acid-binding proteins in human placenta. Placenta 19: 409–415, 1998
32. Campbell FM, Duttaroy AK: Plasma membrane fatty acid-binding protein (FABPpm) is exclusively located in the maternal facing membranes of the human placenta. FEBS Letts 375: 227–230, 1995
33. Campbell FM, Gordon MJ, Duttaroy AK: Plasma membrane fatty acid binding protein from human placenta: Identification and characterization. Biochem Biophys Res Commun 209: 1011–1017, 1995
34. Bradford MM: A rapid and sensitive method for the quantitation of microgramme quantitites of protein utilizing the principle of protein-dye binding. Anal Biochem 72: 248–254, 1976
35. Legrand C, Dubernard V, Kieffer N, Nurden AT: Use of a monoclonal antibody to measure the surface expression of thrombospondin following platelet activation. Eur J Biochem. 171: 393–399, 1988
36. Legrand C, Pidart D, Beiso P, Tenza D, Edelman L: Interaction of monoclonal antibody to glycoprotein IV (CD36) with human platelets and its effect on platelet function. Platelets 2: 99–105, 1991
37. Duttaroy AK, Sinha AK: Purification and properties of prostaglandin E_1/prostacyclin receptors of human blood platelets. J Biol Chem 262: 12685–12891, 1987
38. Yamamoto N, Ikeda H, Tandon NN, Herman J, Tomiyama Y, Mitani T, Sekiguchi S, Lipskey R, Kralisz U, Jamieson GA: A platelet membrane glycoprotein (GP) deficiency in healthy blood donors: NAK^a platelets lack detectable GPIV (CD36). Blood 76: 1698–1703, 1990
39. Bligh EG, Dyer WJ: A rapid method of total lipid extraction and purification. Can J Biochem Physiol 37: 911–917, 1959
40. Crabtree J, Gordon MJ, Campbell FM, Duttaroy AK: Differential distribution and metabolism of arachidonic acid and docosahexaenoic

acids by human placental cells. Mol Cell Biochem 185: 205–212, 1998
41. Major C, Wolf BA: Quantitation of the fatty acids composition of phosphatidic acid by capillary gas chromatography electron capture detection with picomole sensitivity. J Chromat B 658: 233–240, 1994
42. Kerkhoff C, Sorg C, Tandon NN, Nacker W: Interaction of S100A8/S100A9-arachidonic acid complexes with the scavenger receptor CD36 may facilitate fatty acid uptake by endothelial cells. Biochem 40: 241–248, 2001
43. Volf I, Moeslinger T, Cooper J, Schmid W, Koller E: Human platelets exclusively bind oxidised low density lipoprotein showing no specificity for acetylated low density lipoprotein. FEBS Lett 449: 141–145, 1999
44. Mahfouz MM, Kummerow FA: Oxidised low-density lipoprotein (LDL) enhances thromboxane A2 synthesis by platelets, but lysolecithin as a product of LDL oxidation has an inhibitory effect. Prost Other Lipid Mediat 62: 183–200, 2000
45. Chawla A, Boisvert WA, Lee CH, Laffitte BA, Barak Y, Joseph SB, Lia D, Nagy L, Edwards PA, Curtiss LK, Evans RM, Tontonoz P: A PPAR gamma-LXR-ABCA1 pathway in macrophages is involved in cholesterol efflux and atherogenesis. Mol Cell 7: 161–171, 2001
46. Devraj S, Hugou I, Jialal I: α-Tocopherol decreases CD36 expression in human monocyte-derived macrophages. J Lipid Res 42: 521–527, 2001
47. Kalofoutis A, Lekakis J: Changes of platelet phospholipids in diabetes mellitus. Diabetologia 21: 540–543, 1981
48. Jones DB, Carter RD, Haitas B, Mann JI: Low phospholipid arachidonic acid values in diabetic platelets. Br Med J 286: 173–175, 1983
49. Nordy A, Rodset JM: Platelet phospholipids and their function in patients with juvenile diabetes and maturity onset diabetes. Diabetes 19: 698–702, 1970
50. Castaldi PA, Brendt MC, Booth W, Bull H, Greaves M: Evidence for a platelet membrane defect in the myeloproliferative syndromes. Thromb Res 27: 601–609, 1982

Molecular and Cellular Biochemistry **239**: 213–219, 2002.

Sulfo-*N*-succinimidyl esters of long chain fatty acids specifically inhibit fatty acid translocase (FAT/CD36)-mediated cellular fatty acid uptake

Susan L.M. Coort,[1] Jodil Willems,[1] Will A. Coumans,[1] Ger J. van der Vusse,[1] Arend Bonen[2], Jan F.C. Glatz[1] and Joost J.F.P. Luiken[1]

[1]*Department of Physiology, Cardiovascular Research Institute Maastricht (CARIM), Maastricht University, Maastricht, The Netherlands;* [2] *Department of Kinesiology, University of Waterloo, ON, Canada*

Abstract

Sulfo-*N*-succinimidyl esters of LCFAs are a powerful tool to investigate the functional significance of plasmalemmal proteins in the LCFA uptake process. This notion is based on the following observations. First, sulfo-*N*-succinimidyl oleate (SSO) was found to inhibit the bulk of LCFA uptake into various cell types, i.e. rat adipocytes, type II pneumocytes and cardiac myocytes. Second, using cardiac giant membrane vesicles, in which LCFA uptake can be investigated in the absence of mitochondrial β-oxidation, SSO retained the ability to largely inhibit LCFA uptake, indicating that inhibition of LCFA transsarcolemmal transport is its primary action. Third, SSO has no inhibitory effect on glucose and octanoate uptake into giant membrane vesicles derived from heart and skeletal muscle, indicating that its action is specific for LCFA uptake. Finally, SSO specifically binds to the 88 kDa plasmalemmal fatty acid transporter FAT, a rat homologue of human CD36, resulting in an arrest of the transport function of this protein.

In addition to its inhibitory action at the plasma membrane level, evidence is presented for the lack of a direct inhibitory effect on subsequent LCFA metabolism. First, the relative contribution of oxidation and esterification to LCFA uptake is not altered in the presence of SSO. Second, isoproterenol-mediated channeling of LCFAs into oxidative pathways is not affected by sulfo-*N*-succinimidyl palmitate (SSP). As an example of its application, we used SSP to study the role of FAT/CD36 in contraction- and insulin-stimulated LCFA uptake by cardiac myocytes , showing that this transporter is a primary site of regulation of cellular LCFA utilization. (Mol Cell Biochem **239**: 213–219, 2002)

Key words: uptake of long-chain fatty acids, insulin, cellular stimulation, isoproterenol

Abbreviations: LCFAs – long-chain fatty acids; FAT – fatty acid translocase; SSO – sulfo-*N*-succinimidyl oleate; SSP – sulfo-*N*-succinimidyl palmitate

Introduction

Long-chain fatty acids (LCFAs) are not only important substrates for energy production, but also function as precursors for signaling molecules, mediators of gene expression and constituents of membrane phospholipids [1]. The mechanism of LCFA entry into the cell has, however, been controversial. The debate centers around the idea as to whether LCFAs enter the cell via passive diffusion and/or protein-mediated (facilitated) transmembrane uptake mechanisms. Because of their lipophilic properties it has been argued that LCFAs are transported mainly by passive diffusion. In this respect, Hamilton *et al.* proposed a model in which un-ionized fatty acids (FA) permeate the phospholipid bilayer by a rapid flip-flop mechanism independent of the involvement of membrane-associated proteins [2–4]. On the other hand, convincing

Address for offprints: S.L.M. Coort, Department of Physiology, CARIM, Maastricht University, P.O. Box 616, 6200 MD Maastricht, The Netherlands (E-mail: S.Coort@fys.unimaas.nl)

evidence has been gathered for the notion that the bulk of transsarcolemmal LCFA transport is protein-mediated (for review see [5]). First, LCFA uptake into various mammalian cell types exerts saturation kinetics [6–9]. Second, transmembrane LCFA transport across the cellular plasma membrane can be reduced by inhibitors of protein-mediated transport (i.e. trypsin, phloretin and diisothiocyanodisulfonic acid (DIDS)) [6, 7, 9, 10]. Finally, uptake of LCFAs is subject to competitive inhibition by other LCFAs, but not by medium-chain fatty acids or other substrates [7–10]. Despite these recent observations that a protein-mediated mechanism is involved in cellular LCFA uptake, some skepticism remains [4].

In order to investigate the significance of plasma membrane proteins in the LCFA uptake process in mammalian cells, derivatives of LCFAs, which maintain the lipophilic properties of LCFAs but are able to crosslink with proteins, could very well be powerful tools. In particular, LCFAs have been coupled to *N*-hydroxysulfosuccinimides through an ester linkage. In general, sulfo-*N*-succinimidyl esters of carboxylic acids are used as acylating reagents for proteins and as membrane impermeable crosslinkers [11]. Sulfo-*N*-succinimidyl esters were found to react at high yield with proteins present in the plasma membrane, because of their low rate of hydrolysation compared to their rates of reacting with nucleophilic groups in proteins [11, 12].

Harmon *et al.* used sulfo-*N*-succinimidyl esters of FA with different chain lengths in order to identify plasma membrane proteins that may be involved in the LCFA uptake process [13]. They found that sulfo-*N*-succinimidyl oleate (SSO) labeled an 88 kDa rat adipocyte membrane protein, which resulted in a decrease in LCFA uptake in isolated rat adipocytes. They also discovered that in isolated rat adipocytes pretreated with SSO the intracellular water space, glucose transport, and retinoic acid permeation were not significantly altered. These findings gave indirect evidence that SSO acts at the level of the plasma membrane. Furthermore, because of their membrane impermeability intracellular processes are thought not to be directly affected by sulfo-*N*-succinimidyl esters of LCFAs.

In this review we discuss the use of sulfo-*N*-succinimidyl esters of LCFAs for examining LCFA uptake. First, the chemical properties and synthesis of sulfo-*N*-succinimidyl acyl esters are described. Second, we discuss the possible role of an 88 kDa plasma membrane protein, notably fatty acid translocase (FAT/CD36) which was labeled by SSO in rat adipocyte membranes, in the SSO-exerted inhibition of LCFA uptake by heart and skeletal muscles. Subsequently, evidence is presented that SSO specifically binds to FAT/CD36. Finally, examples are given of the application of sulfo-*N*-succinimidyl esters of LCFAs in studies on the regulation of cellular uptake by FAT/CD36.

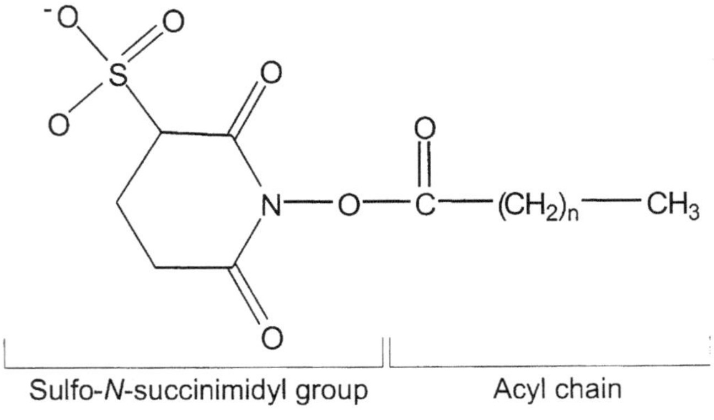

Fig. 1. Structure of sulfo-*N*-succinimidyl esters of fatty acids. The molecule consists of a reactive sulfo-*N*-succinimidyl head group and a acyl chain tail, such as myristate (C14:0), palmitate (C16:0) and oleate (C18:1).

Characterization and synthesis of sulfo-*N*-succinimidyl esters of LCFAs

Sulfo-*N*-succinimidyl acyl esters are composed of a LCFA, such as palmitic acid or oleic acid, coupled to *N*-hydrosulfosuccinimide. These esters of LCFAs can be synthesized as described by Harmon *et al.* [13]. Briefly, in this procedure the fatty acid reacts with a *N*-hydrosulfosuccinimide in the presence of dicyclohexylcarbodiimide (DCC) under anhydrous conditions. The general structure of a sulfo-*N*-succinimidyl ester of a FA is given in Fig. 1.

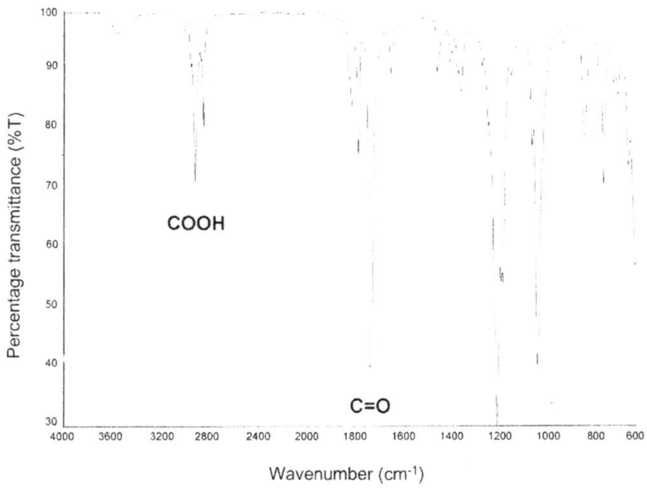

Fig. 2. Infrared (IR) spectrum of sulfo-*N*-succinimidyl palmitate (SSP) is presented as the percentage transmittance (%T) as a function of the wavenumber (cm^{-1}). The absorption at 1750–1735 cm^{-1} corresponds to the C = O bond and that at 3000–2500 cm^{-1} to the COOH group. The region from 1500–400 cm^{-1} is the so-called fingerprint region of SSP.

To identify the synthesized sulfo-*N*-succinimidyl ester of a specific LCFA, mass spectral analysis can be applied [13] as well as infrared (IR) spectroscopy, which is routinely used in our laboratory [7]. Infrared spectra provide information about the presence of functional groups in a molecule [14]. The C = O bond, originating from the fatty acid, is present in all IR spectra of sulfo-*N*-succinimidyl esters of FA and can be recognized with a peak at approximately 1750–1735 cm^{-1}. The region from 1500 to 400 cm^{-1} reveals certain structural features of the sulfo-*N*-succinimidyl esters of a specific FA and is the fingerprint of the molecule. In the IR spectra of sulfo-*N*-succinimidyl palmitate (SSP) there is also a small peak detectable at 3000–2500 cm^{-1}, corresponding to a COOH-group, which indicates that only minor quantities of non-esterified palmitate are present in the SSP preparation (Fig. 2).

Effects of sulfo-*N*-succinimidyl acyl esters on uptake of LCFAs by various cell types

The first study that investigated the effect of sulfo-*N*-succinimidyl esters of LCFAs on substrate utilization was performed by Harmon *et al.* [13]. The uptake of oleate, linoleate and stearate into isolated rat adipocytes treated for 25 min with 200 μM SSO was reduced by about 65%, in each case. These same researchers showed that the length of the carbon chain of the acyl moiety chemically bound to the reactive sulfo-*N*-succinimidyl group was crucial for the function of the ester. Sulfo-*N*-succinimidyl propionate (C3:0) had no effect on the oleate uptake by rat adipocytes while esters of oleate (C18:1), palmitate (C16:0) and myristate (C14:0) had an inhibiting effect on the LCFA uptake by adipocytes. This indicates that only LCFAs coupled to a reactive sulfo-*N*-succinimidyl group

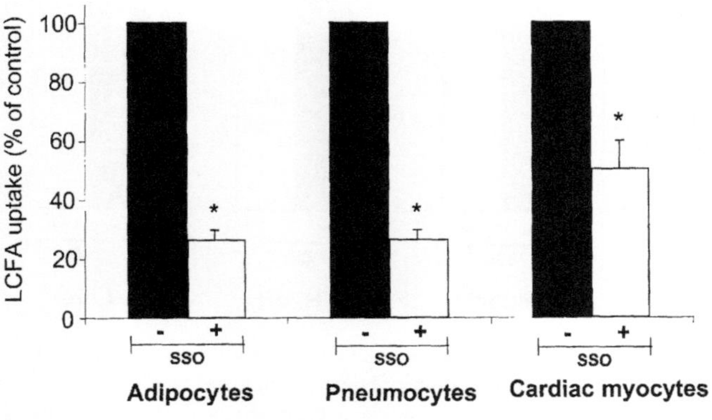

Fig. 3. Effect of SSO on LCFA uptake by various cell types, i.e. rat adipocytes, type II pneumocytes and cardiac myocytes. Data are presented as percentage of control values ± S.E.M. and were obtained from Harmon *et al.* (13]; Guthmann *et al.* [15]; Luiken *et al.* [17]. *Significantly different from control ($p < 0.05$).

affect cellular LCFA uptake, which is in agreement with the notion that the transport system mediating uptake of oleate is specific for LCFAs, and is not involved in uptake of short- and medium chain fatty acids [13].

In rat type II pneumocytes, in which LCFAs are required for production of surfactant, uptake of palmitate was reduced upon pretreatment with SSO, the maximal inhibition being 80% at 1.0 mM [15]. In cardiac myocytes, for which LCFAs are the main source of energy [16], incubation with SSO resulted in an inhibition of palmitate uptake by 50% [17]. Thus, in rat adipocytes, type II pneumocytes and cardiac myocytes, SSO exerts an inhibitory effect of comparable magnitude on LCFA uptake, i.e. between 50 and 80% (Fig. 3).

Effect of sulfo-*N*-succinimidyl esters of LCFA on cardiac LCFA uptake

In view of the crucial role of LCFAs in cardiac energy production [16], a great deal of research with respect to the effect of sulfo-*N*-succinimidyl acyl esters on LCFA uptake has been performed in the heart. In this section we will compare the effect of SSO/SSP on cardiac LCFA uptake in three different models. First, in isolated rat cardiac myocytes, in which LCFA uptake is followed by its rapid metabolism, SSO had an inhibitory effect on LCFA uptake, i.e. 50% [5, 17], as described in the previous section. Second, giant membrane vesicles derived from rat heart tissue were used to investigate cardiac LCFA uptake. Giant membrane vesicles are an experimental model by which LCFA uptake can be investigated in the absence of subsequent metabolism [17, 18]. These vesicles consist of a membrane of pure sarcolemmal origin and contain soluble cytoplasmatic proteins, like cytoplasmic fatty acid-binding protein (FABPc) acting as a sink for sequestered FA. In addition, giant membrane vesicles lack intracellular structures like endoplasmatic reticulum, endosomes and mitochondria which latter absence indicates the inability of giant membrane vesicles to oxidize LCFA [17, 18]. Using giant membrane vesicles derived from heart, we discovered that inhibition of LCFA uptake by SSO occurs in the absence of mitochondrial β-oxidation. The comparable inhibitory effect of SSO in heart giant membrane vesicles (–70%) vs. cardiac myocytes (–50%), i.e. in the absence and presence of LCFA metabolism, implicates that inhibition of LCFA transsarcolemmal transport is the primary action of sulfo-*N*-succinimidyl esters of LCFAs.

Third, the effect of SSP was investigated at total heart level by Tanaka *et al.* They showed that perfusing isolated rat hearts with 10 μM SSP resulted in a decrease in palmitate uptake by approximately 24% [19]. Thus, not only at the cellular level but also at the total tissue level, sulfo-*N*-succinimidyl esters of LCFAs are able to inhibit cardiac LCFA uptake.

Inability of sulfo-*N*-succinimidyl esters of LCFAs to inhibit cellular uptake of substrates other than LCFAs

In order to study the specificity of SSO for inhibition of LCFA uptake, we have used giant membrane vesicles of rat heart and skeletal muscle to examine whether SSO inhibits the uptake of other substrates such as glucose and medium-chain FAs, as well as LCFAs. SSO had no effect on the uptake of glucose into giant membrane vesicles derived from either heart or skeletal muscle whereas phloretin, a non-specific inhibitor of carrier-mediated transport, was able to inhibit glucose uptake into these vesicles by more than 80% [17]. In contrast, in metabolically active cardiac myocytes [7] and in perfused heart [19], the presence of SSO caused a substantial increase in glucose uptake by 40 and 31%, respectively. We speculate that the SSO-induced inhibition of LCFA uptake by the heart lowers the levels of intracellular 'free' (non-esterified) LCFAs. This lower LCFA level is expected to relieve glucose metabolism from inhibition by LCFA-induced Randle-cycle-effects to compensate for the fact that less LCFAs are intracellularly available for energy conversion.

With respect to medium-chain FAs, the uptake of octanoate in heart giant membrane vesicles was not inhibited by SSO [17]. Taken together these findings indicate that the effect of sulfo-*N*-succinimidyl esters of LCFAs are restricted to uptake of LCFAs.

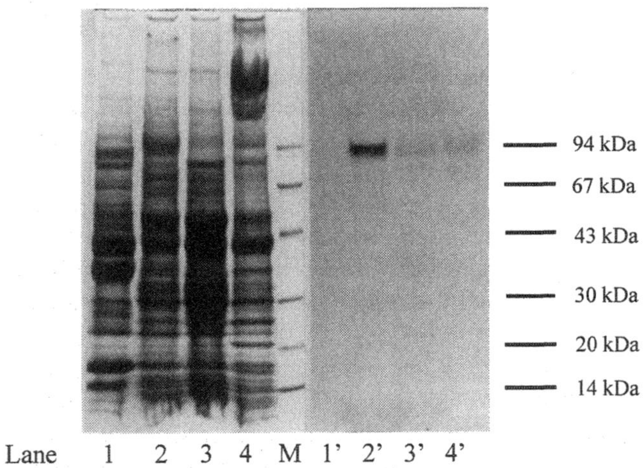

Fig. 4. Presence of [^{3}H]-SSP in myocardial cell fractions obtained from [^{3}H]-SSP perfused rat hearts. SDS-PAGE was performed, followed by autoradiography. The molecular weight standards (lane M) were phosphorylase b (94,000), bovine serum albumin (67,000), ovalbumin (43,000), carbonic anhydrase (30,000), soybean trypsin inhibitor (20,100) and α-lactalbumin (14,000). Lanes 1–4 and 1'–4' respectively present rat heart fractions and autoradiographs. Lanes 1 and 1': cytosolic fraction, lanes 2 and 2': microsomal fraction, lanes 3 and 3': mitochondrial fraction, lanes 4 and 4': cell debris. These data were reproduced with permission from Dr. T. Tanaka, Osaka, Japan [19].

Specific binding of sulfo-*N*-succinimidyl esters of LCFAs to fatty acid translocase (FAT)/CD36

In the literature, three putative plasma membrane proteins are described to be involved in transmembrane LCFA transport, and each shows a characteristic tissue occurrence. These are a 40 kDa plasma membrane fatty acid binding protein, FABPpm [20], an 88 kDa highly glycosylated fatty acid translocase, FAT, a rat homologue of human CD36 [21, 22], and a 62 kDa fatty acid transport protein, FATP [23]. Various cell types, in particular cardiac myocytes, are known to contain all three described plasma membrane proteins.

There are two lines of evidence that sulfo-*N*-succinimidyl esters of LCFAs specifically binds to FAT/CD36, and in that way inhibits the transport of LCFAs across the plasma membrane. First, in plasma membrane fractions of rat adipocytes, in which SSO reduced LCFA uptake by 65% (see earlier section), [^{3}H]-SSO specifically labeled an 88 kDa rat adipocyte membrane protein which was identified as FAT [19]. In addition, Tanaka *et al.* [19] showed, using SDS-PAGE autoradiography, that in myocardial cell fractions from [^{3}H]-SSP perfused rat hearts a single microsomal protein of 80–90 kDa was labeled (Fig. 4). Second, we discovered that, in contrast to giant membrane vesicles derived from heart and skeletal muscle, in which SSO inhibited LCFA uptake by 70 and 50% respectively, SSO had no effect on LCFA uptake into liver giant membrane vesicles (Fig. 5). These liver giant membrane vesicles display a different set of LCFA transporters. Nota-

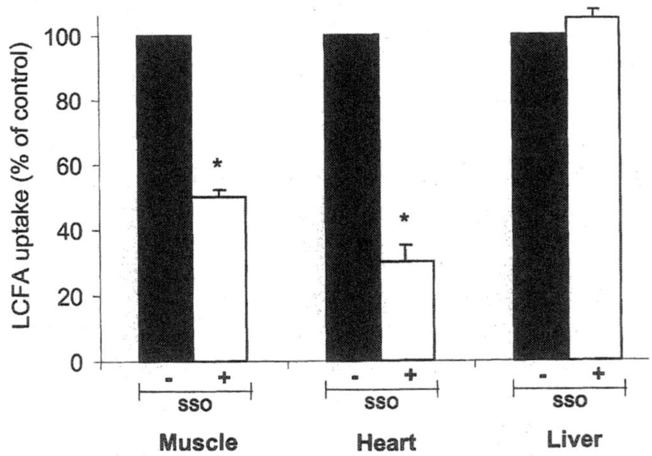

Fig. 5. Effect of sulfo-*N*-succinimidyl oleate (SSO) on palmitate uptake in heart, liver and muscle giant membrane vesicles. Initial ^{3}H-labeled palmitate uptake, 15 sec, was measured in rat heart, liver and muscle giant membrane vesicles. Vesicles were preincubated with 50 μM SSO which was removed prior to assessment of palmitate uptake. Data are presented as means ± S.E.M. of 3 experiments carried out in different giant membrane preparations and published by Luiken *et al.* [18]. *Significantly different from control ($p < 0.05$).

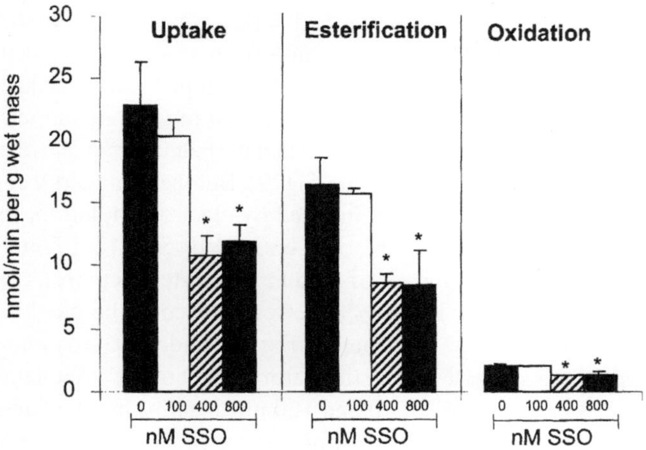

Fig. 6. Cardiac palmitate uptake, oxidation and esterification as function of the concentration of SSO. Uptake and esterification were measured 3 min and $^{14}CO_2$ production, thus oxidation, 30 min after ^{14}C-labeled palmitate addition in isolated cardiac myocytes. These cardiac myocytes were pre-incubated with different concentrations SSO, dissolved in DMSO, for 15 min by 37°C. Data are means ± S.D. of 3 experiments carried out in different cardiac myocytes preparations and published by Luiken *et al.* [7]. *Significantly different from control ($p < 0.05$).

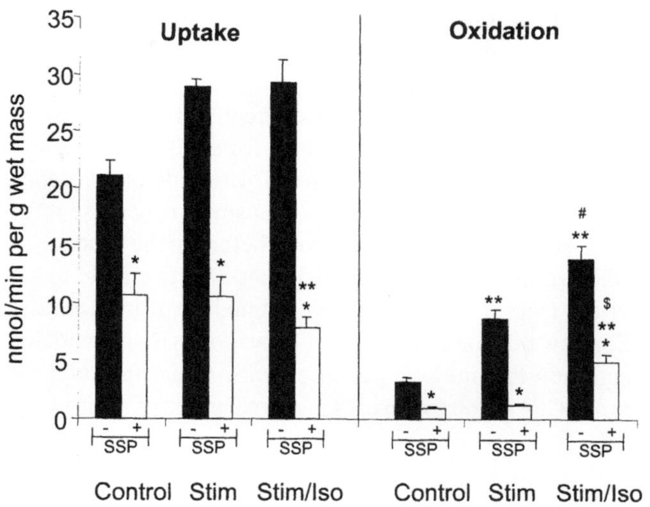

Fig. 7. Effect of sulfo-*N*-succinimidyl palmitate (SSP) on isoproterenol mediated channeling of LCFA into oxidation pathway in electrically stimulated cardiac myocytes. Electrical stimulation of cardiac myocytes was carried out at 200 V/4 Hz, as previously described [7]. Isoproterenol was added at 1 μM 15 min prior to measurement of LCFA uptake or oxidation. SSP was added at 200 μM for 30 min and removed by centrifugation prior to measurement of LCFA uptake or oxidation. Data are presented as means ± S.E.M. of 3 experiments carried out in different cardiac myocytes. *Significantly different from not treated with SSP ($p < 0.05$), **significantly different from control ($p < 0.05$), #significantly different from stimulated without SSP ($p < 0.05$), $significantly different from stimulated with SSP ($p < 0.05$).

bly the absence of FAT/CD36 is inferred on the inability to detect messenger levels of this transporter, whereas FABPpm and FATP have been found to be present. [24]. A lack of inhibitory action of SSO on LCFA uptake in rat liver giant membrane vesicles suggests that SSO specifically interacts with FAT and that FATP and FABPpm play no prominent role. The lesser inhibitory effect of SSO in skeletal muscle compared to heart is related to the lower abundance of FAT/CD36 in the former tissue. Thus, these two lines of evidence bolster the notion that SSO specifically inhibits the LCFA transport function of FAT/CD36.

Metabolic effects of sulfo-*N*-succinimidyl esters of LCFAs

In our studies we have observed that SSO apparently inhibited mitochondrial b-oxidation and esterification in cardiac myocytes (Fig. 6) [7]. The maximal inhibition of initial palmitate uptake in cardiac myocytes pretreated with SSO was observed at a concentration of 0.4 mM or higher and amounted to 50%. Mitochondrial β-oxidation and esterification were also inhibited by SSO, comparable with the amount of inhibition of palmitate uptake, 49 and 35% respectively [7]. Importantly, (i) these inhibitory effects of SSO on palmitate metabolism (esterification and oxidation) are comparable to the SSO induced inhibition of palmitate uptake across the sarcolemma and (ii) the relative (%) partitioning of LCFAs into esterification and oxidation was not changed by SSO. These observations suggests that inhibition of LCFA esterification and oxidation are secondary to inhibition of LCFA uptake.

Further evidence for this notion and for the fact that SSP does not influence the metabolic fate of LCFA comes from recent studies with isoproterenol. This synthetic catecholamine elevates the cellular cAMP levels and is known to increase, at optimal doses, the contraction amplitude of cardiac myocytes by 3-fold [25]. In addition, there is a proportionally enhanced oxygen consumption [25] and an enhanced LCFA oxidation, i.e. from 8.6–13.7 nmol/min per g wet mass (difference 5.1 nmol/min per g wet mass). But we have observed that isoproterenol did not affect LCFA uptake in electrically stimulated cardiac myocytes (Fig. 7). Since the rate of LCFA uptake did not change, the increase in LCFA oxidation in these electrically stimulated cardiac myocytes was accompanied by a decrease in LCFA esterification (data not shown). Thus isoproterenol apparently affects the channeling of incoming LCFA in contracting cardiac myocytes in favor of β-oxidation. Yet, SSP reduced the uptake of LCFAs to the same extent either in the absence or presence of isoproterenol. Studying the influence of SSP on LCFA utilization by these isoproterenol treated and 4 Hz stimulated cardiac myocytes revealed that this specific FAT/CD36 inhibitor reduced the rate of LCFA uptake at 4 Hz stimulation whether in the absence or presence of isoproterenol to the same level as that observed with non-stimulated cardiac myocytes (Fig. 7) In contrast, LCFA oxidation in electrically stimulated myocytes treated with SSP was elevated from 1.2–4.9 nmol/min per g wet mass when isoproterenol was added (difference 3.7 nmol/min per g wet mass). This similar increase in of LCFA oxidation by isoproterenol whether in the absence or in the presence of SSP indicated that SSP is unable to affect the isoproterenol-mediated channeling of intracellular LCFA towards oxidation. This observation is in line with the suggestion that the sole action of sulfo-*N*-succinimidyl esters of LCFAs is on the level of the sarcolemma (plasma membrane).

Application of sulfo-*N*-succinimidyl esters of LCFA in studies on the regulation of cellular LCFA uptake

Recently, we have observed that LCFA uptake in heart and muscle is subject to short-term regulation by cellular contractions [26,27] and by insulin [28]. For these studies sulfo-*N*-succinimidyl esters of LCFAs were crucial to describe the involvement of FAT/CD36 in short term inducible LCFA uptake. The first evidence that FAT/CD36 was involved in the regulation of LCFA uptake came from studies with giant membrane vesicles derived from rat hind limb muscle that had been electrically stimulated. We showed in giant membrane vesicles from contracting muscle that the palmitate uptake rate was increased by 47% ($p < 0.05$) [29]. But the stimulatory effect on palmitate uptake, induced by electrostimulation, in these vesicles could be blocked entirely by SSO [26]. Similarly, electrostimulation of cardiac myocytes increased the LCFA uptake by, i.e. 49%, an effect that could be blocked entirely by SSP, as illustrated in Fig 8. Besides electrostimulation, insulin also had a stimulatory effect on LCFA uptake (i.e. +51 and +56% at 10 and 100 nM insulin) into isolated cardiac myocytes. This insulin-induced increase in LCFA uptake was also blocked completely by SSP (Fig. 8).

These inhibitory actions of SSO indicate that the stimulatory effects of both electrostimulation and insulin are mediated by FAT/CD36. Subsequently, we showed with subcellular fractionation experiments that the contraction- and insulin-stimulated LCFA uptake involves the translocation of FAT/CD36 from intracellular depots to the sarcolemma [26, 30].

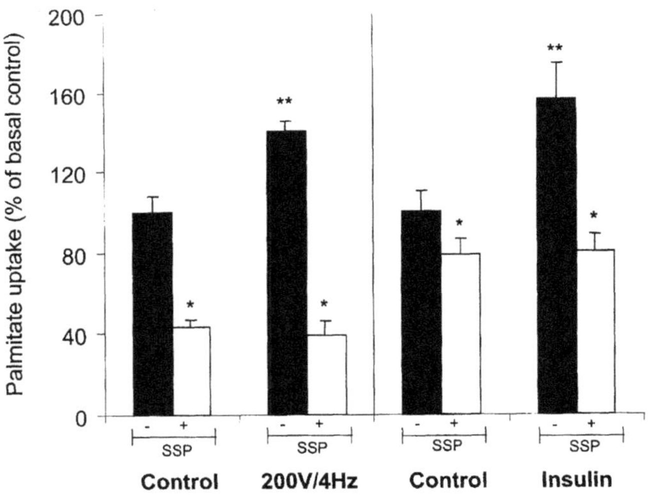

Fig. 8. Effect of sulfo-*N*-succinimidyl palmitate (SSP) on LCFA uptake in cardiac myocytes electrically stimulated or treated with insulin. ^{14}C-palmitate uptake was measured in electrically stimulated cardiac myocytes, derived from fed rats or in cardiac myocytes, derived from fasted rats, treated with insulin, in the absence or presence of SSP. Data are presented as percentage of control values ± S.E.M. *Significantly different from not treated with SSP ($p < 0.05$), **significantly different from control ($p < 0.05$).

Concluding remarks

Sulfo-*N*-succinimidyl esters of LCFAs are useful compounds to study the mechanism and regulation of LCFA uptake by mammalian cells, because (i) they are specific in blocking uptake of LCFAs while having no direct effect on uptake of other substrates, and (ii) they do not affect the intracellular metabolic fate of LCFAs. Furthermore there is convincing evidence that SSO and SSP specifically bind to FAT/CD36, and in doing so inhibit its transport function. When cellular LCFA utilization is altered, sulfo-*N*-succinimidyl esters of LCFAs are useful tools to determine whether or not FAT/CD36 is involved. However, more insight is needed to identify the binding site of sulfo-*N*-succinimidyl esters of LCFAs to FAT/CD36 in order to unravel the molecular mechanism of their inhibitory action on LCFA uptake by this transporter.

Acknowledgements

We wish to thank Dr. T. Tanaka, Osaka, Japan for providing the SDS-PAGE autoradiographic blot presented in Fig. 4 and Dr. M. van Genderen, Technical University, Eindhoven, the Netherlands for providing the infrared spectra of SSP. Our studies were supported by the Netherlands Heart Foundation, grants D98.012 and 2000.156, the Canadian Institutes of Health Research, and the Heart and Stroke Foundation of Ontario.

References

1. van der Vusse GJ, van Bilsen M, Glatz JFC: Cardiac fatty acid uptake and transport in health and disease. Cardiovasc Res 45: 279–293, 2000
2. Hamilton JA, Kamp F: How are free fatty acids transported in membranes? Is it by proteins or by free diffusion through the lipids? Diabetes 48: 2255–2269, 1999
3. Hamilton JA: Fatty acid transport: difficult or easy? J Lipid Res 39: 467–481, 1998
4. Hamilton JA, Johnson RA, Corkey B, Kamp F: Fatty acid transport: The diffusion mechanism in model and biological membranes. J Mol Neurosci 16: 99–108, 2001
5. Glatz JFC, Storch J: Unravelling the significance of cellular fatty acid-binding proteins. Curr Opin Lipidol 12: 267–274, 2001
6. Stremmel W: Fatty acid uptake by isolated rat heart myocytes represents a carrier-mediated transport process. J Clin Invest 81: 844–852, 1988
7. Luiken JJFP, van Nieuwenhoven FA, America G, van der Vusse GJ, Glatz JFC: Uptake and metabolism of palmitate by isolated cardiac myocytes from adult rats: Involvement of sarcolemmal proteins. J Lipid Res 38: 745–758, 1997
8. Abumrad NA, Park JH, Park CR: Permeation of long-chain fatty acid into adipocytes. Kinetics, specificity, and evidence for involvement of a membrane protein. J Biol Chem 259: 8945–8953, 1984
9. Bonen A, Luiken JJFP, Liu S, Dyck DJ, Kiens B, Kristiansen S, Turcotte LP, van der Vusse GJ, Glatz JFC: Palmitate transport and fatty acid transporters in red and white muscles. Am J Physiol 275: E471–E478, 1998
10. Stremmel W, Berk PD: Hepatocellular influx of [^{14}C]oleate reflects membrane transport rather than intracellular metabolism or binding. Proc Natl Acad Sci USA 83: 3086–3090, 1986
11. Anjaneyulu PSR, Staros JV: Reactions of N-hydroxysulfosuccinimide active esters. Int J Pept Prot Res 30: 117–124, 1987
12. Giedroc DP, Puett D, Ling N, Staros JV: Demonstration by covalent cross-linking of a specific interaction between beta-endorphin and calmodulin. J Biol Chem 258: 16–19, 1983
13. Harmon CM, Luce P, Beth AH, Abumrad NA: Labeling of adipocyte membranes by sulfo-*N*-succinimidyl derivatives of long-chain fatty acids: Inhibition of fatty acid transport. J Membr Biol 121: 261–268, 1991
14. Alpert NL, Keiser WE, Szymanski HA: In: IR: Theory and Practice of Infrared Spectroscopy. Plenum Press, New York, 1970
15. Guthmann F, Haupt R, Looman AC, Spener F, Rüstow B: Fatty acid translocase/CD36 mediates the uptake of palmitate by type II pneumocytes. Am J Physiol 277: L191–L196, 1999
16. van der Vusse GJ, Glatz JFC, Stam HCG, Reneman RS: Fatty acid homeostasis in the normoxic and ischemic heart. Physiol Rev 72: 881–940, 1992
17. Luiken JJFP, Turcotte LP, Bonen A: Protein-mediated palmitate uptake and expression of fatty acid transport proteins in heart giant vesicles. J Lipid Res 40: 1007–1016, 1999
18. Luiken JJFP, Schaap FG, van Nieuwehoven FA, van der Vusse GJ, Bonen A, Glatz JFC: Cellular fatty acid transport in heart and skeletal muscle as facilitated by proteins. Lipids 34(suppl): S169–S175, 1999
19. Tanaka T, Kawamura K: Isolation of myocardial membrane long-chain fatty acid-binding protein: Homology with a rat membrane protein implicated in the binding or transport of long-chain fatty acids. J Mol Cell Cardiol 27: 1613–1622, 1995
20. Isola LM, Zhou SL, Kiang CL, Stump DD, Bradbury MW, Berk PD: 3T3 fibroblasts transfected with a cDNA for mitochondrial aspartate aminotransferase express plasma membrane fatty acid-binding protein and saturable fatty acid uptake. Proc Natl Acad Sci USA 92: 9866–9870, 1995
21. Ibrahimi A, Sfeir Z, Magharaie H, Amri EZ, Grimaldi PA, Abumrad NA: Expression of the CD36 homolog (FAT) in fibroblast cells: Effects on fatty acid transport. Proc Natl Acad Sci USA 93: 2646–2651, 1996
22. Abumrad NA, El-Maghrabi MR, Amri EZ, Lopez E, Grimaldi PA: Cloning of a rat adipocyte membrane protein implicated in binding or transport of long-chain fatty acids that is induced during preadipocyte differentiation. Homology with human CD36. J Biol Chem 268: 17665–17668, 1993
23. Schaffer JE, Lodish HF: Expression cloning and characterization of a novel adipocyte long chain fatty acid transport protein. Cell 79: 427–436, 1994
24. Van Nieuwenhoven FA, Verstijnen CPHJ, Abumrad NA, Willemsen PHM, Van Eys GJJM, van der Vusse GJ, Glatz JFC: Putative membrane fatty acid translocase and cytoplasmic fatty acid-binding protein are co-expressed in rat heart and skeletal muscles. Biochem Biophys Res Commun 207: 747–752, 1995
25. Rose H, Strotman KH, Pöpping S, Fisher Y, Kulsch D, Kammermeier H: Simultaneous measurement of contraction and oxygen consumption in cardiac myocytes. Am J Physiol 261: H1329–H1334 , 1991
26. Bonen A, Luiken JJFP, Arumugam Y, Glatz JFC, Tandon NN: Acute regulation of fatty acid uptake involves the cellular redistribution of fatty acid translocase. J Biol Chem 275: 14501–14508, 2000
27. Bonen A, Dyck DJ, Ibrahimi A, Abumrad NN: Muscle contractile activity increases fatty acid metabolism and transport and FAT/CD36. Am J Physiol 276: E642–E649, 1999
28. Dyck DJ, Steinberg G, Bonen A: Insulin increases FA uptake and esterification but reduces lipid utilization in isolated contracting muscle. Am J Physiol Endocrinol Metab 281: E600–E607, 2001
29. Luiken JJFP, Willems J, van der Vusse GJ, Glatz JFC: Electrostimulation enhances FAT/CD36-mediated long-chain fatty acid uptake by isolated rat cardiac myocytes. Am J Physiol Endocrinol Metab 281: E704–E712, 2001
30. Luiken JJFP, Dyck DJ, Han XX, Tandon NN, Arumugam Y, Glatz JFC, Bonen A: Insulin induces the translocation of the fatty acid transporter FAT/CD36 to the plasma membrane. Am J Endocrinol Metab 282: E491–E495, 2002

Molecular and Cellular Biochemistry **239**: 221–225, 2002.

Ex-FABP, extracellular fatty acid binding protein, is a stress lipocalin expressed during chicken embryo development

Fiorella Descalzi Cancedda,[1,2] Beatrice Dozin,[1] Barbara Zerega,[1] Silvia Cermelli,[1] Chiara Gentili[1] and Ranieri Cancedda[1,3]

[1]*Istituto Nazionale per la Ricerca sul Cancro, Centro Biotecnologie Avanzate, Genova;* [2]*Centro di Studio per la Neurofisiologia Cerebrale, Consiglio Nazionale delle Ricerche, Genova;* [3]*Dipartimento di Oncologia, Biologia e Genetica, Universita' di Genova, Italy*

Abstract

Extracellular Fatty Acid Binding Protein (Ex-FABP) is a 21 kDa lipocalin, expressed during chicken embryo development in hypertrophic cartilage, in muscle fibres and in blood granulocyte. The protein selectively binds with high affinity fatty acids, preferably long chain unsaturated fatty acids in chondrocyte and myoblast cultures Ex-FABP expression is increased by inflammatory-agents and repressed by anti-inflammatory-agents. In adult cartilage, Ex-FABP is expressed only in pathological conditions such as in dyschondroplastic and osteoarthritic chicken cartilage. We propose that lipocalin Ex-FABP represents a stress protein physiologically expressed in tissues where active remodelling is taking place during development and also present in tissues characterized by a stress response due to pathological conditions. (Mol Cell Biochem **239**: 221–225, 2002)

Key words: extracellular fatty acid binding protein, lipocalin, hypertrophic cartilage, stress protein

Cultured chick embryo chondrocytes and myoblasts secrete a large amount of a low molecular weight protein of the lipocalin family

Our laboratory has focused on the mechanisms involved in endochondral bone formation. The process is characterized by the differentiation of prechondrogenic mesenchymal cells to resting, proliferating and hypertrophic chondrocytes with subsequent cartilage calcification, erosion, invasion by blood vessels and replacement of hypertrophic cartilage by bone (for a review see [1]). We have developed a cell culture system that recapitulates the chick embryo chondrocyte differentiation process. Starting from a 6 day chick embryo tibia, dissociated cartilage cells expressing type II collagen are grown and expanded in adhesion on plastic dishes, a culture condition that promotes dedifferentiation of cells to a prechondrogenic stage characterized by type I collagen expression. Dedifferentiated cells retain their chondrogenic potential. In fact when grown in suspension culture, they undergo differentiation and express type II collagen and subsequently type X collagen, a specific marker of hypertrophic chondrocytes [2].

We have observed that at late stage of differentiation, at the time type X collagen is expressed, chondrocytes synthesize and secrete large amounts of a protein with an apparent molecular weight of 21 kDa [3] (Fig. 1A). This protein, initially called Ch21, was later renamed Ex-FABP based on its binding properties. Ex-FABP is also expressed by other cells in culture and abundantly released in cell culture medium. '*In vitro*' differentiating chicken myoblasts express a large amount of protein after the onset of myoblast fusion [4], when also apolipoprotein A1, a protein known to be released by differentiated myotubes is expressed [5] (Fig. 1B).

The same protein, named P20K protein, has been described by Bedard *et al.* as a protein secreted by quiescent chicken embryo fibroblasts and chicken heart mesenchymal cells [6]

Address for offprints: F. Descalzi Cancedda, Centro di Biotecnologie Avanzate, Largo Rosanna Benzi, 10, I-16132 Genova, Italy (E-mail: descalzi@cba.unige.it)

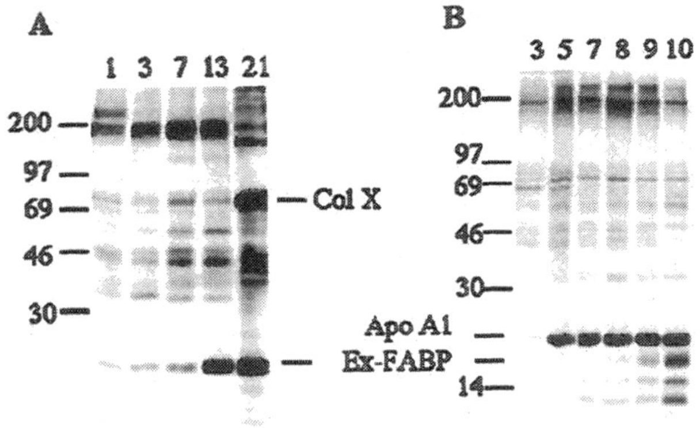

Fig. 1. Proteins secreted by cultured chick embryo chondrocytes and myoblasts. (A) SDS polyacrylamide gel electrophoresis of radioactive proteins released in the culture medium by dedifferentiated chondrocytes transferred in suspension culture. Numbers refer to the days in culture. (B) SDS polyacrylamide gel electrophoresis of radioactive proteins released in the culture medium by differentiating myoblasts. Numbers refer to the days in culture. Bars on the left in panels refer to migration of molecular weight markers. Ex-FABP is indicated in both panels, Col X is type X collagen, ApoA1 is Apolipoprotein A1. The proteins were identified in culture media by immunoprecipitation with specific antibodies.

The authors report that the protein is repressed after mitotic stimulation of these cells [6]. They have included the gene in a set of genes referred to as growth arrest-specific (GAS). The same group has also shown that the expression of the Ex-FABP/p20K gene is depending on the activation of a promoter specific region, quiescence-responsive unit (QRU), possibly by the transcription factor C/EBPβ [7]. We have also observed that confluent hypertrophic chondrocytes enhance EX-FABP expression . In confluent cultures of embryonic chick cells of different origin, we have observed the expression of Ch21/Ex-FABP protein in some cell lines (fibroblasts from the whole body and dedifferentiated chondrocytes), but not in others (skin fibroblasts) [8]. In a quail chondrocyte cell line, when proliferation was stimulated by the constitutive expression of *myc* (QEC-v-myc), the Ex-FABP was abundantly synthesized [9]. Chick chondrocytes proliferate and differentiate also in defined culture medium [10]. In the absence of serum, dedifferentiated chondrocytes can be expanded '*in vitro*' as adherent cells in a medium containing insulin and FGF2. Cells proliferating in these extreme conditions express a large amount of Ex-FABP [10] whereas in the presence of serum, the same factors do not have any effect on the synthesis of the protein (unpublished observation). From these observations ExFABP expression seems to be related to quiescence as well as to other conditions.

The Ch21/Ex-FABP is a not-mannose containing protein, has intrachain disulfide bond and a conformation conferring resistance to limited pepsin digestion. The protein was purified from the culture medium of hypertrophic chondrocytes and characterized [11]. The complete aminoacid sequence was determined at the protein level, the corresponding cDNA was cloned and its nucleotide sequence was determined. Translation of the nucleotide sequence perfectly matched the peptide sequence determined. Computer-assisted comparison of this sequence with sequences existing in data bases revealed that the Ex-FABP belongs to the superfamily of lipocalins [12].

In cultured chondrocytes, the synthesis of the protein is directed by a 1,450-base mRNA and is mostly, if not exclusively, regulated at the transcription level [8].

Unsaturated long chain fatty acids are specific ligands of ExFABP lipocalin

In order to investigate the nature of the ligand for the new lipocalin, the recombinant protein was produced in the baculovirus system and was purified from the cell culture medium by gel filtration and subsequent ion exchange chromatography. By gel shift experiments (Fig. 2A) and binding assay with radioactive compounds (Fig. 2B) we have presented evidence that purified Ex-FABP selectively bind long chain fatty acids [13]. Unsaturated fatty acids like oleic, linoleic and arachidonic acid, are preferentially bound with a kD in the order of 10^{-7} M (Fig. 2B) whereas other hydrophobic molecules like retinol, retinoic acid, progesterone, vitamin D, alcohol and aldehyde derivatives of long chain fatty acids do not bind to the protein. The protein was present in cell culture medium as a monomer and the ligand binding was in a molar ratio of 1:1 with the monomeric protein [13]. By Western blot analysis Ex -FABP was detected also in chicken serum and represents the first example of an extracellular lipocalin able to selectively bind and transport long chain, unsaturated fatty acids in extracellular fluids. In fact fatty acid binding proteins (about 14 kDa), that belong, like lipocalins, to the superfamily of calycins, have been purified only from tissues, including intestine, liver, heart, mammary, Schwann cell and adipose [14–18]. In all cases the localization of the protein was intracellular. It was a general belief that in blood, fatty acids are transported by albumin [19] that binds fatty acids with high capacity, but low affinity. On the basis of these results the originally named Ch21 protein was renamed Ex-FABP. The presence of Ex-FABP in the bone growth plate and in newly formed muscle fibers, suggests an involvement of lipid and fatty acid metabolism in cartilage development, bone formation and myogenesis.

There is a report [7] that in chicken embryo fibroblasts, expressing Ex-FABP/p20K at confluence, the addition of linoleic acid markedly repressed the synthesis of the protein without inducing mitogenesis. This finding suggests a role of this essential fatty acid in the regulation of the Ex-FABP gene.

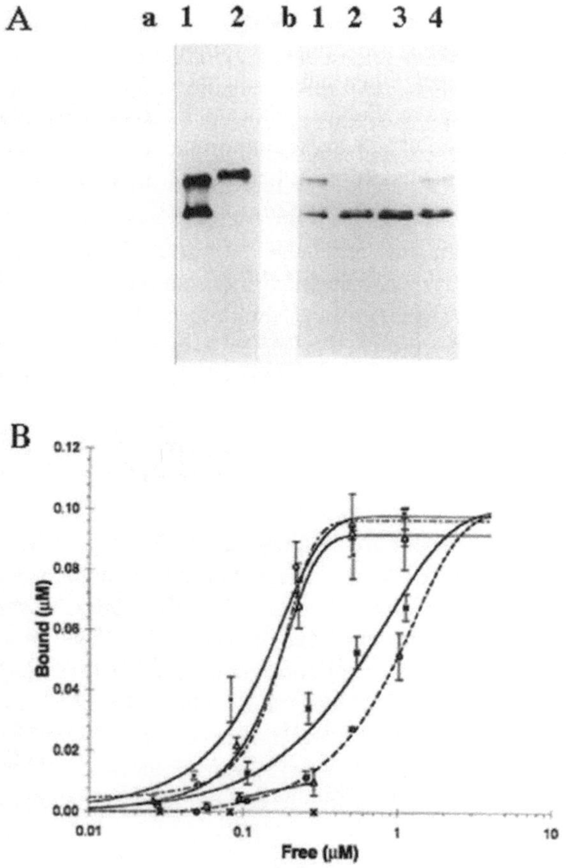

Fig. 2. (A) Association of fatty acids with the purified recombinant Ch21 analyzed by polyacrylamide gel electrophoresis under native conditions. (a) Protein before (lane 1) and after (lane2) extraction with ethyl acetate. (b) Lane1: control; lanes 2–4: protein incubated in PBS with oleic (lane 2), linoleic (lane 3) and stearic acid (lane 4). The purified protein, homogeneous in SDS PAGE, shows two bands by polyacrylamide gel electrophoresis in native conditions. Extraction of the ligand results in a shift to the less anodic band, while incubation with fatty acids results in a shift to the more anodic band. (B) Fatty acid binding to recombinant Ex-FABP. A fixed amount of radioactive hydrophobic ligand (1×10^{-7} M) was incubated with increasing concentration recombinant purified Ex-FABP. Filled circle: myristic acid, K_d 9.8×10^{-7} M; star: retinoic acid; filled diamond: arachidonic acid, K_d 1.6×10^{-7} M; open triangle: oleic acid, K_d 1.7×10^{-7} M; open circle: linoleic acid, K_d 1.7×10^{-7} M; filled square: stearic acid, K_d 5.9×10^{-7} M; open square: progesterone.

Ex-FABP is developmentally regulated during chicken embryogenesis

The synthesis and location of Ex-FABP was investigated in developing chick embryo by immunohistochemistry and '*in situ*' hybridization. Following the expression of the protein in cultured chondrocytes, immunohistochemistry, was first performed in sections of limbs of chickens at different developmental stages. Initially the protein was detected in the calcifying mid-diaphyseal sleeve of 7 day old tibia. Later, it is observed in the matrix and in the hypertrophic chondrocytes along the edges of the growth cone, in prearticular cartilage and in the cartilage matrix and cells surrounding blood vessels [20]. '*In situ*' hybridization on embryo sections confirmed the expression of Ex-FABP by cells of the chondrogenic lineage [6]. By immunohistochemistry also bone trabeculae show intracellular positivity and staining at their edges. Osteoblasts cultures and cultured bone fragments synthesize and secrete the protein into the culture medium [20].

In developing chicken embryo Ex-FABP is present also in newly formed myotubes at early stages of development [4]. Single myoblasts were negative. Localization of the protein and its mRNA was performed by immunohistochemistry and '*in situ*' hybridization respectively. At later stages of development, myofibers still expressed both the mRNA and the protein.

This finding is in agreement with the data obtained by immunohistochemistry in cultured chicken myoblasts showing that multinucleate myotubes, but not immature myotubes and single myoblasts, were clearly stained by antibodies directed against Ex-FABP [4].

Ex-FABP expression was localized also in the developing myocardium and the muscular layer of large blood vessels [4].

By Northern blot analysis of RNAs extracted from several chick embryo tissues, Ex-FABP mRNA was identified also in extracts from tissues other than cartilage such as skin, brain, heart and skeletal muscle [8]. It should also be noted that Ex-FABP is a major translation product in granulocytes [8].

Ex-FABP is modulated by inflammatory-anti-inflammatory agents and expressed in pathological conditions

Ex-FABP synthesis is strongly up-regulated both in '*in vitro*' differentiating chondrocytes and myotubes by inflammatory factors such as LPS or IL-6 while anti-inflammatory drugs repress the synthesis of the protein physiologically expressed and prevent or repress the synthesis of the protein induced by LPS [21] (Figs 3A and 3B). In this respect Ex-FABP can be considered as being expressed in culture as a stress protein. In the adult animal Ex-FABP is expressed in cartilage as a consequence of pathological conditions. In dyschondroplastic chickens a high amount of Ex-FABP was observed in the lesion areas of tibiae at the level of the epiphysis. In osteoarthritic chickens Ex-FABP was expressed in affected articular cartilage. Control cartilages were negative [21].

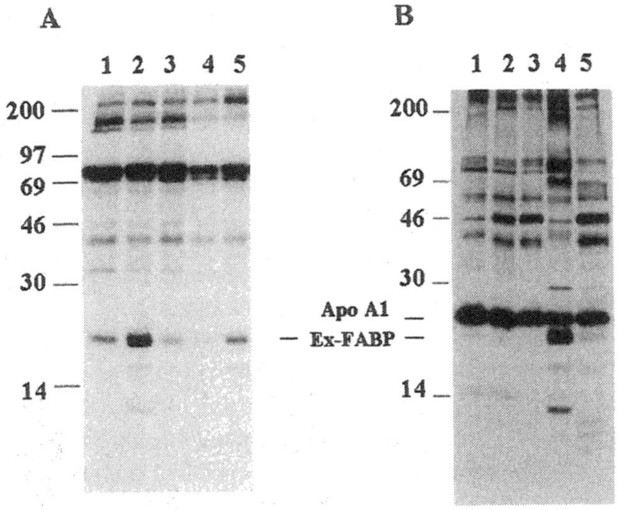

Fig. 3. Proteins released by cultured hypertrophic chondrocytes and myoblasts treated with inflammatory and anti-inflammatory agents. (A) Chicken hypertrophic chondrocytes cultured for 16 days were labeled with [^{35}S]-Methionine. Labeled proteins released into media were analysed on a 15% SDS PAGE. Lane 1: control; lane 2: culture treated with LPS; lane 3: culture treated with Diclofenac (100 μM); lane 4: culture treated with Diclofenac (200 μM); lane 5: culture treated with LPS and Diclofenac (200 μM). (B) Chicken myoblasts cultured for 12 days were labeled with [^{35}S]-Methionine. Labeled proteins released into the media were analysed on a 15% SDS PAGE. Myoblasts cultured for 12 days were labeled after different treatments. Lane 1: control; lane 2: culture treated with Diclofenac (100 μM); lane 3: culture treated with Diclofenac (200 μM); lane 4: culture treated with LPS; lane 5: culture treated with LPS and Diclofenac (200 μM). Bars on the left in panels refer to migration of molecular weight markers. Ex-FABP is indicated in both panels, ApoA1 is Apolipoprotein A1. Both proteins were identified in culture media by immunoprecipitation with specific antibodies.

Therefore Ex-FABP is a stress protein expressed both in tissues where active remodelling is physiologically taking place during development and in tissues characterized by a stress response due to pathological conditions. It is possible that the same applies also to other proteins as proteases, chemotactic and angiogenic factors, etc. We have suggested that some stress responses characteristic of a local inflammatory status are 'physiologically' activated during endochondral bone formation and possibly during initial myogenesis.

Possible function of Ex-FABP protein

Our data indicate that in differentiating embryo tissues undergoing extensive remodelling, Ex-FABP, is expressed by cells that change shape and function, as the endochondral chondrocytes and the fusing myoblasts. In adult mature cartilage (epiphyseal and articular) the protein is not expressed in normal conditions, but its expression is reactivated in pathologies such as tibial dyschondroplasia and osteoarthritis. In culture, the protein is upregulated by inflammatory- and down-regulated by anti-inflammatory agents.

From the data available, it appears that Ex-FABP expression is activated in cell stress conditions of different nature. Stress conditions could be either 'physiological' (chondrocyte differentiation to the hypertrophic stage and subsequent bone formation with extensive extracellular matrix remodelling; myoblast fusion and myofiber formation) or pathological (degenerative diseases). In cultured cells, an increase in Ex FABP expression is observed also in response to inflammatory stimuli and in other stress conditions as serum deprivation and growth arrest.

A possible function of the protein could be suggested by the fact that addition of antibodies directed against Ex-FABP to myoblast cultures resulted in a continuous removal of Ex-FABP from the culture medium, due to the formation of immune complexes. Under these conditions, cells appeared damaged and myotube formation was impaired. Nevertheless a strong enhancement of the synthesis of the same protein was observed in antibody treated cells suggesting that Ex-FABP may exert a feed-back (autocrine) control during myotube differentiation and maturation and may play a role in maintaining cell viability [5]. Additional studies on these important aspects in different cell cultures are necessary.

Acknowledgement

Partially supported by funds from Associazione Italiana per la Ricerca sul Cancro and MURST.

References

1. Cancedda R, Descalzi Cancedda F, Castagnola P: Chondrocyte differentiation. Int Rev Cytol 159: 265–358, 1995
2. Castagnola P *et al.*: Type X collagen synthesis during *in vitro* development of chick embryo tibial chondrocytes. J Cell Biol 102: 2310–2317, 1986
3. Descalzi Cancedda F *et al.*: Developmentally regulated synthesis of a low molecular weight protein (Ch 21) by differentiating chondrocytes. J Cell Biol 107: 2455–2463, 1988
4. Gentili C *et al.*: Expression of the extracellular fatty acid binding protein (Ex-FABP) during muscle fiber formation *in vivo* and *in vitro*. Exp Cell Res 242: 410–418, 1998
5. Ferrari S, Battini R, Cossu G: Differentiation-dependent expression of apolipoprotein A-I in chicken myogenic cells in culture. Dev Biol 140: 430–436, 1990
6. Bedard PA *et al.*: Rapid repression of quiescence-specific gene expression by epidermal growth factor, insulin, and pp60v-src. Mol Cell Biol 9: 1371–1375, 1989
7. Kim S *et al.*: C/EBPbeta (NF-M) is essential for activation of the p20K lipocalin gene in growth-arrested chicken embryo fibroblasts. Mol Cell Biol 19: 5718–5731, 1999

8. Dozin B *et al.*: Expression regulation and tissue distribution of the Ch21 protein during chicken embryogenesis. J Biol Chem 267: 2979–2985, 1992
9. Quarto R *et al.*: Constitutive myc expression impairs hypertrophy and calcification in cartilage. Dev Biol 149: 168–176, 1992
10. Quarto R *et al.*: Modulation of commitment, proliferation and differentiation of chondrogenic cells in defined culture medium. Endocrinology 138: 4966–4976, 1997
11. Descalzi Cancedda F *et al.*: The amino terminal sequence of the developmentally regulated Ch21 protein shows homology with amino terminal sequences of low molecular weight proteins binding hydrophobic molecules. Biochem Biophys Res Commun 168: 933–938, 1990
12. Descalzi Cancedda F *et al.*: Ch21 protein developmentally regulated in chick embryo belongs to the superfamily of hydrophobic molecules transporters. J Biol Chem 265: 19060–19064, 1990
13. Descalzi Cancedda F *et al.*: The developmentally regulated avian Ch21 lipocalin is an extracellular fatty acid-binding protein. J Biol Chem 271: 20163–20169, 1996
14. Ockner RK, Manning JA, Kane JP: Fatty acid binding protein. Isolation from rat liver, characterization, and immunochemical quantification. J Biol Chem 257: 7872–7878, 1982
15. Sacchettini JC *et al.*: Rat heart fatty acid-binding protein is highly homologous to the murine adipocyte 422 protein and the P2 protein of peripheral nerve myelin. J Biol Chem 261: 8218–8223, 1986
16. Bohmer FD *et al.*: Identification of a polypeptide growth inhibitor from bovine mammary gland. Sequence homology to fatty acid- and retinoid-binding proteins. J Biol Chem 262: 15137–15143, 1987
17. Suzuki M *et al.*: The complete amino acid sequence of human P2 protein. J Neurochem 39: 1759–1762, 1982
18. Matarese V, Bernlohr DA: Purification of murine adipocyte lipid-binding protein. Characterization as a fatty acid- and retinoic acid-binding protein. J Biol Chem 263: 14544–14551, 1988
19. Ashbrook JD *et al.*: Long chain fatty acid binding to human plasma albumin. J Biol Chem 250: 2333–2338, 1975
20. Manduca P *et al.*: Synthesis and secretion of Ch 21 protein in embryonic chick skeletal tissues. Eur J Cell Biol 50: 154–161, 1989
21. Cermelli S *et al.*: Extracellular fatty acid binding protein (Ex-FABP) modulation by inflammatory agents: 'Physiological' acute phase response in endochondral bone formation. Eur J Cell Biol 79: 155–164, 2000

Molecular and Cellular Biochemistry **239**: 227–234, 2002.

Plasma concentration of intestinal- and liver-FABP in neonates suffering from necrotizing enterocolitis and in healthy preterm neonates*

Florian Guthmann,[1] Torsten Börchers,[2] Christian Wolfrum,[3] Thomas Wustrack,[1] Sabine Bartholomäus[4] and Friedrich Spener[3]

[1]*Department of Neonatology, Charité Campus Mitte;* [4]*Department of Neonatology, Charité Campus Virchow-Klinikum, Humboldt-Universität zu Berlin;* [2]*Institut für Chemo- und Biosensorik, Münster;* [3]*Department of Biochemistry, University of Münster, Germany*

Abstract

Both early diagnostic and prognostic assessment of the acute abdomen in preterm infants are hampered by the lack of a sensitive and specific parameter for intestinal injury. In this prospective clinical study we wanted to estimate the value of intestinal (I-) and liver (L-) fatty acid binding protein (FABP) in diagnosing necrotizing enterocolitis (NEC). Using highly sensitive and specific sandwich ELISAs which employ recombinant human I- and L-FABP as standard proteins (limit of detection 0.1 ng/ml plasma), the L-FABP concentration (median 7.6 ng/ml) was determined to be about 3 fold that of I-FABP (median 2.52 ng/ml) in plasma of healthy preterm infants. I- and L-FABP concentrations significantly increased with birth weight (1.6 and 5.0 ng/ml per kg, respectively). At onset of symptoms, I-FABP concentration was significantly higher in infants who later developed severe NEC compared to healthy infants and those, whose illness remained confined to stage I or II. L-FABP was significantly elevated compared to the control group at onset of symptoms regardless of the further course of NEC. In conclusion, I-FABP appears to be a specific parameter for early detection of intestinal injury leading to severe NEC stage III. L-FABP, however, is a promising sensitive marker even for stage I of NEC. (Mol Cell Biochem **239**: 227–234, 2002)

Key words: recombinant protein, sandwich ELISA, necrotizing enterocolitis, neonates, preterm infant, fatty acid binding protein

Introduction

The differential diagnosis of acute abdomen of the preterm infant is still a challenge to the neonatologist. This is due to the high frequency of mild abdominal symptoms in preterm neonates like feeding intolerance, abdominal distension, gastric aspirates and bloody stools and, on the other hand, a wide variety of possible clinical courses. These reach from spontaneous remission without recurrence to fulminant intraabdominal catastrophe with sepsis, shock, peritonitis and death. The first symptom of a more severe disorder involving the gut often is gastrointestinal haemorrhage, which shows in the vast majority of infants no identifiable cause [1]. Nevertheless, it may be the presenting symptom in cases of ischemia, such as malrotation with midgut volvulus, intussusception, Hirschsprung disease, Meckel diverticulum, and intestinal duplications. Necrotizing enterocolitis (NEC) is a cause of rectal bleeding that is found more frequently in neonatal intensive care units.

Laboratory evaluation of premature infants suffering from acute abdomen is unspecific and reflects signs of inflammation, including leukocytosis or leukopenia, thrombocytopenia, electrolyte imbalance, acidosis, hypoxia, and hypercapnia. Bacterial growth in blood culture may indicate a primarily

*Results were presented in part at the '4th International Conference on Lipid Binding Proteins' held in Maastricht (NL) from June 10–12, 2001.

Address for offprints: F. Guthmann, Charité Campus Mitte, Department of Neonatology, D-10098 Berlin, FRG (E-mail: florian.guthmann@charite.de)

septic process or may reflect gut barrier failure. The major hindrance for ruling out disorders associated with high or increasing risk of progressive disease is the lack of a sensitive and specific parameter of intestinal damage.

Fatty acid binding proteins (FABPs) comprise a group of cytoplasmic small molecular mass proteins (~ 15 kDa) with high organ specificity. Intestinal (I)-FABP, which is expressed in gut from duodenum to cecum with its maximum at distal jejunum (data from mouse [2]), was shown to be elevated in intestinal ischemia in the rat [3, 4], in an animal model of NEC [5], in cases of human intestinal injury [6], and in a pilot study of NEC [7]. The latter revealed significantly increased plasma concentrations of I-FABP in neonates suffering from NEC stage III compared to stage II (according to the classification of Walsh and Kliegman [8]). However, I-FABP concentrations did not discriminate between stage I and stage II of NEC.

L(liver)-FABP, another member of this protein family is expressed in liver, kidney and intestine. Its anatomical expression along the murine gut peaks proximal of I-FABP [2]. Theoretically, the additional determination of L-FABP should enhance the sensitivity of this evaluation, particularly with regard to spatially limited mucosal injury of the gut. To our knowledge, L-FABP has not been measured in plasma so far.

The aim of our study was (i) to establish highly sensitive and fast ELISAs for detection of I- and L-FABP, (ii) to determine plasma concentration of both the I- and L-type in healthy preterm infants and neonates suffering from different stages of NEC and (iii) to estimate the value of the new parameter L-FABP in diagnosing NEC.

Materials and methods

Study population

Control group: Preterm infants of gestational age < 33 weeks, between days 14 and 28 of age without abdominal symptoms or other clinical signs of illness. There was no evidence of any infectious disease, abdominal symptoms like feeding intolerance, abdominal distension, gastric aspirates or any other conspicuousness. In defining these inclusion criteria, we tried to match some typical properties of the NEC high risk group with respect to immaturity and probable age at onset. Blood (100 μl) was drawn once in the context of an indicated evaluation and processed as described for the patient group.

Patient group: Preterm infants of gestational age < 33 weeks who exhibited abdominal symptoms requiring a septic work up or an abdominal X-ray. Blood (100 μl) was drawn at the initial evaluation (defined as onset of symptoms for purposes of the study). Blood (EDTA) was centrifuged and plasma was stored at –70°C until determination of I- and L-FABP. The protocol was approved by the institutional ethics committee.

Staging of NEC

The staging of Walsh and Kliegman [8] was simplified for purposes of this study. Stage I comprises stages IA and IB of the Walsh and Kliegman classification. If pneumatosis intestinalis was diagnosed on X-ray, the case was assigned to stage II. On occurrence of organ failure (for definition in very low birth weight infants see [12]) in addition to gastrointestinal tract, or on occurrence of peritonitis, ascites, perforated bowel or sepsis/SIRS, subjects were assigned to stage III.

ELISA for human L-FABP

The procedures for production of recombinant human L-FABP (here used as standard protein for generation of calibration curves) as well as for the corresponding sandwich ELISA have been described earlier [9].

Expression and purification of recombinant I-FABP

The cDNA for human I-FABP was cloned by reverse transcription of human intestine mRNA and subsequent polymerase chain reaction using Pfu-proofreading polymerase (Stratagene, Heidelberg, Germany) and primers designed according to the 5′ and 3′ ends of the coding region as published by Sweetser *et al.* [10]. The primers also introduced *Nco*I and *BamH*I restriction sites for directed cloning into the pET3d expression vector (Novagen, Madison, WI, USA). The correct construction of this pEThIFABP (Fig. 1A) was confirmed by cycle sequencing after isolation of the plasmid from *E. coli* DH5α bacteria.

For heterologous expression of recombinant human I-FABP, competent *E. coli* BL21(DE3)pLysS bacteria were transformed with the pEThIFABP expression vector. A 1 L culture of these bacteria in LB-broth supplemented with ampicillin and chloramphenicol was induced with 0.4 mmol/l isopropyl-β-D-thiogalactopyranoside (at 0.6 OD_{600}). After further incubation for 2.5 h, bacteria were harvested, disrupted by sonication, and nucleic acids were precipitated by addition of streptomycin sulphate (final concentration 1.5% w/v). The supernatant after centrifugation was then desalted on a Sephadex G25 column (5 × 20 cm, 10 mM Tris/HCl, pH 7.4) and recombinant human I-FABP was purified by anion exchange chromatography on Q-Sepharose BB (5 × 25 cm, 10 mM Tris/HCl pH 8.0 with a gradient from 0–300 mM NaCl). Final gel filtration was done on a Superdex S200 column (High load prepgrade, 1.6 × 60 cm, 10 mM Tris/HCl, pH 7.4, 150 mM NaCl). All chromatography steps were performed using the FPLC or BioPilot system and chromatography material from Pharmacia (Freiburg, Germany).

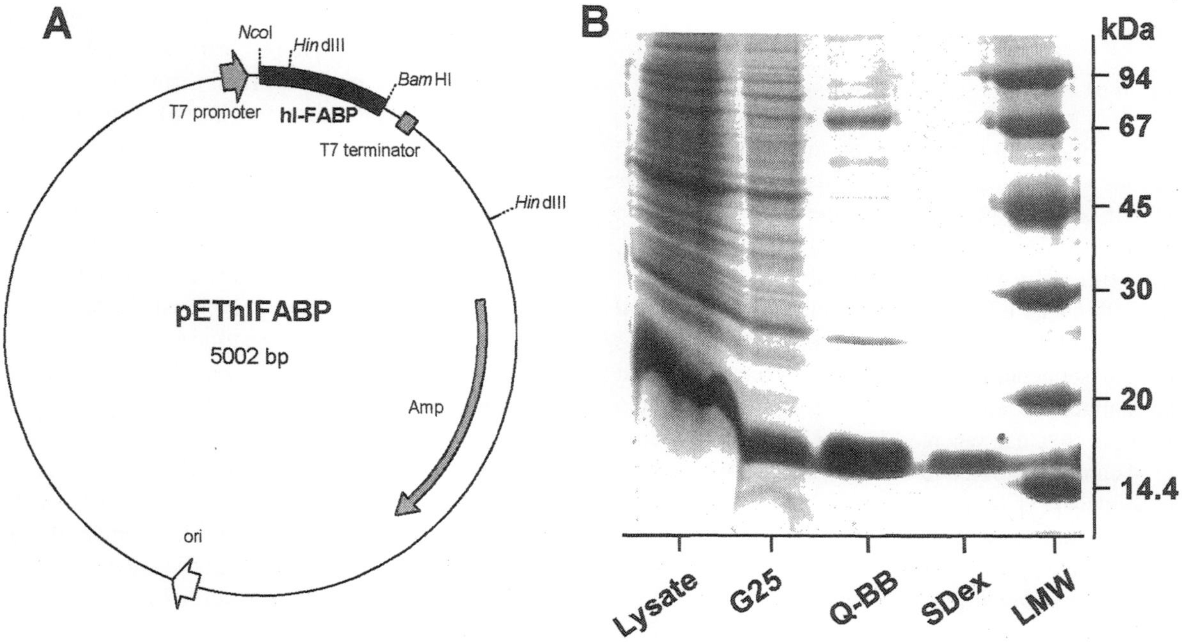

Fig. 1. Heterologous expression of human I-FABP. (A) Expression vector pET-hIFABP, containing the 399 bp coding region of human I-FABP. (B) Progress of purification. Aliquots of fractions of the G25, Q-Sepharose BB and Superdex S200 column were subjected to SDS-PAGE and subsequent Coomassie staining.

Purity of recombinant human I-FABP was assessed by SDS-polyacrylamide gel electrophoresis (SDS-PAGE, 13% gels) and staining with Coomassie Brilliant Blue. Molecular mass marker proteins were obtained from Pharmacia. For quantification of recombinant human I-FABP by UV-spectrometry an extinction coefficient E_{280}(1 mg/ml) of 1.092 (calculated according to Gill and van Hippel [11]) was used.

Affinity purification of antibodies and sandwich ELISA for human I-FABP

Polyclonal antibodies were generated in New Zealand white rabbits by injection of 500 µg recombinant human I-FABP followed by several booster injections. The specific antibodies were affinity-purified from the antiserum using CH-activated Sepharose (Pharmacia) to which the recombinant human I-FABP had been bound covalently and 0.1 M glycine/HCl (pH 2.8) as elution buffer. The affinity purified antibodies were immediately neutralised and dialysed against PBS by ultrafiltration. For preparation of the detector antibody, an aliquot of the affinity purified antibodies was biotinylated by means of sulfosuccinimidyl 6-[biotinamido]hexanoate sodium salt (Sigma, Deisenhofen, Germany). A fraction of the biotinylated antibodies was further purified on rabbit IgG agarose (Sigma).

All incubations of the sandwich ELISA were done in polystyrene microtiter plates (Nunc, Wiesbaden-Biebrich, Germany) at room-temperature in a volume of 100 µl (if not stated otherwise) and were stopped after 1 h by washing 3 times with 10 mM Na-phosphate, 150 mM NaCl, pH 7.4 (PBS) containing 0.05 % (v/v) Tween 20 (PBS-T). Wells were coated with affinity purified antibody (2 µg/ml) in coating buffer (15 mM Na_2CO_3, 35 mM $NaHCO_3$, pH 9.6). For blocking, 200 µl/well of PBS containing 0.5% (w/v) BSA was added directly to the emptied wells (without washing). Samples (diluted 1:2–1:10) and standards were then applied in PBS-T containing 5 g/l BSA (PBS-TB). For specific detection of bound I-FABP, the biotin conjugate of the affinity purified antibody was added (1:250 dilution in PBS-TB, corresponding to ~ 2 µg/ml) followed by incubation with streptavidin-peroxidase (Sigma) diluted 1:5000 in PBS-TB. For colour reaction, o-phenylendiamine (0.8 mg/ml; Sigma) dissolved in phosphate/citrate buffer (50 mM sodium phosphate, 100 mM sodium phosphate, pH 5.0) containing 0.015% H_2O_2 was used. The reaction was terminated after 10–30 min by addition of 50 µl 2 M H_2SO_4 and the absorbance measured at 490 nm with a MR 5000 microplate reader (Dynatech, Denkendorf, Germany).

Statistics

Differences between two groups were assessed using the Mann–Whitney rank test. Statistical calculations were performed with the Statview software program (version 4.57).

Results

Development of a sensitive ELISA for human I-FABP

In order to provide unlimited supply of pure human I-FABP for immunoassay development and as standard protein for generation of calibration curves in sandwich ELISA, we chose to clone the coding part of the human I-FABP cDNA for construction of a bacterial expression system. DNA sequencing after RT-PCR and directed cloning into pET3d revealed 3 nucleotide exchanges compared to the published human I-FABP sequence [10]. The reason for the C216T and G357A exchanges, which are not connected with amino acid substitutions, is not clear, whereas the G163A exchange leading to Thr54 can be attributed to the known Ala/Thr polymorphism at codon 54 [13]. No extra amino acids were introduced into the recombinant protein because the codon for the initiator methionine was part of the *Nco*I restriction site used for construction of the expression vector.

Earlier, good results with heterologous expression of recombinant human heart-type (H-)FABP [14] and other members of this protein family were obtained with the pET3 based expression vectors. The pET vectors are characterised by the strong T7 promoter (Fig. 1 A), and large amounts of recombinant I-FABP were present in the soluble supernatant after induction of *E. coli* BL21(DE3)pLysS bacteria transformed with pEThIFABP. The most effective step of the purification scheme was the anion exchange chromatography, which removed most of the contaminating proteins (Fig. 1 B). The purity of recombinant human I-FABP after the final gel filtration step was found to be clearly above 97%, as in a dilution series of the purified protein from 5–0.15 μg/lane only I-FABP was visible (data not shown). Typical yields of pure protein were approximately 50 mg per litre culture. The purified recombinant protein was used for generation of polyclonal antibodies.

For this study an enzyme-linked immunosorbent assay of the antigen capture type (sandwich ELISA) was developed based on affinity purified polyclonal antibodies of which a portion was biotinylated to be used as detector antibodies. The assay scheme is illustrated in Fig. 2A. In initial experiments an unacceptably high background was observed for the ELISA which could be significantly reduced by passage of the biotinylated antibody over an IgG column (Fig. 2B). After further optimisation, the ELISA was characterised by a limit of detection of at least 0.05 ng/ml, a measuring range from 0.1–4 ng/ml and an interassay coefficient of variation of ~ 5%. The ELISAs were highly selective for their respective antigens, as no signals significantly different from the ELISA blanks were detected for 10 ng/ml recombinant human L-FABP in the I-FABP ELISA and for the same amount of I-FABP in the L-FABP ELISA. Linearity of the ELISA signal was also given for real samples as exemplified in Fig. 3. For various dilutions (from 1:20–1:160) of a plasma sample with a high I-FABP content the same I-FABP concentration (88.1 ± 2.0 ng/ml plasma) was found, indicating the same immunochemical reactivity of native and recombinant human I-FABP. Similarly, addition of increasing amounts (up to 50%

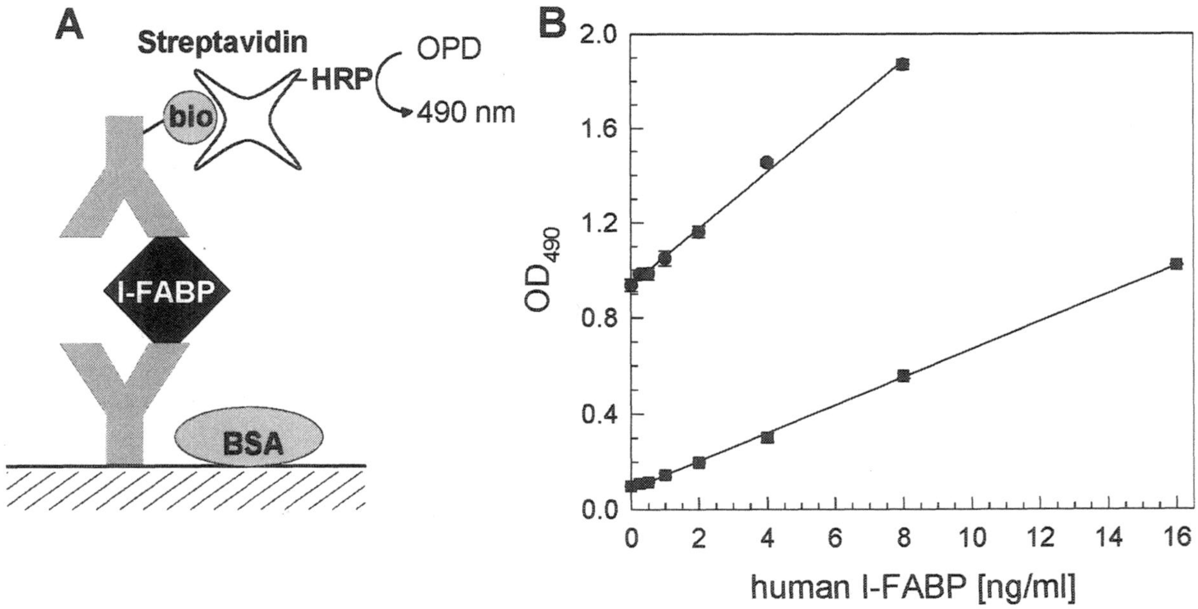

Fig. 2. Sandwich ELISA for human I-FABP. (A) Scheme of the sandwich ELISA; (B) impact of further purification of biotinylated detector antibodies. Biotinylated anti-I-FABP antibodies were used either directly (filled circle) or after passage over a rabbit IgG agarose column (filled square). The concentration of the detector antibodies was 1 μg/ml (5 min colour reaction).

Table 1. Characteristics of control and patient group

Parameters	Control n = 26	Patient n = 14	p
Birth weight (g)	1342	995	< 0.01
Median (range)	830–1865	445–1880	
Gestational age (wk)	30.4	28.6	n.s.
Median (range)	27.7–32.9	24.6–32.9	
Age at onset (days)	24	6	< 0.01
Median (range)	14–28	2–57	

The patient group consisted of infants who developed any stage of NEC within 7 days after onset of symptoms and sample drawing. Significance (p) was calculated by Mann–Whitney rank test.

v/v) of a plasma sample containing 1.2 ng/ml I-FABP to a standard sample with 2 ng/ml I-FABP did not reveal any matrix effects for the ELISA established here.

I- and L-FABP in healthy subjects

Infants born before 33 weeks of gestation (term ≥ 37 weeks) that exhibited an uncomplicated postnatal course were assigned to the control group; the corresponding demographic characteristics are described in Table 1 (control). I- and L-FABP plasma concentrations ranged from 0.46–4.51 ng/ml (median 2.52 ng/ml) and 1.9–13.0 ng/ml (median 7.6 ng/ml), respectively. The 10th and 90th percentiles in this population were 1.4 and 4.1 (I-FABP) and 4.1 and 11.1 ng/ml (L-FABP), respectively. All samples were above the detection limits of 0.1 ng/ml.

Birth weight and age differed significantly between control and patient group (Table 1). To ascertain the impact of these parameters on concentration of FABP, the data were examined for this bias. We found a correlation between birth weight and concentration of I- and L-FABP in healthy infants with a goodness of linear fit of $r = 0.44$ (CI 0.1–0.7) and $r = 0.48$ (CI 0.12–0.74), with slopes of 1.6 ng/ml and 5.0 ng/ml per kg increase in birth weight, respectively, which were different from zero (Fig. 4A, $p = 0.026$; Fig. 4B, $p = 0.013$). There was no significant correlation between age and concentration of FABPs among healthy infants.

I- and L-FABP in subjects exhibiting abdominal symptoms

In order to estimate the clinical relevance of I- and L-FABP, we determined their plasma concentration in preterm infants suffering from abdominal symptoms such as feeding intolerance, abdominal distension, gastric aspirates or bloody stools. Septic work up was done in all infants exhibiting abdominal symptoms. Within seven days after onset of symptoms, the major diagnosis regarding this period was assigned to the infant. As expected, the 14 subjects of this group showed a high diversity with respect to diagnoses and outcome. Diagnostic classifications are shown in Table 2.

The first group, complying with the definition of NEC stage I, was characterised by moderate abdominal symptoms. All children in this group exhibited bloody stools, but the clinical course was mild with complete remission. None of these infants underwent surgery. The infants suffering from NEC stage II presented, aside from more severe symptoms, pneumatosis intestinalis on X-ray. One of these infants underwent laparotomy, which was negative for necrotic lesions. The course of the infants suffering NEC stage III was complicated by the occurrence of septic shock, capillary leak syndrome or paralytic ileus and multiorgan failure. Lethality in this group was 75%. 'Others' comprises 2 infants suffering from incomplete volvulus and meconium ileus.

I-FABP concentrations were significantly higher in stage III compared to each of the other stages ($p < 0.05$) and the control group ($p < 0.002$). None of the infants assigned to

Table 2. Summary of study results

Parameters	Diagnosis NEC I	NEC II	NEC III	Others
n	5	3	4	2
Bloody stools (onset)	4/5	3/3	0/4	0/2
(12–24 h after onset	5/5	2/3	1/4	0/2
Laparotomy	0/5	1/3	3/4	2/2
Lethality	0/5	0/3	3/4	0/2
I-FABP (ng/ml)	2.2	2.5	13.3	1.1
(range)	0.4–3.4	0.4–8.71	11.3–250	0.8–1.4
L-FABP (ng/ml)	16.0	12.0	18.5	7.3
(range)	6.3–24.2	6.9–21.9	18.3–23.4	4.1–10.5

Infants with abdominal symptoms were grouped according to their diagnoses as NEC stage I, II, III and others. Plasma concentrations of I-FABP and L-FABP were determined by ELISA at onset of abdominal symptons and are given as median and range of n patients.

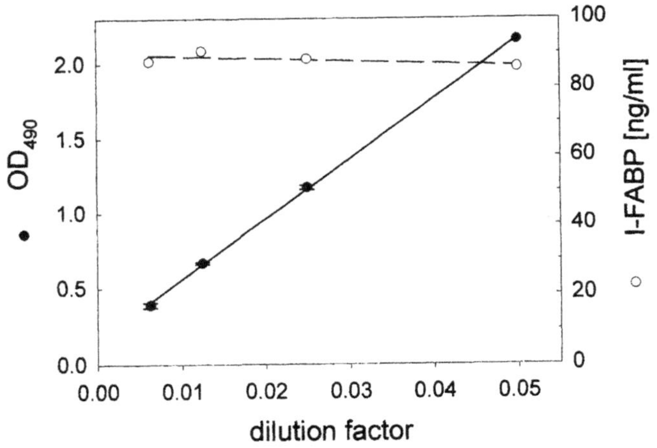

Fig. 3. Equivalence of native and recombinant human I-FABP. Increasing dilutions (1:20–1:160, corresponding to dilution factors 0.05–0.006) of a plasma sample were subjected to the I-FABP ELISA.

stage I, however, exhibited concentrations above the control. Also, at onset of symptoms, I-FABP did not discriminate infants who later developed NEC stage II from infants whose illness remained confined to stage I. In contrary to I-FABP, the liver-type FABP discriminated between control and the very beginning of NEC, stage I ($p < 0.05$). However, development of more severe illness could not be predicted by higher L-FABP concentrations at onset of symptoms, compared to infants whose diagnosis remained restricted to stage I.

In addition to the data from 26 control subjects, Fig. 4 also shows the data from 5 infants with stage I (square), 3 infants with stage II (open circle) and 4 infants with stage III (triangle) NEC. Those 5 out of 12 patients suffering from NEC of stage III were recognised by an elevated I-FABP concentration (Fig. 4A). In comparison, 9 out of the 12 patients were recognised by determination of L-FABP concentration. Two of the 3 undetected patients were suffering from NEC stage I, the third one was assigned to stage II. He underwent surgery whereby only mild findings without necrosis were seen (Fig. 4, open circle at 710 g).

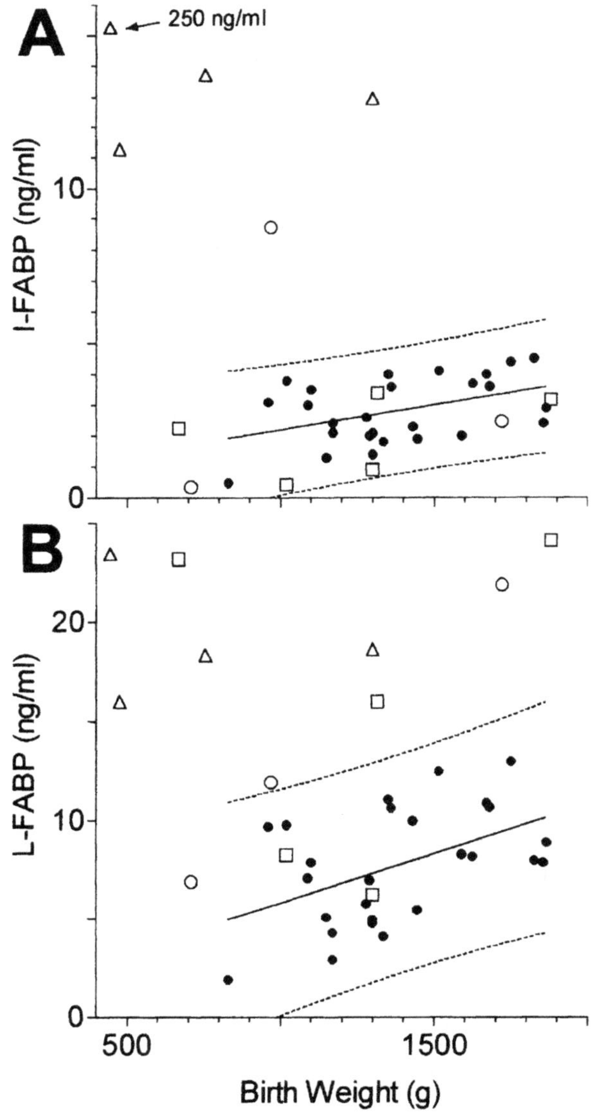

Fig. 4. I-FABP (A) and L-FABP (B) plasma concentration of healthy (filled circle) and sick infants (open symbols) suffering from NEC stage I (square), II (circle) or III (triangle) at onset of symptoms are plotted vs. birth weight. Solid lines represent linear regression of values from healthy subjects with 95% predictive intervals (dashed lines).

Discussion

For the first time, concentrations of I-FABP have been measured in plasma of healthy children. In contrast to previously published data where I-FABP has not been detectable in healthy adult humans or animals [4, 6], we could detect I-FABP in every infant examined. Only ten of 40 infants exhibited I-FABP concentrations below 1.9 ng/ml, the detection limit of the radioimmunoassay used in most of the studies so far [6]. This discrepancy can thus only be partly attributed to the higher sensitivity of our ELISA (detection limit 0.1 ng/ml plasma when a 1:2 dilution is measured). Other reasons may be the use of different standard proteins for calibration or the different characteristics (in particular with respect to matrix effects) of a competitive immunoassay (like the radioimmunoassay of Liebermann *et al.* [6]) compared to a sandwich assay. Interestingly, Kanda *et al.* [15], also employing a sandwich ELISA, generally found considerably higher I-FABP levels in the healthy controls (up to 65 ng/ml) and in adult patient samples (ranging from 115–1496 ng/ml), which coincided with a concomitantly higher limit of detection of 20 ng/ml serum. The reason for these highly differing basal levels of human I-FABP can only be clarified by using standardised I-FABP preparations and/or reference sera. Some of the problems associated with production of recombinant proteins suited as standards for immunoassay (e.g. reproducibility of the production or quantification of the standard protein) have been discussed elsewhere [14].

Among the infants exhibiting abdominal symptoms we found significantly higher concentrations of I-FABP in those who later developed NEC stage III. These data corroborate the findings of Edelson and co-workers [7] who found I-FABP concentrations at onset of symptoms significantly higher in stage III (median 12.5 ng/ml) compared to stages II or I. In contrast to our results, they could not detect I-FABP in infants suspected for NEC (stage I). The highest I-FABP concentration (250 ng/ml) detected in the presented study belongs to an infant who rapidly developed NEC and SIRS necessitating cardio-pulmonary reanimation. Laparotomy of this patient revealed a matt, thickened peritoneum without evidence of bacterial growth or intestinal necrosis.

L-FABP is an interesting new candidate marker for the detection of infants who will develop NEC. In healthy preterm infants, every blood sample examined revealed higher concentration of L-FABP compared to I-FABP. This fact may reflect the physiologic turnover of mucosal cells, provided that L-FABP is expressed at a higher level or more expanded than I-FABP is. However, Poirier *et al.* [16] by Northern blotting found that both, L- and I-FABP are more abundant in the epithelial cells of the villus than of the crypt cells. One explanation may be that L-FABP is not exclusively found in the intestine (as is I-FABP) but is also abundantly expressed in liver. This resembles the situation in the diagnosis of acute myocardial infarction by means of the two markers H-FABP and myoglobin. In that case the latter exhibits a higher basal plasma concentration as well, because it is not only expressed in the heart but also in muscle [17]. The normal turnover of cells from organs is responsible for the basal levels of typical intracellular proteins in the circulation.

The spatial expression pattern of L-FABP in the murine intestine along the duodenum to colon axis differs from that of I-FABP [2]. The original rationale for determination of both I- and L-FABP was to take advantage of this differential expression and to obtain information about the location of the lesion by comparing I-FABP and L-FABP levels. Indeed, samples from infants suffering from abdominal symptoms had a L-FABP/I-FABP ratio between 0.1 and 20.

L-FABP was significantly elevated in infants suspected for NEC (stage I) compared to controls, but statistically, at onset of symptoms there were no differences between the three NEC groups. Because of its initial increase, L-FABP is a promising sensitive marker that detects even stage I of NEC, whereas I-FABP is indicative of the more severe damage occurring in NEC with stage III.

In clinical practice, classification of the patients according to the commonly used grading initially described by Bell [18] and modified by Walsh and Kliegman [8] is difficult. Not every infant fulfils simultaneously all criteria as systemic and intestinal symptoms and radiological signs. Therefore, we used, in this paper, a less subtle classification of NEC confined to three stages. Nevertheless, infants suffering from stage II show a broad spectrum of clinical courses. The occurrence of pneumatosis intestinalis which is an obligatory sign of stage II is according to our experience not a good predictor for surgical therapy or outcome. Our data suggest that the combined determination of L-FABP and I-FABP may improve the value of the clinical classification with respect to treatment of NEC.

Acknowledgement

The authors gratefully acknowledge the contribution of Dr. Carsten Hohoff during the initial phase of the project.

References

1. Crissinger KD, Balistreri WF: Gastrointestinal hemorrhage. In: A.A. Fanaroff, R.J. Martin (eds). Neonatal-Perinatal Medicine: Diseases of the Fetus and Infant, 6th edn. Mosby-Year Book Inc., St. Louis, 1997, pp 1331–1333
2. Sacchettini JC, Hauft SM, Van-Camp SL, Cistola DP, Gordon JI: Developmental and structural studies of an intracellular lipid binding protein expressed in the ileal epithelium. J Biol Chem 265: 19199–19207, 1990
3. Kanda T, Nakatomi Y, Ishikawa H, Hitomi M, Matsubara Y, Ono T, Muto T: Intestinal fatty acid-binding protein as a sensitive marker of intestinal ischemia. Dig Dis Sci 37: 1362–1367, 1992
4. Gollin G, Marks C, Marks WH: Intestinal fatty acid binding protein in serum and urine reflects early ischemic injury to the small bowel. Surgery 113: 545–551, 1993
5. Gollin G, Marks WH: Elevation of circulating intestinal fatty acid binding protein in a luminal contents-initiated model of NEC. J Pediatr Surg 28: 367–370, 1993
6. Lieberman JM, Sacchettini J, Marks C, Marks WH: Human intestinal fatty acid binding protein: Report of an assay with studies in normal volunteers and intestinal ischemia. Surgery 121: 335–342, 1997
7. Edelson MB, Sonnino RE, Bagwell CE, Lieberman JM, Marks WH, Rozycki HJ: Plasma intestinal fatty acid binding protein in neonates with necrotizing enterocolitis: a pilot study. J Pediatr Surg 34: 1453–1457, 1999
8. Walsh MC, Kliegman RM: Necrotizing enterocolitis: treatment based on staging criteria. Pediatr Clin North Am 33: 179–201, 1986
9. Wolfrum C, Buhlmann C, Rolf B, Börchers T, Spener F: Variation of liver-type fatty acid binding protein content in the human hepatoma cell line HepG2 by peroxisome proliferators and antisense RNA affects the rate of fatty acid uptake. Biochim Biophys Acta 1437: 194–201, 1999

10. Sweetser DA, Birkenmeier EH, Klisak IJ, Zollman S, Sparkes RS, Mohandas T, Lusis AJ, Gordon JI: The human and rodent intestinal fatty acid binding protein genes. A comparative analysis of their structure, expression, and linkage relationships. J Biol Chem 262: 16060–16071, 1987
11. Gill SC, von Hippel PH: Calculation of protein extinction coefficients from amino acid sequence data. Anal Biochem 182: 319–326, 1989
12. Sonntag J, Wagner MH, Waldschmidt J, Wit J, Obladen M: Multisystem organ failure and capillary leak syndrome in severe necrotizing enterocolitis of very low birth weight infants. J Pediatr Surg 33: 481–484, 1998
13. Baier LJ, Sacchettini JC, Knowler WC, Eads J, Paolisso G, Tataranni PA, Mochizuki H, Bennett PH, Bogardus C, Prochazka M: An amino acid substitution in the human intestinal fatty acid binding protein is associated with increased fatty acid binding, increased fat oxidation, and insulin resistance. J Clin Invest 95: 1281–1287, 1995
14. Schreiber A, Specht B, Pelsers MMAL, Glatz JFC, Börchers T, Spener F: Recombinant human heart-type fatty acid-binding protein as standard in immunochemical assays. Clin Chem Lab Med 36: 283–288, 1998
15. Kanda T, Fujii H, Tani T, Murakami H, Suda T, Sakai Y, Ono T, Hatakeyama K: Intestinal fatty acid-binding protein is a useful diagnostic marker for mesenteric infarction in humans. Gastroenterology 110: 339–343, 1996
16. Poirier H, Degrace P, Niot I, Bernard A, Besnard P: Localization and regulation of the putative membrane fatty-acid transporter (FAT) in the small intestine. Comparison with fatty acid-binding proteins (FABP). Eur J Biochem 238: 368–373, 1996
17. Van Nieuwenhoven FA, Kleine AH, Wodzig WH, Hermens WT, Kragten HA, Maessen JG, Punt CD, Van Dieijen MP, Van der Vusse GJ, Glatz JFC: Discrimination between myocardial and skeletal muscle injury by assessment of the plasma ratio of myoglobin over fatty acid-binding protein. Circulation 92: 2848–2854, 1995
18. Bell MJ, Ternberg JL, Feigin RD, Keating JP, Marshall R, Barton L: Neonatal necrotizing enterocolitis. Therapeutic decisions based upon clinical staging. Ann Surg 187: 1–7, 1978

Molecular and Cellular Biochemistry **239**: 235–237, 2002.

Index to Volume 239

Developments in Molecular and Cellular Biochemistry

Series Editor: Naranjan S. Dhalla, Ph.D., M.D. (Hon.), FACC

1. V.A. Najjar (ed.): *Biological Effects of Glutamic Acid and Its Derivatives.* 1981 ISBN 90-6193-841-4
2. V.A. Najjar (ed.): *Immunologically Active Peptides.* 1981 ISBN 90-6193-842-2
3. V.A. Najjar (ed.): *Enzyme Induction and Modulation.* 1983 ISBN 0-89838-583-0
4. V.A. Najjar and L. Lorand (eds.): *Transglutaminase.* 1984 ISBN 0-89838-593-8
5. G.J. van der Vusse (ed.): *Lipid Metabolism in Normoxic and Ischemic Heart.* 1989 ISBN 0-7923-0479-9
6. J.F.C. Glatz and G.J. van der Vusse (eds.): *Cellular Fatty Acid-Binding Proteins.* 1990 ISBN 0-7923-0896-4
7. H.E. Morgan (ed.): *Molecular Mechanisms of Cellular Growth.* 1991 ISBN 0-7923-1183-3
8. G.J. van der Vusse and H. Stam (eds.): *Lipid Metabolism in the Health and Diseased Heart.* 1992 ISBN 0-7923-1850-1
9. Y. Yazaki and S. Mochizuki (eds.): *Cellular Function and Metabolism.* 1993 ISBN 0-7923-2158-8
10. J.F.C. Glatz and G.J. van der Vusse (eds.): *Cellular Fatty-Acid-Binding Proteins, II.* 1993 ISBN 0-7923-2395-5
11. R.L. Khandelwal and J.H. Wang (eds.): *Reversible Protein Phosphorylation in Cell Regulation.* 1993 ISBN 0-7923-2637-7
12. J. Moss and P. Zahradka (eds.): *ADP-Ribosylation: Metabolic Effects and Regulatory Functions.* 1994 ISBN 0-7923-2951-1
13. V.A. Saks and R. Ventura-Clapier (eds.): *Cellular Bioenergetics: Role of Coupled Creatine Kinases.* 1994 ISBN 0-7923-2952-X
14. J. Slezák and A. Ziegelhöffer (eds.): *Cellular Interactions in Cardiac Pathophysiology.* 1995 ISBN 0-7923-3573-2
15. J.A. Barnes, H.G. Coore, A.H. Mohammed and R.K. Sharma (eds.): *Signal Transduction Mechanisms.* 1995 ISBN 0-7923-3663-1
16. A.K. Srivastava and J.-L. Chiasson (eds.): *Vanadium Compounds: Biochemical and Therapeutic Applications.* 1995 ISBN 0-7923-3763-8
17. J.M.J. Lamers and P.D. Verdouw (eds.): *Biochemistry of Signal Transduction in Myocardium.* 1996 ISBN 0-7923-4067-1
18. E.-G. Krause and R. Vetter (eds.): *Biochemical Mechanisms in Heart Function.* 1996 ISBN 0-7923-4118-X
19. R. Vetter and E.-G. Krause (eds.): *Biochemical Regulation of Myocardium.* 1996 ISBN 0-7923-4259-3
20. G.N. Pierce and W.C. Claycomb (eds.): *Novel Methods in Molecular and Cellular Biochemistry of Muscle.* 1997 ISBN 0-7923-4387-5
21. F.N. Gellerich and S. Zierz (eds.): *Detection of Mitochondrial Diseases.* 1997 ISBN 0-7923-9925-0
22. P.K. Singal, V. Panagia and G.N. Pierce (eds.): *The Cellular Basis of Cardiovascular Function in Health and Disease.* 1997 ISBN 0-7923-9974-9
23. S. Abdel-aleem and J.E. Lowe (eds.): *Cardiac Metabolism in Health and Disease.* 1998 ISBN 0-7923-8104-1
24. A.K. Srivastava and B. Posner (eds.): *Insulin Action.* 1998 ISBN 0-7923-8113-0
25. V.A. Saks, R. Ventura-Clapier, X. Leverve, A. Rossi and M. Rigoulet (eds.): *Bioenergetics of the Cell: Quantitative Aspects.* 1998 ISBN 0-7923-8118-1
26. G.N. Pierce, H. Rupp, T. Izumi and A. Grynberg (eds.): *Molecular and Cellular Effects of Nutrition on Disease Processes.* 1998 ISBN 0-7923-8171-8
27. K. Ahmed, E. Chambaz and O.G. Issinger (eds.): *Molecular and Cellular View of Protein Kinase CK2.* 1998 ISBN 0-7923-8208-0

Developments in Molecular and Cellular Biochemistry

28. M.V. Cohen, J.M. Downey, R.J. Gelpi and J. Slezák (eds.): *Myocardial Ischemia and Reperfusion*. 1998 ISBN 0-7923-8173-4
29. D.A. Bernlohr and L. Banaszak (eds.): *Lipid Binding Proteins within Molecular and Cellular Biochemistry*. 1998 ISBN 0-7923-8223-4
30. R. Albarez-Gonzalez (ed.): *ADP-Ribosylation Reactions: From Bacterial Pathogenesis to Cancer*. 1998 ISBN 0-7923-8235-8
31. S. Imai and M. Endo (eds.): *Muscle Physiology and Biochemistry*. 1998 ISBN 0-7923-8264-1
32. D.K. Das (ed.): *Stress Adaptation, Prophylaxis and Treatment*. 1999 ISBN 0-7923-8406-7
33. H. Rupp and B. Maisch (eds.): *Control of Gene Expression by Catecholamines and the Renin-Angiotensin System* ISBN 0-7923-7981-0
34. X. Shi, V. Castranova, V. Vallyathan and W.G. Perry (eds.): *Molecular Mechanisms of Metal Toxicity and Carcinogenesis*. 2001 ISBN 0-7923-7498-3
35. K. Ahmed, J.E. Allende and O.-G. Issinger (eds.): *Protein Kinase CK2 – From Structure to Regulation*. 2001 ISBN 0-7923-7666-8
36. R.K. Sharma (ed.): *Guanylate Cyclase*. 2002 ISBN 0-7923-7682-8
37. V. Vallyathan, V. Castranova and S. Shi (eds.): *Oxygen/Nitrogen Radicals: Cell Injury and Disease*. 2002 ISBN 1-4020-7085-3
38. J.F.C. Glatz (ed.): *Cellular Lipid Binding Proteins*. 2002 ISBN 1-4020-7220-1

KLUWER ACADEMIC PUBLISHERS – DORDRECHT / BOSTON / LONDON

MIX
Papier aus verantwortungsvollen Quellen
Paper from responsible sources
FSC® C105338

If you have any concerns about our products,
you can contact us on
ProductSafety@springernature.com

In case Publisher is established outside the EU,
the EU authorized representative is:
Springer Nature Customer Service Center GmbH
Europaplatz 3, 69115 Heidelberg, Germany

Printed by Libri Plureos GmbH
in Hamburg, Germany